Introduction to Medical Virology

Theory and Practice in Tropical Countries

Oluwatayo Israel Olasunkanmi

WILE

Published by John Wiley & Sons, Inc., Hoboken, New Jersey.

Library of Congress Cataloging-in-Publication Data Applied for

Paperback ISBN: 9781394376384

Cover Design: Wiley
Cover Image: © Bakhtiar Zein/Shutterstock

Set in 11.5/13.5pt STIXTwotext by Straive, Pondicherry, India

Dedication

This book is dedicated to all students, healthcare professionals, and scientists who, often under daunting circumstances, confront viral diseases to protect lives and communities.

1. To the students and aspiring virologists who are willing to devote themselves to understanding viruses and developing solutions that make society safer – this book is for you. May it guide your journey, ignite your curiosity, and strengthen your resolve to advance science and public health.
2. It is especially dedicated to the healthcare workers who face immense challenges during pandemics, tirelessly caring for patients, and contributing to the eradication of polioviruses in tropical countries and across the globe. Your courage, dedication, and resilience inspire every page of this work.
3. To scientists, teachers, colleagues, and pioneers whose discoveries illuminate the path of virology who shape the path in virology, and to my family, whose unwavering love and support made this work possible – I dedicate this book with heartfelt gratitude.

To all, may this work honor your efforts, celebrate your perseverance, and inspire a shared commitment to combating viral diseases with knowledge, compassion, and hope globally.

Contents

Forewords

The landscape of global health is increasingly defined by the dynamics of viral pathogens, particularly within the tropical and subtropical regions where the intersection of biodiversity, climate change, and human activity creates a unique crucible for emerging infectious diseases. While the field of virology has advanced remarkably at the molecular level, there has often remained a disconnect between high-level laboratory research and the practical, epidemiological realities faced by clinicians and researchers in tropical countries.

Having spent several decades at the forefront of infectious disease research and academic mentorship, I have witnessed the evolution of our field from both the laboratory bench and the clinical front lines. Throughout my career, I have frequently observed the struggle of students and practitioners who must rely on models that do not always translate to the unique environmental and diagnostic constraints of the tropics. It is from this perspective of long-standing engagement with global health challenges that I recognize the profound necessity of a text that speaks directly to these localized realities.

In "Introduction to Medical Virology for Tropical Countries," Dr. Olasunkanmi Oluwatayo provides a much-needed bridge across this divide. As we grapple with the lessons of recent pandemics and the ongoing challenges of endemic viruses, from the complexities of arboviral transmission to the persistent burden of viral hemorrhagic fevers, the need for a comprehensive, context-specific resource has never been greater.

What distinguishes this work is its meticulous balance of fundamental molecular virology with the clinical and public health nuances specific to tropical ecosystems. Dr. Oluwatayo draws upon extensive expertise in molecular pathogenesis and high-containment laboratory operations to demystify complex viral replication cycles and host-pathogen interactions. Moreover, the text never loses sight of the "tropical" imperative, providing detailed insights into vector biology, diagnostic constraints in resource-limited settings, and the socioeconomic factors that drive viral persistence.

Throughout the chapters, readers will find a rigorous examination of both well-established pathogens and emerging threats. The integration of current research on neuroinflammation, immunology, and novel antiviral strategies ensures that this is not merely a foundational text, but a forward-looking roadmap for the next generation of virologists.

It is a privilege to introduce a work that so clearly reflects a commitment to academic excellence and a profound dedication to improving health outcomes in the regions that bear the heaviest burden of viral disease. I do not doubt that this textbook will become an indispensable resource for students, researchers, and healthcare professionals dedicated to the study and control of viruses in the tropics.

B.M. Okanlawon
Former Interim Head
Department of Medical Lab. Science
Ladoke Akintola University of Technology
Nigeria

The burden of viral infections in tropical countries remains substantial, shaped by unique ecological, epidemiological, and socioeconomic factors. From endemic pathogens to emerging and re-emerging viral threats, the need for context-specific knowledge in medical virology is both urgent and indispensable. In such settings, access to clear, practical, and scientifically sound resources is essential for effective diagnosis, research, and control.

This book has been developed with this need in mind. This textbook provides a comprehensive yet accessible foundation in medical virology, integrating core principles with the realities of practice. It thoughtfully addresses the challenges faced in tropical regions.

One of the notable strengths of this work lies in its emphasis on applicability. The text bridges fundamental virological concepts with diagnostic approaches and public health strategies relevant to tropical countries. In doing so, it equips readers not only with theoretical understanding but also with the practical insight required for real-world problem solving.

This book represents a meaningful contribution to the field of medical virology and to the broader effort of strengthening infectious disease capacity in regions globally.

Oluyinka Oladele Opaleye, PhD
Professor, Department of Microbiology
Ladoke Akintola University of Technology
Nigeria

I can confidently state that this text reflects a career grounded in technical excellence, scholarly rigor, and a profound commitment to advancing health equity in tropical regions.

In Introduction to Medical Virology for Tropical Countries, readers are presented with a thoughtful framework for understanding host–pathogen interactions and the broader complexities of viral disease. The inclusion of advanced topics elevates this work beyond the scope of a conventional introductory text. By combining contemporary scientific concepts with practical realities relevant to tropical settings, the author has produced a resource that will guide teaching, strengthen practice, and expand knowledge.

We are living in a time when the next generation of scientists in tropical countries must be more than recipients of existing knowledge; they must become creators, innovators, and leaders shaping their own scientific future. This textbook provides a strong foundation for that purpose. It is both a pleasure and an honor to recommend this work to students, clinicians, educators, and researchers who will undoubtedly use its pages to advance the future of virology.

Agboola Solomon, PhD
Wayne State University
USA

Preface

Viral infections remain one of the greatest challenges to global health, with their impact being felt across every continent, population, and healthcare system. From emerging and reemerging pathogens to zoonotic threats, viruses affect societies everywhere, irrespective of geography or economic status. Recent pandemics have demonstrated that even the most technologically advanced nations are not immune to widespread viral devastation. At the same time, certain viral infections display distinct patterns of transmission and persistence in tropical regions, where climate, ecology, and vector dynamics play an important role in shaping their epidemiology.

Thus, while virology is fundamentally a global discipline, the study of viral infections in tropical contexts provides unique insights that enrich our understanding of transmission dynamics, diagnostics, prevention, and control. Recognizing these diverse realities, this book, *Introduction to Medical Virology: Theory and Practice in Tropical Countries*, bridges foundational virology concepts with the practical challenges of healthcare and laboratory science in resource-variable settings.

This text is designed for students, medical professionals, and public health practitioners who must not only understand the biology of viruses but also apply that knowledge in real-world contexts to protect communities. By integrating biological, clinical, diagnostic, and public health perspectives, it provides a comprehensive yet practical resource. Readers will find accessible explanations of viral biology alongside hands-on approaches to laboratory diagnosis, outbreak investigation, and infection control. Special attention is given to diagnostic methods adaptable to low-resource environments, making this book especially valuable in settings where advanced technologies may not always be available.

Unlike many traditional virology texts that emphasize theory with limited contextual application, this work balances scientific rigor with practical guidance. It highlights both the viral infections of global concern and those whose burden is particularly evident in tropical regions. In doing so, it underscores that virology is at once a universal science and a discipline shaped by local realities, requiring collaboration, innovation, and adaptability across settings.

The book serves multiple disciplines, including virology, medical microbiology, public health and epidemiology, molecular biology, and clinical laboratory science. It is written at a level suitable for undergraduate and postgraduate students, while also offering insights for practitioners, researchers, and policymakers working to strengthen viral disease surveillance and control.

Ultimately, what distinguishes this book is its contextual focus. It not only provides the foundational knowledge expected of a medical virology text but also addresses the pressing public health realities that define viral infections in tropical regions. It is my hope that it will contribute meaningfully to the training of the next generation of virologists, laboratory scientists, clinicians, and public health professionals – not only in tropical countries but everywhere viral threats pose a challenge. In this way, the book reflects a commitment to advancing knowledge, strengthening healthcare systems, and supporting global health security in the face of emerging and reemerging viral diseases.

The author gratefully acknowledges the use of BioRender in the creation of some of the figures included in this textbook.

Acknowledgments

I sincerely acknowledge Prof. Zhaohua Zhong and Prof. Wenran Zhao from Harbin Medical University for granting permission to use several images featured in this book. I am grateful to the United States Centers for Disease Control and Prevention, Poxvirus and Rabies Branch, for allowing the use of publicly available illustrative images on packing and shipping Category B clinical specimens. Dr. Philip Bryant for providing consent to use images of inclusion bodies in human and animal cells. I also thank Dr. Aroge Temitope for providing valuable images and Hans R. Gelderblom for his contributions.

Special thanks to Dr. Murat Enöz, Ear, Nose, and Throat Specialist (Otorhinolaryngologist), Istanbul, Turkey, for providing images of oral papilloma infections, and to RWD Life Science Inc. for providing incubator images. The electron microscopy and basophilic staining images of Guarnieri bodies typical of smallpox (variola) and vaccinia-infected cells were made possible through Dr. Thieme Verlag, Stuttgart, Germany, Hans R. Gelderblom, and Dr. Philip Bryant, reproduced here with their authorized consent. Additional acknowledgments go to Prof. Friedrich A. Bahmer, Bremen, for providing several stained images of biopsy samples, as well as electron microscopy images of the contents. Negative-stain images of herpes capsules, among others, were kindly provided by Hans R. Gelderblom with authorized consent.

Finally, I am grateful to Dr. Hashem Abu Serhan and colleagues for granting permission to reproduce images of eye influenza cases. Their support and generosity have greatly enriched this work.

List of Abbreviations

A2M Alpha-2-microglobulin
AAV Adeno-associated virus
ACE2 Angiotensin-converting enzyme 2
ADCC Antibody-dependent cellular cytotoxicity
AFP Acute flaccid paralysis
AIDS Acquired immunodeficiency syndrome
AP-1 Activator protein-1
APCs Antigen-presenting cells
APOBEC3G Apolipoprotein B mRNA editing catalytic polypeptide 3G
APP Acute phase proteins
APR Acute phase response
ASGP-R Asialoglycoprotein receptor
ATM Ataxia-telangiectasia mutated
ATR Ataxia-telangiectasia and Rad3-related
BALF1 Bam HI fragment H leftward open reading frame 1 (EBV)
BCA Bicinchoninic acid
BHRF1 Bam HI fragment H rightward open reading frame 1 (EBV)
BLAST Basic Local Alignment Search Tool
bNAbs Broad neutralizing antibodies
BSA Bovine serum albumin
BSCs Biological safety cabinets
BSL Biosafety level
CAR Coxsackievirus and adenovirus receptor
CBC Complete blood count
cccDNA Covalently closed circular DNA
CCR5 C—C Chemokine receptor type 5
CD4 Cluster of Differentiation 4
CDKs Cyclin-dependent kinases
CEF Chick embryo fibroblast
CHO Chinese hamster ovary
CMV Cytomegalovirus
COVID-19 Coronavirus disease 2019
CPE Cytopathic effects
crasiRNA Centromere repeat-associated small RNA
CRISPR Clustered regularly interspaced short palindromic repeats
csrRNA Centromere small RNA
CXCR4 C-X-C Chemokine receptor 4
CXCR4/CCR5 Chemokine receptors
DENV Dengue virus
DEPC Diethyl pyrocarbonate
DETECTR DNA endonuclease-targeted CRISPR trans reporter
DMEM Dulbecco's Modified Eagle Medium
DMSO Dimethyl sulfoxide
DNA Deoxyribonucleic acid
dNTP Deoxynucleotide triphosphates
DR-siRNA Dicer-related siRNA
DSBs Double-strand breaks
dsDNA Double-stranded DNA
dsRNA Double-stranded RNA
E1B-19 K Adenovirus early region 1B 19 kilodalton protein
EAC East African Community
EBV Epstein–Barr virus
EBNA2 Epstein–Barr nuclear antigen 2
EBV Epstein–Barr virus
EC50 Half-Maximal Effective Concentration
EDTA Ethylenediaminetetraacetic acid
EEEV Eastern equine encephalitis virus
EGF Epidermal growth factor

ELISA Enzyme-linked immunosorbent assay
endo-siRNA Endogenous siRNA
EphA2 Ephrin-A receptor 2
ERAD ER-associated degradation
ESR Erythrocyte sedimentation rate
EV-A71 Enterovirus 71
exo-siRNA Exogenous siRNA
FASTA Fast-All (sequence format)
FASTQ Fast-Quality (sequence format with quality scores)
FBC Full blood count
FBS Fetal bovine serum
FcγRIIIa Fc Receptor Gamma IIIa
FDA Food and Drug Administration
FELUDA FnCas9 Editor Linked Uniform Detection Assay
FGF Fibroblast growth factor
FIFO First in, first out
GMEM Glasgow Minimum Essential Medium
GOARN Global Outbreak Alert and Response Network
GP Glycoprotein
gp-340 Glycoprotein-340
GPLN Global Polio Laboratory Network
GPx Glutathione peroxidases
gRNA Guide RNA
HA Hemagglutinin
HA1 Hemagglutinin subunit 1
HAV Hepatitis A virus
HBGA Histo-blood group antigen
HBSS Hanks' Balanced Salt Solution
HBV Hepatitis B virus
HCC Hepatocellular carcinoma
HCV Hepatitis C virus
HEPA High-efficiency particulate air
HEV Hepatitis E virus
HFMD Hand, foot, and mouth disease
HHV-6 Human herpesvirus 6
HiPSCs Human induced pluripotent stem cells
HIV Human immunodeficiency virus
HLA Human leukocyte antigen
hMPV Human metapneumovirus
HPIV Human parainfluenza virus
HR Homologous recombination
HRP Horseradish peroxidase
HSPGs Heparan sulfate proteoglycans
HSR Heat shock response
HSV Herpes simplex virus
HSV-1 Herpes simplex virus type 1
HSV-2 Herpes simplex virus type 2
HTLV Human T-lymphotropic virus
HTLV-1 Human T-cell leukemia virus type 1
HuF Human Foreskin Fibroblast Cells
ICAM-1 Intercellular adhesion molecule 1
ICP0 Infected cell protein 0
ICTV International Committee on Taxonomy of Viruses
IFN Interferon
IFNGR1 Interferon gamma receptor 1
Ig Immunoglobulin
IgG/IgM Immunoglobulin G/M
IL Interleukin
IL10R2 Interleukin-10 receptor subunit 2
IRES Internal ribosome entry site
IRIS Immune reconstitution inflammatory syndrome
ISG Interferon-stimulated gene
ISO International Organization for Standardization
ISOL (ISO, International Organization for Standardization)
JAK–STAT Janus kinase/signal transducer and activator of transcription
JC Virus (JCV) John Cunningham virus
JEV Japanese encephalitis virus

KSHV Kaposi's sarcoma-associated herpesvirus
LAMP Loop-mediated isothermal amplification
LCMV Lymphocytic choriomeningitis virus
LDLR Low-density lipoprotein receptor
LIMS Laboratory Information Management Systems
LLC-MK2 Rhesus monkey kidney epithelial cells
lncRNA-derived small RNAs Long noncoding RNA-derived small RNAs
mAbs Monoclonal antibodies
MAC Membrane attack complex
MAC *Mycobacterium Avium* complex
MDA5 Melanoma differentiation-associated protein 5
MDCK Madin–Darby canine kidney
MDCK-SIAT1 Madin–Darby canine kidney cells expressing SIAT1
MERS Middle East respiratory syndrome
MERS- CoVMiddle East respiratory syndrome coronavirus
MHC Major histocompatibility complex
MICA/MICB MHC class I chain-related proteins A and B
miRNA MicroRNA
MLS/T Medical Laboratory Scientists or Technologists
MOMP Membrane outer mitochondrial permeabilization
MRC-5 Medical Research Council Cell Strain 5
mRNA Messenger RNA
MSDS Material Safety Data Sheets
MVBs Multivesicular bodies
Mxra8 Matrix remodeling-associated protein 8
NA Neuraminidase
NAAT Nucleic acid amplification tests
NAD Nicotinamide adenine dinucleotide
NADPH Nicotinamide adenine dinucleotide phosphate
NCAM Neural cell adhesion molecule
NCBI National Center for Biotechnology Information
NES Nuclear export signal/sequence
NF-κB Nuclear factor kappa B
NGS Next-generation sequencing
NHEJ Non-homologous end joining
NK cells Natural killer cells
NKG2D Natural killer group 2 member D
NKp30 (NCR3) Natural cytotoxicity receptor 3
NKp44 (NCR2) Natural cytotoxicity receptor 2
NNRTI Non-nucleoside reverse transcriptase inhibitor
Nrf2 Nuclear factor erythroid 2-related factor 2
NRTI Nucleoside reverse transcriptase inhibitor
NS1 Nonstructural protein 1
NS5 Nonstructural protein 5
OAS 2,5′-Oligoadenylate synthetase
OPV Oral polio vaccine
ORF Open reading frame
OSR Oxidative Stress Response
p53 Tumor protein p53
PAMPs Pathogen-associated molecular patterns
PBS Phosphate-buffered saline
PCP Pneumocystis Jirovecii pneumonia
PCR Polymerase chain reaction
PDGF Platelet-derived growth factor
Persistent generalized lymphadenopathy (PGL) Enlargement of multiple lymph nodes

PIC Pre-integration complex
piRNA PIWI-interacting RNA
PKR Protein kinase R
PPE Personal protective equipment
pri-miRNA Precursor microRNA
PRRs Pattern recognition receptors
PTPs Protein tyrosine phosphatases
PTX Pentraxins
PV Poliovirus
PVDF Polyvinylidene difluoride
QA Quality assurance
QC Quality control
Rb Retinoblastoma protein
RBC Red blood cell
RDTs Rapid diagnostic tests
RNA Ribonucleic acid
RNAi RNA interference
RNP Ribonucleoprotein complex
RNS Reactive nitrogen species
ROS Reactive oxygen species
RPMI Roswell Park Memorial Institute Medium
rRFs/rsRNAs Ribosomal RNA-derived small RNAs
RRLs Regional reference laboratories
RSV Respiratory syncytial virus
RTEs Recent thymic emigrants
RT-PCR Reverse transcription PCR
RT-qPCR Reverse transcriptase quantitative PCR
SARS Severe acute respiratory syndrome
SARS-CoV-2 Severe acute respiratory syndrome coronavirus 2
scRNA-Seq Single-cell RNA sequencing
sdRNA Small nucleolar RNA-derived RNA
SDS Sodium dodecyl sulfate
SDS-PAGE Polyacrylamide gel with SDS
SHERLOCK Specific High-Sensitivity Enzymatic Reporter UnLOCKing
shRNA Small hairpin RNA
siRNA Small interfering RNA
SIV Simian immunodeficiency virus
siwiRNA Secondary piRNA-Like RNA
SLE Systemic lupus erythematosus
SLEV St. Louis encephalitis virus
snRNA-derived fragments Fragmented small nuclear RNA
SOPs Standard operating procedures
SPI-2 (B13R) Serpin-2 (Vaccinia virus protein)
SSBs Single-strand breaks
ssDNA Single-stranded DNA
STIs Sexually transmitted infections
svRNA Vault RNA-derived small RNA
SYBR No official expansion (proprietary name by Invitrogen for "SYBR dyes," nucleic acid stains)
TAE Tris-acetate-EDTA
TAE buffer Tris-acetate-EDTA buffer
TB Tuberculosis
TBE Tris-borate-EDTA
TBE buffer Tris-borate-EDTA buffer
TBEV Tick-borne encephalitis virus
TBST Tris-buffered saline with Tween 20
TCRs T-cell receptors
TEM Transmission electron microscopy
tiRNA Stress-induced tRNA fragments/tRNA halves
TLR3 Toll-like receptor 3
TNFs Tumor necrosis factors
tRFs/tsRNAs tRNA-derived fragments
UPR Activation of the unfolded protein response
UTR Untranslated region
UV Ultraviolet
VDACs Voltage-dependent anion channels
VDPV Vaccine-derived poliovirus
v-GPCR Viral G protein-coupled receptor

vMIA Viral mitochondria-localized inhibitor of apoptosis (encoded by CMV)
VNTR Variable number of tandem repeat
VP Viral protein
vRNA Viral RNA
vsiRNA Virus-derived siRNA
VSV Vesicular stomatitis virus
VTM Virus transport medium
VZV Varicella-zoster virus
WBC White blood cell
WEEV Western equine encephalitis virus
WHO World Health Organization
WHO-SEARO Southeast Asia Regional Office of the World Health Organization
WPV Wild poliovirus
YsRNA YRNA-derived fragment

1 General Features and Classification of Viruses of Medical Importance

1.1 Introduction

Viruses are obligate intracellular infectious agents. The term "virus" originates from the Latin word "poison." [1]. Structurally, viruses consist of genetic material, either ribonucleic acid (RNA) or deoxyribonucleic acid (DNA) (but not both), encapsulated within a protein coat called a capsid, and in some cases, an outer lipid envelope. Not all viruses have an envelope. Enveloped viruses are nucleocapsids covered by a lipid bilayer membrane, usually derived from the host cell. In contrast, non-enveloped viruses, also known as naked viruses, lack this lipid bilayer.

The size of viruses typically ranges from 20 to 300 nm in diameter, making them too small to be seen with the naked eye and with an ordinary light microscope [2]. Giant viruses are exceptions; they are large viruses that might be visible under very powerful light microscopes, though details would not be clear [3].

Viruses are incredibly diverse, varying in size, shape, and complexity. They can infect all types of organisms, including animals, plants, and even other microorganisms like bacteria. Despite their simplicity, viruses have profound impacts on health, ecology, and evolution, making them a critical focus of study in microbiology and virology. Figure 1.1 illustrates the hypothetical structure of some viruses of medical importance.

Examples of giant viruses:

Mimivirus: One of the largest known viruses, measuring around 400 nm in diameter, the mimivirus is large enough to be seen as a speck under a high-resolution light microscope.

While mimivirus is not considered a typical human pathogen, some studies have suggested its presence in patients with pneumonia. However, the role of mimivirus in causing human disease remains unclear and is debated. It may be associated with respiratory infections [4].

Pandoravirus: Another giant virus, Pandoravirus, is even larger, measuring around 1 μm (1000 nm). This is well within the resolution range of a light

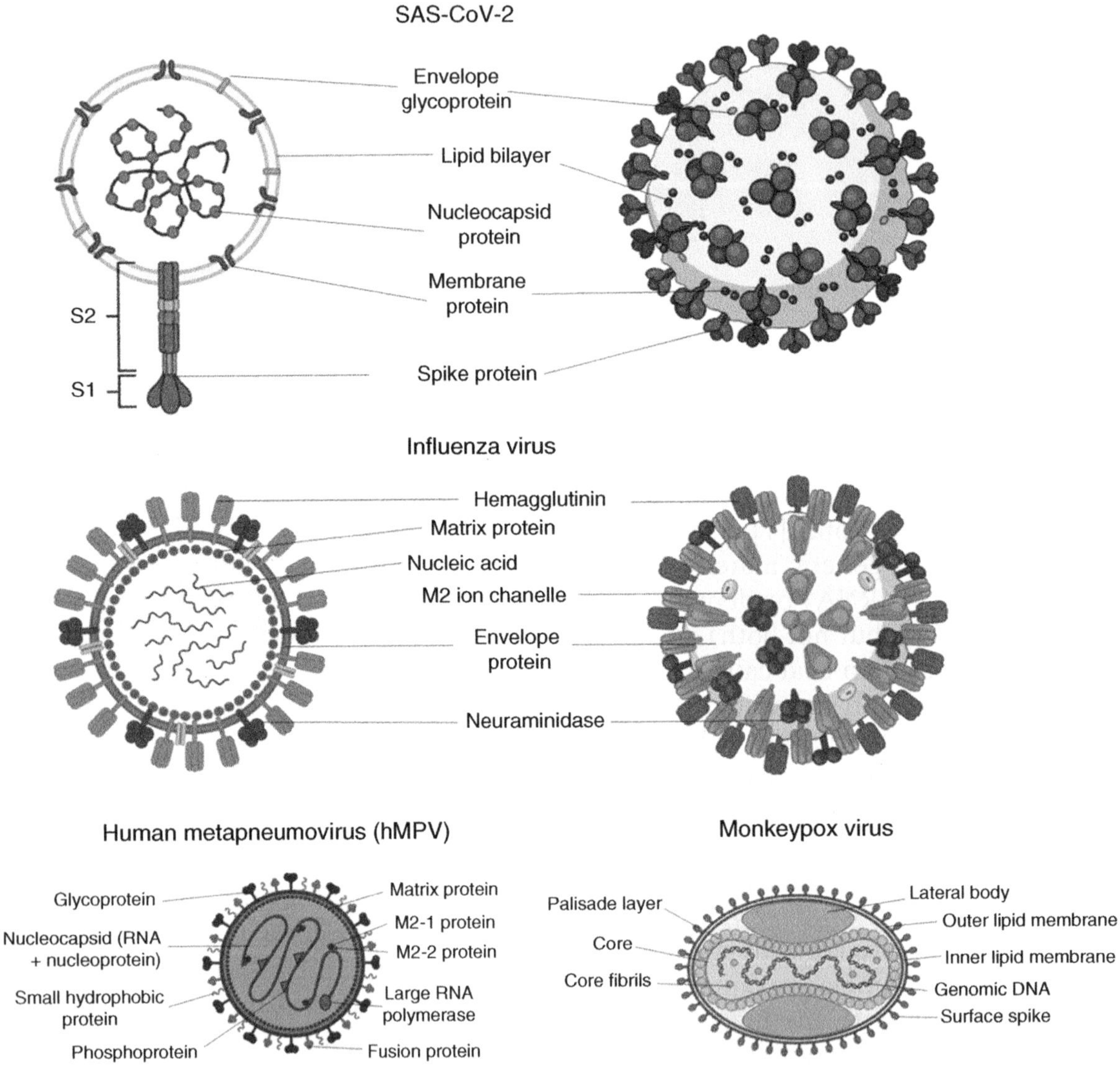

Figure 1.1 Putative structure of some viruses of medical importance.
Source: With permission BioRender.

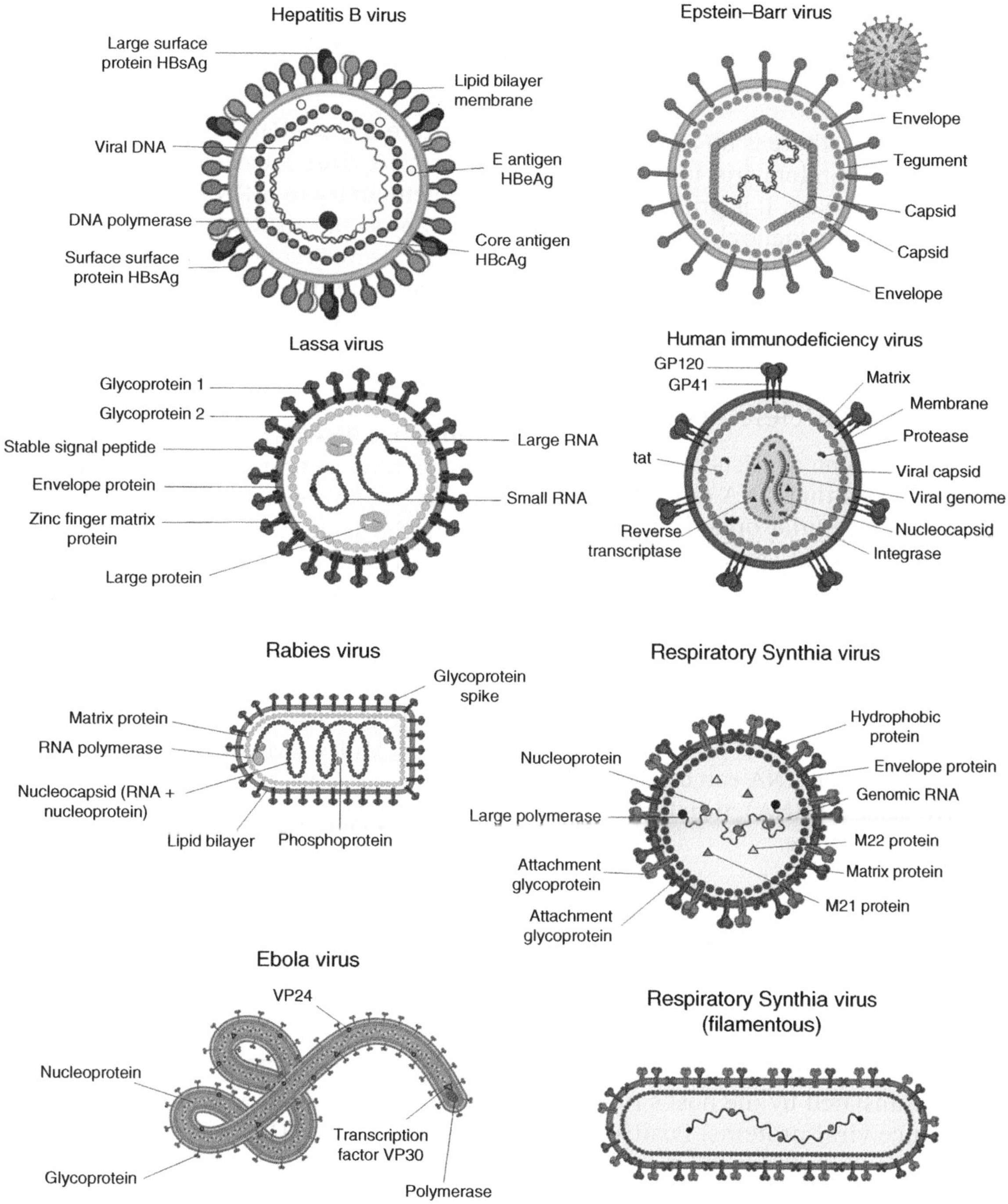

Figure 1.1 (Continued)

microscope, so it can be seen, although its internal structure would still require an electron microscope for detailed observation [5].

Currently, there is no conclusive evidence that Pandoravirus infects humans or causes disease. It is primarily studied for its unique genetic characteristics and its interactions with amoebae.

1.2 Basic Virus Structure

1.2.1 Viral Nucleic Acids

The viral genome consists of nucleic acid, which can be either DNA or RNA, but not both. Unlike cellular organisms or bacterial genomes, viral genomic nucleic acid (RNA or DNA) can be (i) single-stranded or double-stranded, (ii) circular or linear, or (iii) segmented or non-segmented. Segmented means a single virus can have multiple RNA or DNA molecules as part of its genome. Additionally, if a virus is an RNA virus, its genomic RNA can be classified as positive-sense (+RNA), negative-sense (−RNA), or ambisense.

1.2.1.1 RNA Viruses

Positive-Sense RNA Viruses These are viruses whose genetic material functions directly as viral messenger RNA (mRNA), which encodes viral proteins. In other words, the RNA genome can serve as mRNA and be translated by the host's ribosomes to produce viral proteins. Positive-sense RNA viruses make up more than one-third of all known virus genera. Examples include poliovirus, hepatitis C virus (HCV), and coronaviruses such as severe acute respiratory syndrome coronavirus 2 (SARS-CoV-2).

Negative-Sense RNA Viruses Negative-sense RNA viruses have genomes that are complementary to mRNA. Before translation can occur, an RNA-dependent RNA polymerase must convert the negative-sense RNA genome into a positive-sense RNA. Examples of negative-sense RNA viruses include the influenza virus, rabies virus, and Ebola virus.

Ambisense RNA Viruses Ambisense RNA viruses have genomes that contain both positive- and negative-sense sequences within the same RNA molecule. The viral RNA polymerase must first transcribe the negative-sense portion into positive-sense RNA, while the positive-sense portion can be directly translated. Examples of ambisense RNA viruses include some arenaviruses (e.g., Lassa virus) and bunyaviruses (e.g., Rift Valley fever virus) [6, 7].

1.2.1.2 DNA Viruses

The genomes of DNA viruses can be either double-stranded (dsDNA) or single-stranded (ssDNA). In general, DNA viruses are more stable and less prone to mutation compared to RNA viruses, providing a higher degree of genetic stability.

Double-Stranded DNA Viruses Double-stranded DNA (dsDNA) is the most common type of genome in DNA viruses. The genome can be linear or circular. Examples include adenoviruses (linear dsDNA), herpesviruses (linear dsDNA), poxviruses (linear dsDNA with covalently closed ends), and papillomaviruses (circular dsDNA).

Single-Stranded DNA Viruses Single-stranded DNA (ssDNA) viruses are less common and require conversion into dsDNA for replication and transcription. An example is the parvovirus.

Partially dsDNA Viruses Partially dsDNA viruses have circular genomes containing both double-and single-stranded regions. In the case of hepatitis B virus (HBV), one strand is a complete negative-sense strand, while the other is an incomplete positive-sense strand. This "partially" double-stranded genomic structure gives HBV its unique "relaxed" circular DNA structure due to the gap in the positive strand. HBV contains a small portion of RNA (pre-genomic RNA) during certain stages of replication [8].

Clinical Significance Most DNA viruses are genetically more stable than RNA viruses. Also, despite their limited coding capacity, RNA viruses evolve rapidly due to lack of proofreading during replication, contributing to their high mutation rates. This rapid evolution presents challenges for vaccine and antiviral drug development, especially in populations with prior exposure.

Implications of Genetic Stability

Evolutionary dynamics

DNA viruses: The genetic stability of DNA viruses can be beneficial for the development of long-lasting vaccines and antiviral therapies.

RNA viruses: The high mutation rates of RNA viruses drive rapid evolution, which can complicate the development of effective vaccines and treatments. For example, the rapid genetic changes in influenza viruses necessitate regular updates to the flu vaccine.

Pathogenicity and host interaction

DNA viruses: Their stable genomes allow for persistent infections and long-term coevolution with their hosts. Examples include herpesviruses and papillomaviruses.

RNA viruses: Their genetic variability enables quick adaptation to host defenses and environmental changes, often resulting in acute, rapidly spreading infections. Examples include human immunodeficiency virus (HIV), influenza, and coronaviruses.

Reason for stability

Replication fidelity: DNA viruses often replicate using host-cell DNA polymerases or their viral DNA polymerases, which have proofreading and error-correcting mechanisms [9]. This ensures higher fidelity during DNA replication, resulting in fewer mutations. In contrast, RNA viruses typically rely on RNA-dependent RNA polymerases (RdRp) for replication, which lack proofreading capabilities. This leads to higher error rates during RNA synthesis and, consequently, higher mutation rates.

Proofreading mechanisms and error rates: DNA polymerases have much lower error rates compared to the RNA polymerases used by RNA viruses. DNA polymerases, particularly those with proofreading activity, have an error rate of approximately 10^{-8} to 10^{-11} errors per nucleotide. In contrast, RdRp has significantly higher error rates, ranging from approximately 10^{-3} to 10^{-6} errors per nucleotide [10]. This higher error rate results in greater genetic variability in RNA viruses. The lower mutation rates in DNA viruses contribute to their greater genetic stability, allowing these viruses to maintain genome integrity over time.

1.2.2 Virus Capsids and Capsomere Proteins

A capsid is the protein coat that covers, protects, and stabilizes the viral genome. It consists of polypeptide subunits called capsomeres, which form the structural morphology of the virus, giving it a specific shape, such as spherical, rod-like, or conical. Capsid proteins vary phenotypically (in terms of structure and function) and genotypically (in terms of genetics) across different types of viruses. Recent studies have demonstrated that viruses can assume various shapes during different stages of replication. Viral capsids typically exhibit either icosahedral or helical symmetry. Helical capsids can be rigid or flexible, depending on the virus, while icosahedral capsids are generally more rigid [11].

1.2.2.1 Helical Symmetry

Helical symmetry refers to a structural arrangement commonly observed in viruses, where protein subunits, capsomeres, are arranged in a helical structure, creating a rod-like or filamentous shape. This symmetry is primarily found in the nucleocapsids of some viruses, particularly those with single-stranded RNA (ssRNA) genomes, such as influenza and rabies [12].

The helical symmetry of a virus can be mathematically described by the following parameters:

a. **Radius (*r*):** The distance from the center of the helix to the center of a capsomere.
b. **Pitch (*p*):** The vertical distance (along the helix axis) for one complete turn of the helix
c. **Number of capsomeres per turn (*n*):** The number of protein subunits in one complete helical turn.
d. **Axial rise per capsomere (*h*):** The vertical distance between consecutive capsomeres along the helical axis can be calculated as $h = p/n$.
   For example
   Suppose a helical virus has the following properties: Pitch (p) = 5 nm; number of capsomeres per turn (n): 13
   To find the axial rise per capsomere (h): $h = p/n$
   $=5\,\text{nm}/13 \approx 0.385\,\text{nm}$

1.2.2.2 Icosahedral Symmetry

Icosahedral symmetry is a common and efficient structural form adopted by many viruses to build their capsids. An icosahedron consists of 20 equilateral triangular faces, 12 vertices (where 5 triangular faces meet), and 30 edges (shared between 2 triangular faces) [13]. This symmetry allows for the construction of a stable and closed shell using a minimal number of distinct protein subunits, optimizing genetic economy. It is common in many DNA viruses and some RNA viruses.

Triangulation Number (*T*-Number) The triangulation number (T) describes the number of smaller triangular facets into which each of the 20 equilateral triangular faces of the icosahedron is divided. The T-number helps to understand the complexity and size of the viral capsid.

Formula:

$$T = h^2 + hk + k^2,$$

where h and k are nonnegative integers.

The T-number represents the number of quasi-equivalent positions that the capsomeres can occupy on the icosahedron.

Examples of T-numbers:

$T = 1$**:** Simplest form with each triangular face made of a single capsomere. Total of 60 protein subunits (12 vertices × 5 subunits each).

$T = 3$**:** Each triangular face is subdivided into three smaller triangles. Total of 180 protein subunits. Example: poliovirus.

$T = 4$**:** Each triangular face is subdivided into four smaller triangles. Total of 240 protein subunits. Example: HBV.

$T = 7$**:** Each triangular face is subdivided into seven smaller triangles. Total of 420 protein subunits.

Clinical Applications Understanding the structural details of icosahedral capsids aids in vaccine development by targeting specific viral proteins and capsid structures. Drugs can be designed to disrupt capsid assembly or stability, inhibiting viral replication. Additionally, in nanotechnology, the icosahedral symmetry of viruses can be used as a model for designing nanomaterials and nanodevices, leveraging the stability and geometric precision of viral capsids.

1.2.2.3 Complex Symmetry

A virus is said to have complex symmetry when the capsid structure combines elements of both icosahedral and helical symmetry. Some viruses, such as poxviruses and bacteriophages, exhibit complex symmetry [14, 15].

- **Function of the capsids**

  Protection: The capsid protects the viral genome from degradation by nucleases and harsh environmental conditions.

  Attachment and entry: In nonenveloped viruses, capsid proteins often contain receptor-binding sites that facilitate attachment to host cells and entry.

  Delivery of genetic material: Once inside the host, the capsid disassembles to release the viral genome for replication.

The **nucleocapsid** and **capsomere** are both structural components of viruses, but they differ in their composition and function:

Nucleocapsid

- The nucleocapsid is the combination of a virus's genetic material (either DNA or RNA) and the protein coat that surrounds and protects it [16].

- It serves as a protective layer for the viral genome and aids in the process of infection by ensuring the safe delivery of the genetic material into host cells.
- In some viruses, the nucleocapsid is enclosed by an additional lipid envelope, but in others (non-enveloped viruses), the nucleocapsid is the outermost layer.

Capsomere

- A capsomere is a subunit or building block of the capsid, which is the protein shell that surrounds the viral genome.
- Capsomeres are made of protein molecules and self-assemble to form the capsid, which can have various geometric shapes, such as icosahedral or helical.
- The arrangement and number of capsomeres determine the overall shape and symmetry of the viral capsid.

1.2.3 Virus Envelope (Membrane) Protein

The virus envelope is a lipid bilayer, embedded with proteins, that surrounds some virus particles. Most enveloped viruses (except poxviruses) acquire their envelope from the host-cell membrane, such as the Golgi apparatus, endoplasmic reticulum (ER), or plasma membrane, during viral maturation or budding [17]. Aside from the viral envelope, which is the last/outer layer of enveloped viruses, some viruses encode glycoprotein projections called spikes or peplomers, which are present on either the outer layer of the viral capsid or envelope [18]. These projections facilitate virus attachment to host cells, entry, and immune evasion. They are crucial for both viral specificity and infectivity.

The membrane (or envelope) of viruses plays a crucial role during three main stages of the viral life cycle: entry into the host cell, assembly of new virions, and release from the host cell.

1.2.3.1 Virus Membrane Glycoprotein

Virus membrane glycoproteins are crucial components of viral envelopes, playing significant roles in the virus's ability to infect host cells (membrane fusion and entry) and evade the immune system. These glycoproteins can be broadly categorized into two structural classes: Class I and Class II glycoproteins.

Class I Viral Membrane Glycoproteins
Class I glycoproteins are characterized by their prominent alpha-helical transmembrane domains. They typically undergo significant conformational changes during the process of fusion, shifting from a pre-fusion to a post-fusion state. In their functional form, these glycoproteins often form trimeric structures [19].

Examples of viruses with Class I membrane glycoproteins are

- **Influenza virus:** Influenza hemagglutinin (HA1 and 2) is a glycoprotein. It is responsible for the attachment and subsequent fusion of the influenza virus with the host-cell membrane. However, type II glycoproteins are also found on HA1 [20].

- **Human immunodeficiency virus:** HIV envelope glycoprotein (Env) comprises glycoprotein (gp) 120 and gp41 subunits; this glycoprotein is essential for HIV to attach to and fuse with host cells.

Class II Viral Membrane Glycoproteins Class II glycoproteins are characterized by their beta-sheet-rich structures or domains. Unlike Class I glycoproteins, Class II glycoproteins exist as dimers in their pre-fusion state and form a hairpin-like structure upon fusion, facilitating membrane fusion.

Examples of viruses with Class II membrane glycoproteins are

Dengue virus (DENV): DENV envelope glycoprotein (E) is a glycoprotein that mediates DENV entry into host cells by binding to receptors and facilitating membrane fusion [21].

Tick-borne encephalitis virus: Tick-borne encephalitis virus glycoprotein (E), like DENV glycoprotein, also plays a key role in viral attachment and membrane fusion for viral entry [22, 23].

1.2.3.2 *Broad Functions of Viral Membrane Glycoproteins*

Role in Virus Attachment Viral glycoproteins embedded in the envelope, such as hemagglutinin (HA) in influenza viruses or spike (S) proteins in coronaviruses, recognize and bind to specific receptors on the surface of the host cell. This binding determines the host range and tissue tropism of the virus.

Role in Virus Entry and Fusion After attachment, fusion proteins within the viral envelope mediate the merging of the viral envelope with the host-cell membrane. This fusion can occur either directly at the plasma membrane (e.g., HIV) or within endosomes following endocytosis (e.g., influenza virus). Fusion releases the viral nucleocapsid into the host cell's cytoplasm, where viral replication begins.

Role in Virus Release Newly formed enveloped virions are released from the host cell either by budding from the plasma membrane or through exocytosis if they bud into intracellular vesicles first. During budding, the viral nucleocapsid associates with the host membrane at the budding site, where viral proteins and glycoproteins are embedded. The envelope wraps around the nucleocapsid, releasing the mature virion. Viral matrix proteins play a key role in linking the nucleocapsid to the envelope, facilitating non-lytic budding.

Clinical Significance of Viral Membrane Glycoprotein The presence of a lipid envelope has significant implications for the virus's stability and how it interacts with the environment. The lipid envelope makes viruses more susceptible to disinfectants, particularly those containing alcohol and detergents. These substances can disrupt the lipid bilayer, effectively inactivating the virus outside the body. This property is clinically significant for infection control, as proper hand hygiene and surface disinfection can help prevent the spread of some enveloped viruses.

Enveloped viruses are generally less stable outside the host compared to non-enveloped viruses [24]. They are sensitive to environmental conditions such as heat, desiccation, and pH changes. This means that, in most cases, enveloped viruses typically require direct transmission routes such as blood, bodily fluids, or close person-to-person contact. Understanding this structure of the virus can aid in developing strategies for controlling outbreaks and implementing preventive measures.

1.2.4 Matrix Proteins (in Enveloped Viruses)

Matrix proteins are like the tough scaffolding within an enveloped virus [25]. They lie beneath the envelope and link the envelope with the nucleocapsid, playing a crucial role in the structure and assembly of the virus particle. Their structure can vary depending on the specific virus, but they often have regions that bind to lipids in the envelope and regions that interact with the viral RNA or proteins in the nucleocapsid, helping to anchor the nucleocapsid within the viral membrane and maintain the integrity of the virus [26].

Function of the virus matrix protein

In HIV, the matrix protein (MA) plays a crucial role in the assembly and budding of new viral particles.

In influenza viruses, the matrix protein M1 helps maintain the integrity of the viral envelope and interacts with the viral ribonucleoprotein complex (RNP) within the nucleocapsid.

1.2.5 Tegument

The tegument is found in some enveloped viruses, specifically in the space between the capsid and the viral envelope. It is prominent in herpesviruses (e.g., herpes simplex virus) [27]. The tegument is composed of various viral proteins, which vary significantly between different viruses. For example, Epstein–Barr virus (EBV) tegument [28] contains proteins like BGLF and viral protein 16 (VP16) (late tegument protein), which are essential for gene expression, immune modulation, and replication. Also, varicella-zoster virus (VZV) has tegument proteins that play crucial roles in virus replication and immune evasion [29]

1.3 Classification of Viruses of Medical Importance

Classification of viruses is essential for understanding their biology and epidemiology and for developing strategies to combat viral diseases. In the early days of virology, virus classification was primarily based on characteristics observable under an electron microscope and a few other basic properties, such as morphological characteristics, including shape (symmetry of the capsid), size, and stability under certain conditions [30]. This traditional method provided a limited understanding of viral biology.

As technology advanced, electron microscopy allowed for direct visualization of viruses, providing detailed information about their morphology and size. Molecular techniques, such as nucleic acid sequencing, have since provided a much deeper understanding of viral genetics and phylogeny, leading to more accurate and comprehensive classification systems.

1.3.1 Traditional Classification

1.3.1.1 Classification Based on Physicochemical Properties

In this type of classification system, viruses can be classified based on susceptibility to physical or chemical agents (especially ether and detergents), pH stability, thermal stability, solvent stability, cation stability, radiation stability, and buoyant density.

This system of classification is particularly valuable in laboratory research, biosafety, epidemiological surveillance, sterilization protocols, vaccine development, and public health interventions.

a. **Classification based on susceptibility to solvents (e.g., ether, chloroform, and detergents):** Enveloped viruses are generally sensitive to lipid solvents like ether or detergents, which disrupt the viral envelope, leading to inactivation [31]. In contrast, non-enveloped viruses are more resistant [32]. This feature is used to differentiate between these two structural types.
b. **Classification based on pH stability:** Viruses can be classified as acid-stable or acid-labile. That is, some viruses are stable over a wide pH range, while others are inactivated outside narrow pH limits. For example, enteric viruses like picornaviruses, such as poliovirus, are stable in acidic conditions (up to pH 3.0), allowing survival in the stomach [33]. However, rhinovirus is not acid-stable [34]. This characteristic is relevant for understanding oral transmission routes.
c. **Classification based on thermal stability:** This refers to how resistant a virus is to heat. Thermally stable viruses (e.g., hepatitis A and E) can withstand pasteurization-like conditions, whereas others are easily inactivated by mild heat [35]. Exposure to heat at 70–72 °C for 2 minutes substantially lowers the level of infectious virus, though it typically does not achieve a reduction greater than 4 $\log_{10}$ [36]. This property also influences sterilization and vaccine storage protocols.
d. **Classification based on radiation stability:** Most viruses are sensitive to ultraviolet exposure because of the pyrimidine dimer (thymidine) [37]. However, since only DNA and not RNA contains thymidine, DNA viruses are more sensitive to ultraviolet or ionizing radiation than RNA viruses [38]. Ultraviolet radiation, in particular, is used to disinfect surfaces and air in hospital and laboratory settings.
e. **Classification based on cation sensitivity:** Certain viruses require specific ions (e.g., Mg^{2+} or Ca^{2+}) to maintain structural stability or facilitate infectivity. Removing these ions or altering ionic strength can destabilize the virion or affect its ability to infect host cells [39].
f. **Classification based on buoyant density:** Viruses have characteristic buoyant densities in gradient centrifugation media (e.g., cesium chloride). This property is widely used in virus purification and identification techniques.

g. **Classification based on desiccation resistance:** Resistance to drying can vary. For example, non-enveloped viruses such as adenoviruses and enteroviruses are more resistant to desiccation, allowing survival on surfaces for extended periods [40].

h. **Classification based on sensitivity to oxidizing agents and disinfectants:** Some viruses, particularly non-enveloped ones, can survive exposure to alcohols or chlorine-based disinfectants, while enveloped viruses are more readily inactivated [41].

1.3.1.2 Classification Based on Viral Pathogenesis

This involves categorizing viruses according to their ability or affinity to infect specific cells, tissues, or organs, a concept known as "tropism." Classifying viruses based on tissue tropism provides insights into their pathogenicity, clinical manifestations, and epidemiology. These viruses are named based on the part of the body they primarily infect. However, it is important to note that while a virus may have a primary site of infection, it is not exclusively limited to causing symptoms at that site. Following are examples of virus classification based on viral pathogenesis:

Respiratory Viruses These are viruses that primarily infect the respiratory tract, including the nasal passages, throat, trachea, and lungs. Viruses in this category are referred to as respiratory viruses because they predominantly cause clinical symptoms and tissue damage in the respiratory system. Their classification under this nomenclature is based on their tropism for respiratory epithelium, and this tropism is often mediated by specific cellular receptors and cofactors found abundantly on respiratory epithelial cells. These molecular interactions are essential for viral attachment, entry, and replication, which in turn drive their tissue specificity and disease profile.

Examples include

- **Influenza viruses:** Cause respiratory infections, ranging from mild colds to severe pneumonia. Influenza viruses bind to sialic acid residues on glycoproteins of respiratory epithelial cells via their HA proteins, leading to infection of the upper and lower respiratory tract [42].
- **Rhinoviruses:** Major cause of the common cold, infecting the upper respiratory tract. Rhinoviruses target intercellular adhesion molecule-1 (ICAM-1) on nasal mucosa epithelial cells, explaining their restriction to the upper respiratory tract and association with the common cold [43].
- **Respiratory syncytial virus (RSV):** Causes bronchiolitis and pneumonia majorly in infants young children and high-risk adults. RSV uses nucleolin, CX3C chemokine receptor 1, and heparan sulfate proteoglycans (HSPGs) for entry into airway epithelial cells, contributing to bronchiolitis in infants [44, 45].
- **Coronaviruses:** SARS-CoV-2. It uses the angiotensin-converting enzyme 2 (ACE2) receptor, which is expressed in the nasal epithelium, trachea, and alveolar cells, accounting for both mild upper respiratory tract infections and severe lower respiratory tract involvement [46].

- **Human metapneumovirus (hMPV):** hMPV primarily infects respiratory epithelial cells and uses multiple cellular components for entry [47]. Although the exact primary receptor is not fully defined, studies have shown that hMPV binds to HSPGs on the surface of epithelial cells, initiating attachment. In addition, integrins (particularly αVβ1) have been implicated as co-receptors or facilitators of viral entry [48]. These receptor interactions contribute to the virus's tropism for the upper and lower respiratory tract, especially in infants, the elderly, and immunocompromised individuals.

The naming convention, therefore, is not only based on clinical presentation but also reflects virological behavior and host–pathogen interaction. Respiratory viruses may belong to different families (e.g., *Orthomyxoviridae*, *Coronaviridae*, *Paramyxoviridae*), but their shared tissue targeting unifies their classification under this pathogenesis-based nomenclature.

Gastrointestinal Viruses These viruses are grouped under this nomenclature based on their predominant site of infection – gastrointestinal (GI) tract, which includes the stomach, intestines, and associated mucosa. They are referred to as GI or enteric viruses because their clinical manifestations, diarrhea, vomiting, abdominal cramps, and dehydration, result directly from viral replication and damage to intestinal epithelial cells. Their ability to survive harsh gastric conditions, such as low pH and digestive enzymes, is critical to their infectivity and shapes their classification. In particular, they are highly adapted to infect enterocytes (intestinal epithelial cells), and this tropism is mediated by specific cellular receptors and glycans that facilitate attachment, entry, and replication.

Examples

- **Rotaviruses:** One of the leading causes of severe diarrhea in infants and young children globally [49]. Rotavirus targets mature enterocytes in the small intestine. Entry is facilitated by interactions with cell surface glycans, including sialic acid, integrins (α2β1, αVβ3), and heat shock cognate protein 70 (hsc70) [50]. The VP4 and VP7 outer capsid proteins mediate these receptor interactions [51].
- **Noroviruses:** Cause acute gastroenteritis, commonly leading to outbreaks of foodborne illness [52]. They are known for causing acute gastroenteritis in all age groups and are often associated with outbreaks. Noroviruses bind to histo-blood group antigens (HBGAs) expressed on the surface of gut epithelial cells [53]. The variation in HBGA expression among individuals partly explains susceptibility differences.
- **Enteroviruses:** Include coxsackieviruses and echoviruses. They are capable of systemic spread, and initial replication occurs in the oropharynx and intestinal mucosa. Receptors include poliovirus receptor PVR (CD155) for poliovirus and CAR for group B coxsackieviruses [54].

- **Astrovirus:** Infects intestinal epithelial cells, especially in children. The exact receptor remains unclear, but evidence suggests that entry may involve cell surface carbohydrates, and the virus can resist gastric acidity, aiding fecal–oral transmission.

Hepatotropic Viruses Hepatotropic viruses are classified based on their specific affinity for liver tissue. They cause inflammation of the liver (hepatitis) as the primary pathological outcome. The term hepatotropic reflects their tissue tropism – that is, their capacity to target and replicate within hepatocytes (liver cells). This classification is distinct and clinically relevant due to the unique disease course, modes of transmission, and long-term complications associated with liver-specific viral infections. These viruses, despite belonging to different families, share a common pathogenesis – hepatocellular injury, which may result from direct viral cytotoxicity, immune-mediated damage, or chronic inflammation that predisposes the host to fibrosis, cirrhosis, and even hepatocellular carcinoma (HCC).

Examples

- **Hepatitis A virus (HAV):** Causes acute hepatitis, primarily transmitted through ingestion of contaminated food and water. Humans are the only known reservoir [55].
- **HBV:** Can cause both acute and chronic hepatitis, leading to liver cirrhosis and HCC.
- **HCV:** Often leads to chronic hepatitis, with a high risk of progressing to cirrhosis and liver cancer.

Neurotropic Viruses These are viruses that primarily infect the nervous system, including the brain and spinal cord. They are grouped because they share the ability to breach the blood–brain barrier, infect neurons or glial cells, and often cause encephalitis, meningitis, neuropathy, or paralysis.

Examples

- **Rabies virus:** Causes fatal encephalitis, primarily transmitted through animal bites. It particularly causes acute viral encephalitis. Its transmission via animal bites underscores its distinct neuroinvasive pathogenesis
- **Poliovirus:** Infects motor neurons, leading to poliomyelitis and potential paralysis. Poliovirus infects and destroys anterior horn motor neurons, leading to poliomyelitis and flaccid paralysis. Despite initial replication in the gut, its severe outcome defines its nomenclature placement.
- **Herpes simplex virus (HSV):** Can cause encephalitis and meningitis, along with peripheral nervous system infections like cold sores and genital herpes. Both HSV-1 and HSV-2 exhibit neurotropism, establishing latency in sensory ganglia and occasionally causing encephalitis, meningitis, or radiculopathy. Their ability to persist and reactivate in neural tissues contributes to their inclusion in this group.

Lymphotropic Viruses These are viruses that primarily infect the lymphatic system, including lymph nodes and immune cells, specifically targeting T lymphocytes or B lymphocytes.

Examples

- **EBV:** Causes infectious mononucleosis and is associated with certain cancers like Burkitt's lymphoma [56].
- **Human T-lymphotropic virus (HTLV):** Associated with adult T-cell leukemia or lymphoma. Although generally asymptomatic, a few (3–5%) develop adult T-cell leukemia [57].
- **Human immunodeficiency virus (HIV):** Infects T-helper cells, leading to acquired immunodeficiency syndrome (AIDS).

Clinically relevant

- Classifying viruses based on the site of infection or disease presentation (e.g., respiratory, enteric, neurotropic) is highly practical for clinical diagnosis and patient management. It helps healthcare professionals quickly associate symptoms with likely viral agents and guides appropriate testing and treatment.
- This system provides an intuitive and organ-specific framework that is easy to understand and use across disciplines, including medicine, nursing, epidemiology, and public health. Terms like hepatotropic, neurotropic, or dermatotropic viruses immediately signal the type of disease involved.
- Since the classification is tied to disease outcomes and transmission routes, it aids in targeted surveillance, outbreak response, and vaccine deployment. For instance, respiratory viruses can be managed with airborne precautions, while enteric viruses prompt sanitation measures.
- This system supports early outbreak detection based on clinical syndromes, especially where laboratory confirmation is delayed or unavailable. If multiple patients present with meningitis, a neurotropic virus is suspected, prompting faster action.

Overlaps and ambiguity

- Many viruses do not restrict themselves to one organ system. For example, enteroviruses may initially infect the GI tract but later affect the nervous system. This can create confusion in classification and may lead to mislabeling or oversimplification.
- This system does not reflect evolutionary relationships or molecular structure, which are important for understanding virus biology, mutation patterns, and vaccine design. It may group unrelated viruses that happen to affect the same organ.
- In molecular virology and advanced research, pathogenesis-based classification is insufficient. Researchers often require more granular, genetic-based classifications (e.g., Baltimore classification or International Committee on Taxonomy of Viruses [ICTV] taxonomy) to understand replication strategies and genome characteristics.
- The same virus can cause different diseases in different hosts or under different conditions. For instance,

cytomegalovirus (CMV) may be asymptomatic in healthy individuals but neurotropic or hepatotropic in immunocompromised patients, complicating classification.
- Viruses that cause secondary, asymptomatic or mild infections may be under-recognized in pathogenesis-based systems, even if they are epidemiologically significant (e.g., certain blood-borne viruses or emerging zoonoses).

1.3.1.3 Classification Based on Serological Properties

Serological classification of viruses is based on their antigenic properties, specifically how they react to specific antibodies. This method is crucial for understanding virus epidemiology, vaccine development, and diagnosis. Viruses are classified into serotypes and serogroups based on the differences and similarities in their surface antigens (proteins or glycoproteins on the viral envelope or capsid) that are recognized by the immune system.

Virus Serotypes Serotypes are distinct variations within a species of viruses that elicit a specific immune response. Different serotypes of the same virus can cause different immune reactions, as this could lead to type-specific immunity [58].

Examples

- **DENV:** There are four serotypes (DENV-1, DENV-2, DENV-3, and DENV-4) [59]. Infection with one serotype provides lifelong immunity to that serotype but not to the others [60].
- **Poliovirus:** There are three serotypes (PV1, PV2, and PV3), each requiring a specific immune response (minimal heterotypic immunity) [61].

Virus Serogroups A serogroup is a broader classification that includes multiple serotypes sharing common antigenic properties, specifically those that the immune system recognizes.

Examples:

- **Enteroviruses:** Include multiple serogroups such as coxsackieviruses and echoviruses, each with numerous serotypes.
- **Orthomyxoviruses:** Influenza A viruses are categorized into different serogroups based on the HA and neuraminidase (NA) proteins, leading to different subtypes like H1N1 and H3N2.

1.3.1.4 Classification Based on Vector Relationship

Viruses can be classified based on how they are transmitted through vectors. Usually, these viruses are transmitted among animals without ever infecting humans. Humans become incidentally infected by certain arthropods, rodents, and, in some cases, by direct contact with the infected animal.

Arthropod-Borne Viruses

- They are classified as arboviruses. This group of viruses is transmitted through blood-sucking arthropods (especially insects and ticks) [62]. The virus can penetrate and replicate in an insect's gut and salivary gland, where it can be transmitted to another host when bitten. In some instances, the insect is generally unarmed or has no obvious disease. Birds are usually reservoir hosts for some arbovirus. Other reservoir hosts include rodents and nonhuman primates like monkeys. However, there is a high possibility of human-to-human transmission. Examples are listed in Table 1.1.

Rodent-Borne Viruses They are classified as roboviruses [63]. They belong to the *Arenaviridae* and *Hantaviridae* virus families. Rodent-borne viruses are primarily transmitted to humans through contact with rodents or their body fluids, including urine, feces, and saliva. These viruses often have significant implications for human health, causing a range of diseases [63].

Rodents, especially the most diverse order of mammals, serve as reservoirs and vectors for virus transmission. It is rare for reservoir hosts to become chronically infected and overtly diseased. Many robovirus diseases are extremely lethal and contagious and can be acquired in both urban and especially in rural areas among poor people who depend on farming for survival.

Table 1.1 Arboviruses of medical importance based on vector relationship in tropical countries.

Vector type	Virus	Disease
Mosquito-borne viruses	Dengue virus (DENV)	Dengue fever, dengue hemorrhagic fever
	Zika virus	Zika fever, congenital Zika syndrome
	Chikungunya virus	Chikungunya fever
	West Nile virus (WNV)	West Nile fever, a neuroinvasive disease
	Yellow fever virus	Yellow fever
	Japanese encephalitis virus (JEV)	Japanese encephalitis
	Rift Valley fever virus (RVFV)	Rift Valley fever
Tick-borne viruses	Tick-borne encephalitis virus (TBEV)	Tick-borne encephalitis
	Crimean–Congo hemorrhagic fever virus (CCHFV)	Crimean–Congo hemorrhagic fever
	Powassan virus	Powassan encephalitis
Sandfly-borne viruses	Toscana virus	Meningoencephalitis
	Sandfly fever Sicilian virus	Sandfly fever
	Sandfly fever Naples virus	Sandfly fever

Examples are Lassa virus, common in West Africa, transmitted by a multimammate rat (*Mastomys natalensis*) [64]. Transmission of this disease is from rodents through direct handling or contact with infected rodent droppings, saliva, or urine. Other examples are listed in Table 1.2.

Except for the possibility of human-to-human transmission, the distribution of robovirus and arbovirus diseases is generally restricted to the areas inhabited by the vectors or reservoirs.

1.3.2 Baltimore Classification

The Baltimore classification system was developed in 1971 by David Baltimore [65]. It is the most widely used method for classifying viruses. It groups viruses into seven classes based on three key characteristics [66]: (i) Type of nucleic acid – DNA or RNA; (ii) number of strands – single-stranded (ss) or double-stranded (ds); (iii) sense of the RNA genome (for RNA viruses only) – positive-sense (+ssRNA) or negative-sense (–ssRNA). See Table 1.3.

Table 1.2 Rodent-borne viruses of medical importance.

Virus family	Virus name	Disease(s)	Primary rodent host(s)
Hantaviridae	Sin Nombre virus*	Hantavirus pulmonary syndrome (HPS)	Deer mouse (*Peromyscus maniculatus*)
	Hantaan virus*	Hantavirus pulmonary syndrome (HPS) (common in the Americans) and Hemorrhagic fever with renal syndrome (HFRS) (common in Asia)	Striped field mouse (*Apodemus agrarius*)
	Puumala virus*	HFRS	Bank vole (*Myodes glareolus*)
	Seoul virus**	HFRS	Norway rat (*Rattus norvegicus*)
Arenaviridae	Lassa virus*	Lassa fever	Multimammate rat (*Mastomys natalensis*)
	Junin virus	Argentine hemorrhagic fever (AHF)	Drylands vesper mouse (*Calomys musculinus*)
	Machupo virus	Bolivian hemorrhagic fever (BHF)	Large vesper mouse (*Calomys callosus*)
	Lymphocytic choriomeningitis virus (LCMV)	Lymphocytic choriomeningitis (LCM)	House mouse (*Mus musculus*)

* Highlights reported case(s) in tropical countries.
** Reported possible emerging zoonotic risk in tropical countries.

Table 1.3 Baltimore classification of viruses.

Class	Nucleic acid	Replication strategy	Examples
Class I	Double-stranded DNA (dsDNA)	Replicate using host DNA polymerase	Herpesviruses, adenoviruses, poxviruses
Class II	Single-stranded DNA (ssDNA)	Replicate by converting to double-stranded DNA first	Parvoviruses
Class III	Double-stranded RNA (dsRNA)	Replicate using an RNA-dependent RNA polymerase	Rotaviruses
Class IV	Single-stranded RNA (ssRNA, positive-sense)	Directly serves as mRNA for protein synthesis	Picornaviruses, coronaviruses
Class V	Single-stranded RNA (ssRNA, negative-sense)	Replicate via an RNA-dependent RNA polymerase	Influenza viruses, rabies virus
Class VI	ssRNA with reverse transcription	Convert RNA to DNA via reverse transcriptase, then integrate into host DNA	Retroviruses (e.g., HIV)
Class VII	dsDNA with reverse transcription	Replicate via an RNA intermediate and reverse transcription	Hepadnaviruses (e.g., HBV)

Source: Adapted from [65].

1.3.3 Classification Based on Genomic Organization and Molecular Processes

The classification of viruses based on genomic organization and molecular processes is also a more common approach that provides deeper insights into viral biology, replication mechanisms, and evolutionary relationships. This method involves analyzing the structure and function of the viral genome as well as the molecular processes involved in virus replication, transcription, and translation. See Tables 1.4 and 1.5.

1.3.4 International Committee on Taxonomy of Viruses (ICTV) Classification of Viruses

The ICTV was established in 1966, and it is the generally accepted form of virus classification [67]. The ICTV develops and maintains a universal system for naming and categorizing viruses based on a combination of genetic, structural, and biological properties and evolutionary relationships [67]. In this system of classification, viruses are classified in hierarchical order related

Table 1.4 Classification of viruses based on genomic organization and molecular processes.

Properties	Classification
Type of viral genome	DNA or RNA
Number of strands	Single-stranded (ss), double-stranded (ds), or partially double-stranded
Size of the viral genome	Large-genome DNA viruses or small-genome DNA viruses
Genome segment	Segmented or non-segmented
Genome sense	Positive-sense RNA viruses, negative-sense RNA viruses, or RNA ambisense
Open reading frame	Number and position of open reading frame(s)
Other features	Guanine + cytosine (G+C) contents, 5′ terminal cap, 3′terminal poly (A) tail

Table 1.5 Morphological classification of viruses.

Morphological properties	Classification	Examples
Membrane (envelope)	Enveloped viruses	DNA envelope viruses. Examples include herpesviruses, hepadnaviruses (now split into two separate families: *Papillomaviridae* [which includes human papillomaviruses, or HPV] and *Polyomaviridae*)
		RNA enveloped virus. Examples include filovirus, paramyxoviruses, rhabdovirus, and coronaviruses
	Naked or non-enveloped viruses	DNA non-enveloped viruses, e.g., adenoviruses, papovavirus
		RNA non-enveloped virus e.g., picornaviruses, hepatitis A virus
Capsid structure (viral symmetry)	Icosahedral	Non-enveloped icosahedral: poliovirus, hepatitis A virus.
		Enveloped virus Icosahedral: Yellow fever virus, HIV-1, herpes simplex viruses
	Helical	Enveloped virus: measles virus, mumps virus
		Non-enveloped virus: tobacco mosaic virus
Complex structure	Combination of icosahedral and helical capsid	Poxviruses

to the classification system of other cellular organisms – binomial nomenclature. The latest update on viral nomenclature can be found at `https://ictv.global`. This classification is based on the following:

- **Morphology**: Includes the shape of the virus particle, the presence of an envelope, and the symmetry of the capsid.
  - **Genetic material**: Includes the type of nucleic acid (DNA or RNA) and its structure (single or double-stranded, segmented, etc.).
  - **Host range**: Considers the types of organisms that the virus can infect.
- **Phylogenetics**: Considers the evolutionary relationships between viruses.

According to ICTV, virus classification starts from the most diverse realm and continues to a more specific nomenclature – species. This universally accepted classification of viruses employs hierarchical levels (suffix in brackets) from Realm (*-viria*), Subrealm (*-vira*), Kingdom (*-virae*), Subkingdom (*-virites*), Phylum (*-viricota*), Subphylum (*-viricota*), Class (*-viricete*), Subclass (*-viricetidae*), Order (*-virales*), Suborder (*-virineae*), Family (*-viridae*), Subfamily (*-virinae*), Genus (*-virus*), Subgenus (*-virus*), and to species (Tables 1.6 and 1.7).

Table 1.6 Classification of viruses of medical importance based on ICTV classification.

Families	Examples	Common name	Clinical observations/ disease/symptoms
RNA viruses			
Flaviviridae	Flaviviruses	Yellow fever virus	Severe hemorrhagic fever
	Hepatitis C viruses		
Picornaviridae	Enteroviruses	Poliovirus	
	Hepatoviruses	Hepatitis A virus	
Caliciviridae	Caliciviruses		Gastroenteritis
Matonaviridae	Rubiviruses	Rubella virus	Maculopapular rash, often misdiagnosed as measles or scarlet fever
Togaviridae	Alphaviruses	Chikungunya virus	Rash, arthritis
		O'nyong'nyong virus	Rash, arthritis
		Igbo-Ora virus	
Astroviridae	Mamastroviruses		Infantile gastrointestinal infection

(*Continued*)

Table 1.6 (Continued)

Families	Examples	Common name	Clinical observations/ disease/symptoms
Reoviridae	Rotaviruses	Rotavirus 1	Gastroenteritis, vomiting, and fever
Orthomyxoviridae	Influenza A and B viruses	Human influenza virus	Respiratory infection, common cold and flu
Paramyxoviridae	Rubulaviruses	Newcastle disease and mumps virus	Respiratory infections and mumps
	Measles viruses	Measles virus	Measles
Filoviridae	Filovirus	Ebola virus	Severe hemorrhagic fever
		Marburg virus	Severe hemorrhagic fever
Coronaviridae	Coronaviruses	Severe acute respiratory syndrome (SARS)	Respiratory infections and common cold
		Coronavirus disease 2019 (COVID-19)	Respiratory infections and common cold
		Middle East respiratory syndrome coronavirus (MERS-CoV)	Respiratory infections and common cold
Arenaviridae	Arenaviruses	Lassa virus	Severe hemorrhagic fever
Rhabdoviridae	Lyssaviruses	Rabies virus	Encephalitis
		Mokola virus	Fever, convulsion
	Vesiculoviruses		Flu-like symptoms
Retroviridae	Retroviruses	Human T-lymphocyte virus 1	
	Lentiviruses	Human immunodeficiency virus (HIV) 1 and 2	AIDS
Hantaviridae	Orthohantaviruses	Hantavirus	Hantavirus hemorrhagic fever with renal syndrome
Phlenuiviridae	Phleboviruses	Rift Valley fever phlebovirus	Fever, headaches, liver complications
DNA viruses			
Poxviridae	Molluscipoxviruses	Molluscum contagiosum virus	Skin disease
	Orthopoxviruses	Variola (smallpox) virus*	Fever, headache, fatigue

Table 1.6 (Continued)

Families	Examples	Common name	Clinical observations/ disease/symptoms
Adenoviridae	Mastadenoviruses	Human adenovirus	Respiratory and gastrointestinal infection. Keratoconjunctivitis
Papillomaviridae	Papillomaviruses	Human papillomavirus	Warts or papilloma, precancerous lesions
Polyomaviridae	Polyomavirus	Polyomavirus hominis 1 (BK virus)	Nephropathy
		Human polyomavirus 2 (JK virus)	Progressive multifocal leukoencephalopathy
Parvoviridae	Erythroparvoviruses	Human parvovirus B19	*Erythema infectiosum* (fifth disease), arthritis, and arthralgia
	Dependoparvoviruses	Adeno-associated virus	
Hepadnaviridae	*Orthohepadnaviruses*	Hepatitis B	Liver cirrhosis, hepatocellular carcinoma
Herpesviridae	Varicelloviruses	Varicella-zoster virus (chickenpox)	Shingles and fever
	Cytomegaloviruses	Human cytomegalovirus	Cytomegalic inclusion disease and cerebral calcification
	Lymphocryptoviruses	Epstein–Barr virus	Swollen glands in the neck, fever, and rash
	Simplex viruses	Herpes simplex virus 1 and 2	Mucosal ulcers, encephalitis, skin vesicles, and meningitis
	Roseola viruses	Roseola infantum (sixth disease)	Febrile seizures, fever, and rash
	Rhadinoviruses	Kaposi's sarcoma-associated herpesvirus	Kaposi's sarcoma, cancer (common in AIDS patients), and primary effusion lymphoma

* Eradicated worldwide.

Table 1.7 List of viruses of medical importance and the disease(s) they cause.

Virus name	Family	Diseases
Human immunodeficiency virus (HIV)	*Retroviridae*	Acquired immunodeficiency syndrome (AIDS)
Hepatitis B virus (HBV)	*Hepadnaviridae*	Hepatitis B, liver cirrhosis, hepatocellular carcinoma
Hepatitis C virus (HCV)	*Flaviviridae*	Hepatitis C, liver cirrhosis, hepatocellular carcinoma
Influenza virus	*Orthomyxoviridae*	Influenza (flu)
Measles virus	*Paramyxoviridae*	Measles
Mumps virus	*Paramyxoviridae*	Mumps
Rubella virus	*Togaviridae*	Rubella (German measles)
Human papillomavirus (HPV)	*Papillomaviridae*	Cervical cancer, genital warts
Varicella-zoster virus (VZV)	*Herpesviridae*	Chickenpox, shingles
Epstein–Barr virus (EBV)	*Herpesviridae*	Infectious mononucleosis, Burkitt's lymphoma
Herpes simplex virus (HSV)	*Herpesviridae*	Herpes simplex (cold sores, genital herpes)
Cytomegalovirus (CMV)	*Herpesviridae*	Retinitis, pneumonia, encephalitis, and gastrointestinal disease.
Zika virus	*Flaviviridae*	Zika fever, congenital Zika syndrome
Dengue virus (DENV)	*Flaviviridae*	Dengue fever
Severe acute respiratory syndrome coronavirus (SARS-CoV)	*Coronaviridae*	Severe acute respiratory syndrome (SARS)
Middle East respiratory syndrome coronavirus (MERS-CoV)	*Coronaviridae*	Middle East respiratory syndrome (MERS)
SARS-CoV-2	*Coronaviridae*	Coronavirus disease 2019 (COVID-19)
Norovirus	*Caliciviridae*	Acute gastroenteritis
Rotavirus	*Reoviridae*	Severe diarrhea in infants and young children
Rhinovirus	*Picornaviridae*	Common cold
Enterovirus	*Picornaviridae*	Hand, foot, and mouth disease, poliomyelitis
Rabies virus	*Rhabdoviridae*	Rabies

Table 1.7 (Continued)

Virus name	Family	Diseases
Ebola virus	*Filoviridae*	Ebola virus disease
Marburg virus	*Filoviridae*	Marburg virus disease
Human T-cell leukemia virus (HTLV)	*Retroviridae*	Adult T-cell leukemia/lymphoma
Human parainfluenza virus (HPIV)	*Paramyxoviridae*	Parainfluenza
Respiratory syncytial virus (RSV)	*Paramyxoviridae*	Respiratory syncytial virus infection
Hantavirus	*Hantaviridae*	Hantavirus pulmonary syndrome, hemorrhagic fever with renal syndrome
Lassa virus	*Arenaviridae*	Lassa fever
Chikungunya virus	*Togaviridae*	Chikungunya fever
Yellow Fever virus	*Flaviviridae*	Yellow fever
West Nile virus	*Flaviviridae*	West Nile fever
Hepatitis A virus (HAV)	*Picornaviridae*	Hepatitis A
Hepatitis E virus (HEV)	*Hepeviridae*	Hepatitis E
Human T-lymphotropic virus 1 (HTLV-1)	*Retroviridae*	Adult T-cell leukemia/lymphoma
Human T-lymphotropic virus 2 (HTLV-2)	*Retroviridae*	HTLV-2 associated myelopathy
Human parvovirus B19	*Parvoviridae*	Fifth disease (erythema infectiosum)
Human herpesvirus 6 (HHV-6)	*Herpesviridae*	Roseola infantum (sixth disease)
Human herpesvirus 7 (HHV-7)	*Herpesviridae*	Roseola-like symptoms
Human herpesvirus 8 (HHV-8)	*Herpesviridae*	Kaposi's sarcoma
John Cunningham virus (JC Virus) or human polyomavirus 2	*Polyomaviridae*	Progressive multifocal leukoencephalopathy (PML)
BK Virus	*Polyomaviridae*	Nephropathy
Simian virus 40 (SV40)	*Polyomaviridae*	Potentially linked to certain cancers
Coxsackievirus	*Picornaviridae*	Hand, foot, and mouth disease, myocarditis
Echovirus	*Picornaviridae*	Aseptic meningitis, myocarditis
Rhinovirus	*Picornaviridae*	Common cold

(*Continued*)

Table 1.7 (Continued)

Virus name	Family	Diseases
SARS-CoV	*Coronaviridae*	Severe acute respiratory syndrome (SARS)
MERS-CoV	*Coronaviridae*	Middle East respiratory syndrome (MERS)
SARS-CoV-2	*Coronaviridae*	COVID-19
Crimean–Congo hemorrhagic fever virus	*Nairoviridae*	Crimean–Congo hemorrhagic fever
Sin Nombre virus	*Hantaviridae*	Hantavirus pulmonary syndrome
Rift Valley fever virus	*Bunyaviridae*	Rift Valley fever
Toscana virus	*Bunyaviridae*	Toscana virus infection
Nipah virus	*Paramyxoviridae*	Nipah virus infection
Hendra virus	*Paramyxoviridae*	Hendra virus infection
Ross river virus	*Togaviridae*	Ross River fever
Eastern equine encephalitis virus	*Togaviridae*	Eastern equine encephalitis
Western equine encephalitis virus	*Togaviridae*	Western equine encephalitis
Venezuelan equine encephalitis virus	*Togaviridae*	Venezuelan equine encephalitis
La Crosse virus	*Bunyaviridae*	La Crosse encephalitis
Powassan virus	*Flaviviridae*	Powassan virus disease
Human metapneumovirus (hMPV)	*Paramyxoviridae*	Respiratory tract infections
Human bocavirus	*Parvoviridae*	Respiratory tract infections
Astrovirus	*Astroviridae*	Gastroenteritis
Sapovirus	*Caliciviridae*	Gastroenteritis
Colorado tick fever virus	*Reoviridae*	Colorado tick fever
Toscana virus	*Bunyaviridae*	Toscana virus infection
Monkeypox virus	*Poxviridae*	Monkeypox
Molluscum contagiosum virus	*Poxviridae*	Molluscum contagiosum

Note: The list is inexcusable.

1.4 How Will Newly Identified Viruses Be Named?

Naming newly identified viruses involves a systematic process overseen by the ICTV. The goal is to ensure that the naming is standardized, informative, and avoids confusion. Here is a step-by-step overview of how newly identified viruses are named:

1. **Initial identification and characterization**
   Detection: New viruses are typically identified through research involving clinical samples, environmental samples, or laboratory investigations.
   Characterization: Scientists characterize the virus based on its genetic material, structure, replication mechanism, and pathogenic properties. Specifically, by determining whether it is DNA or RNA, its symmetry, the presence of an envelope, and other biological properties. If unknown, the information, such as the origin of the virus (animal or human origin) and the host range (types of cells or organisms the virus can infect), must be determined. Also, information regarding the disease association or any observed disease or symptoms linked to the virus must be determined or reported.
2. **Preliminary classification**
   Once sufficient virological data have been gathered – including genetic, structural, and biological characteristics – a preliminary classification is typically established for provisional scientific communication and further study. At this stage, the virus may be tentatively assigned to an existing family based on its key features or classified into a new family if it represents a novel group. Within that family, the virus is further placed into an appropriate genus, or a new genus may be proposed if required. The species designation follows, informed by a combination of sequence identity, host range, replication strategy, and pathogenicity.
3. **Submission to ICTV**
   Proposal: Following this initial classification, researchers proceed with a formal submission to the ICTV. This involves preparing a comprehensive proposal that outlines the proposed taxonomic placement, supported by detailed genomic sequences, structural characteristics, and relevant biological or pathological data.
   Review: The ICTV evaluates these proposals through peer review before ratifying any official changes to virus taxonomy. ICTV may set up specialized committees that review these proposals. This process involves peer review and may include consultation with other experts in virology.
4. **Naming convention**
   Virus family and genus: Names for new viruses are generally chosen based on established conventions. Families often have names that reflect shared properties, while genera are named to reflect more specific relationships.

Species names: New species names usually reflect the disease they cause or other relevant characteristics. The names are chosen to be descriptive and avoid overlap with existing names.
Virus names: The specific name of the virus itself can be derived from the disease it causes or distinctive features of the virus. Recently, naming viruses based solely on their place of discovery is no longer acceptable. The current approach emphasizes more descriptive, systematic, and sometimes neutral names.

5. **Reasons for changing previous naming practices**
Avoiding stigmatization
Naming viruses after geographic locations can sometimes lead to stigma or blame. For example, the early names of viruses such as the "Spanish flu" or "Wuhan coronavirus" can unfairly associate the disease with a specific region or population, potentially causing social or economic issues.
Scientific precision
Modern naming practices aim to reflect the virus's biological and genetic characteristics more accurately. This helps in better understanding and communicating about the virus's nature, replication, and classification.
Globalization and spread
Viruses can spread rapidly across the globe, making it less relevant to name them after a specific location. For instance, the name "Wuhan coronavirus" is unacceptable. The virus quickly became a pandemic infection, and the name was updated to "SARS-CoV-2" to reflect its relationship to SARS.
Also, in classifying viruses, it is important to avoid people's names to prevent any unintended offense or confusion. Similarly, avoiding the names of animals or food sources is important to prevent unnecessary concern or negativity toward certain animals or food items.
6. **Approval and publication**
Approval: The ICTV's executive committee reviews and approves the proposed names. This step ensures that the name follows the taxonomic rules and does not conflict with existing names.
Publication: Once approved, the names are published in the scientific literature and official ICTV resources. This includes databases, taxonomic lists, and other reference materials.
7. **Ongoing updates:**
As more information becomes available, the classification and naming of viruses may be updated. This ensures that the names remain accurate and relevant to the virus's characteristics and relationships.

1.5 Study Questions

Case Study Questions

1.1 A 30-year-old female patient presents with symptoms of fever, headache, and rash. Laboratory tests reveal the presence of a virus with a single-stranded RNA genome and a lipid envelope. Based on this structural information:

(a) What type of virus is most likely causing the infection?

(b) Explain how the presence of a viral envelope might influence the virus's susceptibility to disinfectants and environmental conditions.

1.2 A 24-month-old female is diagnosed with a severe respiratory infection. Electron microscopy reveals the presence of spike proteins, and molecular studies reveal an RNA genome. Considering the basic structure of this virus:

(a) What kind of virus can this be, and why?

(b) How could this impact public health since it is an RNA virus?

(c) What implications does the virus's RNA genome have for its replication strategy and potential treatments?

1.3 A healthcare worker is accidentally exposed to a blood sample spilled on the bench from a patient previously diagnosed to be infected with a viral infection while working on a laboratory bench. The virus is known to have a helical capsid structure and an RNA genome. Based on the virus structure:

(a) What precautions should be taken to prevent further exposure or infection?

(b) Discuss the role of the helical capsid in the virus's ability to infect host cells.

(c) How might the RNA genome affect the mutation rate and the development of antiviral resistance?

1.4 A 32-year-old European man presents with symptoms of fever, myalgia, severe headache, and abdominal pain. A few days later, he develops jaundice, and his laboratory tests show elevated liver enzymes. This is his first time traveling to a tropical region where mosquitoes are prevalent. Considering viral pathogenesis, which class of viruses is most likely responsible for his symptoms, and what is the significance of understanding the pathogenesis of this viral class?

1.5 A 45-year-old woman presents with a persistent cough, chest pain, and shortness of breath. Imaging shows bilateral lung infiltrates. Laboratory results confirm a positive test for a virus with an RNA genome. Considering the classification based on viral pathogenesis, which class of virus is likely responsible, and why is it important to classify this virus based on the disease it causes?

1.6 A 5-year-old child presents to the emergency department with a high fever, severe cough, and difficulty breathing. The child's condition rapidly deteriorated and was admitted to the intensive care unit. A respiratory sample was taken, and laboratory results revealed the presence of a virus with a single-stranded, negative-sense RNA genome. To which Baltimore group does this virus belong? What could be a potential virus responsible?

1.7 A research laboratory was studying a newly discovered virus that causes hemorrhagic fever syndrome in humans. Initial genetic analysis shows that the virus has a single-stranded, positive-sense RNA genome. How would this virus be categorized? What is an example of a known virus in this group, and how does its classification inform the strategies for vaccine development?

Objectives

1.1 **Which of the following is a characteristic of enveloped viruses?**
(a) Increased resistance to environmental stress
(b) Sensitivity to detergents and alcohol-based disinfectants
(c) Replication exclusively in the nucleus
(d) Presence of a helical capsid

1.2 **What structural component of a virus determines its shape?**
(a) Lipid envelope
(b) Nucleic acid type
(c) Capsid
(d) Spike proteins

1.3 **Which of the following statements about RNA viruses is true?**
(a) They are always enveloped
(b) They have a lower mutation rate compared to DNA viruses
(c) They typically replicate in the host cell's cytoplasm
(d) They have a double-stranded RNA genome

1.4 **In which way does the presence of a viral envelope affect the mode of transmission of a virus?**
(a) Increases resistance to desiccation
(b) Enhances transmission via the fecal–oral route
(c) Requires close contact or fluid exchange for transmission
(d) None of the above

1.5 **Which part of a virus is primarily responsible for its ability to attach and enter a host cell?**
(a) Nucleic acid
(b) Capsid
(c) Envelope glycoproteins
(d) Matrix proteins

1.6 **The presence of a double-stranded DNA genome is a characteristic of which virus family?**
(a) *Flaviviridae*
(b) *Herpesviridae*
(c) *Picornaviridae*
(d) *Retroviridae*

1.7 **Which of the following is a major reason why RNA viruses tend to have higher mutation rates than DNA viruses?**
(a) RNA viruses replicate faster
(b) RNA-dependent RNA polymerases lack proofreading ability
(c) RNA viruses are more complex
(d) RNA viruses infect more host species

1.8 **Which of the following virus families is primarily transmitted by arthropod vectors?**
(a) *Herpesviridae*
(b) *Papillomaviridae*
(c) *Flaviviridae*
(d) *Adenoviridae*

1.9 **What type of vector is commonly associated with the transmission of viruses in the *Togaviridae* family?**
(a) Mosquitoes
(b) Ticks
(c) Fleas
(d) Lice

1.10 **Which of the following is a key factor in the pathogenesis of viral infections?**
(a) The virus's ability to evade the immune system
(b) The host's blood type

(c) The ambient temperature
(d) The availability of antibiotics

1.11 Which mechanism is commonly used by viruses to spread from one host cell to another within a host?
(a) Direct cell-to-cell contact
(b) Humoral immunity
(c) Phagocytosis
(d) Extracellular matrix degradation

References

1. Mammas, I.N., et al., *Viruses, vaccinations and RSV: exploring terminology in pediatric virology*. Exp. Ther. Med., 2020. **20**(6): p. 300.
2. Louten, J., *Virus Structure and Classification. Essential Human Virology*. 2016. p. 19–29. 10.1016/B978-0-12-800947-5.00002-8 Epub 2016 May 6. PMCID: PMC7150055.
3. Simón, D., et al., *Two decades ago, giant viruses were discovered: the fall of an old paradigm*. Front. Microbiol., 2024. **15**: 1356711.
4. Saadi, H., et al., *First isolation of Mimivirus in a patient with pneumonia*. Clin. Infect. Dis., 2013. **57**(4): p. e127–e134.
5. Bisio, H., et al., *Evolution of giant pandoravirus revealed by CRISPR/Cas9*. Nat. Commun., 2023. **14**(1): p. 428.
6. Shalamova, L., et al., *Nucleocapsids of the Rift Valley fever virus ambisense S segment contain an exposed RNA element in the center that overlaps with the intergenic region*. Nat. Commun., 2024. **15**(1): p. 7602.
7. Sabsay, K.R. and Velthuis, A.J.W.t., *Negative and ambisense RNA virus ribonucleocapsids: more than protective armor*. Microbiol. Mol. Biol. Rev., 2023. **87**(4): e00082-23.
8. Zhao, Q., et al., *Pregenomic RNA launch hepatitis B virus replication system facilitates the mechanistic study of antiviral agents and drug-resistant variants on covalently closed circular DNA synthesis*. J. Virol., 2022. **96**(24): e0115022.
9. Zeltzer, S., et al., *Host translesion polymerases are required for viral genome integrity*. PNAS, 2022. **119**(33): e2203203119.
10. Domingo, E., et al., *Mutation rates, mutation frequencies, and proofreading-repair activities in RNA virus*. Genetics., 2021. **13**(9): p. 1882.
11. Parvez, M.K., *Geometric architecture of viruses*. World J. Virol., 2020. **9**(2): p. 5–18.
12. Luo, M., Terrell, J.R., and Mcmanus, S.A., *Nucleocapsid structure of negative strand RNA virus*. Viruses, 2020. **12**(8): p. 835.
13. Prasad, B.V. and Schmid, M.F., *Principles of virus structural organization*. Adv. Exp. Med. Biol., 2012. **726**: p. 17–47.
14. Liu, J., et al., *The palisade layer of the poxvirus core is composed of flexible A10 trimers*. Nat. Struct. Mol. Biol., 2024. **31**(7): p. 1105–1113.
15. Condit, R.C., Moussatche, N., and Traktman, P., *In a nutshell: structure and assembly of the vaccinia virion. Adv. Virus Res*., 2006. **66**: p. 31–124. doi: 10.1016/S0065-3527(06)66002-8. PMID: 16877059.
16. Surjit, M. and Lal, S.K., *The nucleocapsid protein of the SARS coronavirus: structure, function and therapeutic potential*, in *Molecular Biology of the SARS-Coronavirus*. Vol. 22, S. Lal, Editor. 2009, Springer. p. 129–151. doi: 10.1007/978-3-642-03683-5_9.
17. Klasse, P.J., Bron, R., and Marsh, M., *Mechanisms of enveloped virus entry into animal cells*. Adv. Drug Delivery Rev., 1998. **34**(1): p. 65–91.
18. Bosch, B.J., et al., *Spike protein assembly into the coronavirion: exploring the limits of its sequence requirements*. Virology, 2005. **334**(2): p. 306–318.
19. Liu, H.Y. and Yang, P.L., *Small-molecule inhibition of viral fusion glycoproteins*. Annu. Rev. Virol., 2021. **8**(1): p. 459–489.
20. Nakamura, K. and Compans, R.W., *Glycopeptide components of influenza viral glycoproteins*. Virology, 1978. **86**(2): p. 432–442.

21. Huang, C.Y.H., et al., *The dengue virus type 2 envelope protein fusion peptide is essential for membrane fusion*. Virology, 2010. **396**(2): p. 305–315.
22. Stiasny, K., et al., *Membrane interactions of the tick-borne encephalitis virus fusion protein E at low pH*. J. Virol., 2002. **76**(8): p. 3784–3790.
23. Garry, R.F. and Dash, S., *Proteomics computational analyses suggest that hepatitis C virus E1 and pestivirus E2 envelope glycoproteins are truncated class II fusion proteins*. Virology, 2003. **307**(2): p. 255–265.
24. Zeng, L., et al., *Environmental stability and transmissibility of enveloped viruses at varied animate and inanimate interfaces*. Environ. Health (Wash), 2023. **1**(1): p. 15–31.
25. Saletti, D., et al., *The Matrix protein M1 from influenza C virus induces tubular membrane invaginations in an in vitro cell membrane model*. Sci. Rep., 2017. **7**: p. 40801.
26. Batishchev, O.V., *Physico-chemical mechanisms of the functioning of membrane-active proteins of enveloped viruses*. Biochem. (Mosc) Suppl. Ser. A Membr. Cell Biol., 2022. **16**(4): p. 247–260.
27. Kinchington, P.R., et al., *Herpes simplex virus and varicella zoster virus, the house guests who never leave*. Herpesviridae, 2012. **3**(1): p. 5.
28. Murata, T., *Epstein–Barr virus: the molecular virology and the associated diseases*. Fujita Med. J., 2023. **9**(2): p. 65–72.
29. He, H.-P., et al., *Structure of Epstein–Barr virus tegument protein complex BBRF2-BSRF1 reveals its potential role in viral envelopment*. Nat. Commun., 2020. **11**(1): p. 5405.
30. Fauquet, C.M., *Taxonomy, classification and nomenclature of viruses*, in *Encyclopedia of Virology*. 2008. p. 9–23. doi: 10.1016/B978-012374410-4.00509-4 Epub 2008 Jul 30. PMCID: PMC7150148.
31. Merettig, N. and Bockmühl, D.P., *Virucidal efficacy of laundering*. Pathogens, 2022. **11**(9): p. 993.
32. Narula, P., et al., *Chemical inactivation of two non-enveloped viruses results in distinct thermal unfolding patterns and morphological alterations*. BMC Microbiol., 2024. **24**(1): p. 413.
33. Oliveira, A.C., et al., *Low temperature and pressure stability of picornaviruses: implications for virus uncoating*. Biophys. J., 1999. **76**(3): p. 1270–1279.
34. Jacobs, S.E., et al., *Human rhinoviruses*. Clin. Microbiol. Rev., 2013. **26**(1): p. 135–162.
35. Pitino, M.A., et al., *The impact of thermal pasteurization on viral load and detectable live viruses in human milk and other matrices: a rapid review*. Appl. Physiol. Nutr. Metab., 2021. **46**(1): p. 10–26.
36. Johne, R., Scholz, J., and Falkenhagen, A., *Heat stability of foodborne viruses – findings, methodological challenges and current developments*. Int. J. Food Microbiol., 2024. **413**: 110582.
37. Lytle, C.D. and Sagripanti, J.L., *Predicted inactivation of viruses of relevance to biodefense by solar radiation*. J. Virol., 2005. **79**(22): p. 14244–14252.
38. Rauth, A.M., *The physical state of viral nucleic acid and the sensitivity of viruses to ultraviolet light*. Biophys. J., 1965. **5**(3): p. 257–273.
39. Zhou, Y., Frey, T.K., and Yang, J.J., *Viral calciomics: interplays between Ca2+ and virus*. Cell Calcium., 2009. **46**(1): p. 1–17.
40. Vasickova, P., et al., *Issues concerning survival of viruses on surfaces*. Food Environ. Virol., 2010. **2**(1): p. 24–34. doi: 10.1007/s12560-010-9025-6 Epub 2010 Feb 4.
41. Wanguyun, A.P., Oishi, W., and Sano, D., *Sensitivity evaluation of enveloped and non-enveloped viruses to ethanol using machine learning: a systematic review*. Food Environ. Virol., 2024. **16**(1): p. 1–13.
42. Zhao, C. and Pu, J., *Influence of host sialic acid receptors structure on the host specificity of influenza viruses*. Viruses, 2022. **14**(10): p. 2141.

43. Bianco, A., et al., *Expression of intercellular adhesion molecule-1 (ICAM-1) in nasal epithelial cells of atopic subjects: a mechanism for increased rhinovirus infection?* Clin. Exp. Immunol., 2000. **121**(2): p. 339–345.
44. Tayyari, F., et al., *Identification of nucleolin as a cellular receptor for human respiratory syncytial virus.* Nat. Med., 2011. **17**(9): p. 1132–1135.
45. Feng, Z., Xu, L., and Xie, Z., *Receptors for respiratory syncytial virus infection and host factors regulating the life cycle of respiratory syncytial virus.* Front. Cell. Infect. Microbiol., 2022. **12**: 858629.
46. Beyerstedt, S., Casaro, E.B., and Rangel, B.E., *COVID-19: angiotensin-converting enzyme 2 (ACE2) expression and tissue susceptibility to SARS-CoV-2 infection.* Eur. J. Clin. Microbiol. Infect. Dis., 2021. **40**(5): p. 905–919.
47. Babawale, P.I. and Guerrero-Plata, A., *Differential responses of pediatric and adult primary epithelial cells to human metapneumovirus and respiratory syncytial virus infection.* Viruses, 2025. **17**(3): p. 380.
48. Klimyte, E.M., et al., *Inhibition of human metapneumovirus binding to heparan sulfate blocks infection in human lung cells and airway tissues.* J. Virol., 2016. **90**(20): p. 9237–9250.
49. Crawford, S.E., et al., *Rotavirus infection.* Nat. Rev. Dis. Primers, 2017. **3**: p. 17083.
50. Amimo, J.O., et al., *Rotavirus interactions with host intestinal epithelial cells.* Front. Immunol., 2021. **12**: 793841.
51. Taube, S., Jiang, M., and Wobus, C.E., *Glycosphingolipids as receptors for non-enveloped viruses.* Viruses, 2010. **2**(4): p. 1011–1049.
52. Koo, H.L., et al., *Noroviruses: the leading cause of gastroenteritis worldwide.* Discov. Med., 2010. **10**(50): p. 61–70.
53. Schroten, H., Hanisch, F.G., and Hansman, G.S., *Human norovirus interactions with histo-blood group antigens and human milk oligosaccharides.* J. Virol., 2016. **90**(13): p. 5855–5859.
54. Pinkert, S., et al., *The coxsackievirus and adenovirus receptor: glycosylation and the extracellular D2 domain are not required for coxsackievirus B3 infection.* J. Virol., 2016. **90**(12): p. 5601–5610.
55. Verma, R. and Khanna, P., *Hepatitis A vaccine should receive priority in National Immunization Schedule in India.* Hum. Vaccines Immunother., 2012. **8**(8): p. 1132–1134.
56. Dunmire, S.K., Hogquist, K.A., and Balfour, H.H., *Infectious mononucleosis.* Curr. Top. Microbiol. Immunol., 2015. **390**(Pt 1): p. 211–240.
57. Giam, C.Z. and Semmes, O.J., *HTLV-1 infection and adult T-cell leukemia/lymphoma-a tale of two proteins: tax and HBZ.* Viruses, 2016. **8**(6).
58. Kim, J.H., Jang, J.Y., and Jang, Y.J., *Human rhinovirus serotypes induces different immune responses.* Virol. j., 2021. **18**(1): p. 232.
59. Yung, C.F., et al., *Dengue serotype-specific differences in clinical manifestation, laboratory parameters and risk of severe disease in adults, Singapore.* Am. J. Trop. Med. Hyg., 2015. **92**(5): p. 999–1005.
60. Ross, T.M., *Dengue virus.* Clin. Lab Med., 2010. **30**(1): p. 149–160.
61. Bachmann, M.F., Mohsen, M.O., and Speiser, D.E., *The impact of viral evolution on vaccine development for SARS-CoV-2.* Curr. Opin. Immunol., 2025. **96**: 102612.
62. Huang, Y.S., Higgs, S., and Vanlandingham, D.L., *Arbovirus-mosquito vector-host interactions and the impact on transmission and disease pathogenesis of arboviruses.* Front. Microbiol., 2019. **10**: p. 22.
63. Rodas, J.D., Londoño, A.F., and Solari, S., *Epidemiological surveillance of rodent-borne viruses (roboviruses).* Methods Mol. Biol., 2018. **1604**: p. 101–109.

64. Lecompte, E., et al., *Mastomys natalensis and Lassa fever, West Africa*. Emerg. Infect. Dis., 2006. **12**(12): p. 1971–1974.
65. Koonin, E.V., Krupovic, M., and Agol, V.I., *The baltimore classification of viruses 50 years later: how does it stand in the light of virus evolution?* Microbiol. Mol. Biol. Rev., 2021. **85**(3): e0005321.
66. Mahmoudabadi, G. and Phillips, R., *A comprehensive and quantitative exploration of thousands of viral genomes*. Elife, 2018. **7**: e31955.
67. Siddell, S.G., et al., *Virus taxonomy and the role of the International Committee on Taxonomy of Viruses (ICTV)*. J. Gen. Virol., 2023. **104**(5): 001840.

2 Molecular Mechanisms of Virus Replication

The molecular basis of virus replication varies significantly among different virus types, reflecting their diverse genome structures, replication strategies, and interactions with host cellular machinery. For instance, DNA viruses, RNA viruses, retroviruses, and other virus families each have unique strategies for replication. Hence, understanding these molecular processes is essential not only for understanding the pathogenesis of viral infections but also for improving diagnostic approaches and guiding the development of effective antiviral therapies and vaccines. The replication cycle can generally be summarized as follows:

2.1 Virus Attachment

All virus replication cycle begins with the attachment of the virus to a specific receptor (s) on the surface of a host cell. This initial binding allows the virus to enter the host cell. Although the process varies significantly among different viruses depending on receptors, viral proteins (ligands), and/or co-receptors (see Table 2.1 for an overview of different viruses and their corresponding host cell receptors), these interactions are largely specific and play a crucial role in determining the host range and "tissue tropism" of the virus.

2.1.1 Tissue Tropism

Tissue tropism refers to the specific tissues or cell types within a host organism that a virus can infect and replicate in. Viruses exhibit diverse tissue tropism, which is largely determined by the interaction between viral surface proteins and host cell receptors, as well as other intracellular environments that support viral replication.

1. **Factors influencing tissue tropism**
   - **Receptor availability:** One of the most critical determinants of viral tissue tropism is the availability of specific host cell receptors. Viruses must attach to and enter host cells to initiate infection, and this process typically requires interaction between viral surface proteins and specific cellular receptors. The distribution of these receptors across different tissues and

Table 2.1 Examples of virus ligands and host receptors required for attachment.

Viruses	Receptors	Human cells (tropism)	Viral ligands
HIV	CD4	T-helper cells, macrophages, dendritic cells	Glycoprotein (gp)120
Influenza virus	Sialic acid	Epithelial cells of the respiratory tract	Hemagglutinin (HA)
SARS-CoV-2	Angiotensin-converting enzyme 2 (ACE2)	Epithelial cells of the respiratory tract, lung, heart, kidney, intestine	Spike (S) protein
Rhinovirus	Intracellular adhesion molecule 1 (ICAM-1)	Epithelial cells of the respiratory tract	Viral protein 1 (VP1)
Epstein–Barr virus (EBV)	CD21 Ephrin receptor A2	B-cells	gp350/220
Rabies virus	Nicotinic acetylcholine receptor (nAchR); neuronal cell adhesion molecule (NCAM); integrin beta 1 (ITGB1)	Neurons in the central and peripheral nervous systems	Rabies virus glycoprotein (RVG)
Hepatitis C virus (HCV)	CD81; scavenger receptor class B type 1 (SR-B1); low-density lipoprotein receptor (LDLR); CD81 tetraspanin; mannose-binding lectins DC-SIGN and L-SIGN; heparan sulfate proteoglycans	Hepatocytes (liver cells)	E2 glycoprotein
Measles virus	Signaling lymphocyte activation molecule (SLAM), CD46 or nectin-4	Epithelial cells, immune cells	Envelope glycoproteins hemagglutinin (H) and fusion (F) protein
Coxsackievirus	Coxsackievirus and adenovirus receptor (CAR)	Epithelial cells, endothelial cells, heart muscle cells	VP1
Cytomegalovirus (CMV)	Heparin sulfate proteoglycans (HSPGs); platelet-derived growth factor receptor (PDGFRα); OR14I1: defines viral epithelial cell tropism	Various cells including epithelial, endothelial, and fibroblasts	In fibroblasts, glycoprotein H (gH), gL, and gO form a trimer that binds to PDGFRα

Table 2.1 (Continued)

Viruses	Receptors	Human cells (tropism)	Viral ligands
Ebola virus	NPC1 (Niemann–Pick C1); T-cell immunoglobulin and mucin domain (TIM) family proteins	Various cells, including macrophages, dendritic cells, endothelial cells	gp; phosphatidylserine (PS) on the surface of EBOV
Marburg virus	Asialoglycoprotein receptor (ASGP-R): a potential liver-specific receptor;TIM-1 serves as receptor for both Zaire ebolavirus and Lake Victoria Marburg virus	Various cells, including macrophages, dendritic cells, endothelial cells	gp
Lassa virus	Alpha-dystroglycan	Various cells, including endothelial cells, macrophages	gp1
Rift Valley fever virus	Heparan sulfate	Various cells, including liver cells, endothelial cells	gN
Dengue virus	DC-SIGN (dendritic cell-specific intercellular adhesion molecule-3-grabbing non-integrin); human mannose-binding receptor (MR) on macrophages	Dendritic cells, macrophages	E (Envelope) protein
Chikungunya virus	Matrix remodeling-associated protein 8 (Mxra8)	Various cells, including fibroblasts, skeletal muscle cells	E2 (Envelope) protein
Yellow fever virus	Heparan sulfate proteoglycans (HSPGs)	Hepatocytes, endothelial cells, dendritic cells	E (Envelope) protein
Zika virus	AXL, Tyro3	Neural progenitor cells, epidermal keratinocytes, fibroblasts	E (Envelope) protein
Human herpesvirus 8 (HHV-8/Kaposi"s sarcoma-associated herpesvirus)	Ephrin-A receptor 2 (EphA2)	Endothelial cells, B-cells, epithelial cells	gB, gH/gL
Crimean–Congo hemorrhagic fever virus (CCHFV)	Uncertain; possibly low-density lipoprotein receptor (LDLR) or heparan sulfate [1]	Various cells, including endothelial cells, hepatocytes	gN (Glycoprotein)

organs largely dictates which cell types a virus can infect. For instance, the human immunodeficiency virus (HIV) targets cluster of differentiation 4 ($CD4^+$) T-helper cells because it binds specifically to the CD4 receptor and a co-receptor, such as C-C chemokine receptor 5 (CCR5) or C-X-C chemokine receptor 4 (CXCR4) that is predominantly expressed on these immune cells [2, 3].

- **Co-receptor:** In many cases, viruses also rely on co-receptors or attachment factors in addition to the primary receptor. These co-receptors, like the primary receptor, can enhance viral binding, facilitate membrane fusion, or enable internalization. In some cases, the absence of this co-receptor may prevent virus entry even in the presence of the primary receptor. For example, HIV requires both the CD4 receptor and either co-receptor CCR5 or CXCR4 or both for successful entry into immune cells. The specific co-receptor required not only affects which cells the virus infects but also influences disease progression [4].

However, receptor and co-receptor expression alone do not fully explain tissue tropism and susceptibility to virus infection. Other contributing factors include:

- **Viral surface proteins:** The interaction between viral surface proteins and host cell receptors is a primary determinant of tissue specificity. These proteins, often glycoproteins embedded in the viral envelope or capsid, mediate attachment and entry into host cells. For example, the spike (S) protein of severe acute respiratory syndrome coronavirus 2 (SARS-CoV-2) has a high binding affinity for the angiotensin-converting enzyme 2 (ACE2) receptor, which is abundantly expressed in the respiratory tract and gastrointestinal system [5], explaining the broad tissue tropism observed with coronavirus disease 2019 (COVID-19). Mutations in viral surface proteins can expand or restrict tissue targeting by altering receptor affinity or specificity [6, 7].
- **Host immune system:** Immune defenses of various tissues may restrict viral replication and limit tissue tropism.

2. **Other host factors include**
   - **Age:** Children are more susceptible to certain viruses [8], such as poliovirus and coxsackievirus, due to their developing immune systems. In contrast, older adults may experience more severe disease due to immune senescence [9, 10].
   - **Nutritional status:** Adequate nutrition is essential for maintaining effective immune responses. Malnutrition can weaken the immune system, making individuals more vulnerable to viral

infections. For instance, deficiencies in micronutrients like vitamins A, C, and D and zinc can impair immune responses, increasing susceptibility to viral diseases like measles or respiratory infections [11].

- **Gender:** Gender can influence viral infection outcomes. For example, human papillomavirus (HPV) targets different tissues in males and females, predominantly infecting the cervix in females and the penile epithelium in males. Hormonal differences also play a role in susceptibility and immune response to certain viruses. Estrogen and testosterone modulate immune responses differently, contributing to gender-based differences in susceptibility and disease outcomes. Additionally, some viruses may exploit hormone-regulated gene expression to establish infection in a sex-specific manner.
- **Genetic background of the host:** Host genetic factors, such as variations in immune system genes (e.g., major histocompatibility complex [MHC] or human leukocyte antigen [HLA]), can influence susceptibility to specific viruses [12]. Some individuals may have a genetic predisposition to either heightened resistance or increased vulnerability to infections like HIV or hepatitis [13–15].
- **Humoral and cellular-mediated immune response:** The strength and specificity of the host's humoral (antibody-mediated) and cellular immune responses are critical in determining the outcome of viral infections. An effective immune response can limit viral spread and replication, while a compromised immune system (e.g., due to HIV infection) can lead to severe or chronic viral diseases [16, 17].

3. **Clinical implications of tissue tropism**
   - **Disease manifestation:** Tissue tropism directly impacts the clinical manifestations of viral infections. For example, respiratory viruses such as influenza primarily infect the respiratory tract, leading to symptoms like cough, fever, and congestion. In contrast, neurotropic viruses like rabies and herpes simplex virus (HSV) target the nervous system, leading to neurological symptoms such as paralysis or encephalitis.
   - **Transmission:** Tissue tropism influences the mode of viral transmission [18]. For example, viruses that infect the respiratory tract (e.g., influenza, SARS-CoV-2) are typically spread through airborne droplets, whereas viruses with gastrointestinal tropism (e.g., rotavirus, norovirus) are transmitted through fecal–oral routes.
   - **Viral spread within the host:** Tissue tropism helps explain how viruses move from one organ

system to another within the host. For instance, HIV exhibits a tropism for immune cells such as $CD4^+$ T cells, macrophages, and dendritic cells, leading to widespread immune suppression. This tissue-specific replication affects both the clinical progression of the disease and the immune response.

- **Persistent infections:** Some viruses persist in specific tissues, often evading immune detection. For instance, herpesviruses can remain latent in nerve cells for long periods, reactivating later to cause disease (e.g., shingles) [19]. Understanding tissue tropism helps researchers identify potential reservoirs of viral persistence within the host.

Examples of receptor binding and host specificity during virus infection:

a. **Influenza virus**
Receptor: Influenza viruses attach to sialic acid residues present on host cell surface glycoproteins and glycolipids. The specific linkage of sialic acid (α-2,3 or α-2,6) determines the host range of the virus [20].
Viral protein: The influenza virus hemagglutinin (HA) protein binds to sialic acid receptors. After binding, the virus is endocytosed into the host cell.
Co-receptor: No specific co-receptor is required for influenza virus entry.

b. **Human immunodeficiency virus**
Receptor: HIV primarily targets CD4, a receptor protein found on the surface of T-cells, macrophages, and dendritic cells. Additionally, HIV binds to a co-receptor, either CCR5 or CXCR4, which determines viral tropism [21].
Viral protein: The viral glycoprotein gp120 on the HIV envelope interacts with CD4 and the co-receptor. This interaction triggers conformational changes that allow viral entry.
Co-receptor: Depending on the co-receptor used (CCR5 or CXCR4), HIV can infect different types of immune cells.

c. **Herpes simplex virus**
Receptor: HSV uses various receptors, including heparan sulfate proteoglycans on the cell surface and specific cellular receptors such as herpesvirus entry mediator (HVEM) or nectin-1 [22].
Viral protein: The HSV glycoproteins gB, gD, and gH/gL are involved in attachment and entry [23]. gD binds to the cellular receptors, triggering the fusion of the viral envelope with the host cell membrane.
Co-receptor: For efficient entry, HSV often requires co-receptors like nectin-2 or 3-O-sulfated heparan sulfate.

d. **Adenovirus**
Receptor: Adenoviruses attach to host cells through interactions with the coxsackievirus and adenovirus receptor (CAR), a cell adhesion molecule.
Viral protein: The adenovirus fiber knob protein binds to CAR on the

host cell surface, initiating viral entry [24].
Co-receptor: No specific co-receptor is required for adenovirus entry.

e. **Measles virus**
Receptor: Measles virus attaches to host cells using the CD150 receptor (also known as SLAMF1) found on immune cells [25].
Viral protein: The measles virus HA protein interacts with the CD150 receptor to initiate entry into host cells.
Co-receptor: No specific co-receptor is required for measles virus entry.

f. **SARS-CoV-2 (severe acute respiratory syndrome coronavirus 2)**
Receptor: SARS-CoV-2 attaches to ACE2 receptors found on the surface of human respiratory epithelial cells, particularly in the nasal cavity, lungs, and other organs. This interaction facilitates viral entry and is a key determinant of tissue tropism and pathogenicity.
Viral protein: The spike protein (S protein) on the virus binds to ACE2, facilitating viral entry.
Co-receptor: Some evidence suggests that co-receptors like neuropilin-1 may enhance SARS-CoV-2 entry.

g. **Dengue virus (DENV)**
Receptor: Dengue virus attaches to host cells by interacting with heparan sulfate proteoglycans and the attachment receptor, dendritic cell-specific intercellular adhesion molecule-3-grabbing non-integrin (DC-SIGN).
Viral protein: The envelope (E) and membrane (M) proteins of DENV are involved in attachment and entry.
Co-receptor: While DENV primarily uses heparan sulfate proteoglycans and DC-SIGN as attachment receptors, there is evidence suggesting that TAM receptors (Tyro3, Axl, Mer) may act as co-receptors, facilitating viral entry into host cells. TAM receptors, when engaged by DENV, can enhance infection by promoting viral entry and replication.

h. **Rhinovirus**
Receptor: Rhinoviruses bind to intercellular adhesion molecule 1 (ICAM-1) as the primary receptor.
Viral protein: The viral capsid protein viral protein 1 (VP1) mediates attachment to ICAM-1, allowing entry.

i. **Ebola virus**
Receptor: Ebola virus binds to host cells through using receptors such as T-cell immunoglobulin and mucin domain 1 (TIM-1).
Viral protein: The Ebola virus glycoprotein (gp) facilitates viral entry by interacting with attachment receptors and co-factors.

j. **Norovirus**
Receptor: Noroviruses use histo-blood group antigens (HBGAs) as attachment receptors.
Viral protein: The viral capsid protein VP1 recognizes and binds to HBGAs for entry.
Co-receptor: There is no defined co-receptor in this context. HBGAs alone are generally sufficient for norovirus entry into host cells.

2.2 Entry and Uncoating

The entry and uncoating steps in the viral replication cycle are another crucial process that allows the virus to release its genetic material from its protective capsid or envelope, making the viral genome accessible for further replication within the host cell. The mechanisms involved in entry and uncoating can vary significantly among different viruses, depending on their genetic material, structure, and entry mechanisms (Table 2.2).

1. **Enveloped RNA viruses (e.g., influenza virus)**
   Enveloped RNA viruses, like the influenza virus, typically enter host cells through endocytosis, specifically clathrin- and caveolin-independent pathways. Once inside endosomes, a decrease in pH triggers a conformational change in the viral HA protein. This change allows the viral envelope to fuse with the endosomal membrane, releasing the viral RNA into the host cell cytoplasm. Uncoating involves the disruption of the viral envelope.
2. **Enveloped DNA viruses (e.g., HSV)**
   Enveloped DNA viruses, such as Herpes simplex virus (HSV), enter host cells by fusion (see Figure 2.1) [26]. Once the viral envelope fuses with the host cell membrane, the viral capsid is released into the cytoplasm. Subsequently, the viral capsid undergoes

Table 2.2 Virus entry mechanisms.

Mechanism	Description	Examples of viruses
Receptor-mediated endocytosis	Virus binds to specific host cell surface receptors and is internalized via endocytosis. Entry may be clathrin- or caveolin-dependent.	Influenza virus (clathrin-mediated), dengue virus, adenovirus
Direct membrane fusion	Viral envelope fuses directly with the host cell membrane, releasing viral contents into the cytoplasm.	Herpes simplex virus, measles virus
Clathrin-mediated endocytosis	A form of receptor-mediated entry where viruses are taken in via clathrin-coated pits.	Primary entry for many viruses
Caveolin-mediated endocytosis	Entry via caveolae, small, cholesterol-rich invaginations in the plasma membrane.	HPV31
Macropinocytosis	Virus induces large-scale engulfment of extracellular fluid and particles.	Ebola virus, vaccinia virus
Phagocytosis (rare for viruses)	Used mainly by large viruses or virus-infected apoptotic bodies in immune cells.	Mimivirus (in amoebae)
pH-dependent fusion (endosomal escape)	After endocytosis, acidification in endosomes triggers fusion of the viral envelope with the endosomal membrane.	Influenza virus, flaviviruses (e.g., Zika)

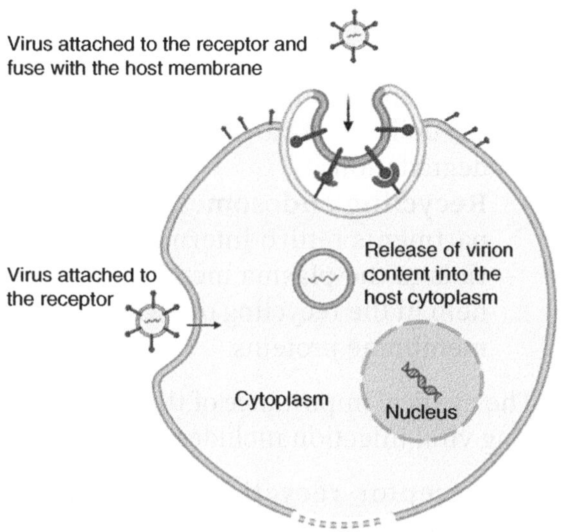

Figure 2.1 Systematic illustration of virus fusion.

uncoating, releasing the viral DNA to enter the host nucleus.

3. **Non-enveloped RNA viruses (e.g., poliovirus)**

   Non-enveloped RNA viruses, like poliovirus, enter host cells by receptor-mediated endocytosis (see Figure 2.2). In the endosome, the low-pH environment triggers conformational changes in the viral capsid proteins, leading to the release of viral RNA into the host cell cytoplasm [27].

4. **Non-enveloped DNA viruses (e.g., adenovirus)**

   Non-enveloped DNA viruses, such as adenovirus, enter host cells through receptor-mediated endocytosis. Once inside the endosome, the acidic pH environment induces conformational changes in the viral capsid, leading to capsid disassembly and the release of viral DNA into the host cell nucleus [27].

5. **Retroviruses (e.g., HIV)**

   Retroviruses, like HIV, have an outer envelope derived from the host cell membrane. Following fusion with the host cell membrane receptor, the viral core is released into the host cell cytoplasm. Uncoating involves the disassembly of the viral core, releasing the viral RNA genome.

Clathrin is a cytoplasmic protein that plays a crucial role in the formation of

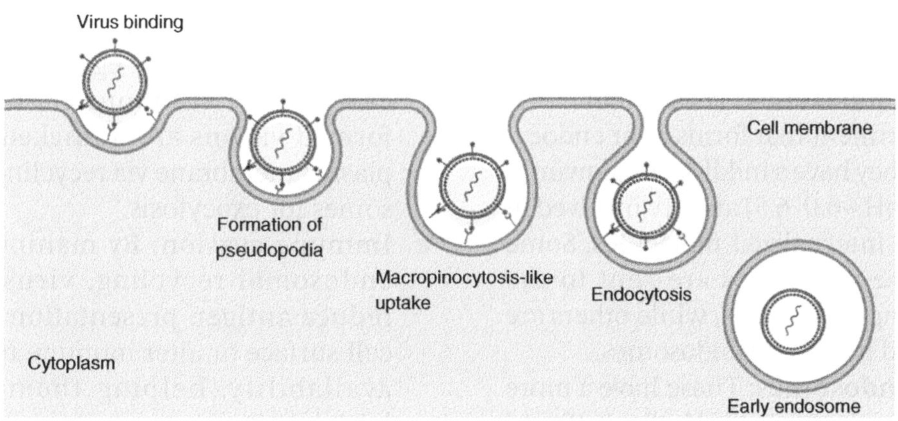

Figure 2.2 Systematic illustration of receptor-mediated endocytosis of virus entry.

clathrin-coated vesicles during endocytosis. It assembles into a triskelion shape that forms a lattice-like coat around vesicles, allowing the cell to internalize various molecules, including viruses. Many viruses exploit this clathrin-mediated pathway to gain entry into host cells after binding to specific surface receptors.

Caveolin is an integral membrane protein essential for the formation and function of caveolae, which are small, flask-shaped invaginations of the plasma membrane. These structures are rich in cholesterol and sphingolipids and serve as platforms for signal transduction and endocytosis. Certain viruses use caveolin-mediated endocytosis to enter cells, often avoiding degradation pathways and targeting specific cellular compartments.

What are endosomes?

Endosomes are membrane-bound compartments within eukaryotic cells that function as part of the endocytic pathway. They play a crucial role in the sorting and trafficking of internalized molecules, including nutrients, membrane proteins, and virus particles.

Types of endosomes:

- **Early endosomes**: These are the first compartment that forms after endocytosis. They have a mildly acidic environment (pH ~6.0–6.5) and are involved in sorting internalized molecules. Some of these molecules are sent to the recycling endosomes, while others are directed to the late endosomes.
- **Late endosomes**: These have a more acidic environment (pH ~5.0–6.0) and are often referred to as multivesicular bodies (MVBs). They mature from early endosomes and are involved in the delivery of cargo to lysosomes for degradation.
- **Recycling endosomes**: These compartments return internalized materials to the plasma membrane. They help in the recycling of receptors and membrane proteins.

The medical importance of the endosome during viral infection includes:

- Receptor recycling for re-entry: Viruses such as influenza and coronaviruses may rely on the recycling of their entry receptors to maintain infection cycles. Recycling endosomes return these receptors to the surface, ensuring that new virions can bind and infect additional cells.
- Intracellular trafficking: Certain viruses use recycling endosomes as trafficking hubs to navigate the intracellular environment. For example, HPV and adenoviruses hijack this pathway to reach the perinuclear region, where replication can occur.
- Viral assembly and egress: In some cases, viral components or fully formed virions are trafficked to the plasma membrane via recycling endosomes for exocytosis.
- Immune evasion: By manipulating endosomal recycling, viruses can reduce antigen presentation on the cell surface or alter immune receptor availability, helping them evade host defenses.

2.3 Transcription of the Viral Genome

Transcription is the process by which viral genomic materials are copied, enabling the production of viral proteins necessary for replication and assembly within the host cell, either in the cytoplasm or the nucleus. This replication occurs using the host cell's machinery. The exact mechanisms vary between different types of viruses. Most DNA viruses typically rely on the host cell's DNA polymerase, while RNA viruses may encode their own RNA polymerases.

2.3.1 Transcription of the Viral RNA Genome in the Cytoplasm

The genome replication of RNA viruses in the cytoplasm is a crucial step in their life cycle. This process involves the synthesis of new viral RNA molecules, which can be positive-sense or negative-sense, depending on the type of RNA virus (see Figure 2.3). The process typically involves the following steps as described below:

a. **Template selection:** For RNA viruses that replicate in the cytoplasm, the viral RNA genome serves as a template for replication. In the case of positive-sense RNA viruses, this genome can be directly used as a template for replication, while negative-sense RNA viruses require the synthesis of a complementary positive-sense RNA intermediate.
b. **RNA polymerase:** RNA viruses typically encode their RNA-dependent RNA polymerase (RdRP), which is responsible for replicating the viral

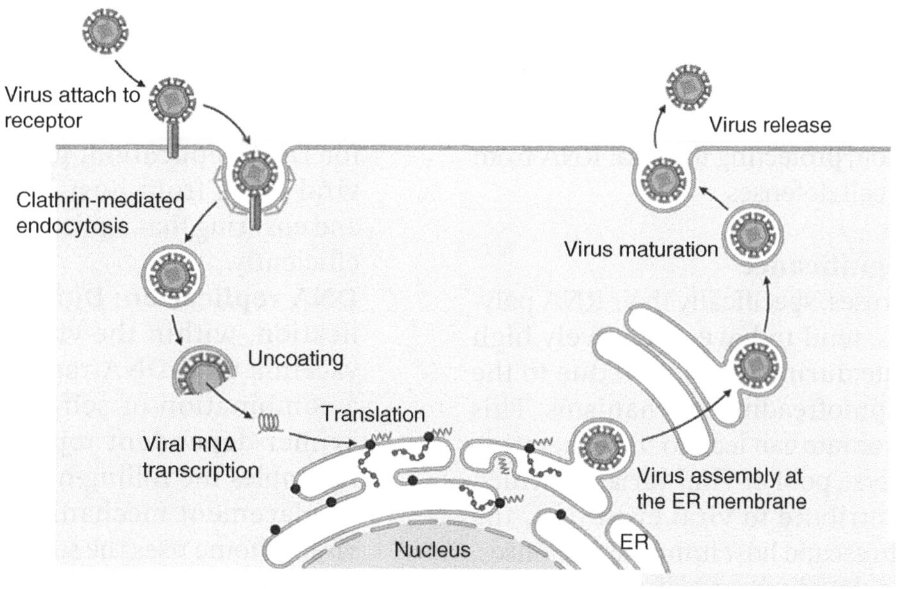

Figure 2.3 Systematic illustration of RNA virus replication in the cytoplasm.

genome [28]. This enzyme is critical for synthesizing new RNA strands based on the existing viral RNA template. Many single-stranded negative-sense RNA viruses require RdRP to first synthesize a complementary positive-sense RNA strand (antigenome synthesis) based on the negative-sense genome. This positive-sense RNA strand then serves as a template to produce more negative-sense RNA genomes. For positive-sense RNA viruses, the RdRP directly uses the viral RNA genome as a template to produce additional copies of the positive-sense RNA. This newly synthesized positive-sense RNA can be used for both translation into viral proteins and as a template for further genome replication.

c. **Formation of replication complexes:** Some RNA viruses often induce the formation of specialized replication complexes within host cells. These complexes provide a controlled environment for viral RNA replication, protecting the viral RNA from host cell defenses.

Clinical significance

RNA viruses, specifically their RNA polymerases, tend to have a relatively high error rate during replication due to the lack of proofreading mechanisms. This high error rate can lead to the generation of a diverse pool of viral variants, which may contribute to viral evolution, the ability to escape host immune responses, and the development of resistance to antiviral agents.

2.3.2 Transcription of Viral DNA in the Cytoplasm

Some DNA viruses replicate their genomes in the cytoplasm, as they lack access to the host cell nucleus [29]. Once inside the cytoplasm, the viral DNA is transcribed by viral enzymes, leading to the expression of early genes. These early genes often encode enzymes necessary for DNA replication, termed early gene expression. Poxviruses, such as the variola virus (which causes smallpox), are well-known examples of cytoplasmic DNA viruses that employ this unique replication strategy [30].

a. **Formation of replication complexes:** Poxvirus, a cytoplasmic DNA virus, induces the formation of replication complexes within the host cell's cytoplasm (also known as the virus factory) [31]. These complexes are composed of viral and host cell proteins that work together to facilitate efficient genome replication. They also provide a controlled environment for DNA replication, protecting the viral DNA from host cell defenses and ensuring that replication proceeds efficiently.
b. **DNA replication:** During virus replication, within the virus factories, vaccinia viral DNA replicates using a combination of self-priming and primer-dependent replication that resembles the rolling-hairpin strand-displacement mechanism [32]. The viral genome uses the strand displacement model that utilizes the terminal hairpin loop of the genome to initiate

DNA synthesis. A nick occurs at one end, which leads to the generation of a 3′-hydroxyl group that then serves as a primer for DNA polymerase to replicate DNA. As DNA synthesis proceeds, the viral replication mechanism displaces the existing strand to generate long double-stranded DNA molecules called concatemers [33, 34].

c. **Resolution and packaging:** After DNA replication, the concatemers are formed. These concatemers are not single, discrete molecules but long double-stranded DNA intermediates. They are subsequently resolved by the viral resolvase into unit-length double-stranded genomes, which serve as the mature replicative form.
d. **Late gene expression and assembly:** Once the viral DNA is replicated and packaged in the cytoplasm, late gene expression is initiated. Late genes encode structural proteins required for the assembly of new viral particles.

Note:
Some other examples of *Poxviridae* family members that replicate in the cytoplasm include Molluscum contagiosum virus. While *Poxviridae* family members are perhaps the most notable group of DNA viruses that replicate in the cytoplasm, they are not the only ones.

2.3.3 Transcription of Some RNA Viruses (Non-retrovirus) in the Nucleus

Replication of RNA viruses in the host cell's nucleus is relatively uncommon, as most RNA viruses replicate in the cytoplasm. However, there are a few non-retrovirus RNA viruses that have evolved the ability to replicate their genomes or transcription intermediates within the host cell nucleus. This nuclear replication strategy provides them with unique advantages in terms of accessing host cell resources and avoiding host cell defenses. For example,

- **Influenza virus (*Orthomyxoviridae*):** Influenza virus is a well-known example of an RNA virus that replicates in the nucleus [35]. Influenza replicates its RNA entirely in the nucleus by utilizing the host cell's RNA polymerase II to transcribe and replicate its RNA. The viral RNA polymerase, composed of viral proteins, assists in both transcription and replication. Importantly, the virus uses a unique "cap-snatching" mechanism to "steal" 5′ ends of cellular pre-messenger RNA (mRNA) molecules (including the cap structure) to extend viral coding sequences [36, 37].
- **Borna disease virus (*Bornaviridae*):** Borna disease virus (BDV) is an RNA virus that belongs to the *Bornaviridae* family. BDV has a non-segmented negative-sense RNA genome and replicates within the nucleus of infected cells [38, 39]. Unlike influenza virus, it encodes a viral polymerase that enters the nucleus and catalyzes RNA replication and transcription [39].
- **Hepatitis delta virus (Delta virus):** Hepatitis delta virus (HDV) is a satellite virus that requires the helper function of the hepatitis B virus (HBV) to supply surface antigens (HBsAg) necessary for viral assembly and

infectivity. HDV contains a circular, single-stranded, negative-sense RNA genome with a ribozyme that allows self-cleavage. The ribozyme cleaves and ligates HDV RNA in the nucleus to generate the circular RNA genome, using host RNA polymerase II, in the nucleus of an infected hepatocyte [40].

2.3.4 Transcription of Some DNA Viruses in the Nucleus

Most DNA viruses replicate in the host nucleus. They have evolved various strategies to replicate their genetic material within the host cell nucleus by utilizing the host's nuclear replication machinery to their advantage. The replication mechanisms and the interactions between viral and host cell factors can vary among different virus families (Figure 2.4).

- ***Herpesviridae* family:** Herpesviruses, HSV, varicella-zoster virus (VZV), and Epstein–Barr virus (EBV), have large, linear double-stranded DNA genomes that replicate in the nucleus [41]. Herpesvirus replication involves the formation of viral DNA replication compartments in the nucleus, known as replication compartments [42], and viral enzymes facilitate genome replication and gene expression.
- ***Papillomaviridae* family:** HPVs, which are responsible for a range of conditions, including warts and some cancers, have small, circular double-stranded DNA genomes that replicate in the host cell nucleus. The viral genome replicates and is maintained as an episome (meaning an extrachromosomal element) in the nucleus of infected cells [43].

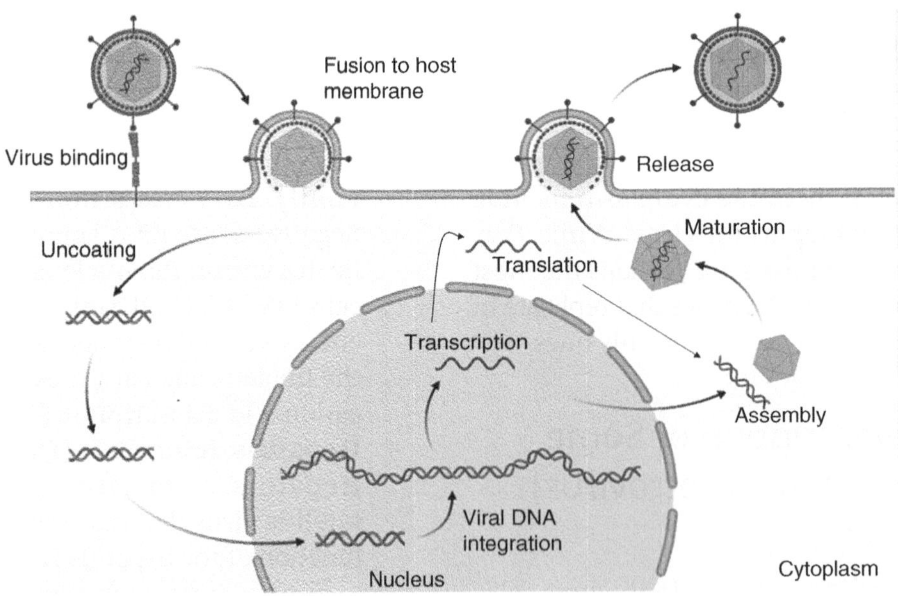

Figure 2.4 Systematic illustration of DNA virus replication in the nucleus.

- ***Polyomaviridae* family:** Polyomaviruses, including simian virus 40 (SV40) and human polyomaviruses like John Cunningham (JC) virus, have small, circular double-stranded DNA genomes [44]. They replicate in the nucleus of host cells. The viral DNA undergoes bidirectional replication, and the resulting multiple copies are maintained as episomes in the nucleus.
- ***Adenoviridae* family:** Adenoviruses, such as human adenovirus, possess linear double-stranded DNA genomes that replicate in the host cell nucleus [45]. The viral genome is replicated by the host cell DNA polymerase and associated viral proteins in specialized nuclear replication centers.

For DNA viruses that replicate in the nucleus, viral DNA replication and transcription occur within the host cell nucleus. However, the resulting viral mRNAs can be exported to the cytoplasm, where host ribosomes translate them into viral proteins. In some cases, such as with herpesviruses, structural proteins, including capsid components, synthesized in the cytoplasm return to the nucleus to facilitate virion assembly.

The influenza virus is a notable exception to RNA viruses that replicate in the nucleus; viral mRNAs and newly synthesized proteins are then transported to the cytoplasm for translation and particle assembly. Retroviruses, such as HIV, follow a distinct replication strategy. After entering the host cell, their RNA genome is reverse-transcribed into DNA in the cytoplasm. This DNA then enters the nucleus, where it integrates into the host genome and is transcribed into viral RNA. The newly formed RNA is exported to the cytoplasm for translation into viral proteins and packaging into new virions. Therefore, most viruses that use the DNA in their replicatory strategies typically rely on a special protein called the nuclear export signal or sequence (NES) [46].

Nuclear Export of the Synthesized Viral Replication Component

After the viral genome has been synthesized in the nucleus, it often needs to be transported to the cytoplasm, where replication continues for certain viruses. This is a crucial step in the replication of viruses. These NESs are recognized by cellular export machineries, such as exportins or nuclear export receptors, which facilitate the transport of cargo from the nucleus to the cytoplasm (see Table 2.3). The nuclear export of viral proteins often relies on the presence of specific sequences.

Examples of Viruses with Nuclear Export of Viral Proteins

- **Influenza virus (*Orthomyxoviridae*):** Influenza virus, a negative-sense RNA virus, replicates its viral RNA (vRNA) in the nucleus with nuclear export protein (NEP) [47]. After transcription, the newly synthesized viral genomic segments, along with viral ribonucleoprotein (vRNP) complexes, are exported from the nucleus to the cytoplasm. This export is mediated by a viral protein called NEP, a nonstructural protein 2 (NS2), which contains

Table 2.3 Examples of viral proteins that contain nuclear export signals.

Viral nuclear export proteins	Virus name
Nonstructural protein 5 (NS5)	Zika virus
NS5A protein	Hepatitis C virus (HCV)
Rev protein	Human immunodeficiency virus (HIV)
HBx protein	Hepatitis B virus (HBV)
Tax protein	Human T-cell leukemia virus (HTLV-1)
Viral matrix (M) protein	Vesicular stomatitis virus (VSV)
Matrix protein 1 (M1) and Nonstructural protein 2 (NS2)	Influenza A virus
E1A protein	Adenovirus
Core protein	Hepatitis D virus (HDV)

an NES. NEP binds to the cellular export receptor CRM1 (chromosome region maintenance 1) to facilitate the export of vRNPs to the cytoplasm, where they are involved in virion assembly [35, 47].

- **Hepatitis B virus (*Hepadnaviridae*):** HBV is a partially double-stranded DNA virus that replicates in the nucleus of liver cells (hepatocytes) [48]. During replication, the viral pregenomic RNA (pgRNA) is synthesized in the nucleus and exported to the cytoplasm. In the cytoplasm, the viral capsid protein, also known as the core protein, is synthesized. The core protein encapsulates the pgRNA, along with the viral polymerase, forming a nucleocapsid. Inside this nucleocapsid, the reverse transcription of pgRNA into relaxed circular DNA (rcDNA) occurs. The nucleocapsid can either be transported back to the nucleus for covalently closed circular DNA (cccDNA) formation (and establish a persistent infection) or be enveloped and exported out of the cell as a mature virion. The export of the viral nucleocapsid and the formation of the mature virion are facilitated by the host's cellular machinery.
- **Hepatitis C virus (Flavivirus):** Hepatitis C virus (HCV) is an RNA virus that primarily replicates in the cytoplasm. However, during the viral life cycle, some viral proteins, such as the HCV core protein, are known to shuttle between the nucleus and cytoplasm [46]. The core protein contains an NES that allows it to be exported from the nucleus to the cytoplasm. The precise role of nuclear export of the HCV core protein is not fully understood but may be related to virion assembly.

Virus Interaction with the Nucleolus and Other Subnuclear Compartments

Viruses can also interact with other various subnuclear compartments, including the nucleolus, to facilitate their replication, transcription, and other aspects of their life cycle. These interactions are often complex and can involve the modulation of host cell processes for the benefit of the virus.

- **Nucleolus and virus replication:** The nucleolus is a subnuclear compartment responsible for ribosome biogenesis and is involved in numerous cellular processes. Some viruses, like herpesviruses (e.g., Epstein–Barr virus), adenoviruses, HIV, and coronaviruses, have been found to interact with the nucleolus [49, 50]. Viruses may target the nucleolus to sequester host cell proteins or factors for their replication, modulate the cell cycle, or influence the host's antiviral response.
- **Subnuclear compartments and virus replication:** Viruses can utilize subnuclear compartments to create replication niches within the host cell. In the case of the HBV, the viral genome is maintained as a cccDNA episome within the nucleoplasm. The cccDNA serves as a template for transcription and replication.
- **Nuclear speckles and virus replication:** Nuclear speckles are subnuclear compartments involved in splicing and transcription. Some viruses, like herpesviruses, have been found to interact with nuclear speckles, potentially influencing host cell gene expression and splicing processes [51, 52].

Host Cell Shutoff Mechanisms and Antiviral Evasion During Nuclear Replication

Host cell shutoff mechanisms involve the partial reduction or complete shutdown of certain cellular pathways or metabolic processes, whether essential or nonessential, by the host cell. These strategies enable viruses to (i) evade the antiviral strategy of the cells and (ii) manipulate host cell processes to ensure the successful completion of their replication cycle within the host cell nucleus, particularly to establish components of a virus's arsenal during nuclear replication.

Virus-Induced Host Cell Shutoff Mechanisms Host cell shutoff mechanisms are strategies employed by viruses to disrupt host cell gene expression and protein synthesis. These mechanisms often involve mechanisms such as degradation of host cell mRNAs and suppression of host cell gene transcription. Further key features of host cell shutoff mechanisms include:

a. **Degradation of host cell mRNAs:** Many viruses, including herpesviruses, adenoviruses, and cytomegalovirus (CMV), encode viral proteins that target host cell mRNAs for degradation [53, 54]. This depletes the cellular pool of mRNAs, reducing the production of host cell proteins [55].
b. **Suppression of host cell transcription:** Some viruses can inhibit host cell transcription by interfering with the activity of host cell RNA polymerases. For instance, herpesviruses encode proteins that inhibit host cell RNA polymerase II (Pol II) activity [56].
c. **Reprogramming of translation:** Viruses may redirect host cell translation machinery to prioritize the translation of viral mRNAs over host cell mRNAs [57]. This ensures the efficient production of viral proteins.

Antiviral Evasion During Replication During replication, viruses need to evade host cell defenses that recognize and combat their successfully replication within the targeted cell. Antiviral evasion strategies employed by viruses in the nucleus include:

a. **Subversion of immune sensors:** Viruses often target cellular immune sensors and signaling pathways that detect viral infections. For example, viruses may inhibit interferon production or interfere with the function of pattern recognition receptors (PRRs) [58].
b. **Inhibition of apoptosis:** Apoptosis, or programmed cell death, is a common host defense mechanism. Many viruses, like herpesviruses, encode proteins that inhibit apoptosis, ensuring the survival of infected cells [59].
c. **Modulation of host cell proteins:** Viruses can manipulate host cell proteins to avoid immune detection. For instance, some herpesviruses produce proteins that modify host cell histones to regulate the host cell immune response [60].
d. **Concealing viral DNA:** Certain DNA viruses maintain their genomes as episomes within the host cell nucleus to evade immune detection [61]. HBV and EBV are examples of viruses that use this strategy.
e. **Interference with host defense factors:** Viruses may produce proteins that antagonize host cell defense factors, such as proteins that interfere with the activity of interferon-stimulated genes (ISGs) [62]. For example:
   - Influenza A virus NS1 protein. NS1 is known to inhibit the activity of several ISGs, such as protein kinase R (PKR) and 2'-5'-oligoadenylate synthetase (OAS), which play essential roles in antiviral responses [63].
   - Hepatitis C Virus NS5A protein. NS5A is involved in multiple functions, including interfering with ISG activity. It can inhibit the antiviral effects of interferon and promote viral replication [64].
   - Hepatitis B virus HBx protein. HBx can interfere with the induction of certain ISGs, including those involved in the innate immune response against viral infections [65].
   - HIV viral protein R (Vpr) and viral infectivity factor (Vif) proteins. HIV-1 encodes the accessory proteins Vpr and Vif, which are involved in immune evasion [66, 67]. Vpr can counteract the activity of certain ISGs, while Vif targets host antiviral factors like apolipoprotein B mRNA editing catalytic polypeptide (APOBEC3G) for degradation [68].
   - Herpes simplex virus (*Herpesviridae*) infected cell protein 0 (ICP0): HSV produces the ICP0. ICP0 is known to counteract ISGs by inhibiting their expression and functions [69, 70].

The effectiveness of these antiviral evasion mechanisms can vary among different viruses, depending on their specific replication strategies and the host's immune response.

2.3.5 Transcription of Retrovirus RNA

Retrovirus replication is a unique process that involves reverse transcription, which converts viral RNA into complementary DNA (cDNA) by the viral enzyme reverse transcriptase (RT). This process involves several steps (see Figure 2.5):

- **Synthesis of (−) DNA strand**: RT uses the viral single-stranded RNA (ssRNA) as a template to synthesize a complementary (−) DNA strand. The retrovirus carries RT within its capsid, as it needs to synthesize DNA from its RNA genome.
- **Degradation of RNA**: The RNA strand is then degraded by the RNase H activity of RT, leaving a single-stranded DNA.
- **Synthesis of (+) DNA strand**: A second DNA strand (+) is synthesized using the newly formed (−) DNA strand as a template, resulting in a double-stranded DNA (dsDNA) molecule. This newly synthesized DNA is called proviral DNA.
- **Nuclear import:** The newly synthesized proviral DNA is transported into the host cell nucleus, typically during cell division when the nuclear membrane is broken down, or through nuclear pores in nondividing cells.
- **Integration:** Once inside the nucleus, the viral enzyme integrase integrates the proviral DNA into the host genome at a random location. The integrated viral

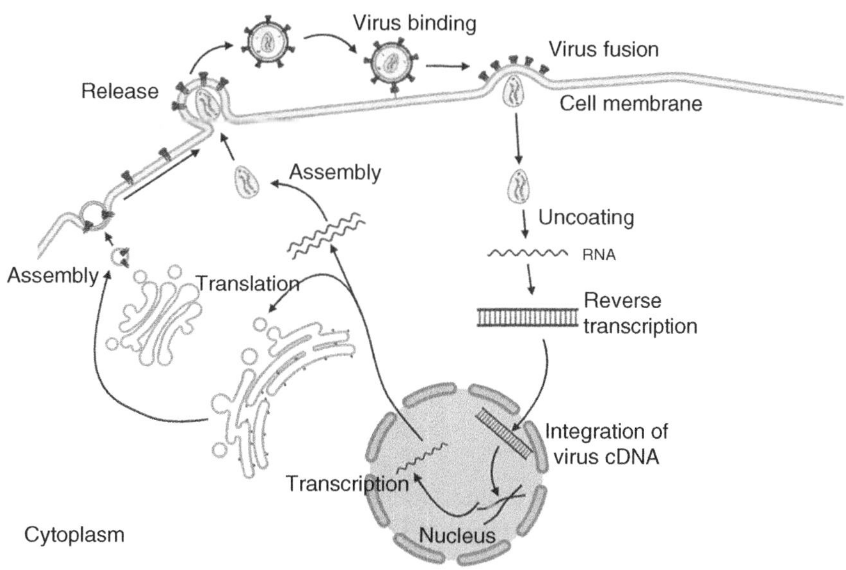

Figure 2.5 Systematic illustration of retrovirus replication.

DNA becomes a provirus, which is now a permanent part of the host's genome.

- **Transcription and translation:** Host RNA polymerase transcribes the integrated proviral DNA into viral mRNA. Viral mRNAs are exported from the nucleus to the cytoplasm, where they are translated into viral proteins by the host's ribosomes. Some of these transcripts serve as full-length viral RNA genomes for new virions.

2.4 Translation and Posttranslational Modifications

Viral RNA translation is the process by which a host cell's ribosomes decode viral RNA to produce viral proteins. This step is essential in the viral life cycle, as the newly synthesized proteins are needed for replication, assembly, and infection of new cells. Depending on the virus, translation can occur directly from the viral RNA (for positive-sense RNA viruses) or after transcription into messenger RNA (for negative-sense RNA or DNA viruses). Subsequently, posttranslational modifications (PTMs) of viral protein occur. These are crucial processes that ensure their successful replication and propagation within host cells.

2.4.1 Viral RNA Translation

a. **Initiation of translation:** Viruses typically hijack the host cell's ribosomes and translation initiation factors to begin protein synthesis. They often use specialized elements in their viral RNA or mRNA to recruit ribosomes.
b. **Ribosome shunting:** Some viruses employ mechanisms such as ribosome shunting or internal ribosome entry sites (IRESs) to bypass the cap-dependent translation initiation used by host cells. This allows the virus to initiate protein synthesis more efficiently [71].
c. **Frameshifting:** Viruses may use frameshifting to produce multiple proteins from a single RNA transcript. This can increase the diversity of viral proteins and their functions [72].
d. **Synthesis of viral polyproteins:** Many RNA viruses, like picornaviruses, encode large polyproteins at once, which are later processed into individual functional proteins by viral proteases [73]. This processing is crucial for the formation of viral replication complexes.
e. **Cap snatching (in RNA viruses):** In the case of influenza A virus, the viral polymerase "snatches" the 5' cap from host cell mRNAs, which is then used to prime the synthesis of viral mRNA. This process allows the virus to acquire host cell-derived RNA caps for its transcripts [74].

2.4.2 Posttranslational Modifications (PTMs) of Viral Proteins

The PTMs of viral proteins refer to the chemical changes made to viral proteins after they are synthesized by host ribosomes.

If this does not occur, the viral protein may not become functional or effective. Viruses can manipulate cellular PTM events to modify their proteins and influence host cell processes. Key PTMs relevant to viral replication include:

a. Phosphorylation
   Phosphorylation involves the addition of phosphate groups, usually to serine, threonine, or tyrosine residues of proteins, mediated by host or viral kinases. This modification can induce conformational changes that affect protein stability, enzymatic activity, subcellular localization, or interactions with host or viral partners.

   Many viruses rely on phosphorylation to activate viral proteins and regulate various stages of the replication cycle. For example, in HIV, phosphorylation of the Rev and Tat proteins is essential for their nuclear localization and transactivation functions, respectively [75]. Similarly, phosphorylation of the matrix and Vpr, Gag proteins influences virus assembly, nuclear import, and infection of nondividing cells [76]. By hijacking host kinase pathways, viruses optimize their replication, evade immune responses, and modulate host signaling networks to favor infection [77].
b. Glycosylation
   Glycosylation is a PTM in which carbohydrate chains (glycans) are covalently attached to specific amino acid residues of proteins, typically asparagine (N-linked) or serine/threonine (O-linked). Glycosylation is essential for the proper folding and function of many viral envelope proteins [78]. It can also play a role in immune evasion by masking epitopes recognized by the host immune system.
   - **HIV:** The envelope glycoprotein of HIV, known as gp120, undergoes extensive glycosylation [79]. These glycans help shield vulnerable epitopes on the viral surface, making it challenging for the immune system to recognize and neutralize the virus [80]. Env is vastly overglycosylated compared to host glycoproteins. HIV's high glycan density and variability contribute to its ability to evade the host immune response and infectivity.
   - **Influenza A virus:** The HA and neuraminidase (NA) proteins of influenza A virus are glycosylated. Glycosylation of HA plays a role in immune evasion by masking key antigenic sites on the virus [81]. The high mutation rate in the HA gene also contributes to the virus's ability to escape immune recognition.
   - **HCV:** HCV envelope glycoproteins, E1 and E2, are heavily glycosylated [82]. Glycosylation of these proteins is involved in immune evasion and aids in shielding epitopes that might otherwise be targeted by the host immune system.
   - **HSV (*Herpesviridae*):** Herpesviruses, such as HSV, produce glycoproteins that play critical roles in viral entry and immune evasion. Glycosylation of these

glycoproteins can hinder immune recognition, contributing to the virus's ability to establish latency and evade host defenses.

- **HPV:** HPV encodes L1 and L2 major and minor capsid proteins, respectively. L1 capsid protein can be glycosylated, while the L2 capsid protein is generally not considered to be significantly glycosylated [83]. They are critical for viral assembly and entry. Glycosylation may influence the virion's interaction with the host cell and may also contribute to immune evasion by concealing epitopes.

c. Ubiquitination

Ubiquitination is a PTM that involves the attachment of ubiquitin molecules to target proteins, leading to various outcomes, including protein degradation, signal transduction, and protein trafficking. Some viruses have evolved to manipulate the ubiquitination process to their advantage [84].

- **HIV:** HIV is known to manipulate the ubiquitination process in multiple ways. The viral protein u (Vpu) targets the host cell protein CD4 for ubiquitin-dependent degradation [85, 86]. This degradation reduces the display of CD4 on the cell surface and promotes the release of new virus particles from infected cells.
- **HPV:** High-risk HPV strains, such as HPV 16 and 18, produce the E6 and E6-associated protein (E6AP). The E6 protein interacts with E6AP to target the tumor suppressor protein p53 for degradation [87]. This degradation allows HPV-infected cells to evade the host's immune response and continue to grow and divide uncontrollably. During HPV infection, the E6 protein interacts with E6AP to promote the ubiquitination of p53. Ubiquitination marks p53 for degradation by proteasomes, a cellular machinery that breaks down proteins [88]. This degradation is essential for HPV-infected cells to evade the host's immune response and promote cell growth and division.
- **Adenovirus:** Adenoviruses encode E1B-55K and E4orf6 proteins that function as an E3 ubiquitin ligase complex [89]. These complexes target and ubiquitinate host cell proteins such as p53, leading to their proteasomal degradation [90]. This degradation supports viral replication and prevents the activation of p53-mediated antiviral responses.
- **EBV:** EBV encodes the viral protein, Epstein–Barr virus immediate-early protein (BZLF1), which promotes the ubiquitination and subsequent degradation of host cell proteins such as the promyelocytic leukemia (PML) protein [91, 92].
- **Kaposi's sarcoma-associated herpesvirus (KSHV):** KSHV encodes a viral ubiquitin ligase, K3, that can ubiquitinate host cell MHC Class I molecules for

degradation [93]. This ubiquitination leads to the internalization and degradation of MHC Class I, impairing the host cell's ability to present viral antigens to cytotoxic T cells [94].

d. SUMOylation
SUMOylation is a PTM in which a Small Ubiquitin-like Modifier (SUMO) protein is covalently attached to target proteins, thereby altering their activity, cellular localization, or interactions with other proteins. This process is mediated by an enzymatic cascade involving E1-activating, E2-conjugating, and E3-ligating enzymes. Several viruses have evolved strategies to manipulate SUMOylation processes, affecting both host cell and viral proteins [95].
   - **CMV:** CMV encodes viral proteins, such as IE1 and UL69, that target the SUMOylation machinery [96]. They can interact with SUMO and SUMO-conjugating enzymes to regulate host cell processes.
   - **EBV:** EBV-encoded latent membrane protein 1 (LMP1) is known to dysregulate the cellular SUMOylation process. LMP1 influences the SUMOylation of host cell proteins, including its ubiquitination, which can have effects on viral gene expression and signaling [97].
   - **HPV**: Certain high-risk HPV types (HPV-16 L2 protein) have been shown to serve as substrates for SUMOylation processes [98]. The HPV E2 protein can compete with host cell proteins for SUMOylation, affecting viral transcription and replication.
   - **HSV** – *Herpesviridae*: HSV-1 ICP0 is known to target host cell proteins involved in SUMOylation [99]. This activity can have multiple effects on viral gene expression and immune evasion.

e. Acetylation and methylation
Histone acetylation and methylation can modify histones and regulate viral gene expression [100]. For example, EBV encodes proteins, particularly the protein known as BKRF4, that modify histones to modulate viral gene transcription [101].

f. Myristoylation and palmitoylation
Myristoylation involves the attachment of myristic acid (a 14-carbon saturated fatty acid) to the N-terminal glycine residue of a protein [102]. Palmitoylation involves the attachment of palmitic acid (a 16-carbon saturated fatty acid) to a cysteine residue within a protein[103].

These lipid modifications are critical for the membrane association and trafficking of many viral proteins. They are often involved in viral assembly and release.

Myristoylation in viral replication:

- **Human immunodeficiency virus-1:** The HIV-1 Gag protein is myristoylated [104]. Myristoylation is essential for the targeting of Gag to the plasma membrane, where viral assembly takes place.
- **Moloney murine leukemia virus (MoMLV):** The Gag protein of MoMLV

is myristoylated [105]. Myristoylation is involved in virus assembly and budding.

- **Picornaviruses (e.g., poliovirus):** Certain picornavirus proteins, such as VP4, are myristoylated. Specifically, the N-terminal glycine residue of VP4 is N-myristoylated [106]. This myristoylation is important for anchoring the viral capsid to the host cell membrane during entry.

Palmitoylation in viral replication:

- **HCV:** The HCV core protein undergoes palmitoylation, which plays a role in the formation of the viral nucleocapsid and contributes to virion assembly [107].
- **Influenza A virus:** The influenza A virus HA and NA proteins are palmitoylated. Palmitoylation helps anchor these viral glycoproteins in the host cell membrane [108].
- **Retroviruses (e.g., HIV-1):** Cytoplasmic domain of the HIV-1 envelope glycoprotein (gp160) contains two palmitoylated [109]. Palmitoylation is involved in membrane binding and virus assembly at the plasma membrane.
- **Varicella-zoster virus:** Some VZV glycoproteins, such as gE and gI, undergo palmitoylation [110]. Palmitoylation is important for the assembly and egress of VZV virions.

2.5 Assembly

Viral assembly and maturation are also critical steps in the viral life cycle. This is where new viral particles are "constructed" within the host cell. This process involves the packaging of replicated genetic material, assembly of structural components, and sometimes the acquisition of an envelope.

2.5.1 Packaging of Genetic Material

Once the viral genome has been replicated within the host cell, it needs to be packaged into a protective structure. This is usually a capsid, which is a protein coat that encapsulates the viral genetic material. Some viruses use a procapsid structure that undergoes maturation and becomes the mature capsid. Viral genomes are often packed with great precision. This process is driven by specific interactions between viral proteins and the genomic material, ensuring that the right amount of genetic material is packaged into each virion (viral particle). The capsid, or protein coat, is typically composed of multiple protein subunits that assemble into a defined structure. This self-assembly process is driven by protein–protein interactions and can be quite complicated, leading to the formation of a highly ordered and protective shell around the viral genome.

2.5.2 Synthesis of Structural Proteins

In parallel with genome replication and packaging, within the host cell, the synthesized structural proteins required for the assembly of the new viral particles are also produced. These proteins include the capsid, as well as any other structural elements needed by the virus. The expression and synthesis of structural proteins are also tightly regulated to ensure the correct stoichiometry and structure of the virion components.

2.5.3 Acquisition of an Envelope (If Applicable)

Some viruses, such influenza virus or HIV, acquire a lipid bilayer envelope as they bud off from the host cell membrane. This envelope contains viral glycoproteins and lipids derived from the host cell membrane. Enveloped viruses can be released from the host cell by budding, a process where they take a portion of the host cell membrane as their envelope. During assembly, the envelope protein(s) assemble on the portion of the host cell membrane, facilitating maturation. The synthesis and assembly of viral membrane proteins vary depending on the specific viral strategy for acquiring their envelope from different cellular membranes, such as the endoplasmic reticulum (ER), Golgi, or plasma membrane.

i. **Viruses that derive their membrane from the Golgi**
   Viruses that derive their envelopes from the Golgi apparatus (Golgi-derived viruses) rely on an interaction between viral and cellular machinery to synthesize and incorporate their membrane proteins into the viral envelope. Examples include EBV and CMV [111].
   a. **Transcription and translation:** Like cellular proteins, viral membrane proteins are synthesized by ribosomes in the cytoplasm using viral mRNA. These ribosomes may be free or associated with the ER membrane, depending on the specific virus.
   b. **Targeting to the ER:** Viral membrane proteins often contain N-terminal signal sequences that target them to the ER [112]. These sequences are like those found in cellular proteins destined for the secretory pathway. The signal recognition particle (SRP) in the cytoplasm might play a limited role in some cases, although it is not the primary mechanism for Golgi-derived viruses.
   c. **Translocation and modifications:** The signal sequence of the viral protein can be recognized by the Sec61 translocon complex in the ER membrane [113]. The protein is then co-translationally (as it is being synthesized) translocated into the ER lumen (interior). Within the ER, viral membrane proteins undergo various PTMs, including glycosylation (the addition of sugars) and folding into their final three-dimensional structure. These modifications are crucial for proper function and targeting.

   While still in the ER, viral membrane proteins might begin to interact with each other or with other viral components like the viral nucleocapsid (core). Once processed in the ER, viral membrane proteins are packaged into transport vesicles, such as the coat protein. Some viruses have reportedly used coat protein complex II (COPII) proteins to form a replication site [114]. COPII vesicles specifically bud from specific ER exit sites, ensuring the trafficking and proper sorting of viral proteins [115].

d. **Vesicular transport:** The vesicles containing viral membrane proteins then navigate through the cellular secretory pathway. They move through the ER–Golgi intermediate compartment (ERGIC) and potentially through the cis-Golgi (the entry point of the Golgi apparatus). During this journey, viral proteins might undergo further modifications, like glycosylation, within the Golgi compartments [116].

e. **Viral assembly and budding:** At the Golgi membrane, viral assembly also takes place [116]. Viral nucleocapsids (containing the viral genome) may interact with the Golgi membrane, enriched with viral membrane proteins. This interaction triggers viral budding, where a new viral envelope is formed by pinching off a portion of the Golgi membrane containing the viral proteins. The newly formed virus particle is then released from the host cell through exocytosis.

ii. **Viruses that derive their membrane from the ER**

Many enveloped viruses rely on the host cell's ER as a source for their viral envelope [117]. In this case, these viruses encode envelope proteins that get integrated into the ER membrane during biosynthesis and then become part of the newly formed viral particle when the virus buds from the ER. Examples include DENV and HCV (although some viral envelope components are derived from the Golgi) [118].

a. **Transcription and translation:** Viral genes need to be transcribed into mRNA within the host cell. Translation usually begins on free ribosomes in the cytoplasm. For membrane proteins, translation is then directed to the ER. The ribosome becomes a membrane-bound ribosome, and the nascent protein is translocated into the ER lumen or membrane.

b. **Glycosylation and processing:** Virus protein (for example, nonstructural protein [NSPs] of coronaviruses) can undergo N-linked glycosylation (N- most common form of protein modification), or other modifications(s) within the ER [119]. Misfolded proteins are targeted for degradation, while correctly folded proteins are transported further.

c. **Viral assembly and budding:** After modifications, viral membrane proteins become concentrated in specific regions of the ER membrane, often interacting with other viral components like the nucleocapsid (containing the viral genome). The viral particle assembles at these ER membrane sites, with the viral membrane proteins forming the viral envelope surrounding the nucleocapsid. The assembled virus particle then buds from the ER membrane, acquiring its envelope derived from the host cell's ER membrane.

d. **Transport and release:** Where necessary, enveloped virions may be transported to the Golgi apparatus via vesicles. In the Golgi, proteins may undergo additional modifications. Finally, virions are transported to the plasma membrane and released via exocytosis.

iii. **Viruses that derive their membrane from the plasma membrane** Some enveloped viruses may also acquire their membranes by hijacking the host cell's plasma membrane. Examples include retroviruses, togaviruses, and rhabdoviruses.

a. **Viral protein synthesis:** The virus utilizes the host cell's ribosomes and translation machinery to synthesize its viral proteins, including the membrane proteins. The viral mRNA, which encodes the viral proteins, can be localized in the cytoplasm or associated with specific cellular compartments depending on the virus.

b. **Formation of viral replication vesicles:** For many enveloped viruses, viral replication occurs within specialized membrane-bound compartments known as viral replication vesicles. These vesicles are derived from intracellular membranes, such as the ER or Golgi apparatus, and serve to protect viral RNA replication from host immune detection. The viral RNA genome is replicated within these vesicles, providing an optimal environment for efficient genome amplification and assembly of viral proteins. The replication vesicles also help coordinate the synthesis of viral proteins, ensuring they are directed to the correct cellular locations, such as the ER or plasma membrane.

c. **Translocation and early interactions:** Viral membrane proteins are likely translocated into the ER lumen (interior) via the Sec61 translocon, a protein channel in the ER membrane.

During or after translocation, some viral proteins might interact with host cell chaperone proteins in the ER lumen. These chaperones assist in proper protein folding and prevent aggregation.

d. **Modifications and trafficking:** Once inside the ER, as mentioned earlier, viral membrane proteins may undergo various modifications, like glycosylation (adding sugar groups) or acylation (adding fatty acid groups). These modifications can affect protein stability, function, and interaction with other viral components. While in the ER, viral membrane proteins might begin to interact with each other or with other viral components, like the viral nucleocapsid (core).

e. **Exit from the ER and Golgi apparatus:** All viral membrane proteins typically do not traffic through the entire secretory pathway used by cellular proteins destined for secretion. Specific mechanisms, which are not fully understood for all viruses, guide the exit of viral membrane proteins from the ER. This might involve interacting with specific host cell factors or viral proteins that act as retrieval signals.

The Golgi apparatus, a major sorting station in the secretory pathway, might not be heavily involved in some viruses. However, some viruses might utilize specific Golgi enzymes for further modifications of their membrane proteins.

f. **Delivery to the plasma membrane:** Viral membrane proteins are transported to specific regions of the plasma membrane. The exact mechanisms directing this transport are still under study, but it likely involves interactions with the cytoskeleton and specific host cell factors.
g. **Assembly and release:** At the plasma membrane, viral membrane proteins accumulate and interact with other viral components like the nucleocapsid. This interaction triggers the formation of new viral particles, which are released through a process known as budding. During budding, a portion of the host cell's plasma membrane is incorporated into the viral envelope, encapsulating the viral components.

2.6 Maturation

After assembly, many viruses undergo a maturation process. During maturation, the virion may undergo structural changes, enzymatic cleavages, or other modifications that make it fully infectious and ready to exit the host cell. This is crucial for ensuring that the new viral particles are functional and capable of infecting new host cells. It often involves the activation or inactivation of specific viral proteins, ensuring that the virus can complete its life cycle. This maturation process is particularly important for certain types of viruses, such as retroviruses (e.g., HIV) and some enveloped viruses (e.g., influenza), where the virus undergoes structural and functional changes before it can successfully infect new host cells.

Viruses can undergo some of the following maturation steps:

1. **Structural changes:** During the maturation phase, the newly assembled viral particles may undergo structural changes. In the case of retroviruses, for example, the immature viral particle initially forms a spherical structure called the immature virion. This immature virion contains many copies of the viral structural protein, often in a precursor form. These immature virions are not yet infectious [120, 121]. They contain loosely packed genetic material and are unable to attach to and enter new host cells.
2. **Enzymatic cleavages:** Maturation also involves the enzymatic cleavage

of specific viral proteins. For retroviruses like HIV, the viral protease plays a critical role in this process. The viral protease cleaves the precursor proteins into their mature, functional forms. The protease cleaves the Gag and Gag-Pol polyproteins into individual proteins, allowing them to assemble into a mature, infectious virion [122].

3. **Conformational changes:** Enzymatic cleavages lead to conformational changes in the viral proteins. These structural alterations are essential for the virus to become infectious. The newly cleaved proteins adopt their final, functional shapes.
4. **Budding and envelope modification (if applicable)**
   Some enveloped viruses acquire their lipid bilayer envelope during the maturation process. This envelope may be derived from the host plasma membrane and contains viral glycoproteins. For these viruses, budding from the host cell membrane results in the acquisition of the envelope. The maturation process ensures that the viral envelope is properly structured and functional.
5. **Functional readiness**
   Once the virus has undergone maturation, it becomes functionally active and ready to infect new host cells. Its structural changes, proteolytic cleavages, and envelope modifications ensure that it can effectively engage with host cell receptors, enter the host cell, and initiate the viral replication cycle.

2.7 Release or Exit

As the assembled viral particles undergo maturation, they are subsequently ready to be released from the host cell. This release/exit can occur through cell lysis, where the host cell is destroyed, or through a budding process, where the virus acquires an envelope from the host cell's membrane and is released. During this process, the virus may or may not destroy the cell (See Figure 2.6).

2.7.1 Cell Lysis

In this type of release, the matured viral particles essentially "break out" of the host cell by destroying it. The virus induces changes in the host cell's structure and function, leading to cell rupture or even death. This process is often referred to as cytolysis. Cell lysis is common in non-enveloped viruses and some enveloped viruses that do not utilize a budding mechanism. Notable examples include many bacteriophages, adenoviruses, and some picornaviruses [123].

The host cell's death typically triggers an immune response, and it can lead to inflammation and tissue damage. In some cases, this results in observable symptoms of infection.

2.7.2 Budding

Budding is an alternative and less destructive release mechanism used by some enveloped viruses. During this process, the viral particle acquires an envelope derived from the host cell's membrane. The virus typically assembles and matures at the host cell's plasma membrane, incorporating viral glycoproteins and

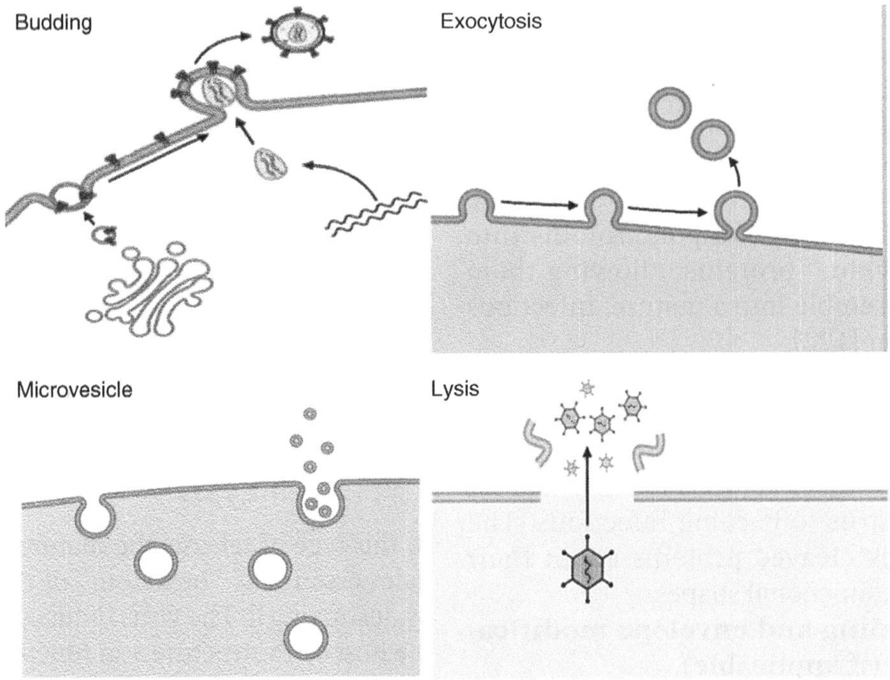

Figure 2.6 Illustration of different mechanisms of virus exit.

lipids from the cell membrane into the virion's envelope. This process is less destructive to the host cell, and the cell often remains viable after budding. Examples of viruses that use budding include HIV, influenza virus, and herpesviruses [124, 125].

Budding is often a stealthier way for the virus to exit the host cell, as it may not immediately trigger a host immune response. Initially, the enveloped virus particle can resemble a normal host cell membrane, making it less conspicuous to the immune system.

Beyond the well-known mechanisms of viral release through budding and cell lysis, there are other, less common ways that viruses can exit host cells. These alternative methods of viral release demonstrate the diversity of strategies that viruses have evolved to exit host cells and infect new ones. The specific mechanism used by a virus can depend on its type, structure, and the nature of its host cell. Here are a few additional methods:

2.7.3 Exocytosis

Some viruses, particularly large DNA viruses like poxviruses, can exit host cells through exocytosis [126]. This process involves the virus being enclosed within a vesicle or membrane-bound structure. The vesicle then fuses with the host cell's plasma membrane, releasing the enclosed virus into the extracellular space. Exocytosis is also a nondestructive process that allows the host cell to survive and continue producing new virus particles. Poxviruses are a notable example of viruses that use exocytosis for release.

2.7.4 Syncytia Formation

Certain viruses, like respiratory syncytial virus (RSV) and some paramyxoviruses (e.g., measles virus), can induce the formation of syncytia, which are large, multinucleated cells resulting from the fusion of multiple host cells. This fusion process often leads to the release of virus particles, as the syncytium ruptures to release new virions. Syncytia formation can be damaging to host tissues and plays a role in the pathogenesis of these viruses [127].

2.7.5 Apoptosis-Induced Release

Some viruses can manipulate the host cell's apoptotic pathway, leading to programmed cell death (apoptosis). During apoptosis, the host cell undergoes controlled disintegration, which can release virions into the surrounding environment. This mechanism can be observed with some viruses, including adenoviruses [128].

2.7.6 Microvesicle Release

Some enveloped viruses can be released from host cells in small vesicles known as microvesicles or exosomes. These vesicles contain viral particles and can be released without causing immediate host cell lysis. This process allows the virus to evade the host's immune system. Examples of viruses include enteroviruses [129]. Poliovirus can exit through autophagosome-like vesicles [130].

Impact of Virus Release on the Host Cell

The choice between cell lysis and budding can have different effects on the host cell. Cell lysis causes host cell death and may trigger immediate inflammation and immune responses. In contrast, budding allows the host cell to survive and continue producing new virus particles. In cases of persistent viral infections where the host cell remains alive and keeps producing virus, the cell can serve as a long-term reservoir for the virus. This is often observed in chronic infections, such as those caused by HIV.

Clinical Impact of Virus Release on the Host Cell to Acute Flaccid Paralysis (AFP)

Acute flaccid paralysis is a medical condition marked by the sudden onset of weakness or paralysis, usually affecting the limbs. AFP is most commonly linked to poliovirus infection, but it can also result from other enteroviruses and neurotropic viruses. The effect of viral release on host cells, especially regarding poliovirus and AFP, is a key factor in how paralysis develops.

1. **Cytopathic effects on neural cells**
   Poliovirus and other enteroviruses that cause AFP infect motor neurons in the anterior horn of the spinal cord. During the viral life cycle, after replication inside the host cell, new virions are assembled and eventually released. This release typically involves cell lysis, a process in which the host cell membrane is disrupted,

leading to the death of the infected neuron [131].

- **Lysis of motor neurons**: The destruction of these neurons due to viral release results in the inability of the affected nerve cells to transmit signals to muscles, leading to muscle weakness or paralysis, a hallmark of AFP or polio.

2. **Viral budding and immune evasion** While poliovirus primarily uses a lytic cycle for release, some other viruses that can contribute to AFP may use non-lytic mechanisms, such as budding, to exit the host cell. This type of viral release can allow the virus to evade immediate detection by the immune system, prolonging the infection.
   - **Chronic inflammation**: Even without immediate cell death, the release of virions via budding can trigger chronic inflammation in the nervous system. Inflammation can contribute to neuronal damage over time.

2.8 Study Questions

Case Study

2.1 A community experiences an outbreak of viral gastroenteritis. The virus involved is known to attach to specific glycoproteins on the surface of intestinal cells. What role does the virus attachment play in the spread and severity of gastrointestinal symptoms during an outbreak? How could interrupting virus attachment reduce the spread of infection?

2.2 A pregnant woman in her second trimester contracts Zika virus infection. Subsequent ultrasound reveals microcephaly in the developing fetus. Discuss how the tissue tropism of Zika virus for neural progenitor cells contributes to congenital abnormalities. What are the clinical implications of Zika virus's neurotropism for pregnant women and their developing fetuses?

2.3 A 45-year-old male presents with jaundice, fatigue, and elevated liver enzymes. After further investigation, he is diagnosed with chronic HBV. Discuss the concept of tissue tropism in the context of HBV. Why does the virus predominantly infect liver cells, and what are the clinical implications of this tropism?

2.4 A 10-month-old girl presents to the clinic with a history of fever, sore throat, and muscle weakness that began a week ago. His parents report that he initially had mild respiratory symptoms like a common cold. However, in the past few days, he developed significant weakness in his legs. Physical examination reveals flaccid paralysis of the lower limbs. Laboratory tests confirm the presence of poliovirus. Meanwhile, in the same clinic, a 15-year-old girl presents with a high fever, cough, and difficulty breathing. Her symptoms started two days ago. On examination, she shows signs of a respiratory infection, but there is no evidence of muscle weakness or paralysis. Further testing reveals an infection with a respiratory virus, likely influenza.

How does the mechanism of cell damage differ between poliovirus and the respiratory virus in these cases, and what are the implications for long-term outcomes for each patient?

Objectives

2.1 **Which of the following steps is NOT involved in the replication cycle of RNA viruses?**
(a) Transcription of viral RNA into mRNA
(b) Reverse transcription of DNA into RNA
(c) Translation of viral proteins
(d) Integration of viral DNA into the host genome

2.2 **Which enzyme is crucial for the replication of retroviruses?**
(a) DNA polymerase
(b) RNA polymerase
(c) Reverse transcriptase
(d) Helicase

2.3 **During which phase of the viral replication cycle do viral proteins and nucleic acids form into new virions?**
(a) Attachment
(b) Penetration
(c) Replication
(d) Assembly

2.4 **In which part of the host cell does the replication of most DNA viruses typically occur?**
(a) Cytoplasm
(b) Nucleus
(c) Mitochondria
(d) Endoplasmic reticulum

2.5 **Which viral protein is often targeted by antiviral drugs that inhibit viral replication?**
(a) Structural proteins
(b) Enzymes required for replication
(c) Host cell receptors
(d) Transport proteins

2.6 **What is the role of the viral polymerase enzyme in the replication of DNA viruses?**
(a) To synthesize new viral DNA from RNA
(b) To copy the viral genome for new virion assembly
(c) To degrade host DNA
(d) To transport viral nucleic acids into the nucleus

2.7 **Which of the following processes is NOT essential for the release of new viral particles from an infected cell?**
(a) Fusion with the host cell membrane
(b) Lysosomal degradation
(c) Endocytosis
(d) Apoptosis

References

1. Xu, Z.-S., et al., *LDLR is an entry receptor for Crimean-Congo hemorrhagic fever virus*. Cell Res., 2024. **34**(2): p. 140–150.
2. Masenga, S.K., et al., *HIV-host cell interactions*. Cells, 2023. **12**(10): p. 1351.
3. Faivre, N., Verollet, C., and Dumas, F., *The chemokine receptor CCR5: multi-faceted hook for HIV-1*. Retrovirology, 2024. **21**(1): p. 2.
4. Cilliers, T., et al., *The CCR5 and CXCR4 coreceptors are both used by human immunodeficiency virus type 1 primary isolates from subtype C*. J. Virol., 2003. **77**(7): p. 4449–4456.

5. Bourgonje, A.R., et al., *Angiotensin-converting enzyme 2 (ACE2), SARS-CoV-2 and the pathophysiology of coronavirus disease 2019 (COVID-19)*. J. Pathol., 2020. **251**(3): p. 228–248.
6. Maginnis, M.S., *Virus-receptor interactions: the Key to cellular invasion*. J. Mol. Biol., 2018. **430**(17): p. 2590–2611.
7. Khan, M.Z.I., et al., *An overview of viral mutagenesis and the impact on pathogenesis of SARS-CoV-2 variants*. Front. Immunol., 2022. **13**: p. 1034444.
8. Yang, J. and Feng, J., *Editorial: Viruses and immune response in pediatric infection*. Front. Cell. Infect. Microbiol., 2023. **13**: 1257807.
9. Blake, I.M., et al., *The role of older children and adults in wild poliovirus transmission*. Proc. Natl. Acad. Sci. U. S. A., 2014. **111**(29): p. 10604–10609.
10. Donbraye, E., et al., *Abundance of enterovirus C in RD-L20B cell culture-negative stool samples from acute flaccid paralysis cases in Nigeria is geographically defined*. J. Med. Microbiol., 2018. **67**(6): p. 854–865.
11. Erickson, K.L., Medina, E.A., and Hubbard, N.E., *Micronutrients and innate immunity*. J. Infect. Dis., 2000. **182**(Supplement_1): p. S5–S10.
12. Elhabyan, A., et al., *The role of host genetics in susceptibility to severe viral infections in humans and insights into host genetics of severe COVID-19: a systematic review*. Virus Res., 2020. **289**: 198163.
13. Naranjo-Covo, M.M., et al., *Natural resistance to HIV infection: role of immune activation*. Immun. Inflamm. Dis., 2025. **13**(2): e70138.
14. Marmor, M., et al., *Resistance to HIV infection*. J. Urban Health, 2006. **83**(1): p. 5–17.
15. Zhang, Z., et al., *Host genetic determinants of hepatitis B virus infection*. Front. Genet., 2019. **10**: p. 696.
16. Alcami, A. and Koszinowski, U.H., *Viral mechanisms of immune evasion*. Trends Microbiol., 2000. **8**(9): p. 410–418.
17. Virgin, H.W., Wherry, E.J., and Ahmed, R., *Redefining chronic viral infection*. Cell, 2009. **138**(1): p. 30–50.
18. Brierley, L., Pedersen, A.B., and Woolhouse, M.E.J., *Tissue tropism and transmission ecology predict virulence of human RNA viruses*. PLoS Biol., 2019. **17**(11): e3000206.
19. Grinde, B., *Herpesviruses: latency and reactivation – viral strategies and host response*. J. Oral Microbiol., 2013. **5**: 22766.
20. Zhao, C. and Pu, J., *Influence of host sialic acid receptors structure on the host specificity of influenza viruses*. Viruses, 2022. **14**(10): p. 2141.
21. Alkhatib, G., *The biology of CCR5 and CXCR4*. Curr. Opin. HIV AIDS, 2009. **4**(2): p. 96–103.
22. Shieh, M.T., et al., *Cell surface receptors for herpes simplex virus are heparan sulfate proteoglycans*. J. Cell Biol., 1992. **116**(5): p. 1273–1281.
23. Jambunathan, N., et al., *Two sides to every story: herpes Simplex Type-1 Viral Glycoproteins gB, gD, gH/gL, gK, and cellular receptors function as key players in membrane fusion*. Viruses, 2021. **13**(9): p. 1849.
24. Nicklin, S.A., et al., *The influence of adenovirus fiber structure and function on vector development for gene therapy*. Mol. Ther., 2005. **12**(3): p. 384–393.
25. Yigit, B., et al., *SLAMF6 in health and disease: implications for therapeutic targeting*. Clin. Immunol., 2019. **204**: p. 3–13.
26. Weed, D.J. and Nicola, A.V., *Herpes simplex virus Membrane Fusion*. Adv. Anat. Embryol Cell Biol., 2017. **223**: p. 29–47.
27. Hogle, J.M., *Poliovirus cell entry: common structural themes in viral cell entry pathways*. Annu. Rev. Microbiol., 2002. **56**: p. 677–702.
28. Ferrer-Orta, C., Ferrero, D., and Verdaguer, N., *RNA-dependent RNA polymerases*

of picornaviruses:from the structure to regulatory mechanisms. Viruses, 2015. **7**(8): p. 4438–4460.
29. Schmid, M., et al., *DNA virus replication compartments*. J. Virol., 2014. **88**(3): p. 1404–1420.
30. Moss, B., *Poxvirus DNA replication*. Cold Spring Harbor Perspect. Biol., 2013. **5**(9): a010199.
31. Condit, R.C., *Vaccinia, Inc.—probing the functional substructure of poxviral replication factories*. Cell Host Microbe, 2007. **2**(4): p. 205–207.
32. Sèle, C., et al., *Low-resolution structure of vaccinia virus DNA replication machinery*. J. Virol., 2013. **87**(3): p. 1679–1689.
33. Cotmore, S., Tattersall, P., and DePamphilis, M., *DNA replication and human disease*. 2006, Cold Spring Harbor Laboratory Press Cold Spring Harbor: NY.
34. Fields, B.N., *Fields' Virology*. Vol. 1. 2007, Lippincott Williams & Wilkins.
35. Dou, D., et al., *Influenza A virus cell entry, replication, virion assembly and movement*. Front. Immunol., 2018. **9**: p. 1581.
36. Koppstein, D., Ashour, J., and Bartel, D.P., *Sequencing the cap-snatching repertoire of H1N1 influenza provides insight into the mechanism of viral transcription initiation*. Nucleic Acids Res., 2015. **43**(10): p. 5052–5064.
37. Decroly, E., et al., *Conventional and unconventional mechanisms for capping viral mRNA*. Nat. Rev. Microbiol., 2011. **10**(1): p. 51–65.
38. Cubitt, B. and de la Torre, J.C., *Borna disease virus (BDV), a nonsegmented RNA virus, replicates in the nuclei of infected cells where infectious BDV ribonucleoproteins are present*. J. Virol., 1994. **68**(3): p. 1371–1381.
39. Honda, T. and Tomonaga, K., *Nucleocytoplasmic shuttling of viral proteins in borna disease virus infection*. Viruses, 2013. **5**(8): p. 1978–1990.
40. Netter, H.J., et al., *Hepatitis delta virus (HDV) and delta-like agents: insights into their origin*. Front. Microbiol., 2021. **12**: 652962.
41. Copeland, A.M., Newcomb, W.W., and Brown, J.C., *Herpes simplex virus replication: roles of viral proteins and nucleoporins in capsid-nucleus attachment*. J. Virol., 2009. **83**(4): p. 1660–1668.
42. Kobiler, O. and Weitzman, M.D., *Herpes simplex virus replication compartments: from naked release to recombining together*. PLoS Pathog., 2019. **15**(6): e1007714.
43. Rossi, N.M., et al., *Extrachromosomal amplification of human papillomavirus episomes is a mechanism of cervical carcinogenesis*. Cancer Res., 2023. **83**(11): p. 1768–1781.
44. Moens, U., Ludvigsen, M., and Van Ghelue, M., *Human polyomaviruses in skin diseases*. Patholog. Res. Int., 2011. **2011**: 123491.
45. Hidalgo, P., et al., *The protein composition of human adenovirus replication compartments*. mBio, 2025. **16**(1): e0214424.
46. Suzuki, R., et al., *Molecular determinants for subcellular localization of hepatitis C virus core protein*. J. Virol., 2005. **79**(2): p. 1271–1281.
47. Yang, C.-C., et al., *CRM1-spike-mediated nuclear export of hepatitis B virus encapsidated viral RNA*. Cell Rep., 2022. **38**(10): 110472.
48. Liang, T.J., *Hepatitis B: the virus and disease*. Hepatology, 2009. **49**(5 Suppl): p. S13–S21.
49. Matthews, D., Emmott, E., and Hiscox, J., *Viruses and the nucleolus*. Nucleolus., 2011. **15**: p. 321–345. doi: 10.1007/978-1-4614-0514-6_14 eCollection 2011.
50. Ulloa-Aguilar, J.M., et al., *The nucleolus and its interactions with viral proteins required for successful infection*. Cells, 2024. **13**(18): p. 1591.
51. Salsman, J., et al., *Genome-wide screen of three herpesviruses for protein subcellular*

localization and alteration of PML nuclear bodies. PLoS Pathog., 2008. **4**(7): e1000100.
52. Li, K. and Wang, Z., *Speckles and paraspeckles coordinate to regulate HSV-1 genes transcription*. Commun. Biol., 2021. **4**(1): p. 1207.
53. Perez-Parada, J., Saffran, H.A., and Smiley, J.R., *RNA degradation induced by the herpes simplex virus vhs protein proceeds 5' to 3' in vitro*. J. Virol., 2004. **78**(23): p. 13391–13394.
54. Abernathy, E., et al., *Viral nucleases induce an mRNA degradation-transcription feedback loop in mammalian cells*. Cell Host Microbe, 2015. **18**(2): p. 243–253.
55. Jasinski-Bergner, S., Mandelboim, O., and Seliger, B., *Molecular mechanisms of human herpes viruses inferring with host immune surveillance*. J. Immunother. Cancer, 2020. **8**(2): e000841.
56. Lyles, D.S., *Cytopathogenesis and inhibition of host gene expression by RNA viruses*. Microbiol. Mol. Biol. Rev., 2000. **64**(4): p. 709–724.
57. Bercovich-Kinori, A., et al., *A systematic view on influenza induced host shutoff*. Elife, 2016: p. 5.
58. Bowie, A.G. and Unterholzner, L., *Viral evasion and subversion of pattern-recognition receptor signalling*. Nat. Rev. Immunol., 2008. **8**(12): p. 911–922.
59. You, Y., et al., *The suppression of apoptosis by α-herpesvirus*. Cell Death Dis., 2017. **8**(4): e2749.
60. Hu, B., et al., *Cellular responses to HSV-1 infection are linked to specific types of alterations in the host transcriptome*. Sci. Rep., 2016. **6**(1): p. 28075.
61. Weitzman, M.D. and Fradet-Turcotte, A., *Virus DNA replication and the host DNA damage response*. Annu. Rev. Virol., 2018. **5**(1): p. 141–164.
62. García-Sastre, A., *Ten strategies of interferon evasion by viruses*. Cell Host Microbe, 2017. **22**(2): p. 176–184.
63. Hale, B.G., et al., *The multifunctional NS1 protein of influenza A viruses*. J. Gen. Virol., 2008. **89**(Pt 10): p. 2359–2376.
64. Polyak, S.J., et al., *Hepatitis C virus nonstructural 5A protein induces interleukin-8, leading to partial inhibition of the interferon-induced antiviral response*. J. Virol., 2001. **75**(13): p. 6095–6106.
65. Wang, F., et al., *Role of hepatitis B virus non-structural protein HBx on HBV replication, interferon signaling, and hepatocarcinogenesis*. Front. Microbiol., 2023. **14**: 1322892.
66. Okumura, A., et al., *HIV-1 accessory proteins VPR and Vif modulate antiviral response by targeting IRF-3 for degradation*. Virology, 2008. **373**(1): p. 85–97.
67. Mouzakis, A., et al., *Mechanisms of immune evasion in HIV-1: the role of virus-host protein interactions*. Curr. Issues Mol. Biol., 2025. **47**(5): p. 367.
68. Cadima-Couto, I. and Goncalves, J., *Towards inhibition of Vif-APOBEC3G interaction: which protein to target?* Adv. Virol., 2010. **2010**: 649315.
69. Everett, R.D. and Orr, A., *Herpes simplex virus type 1 regulatory protein ICP0 aids infection in cells with a preinduced interferon response but does not impede interferon-induced gene induction*. J. Virol., 2009. **83**(10): p. 4978–4983.
70. Lanfranca, M.P., Mostafa, H.H., and Davido, D.J., *HSV-1 ICP0: An E3 ubiquitin ligase that counteracts host intrinsic and innate immunity*. Cells, 2014. **3**(2): p. 438–454.
71. Sorokin, I.I., et al., *Non-canonical translation initiation mechanisms employed by eukaryotic viral mRNAs*. Biochemistry (Mosc), 2021. **86**(9): p. 1060–1094.
72. Penn, W.D., et al., *Regulators of viral frameshifting: more than RNA influences translation events*. Annu. Rev. Virol., 2020. **7**(1): p. 219–238.

73. Francisco-Velilla, R., et al., *Picornavirus translation strategies*. FEBS Open Bio., 2022. **12**(6): p. 1125–1141.
74. De Vlugt, C., Sikora, D., and Pelchat, M., *Insight into influenza: a virus cap-snatching*. Viruses, 2018. **10**(11): p. 641.
75. Ivanov, A., et al., *HIV-1 Tat phosphorylation on Ser-16 residue modulates HIV-1 transcription*. Retrovirology, 2018. **15**(1): p. 39.
76. Kudoh, A., et al., *The phosphorylation of HIV-1 Gag by atypical protein kinase C facilitates viral infectivity by promoting Vpr incorporation into virions*. Retrovirology, 2014. **11**: p. 9.
77. Li, G., et al., *Cell cycle G2/M arrest through an S phase-dependent mechanism by HIV-1 viral protein R*. Retrovirology, 2010. **7**(1): p. 59.
78. Feng, T., et al., *Glycosylation of viral proteins: Implication in virus-host interaction and virulence*. Virulence, 2022. **13**(1): p. 670–683.
79. Rathore, U., et al., *Glycosylation of the core of the HIV-1 envelope subunit protein gp120 is not required for native trimer formation or viral infectivity*. J. Biol. Chem., 2017. **292**(24): p. 10197–10219.
80. Doores, K.J., *The HIV glycan shield as a target for broadly neutralizing antibodies*. Febs. J, 2015. **282**(24): p. 4679–4691.
81. Kim, P., et al., *Glycosylation of hemagglutinin and neuraminidase of influenza A virus as signature for ecological spillover and adaptation among influenza reservoirs*. Viruses, 2018. **10**(4): p. 183.
82. Tong, Y., et al., *Role of hepatitis C virus envelope glycoprotein E1 in virus entry and assembly*. Front. Immunol., 2018. **9**: p. 1411.
83. Liu, W., et al., *Mutation profiles, glycosylation site distribution and codon usage bias of human papillomavirus type 16*. Viruses, 2021. **13**(7): p. 1281.
84. Valerdi, K.M., et al., *The role of the host ubiquitin system in promoting replication of emergent viruses*. Viruses, 2021. **13**(3): p. 369.
85. Douglas, J.L., et al., *Vpu directs the degradation of the human immunodeficiency virus restriction factor BST-2/Tetherin via a {beta} TrCP-dependent mechanism*. J. Virol., 2009. **83**(16): p. 7931–7947.
86. Soper, A., et al., *Various plus unique: Viral protein U as a plurifunctional protein for HIV-1 replication*. Exp. Biol. Med. (Maywood), 2017. **242**(8): p. 850–858.
87. Talis, A.L., Huibregtse, J.M., and Howley, P.M., *The role of E6AP in the regulation of p53 protein levels in human papillomavirus (HPV)-positive and HPV-negative cells*. J. Biol. Chem., 1998. **273**(11): p. 6439–6445.
88. Đukić, A., et al., *HPV oncoproteins and the ubiquitin proteasome system: a signature of malignancy?* Pathogens, 2020. **9**(2): p. 133.
89. Dallaire, F., et al., *The human adenovirus type 5 E4orf6/E1B55K E3 ubiquitin ligase complex enhances E1A functional activity*. mSphere, 2016. **1**(1): 00014-15.
90. Harada, J.N., et al., *Analysis of the adenovirus E1B-55K-anchored proteome reveals its link to ubiquitination machinery*. J. Virol., 2002. **76**(18): p. 9194–9206.
91. Mauser, A., et al., *The Epstein-Barr virus immediate-early protein BZLF1 regulates p53 function through multiple mechanisms*. J. Virol., 2002. **76**(24): p. 12503–12512.
92. Bowling, B.L. and Adamson, A.L., *Functional interactions between the Epstein-Barr Virus BZLF1 protein and the promyelocytic leukemia protein*. Virus Res., 2006. **117**(2): p. 244–253.
93. González Carlos, M., Wang, L., and Damania, B., *Kaposi's sarcoma-associated herpesvirus encodes a viral deubiquitinase*. J. Virol., 2009. **83**(19): p. 10224–10233.
94. Mansouri, M., et al., *Kaposi sarcoma herpesvirus K5 removes CD31/PECAM from endothelial cells*. Blood, 2006. **108**(6): p. 1932–1940.

95. Imbert, F. and Langford, D., *Viruses, SUMO, and immunity: the interplay between viruses and the host SUMOylation system*. J. Neurovirol., 2021. **27**(4): p. 531–541.
96. Nevels, M., Brune, W., and Shenk, T., *SUMOylation of the human cytomegalovirus 72-kilodalton IE1 protein facilitates expression of the 86-kilodalton IE2 protein and promotes viral replication*. J. Virol., 2004. **78**(14): p. 7803–7812.
97. Bentz, G.L., Shackelford, J., and Pagano, J.S., *Epstein-Barr virus latent membrane protein 1 regulates the function of interferon regulatory factor 7 by inducing its sumoylation*. J. Virol., 2012. **86**(22): p. 12251–12261.
98. Marusic, M.B., et al., *Modification of human papillomavirus minor capsid protein L2 by sumoylation*. J. Virol., 2010. **84**(21): p. 11585–11589.
99. Everett, R.D., et al., *Sequences related to SUMO interaction motifs in herpes simplex virus 1 protein ICP0 act cooperatively to stimulate virus infection*. J. Virol., 2014. **88**(5): p. 2763–2774.
100. Lusic, M., et al., *Regulation of HIV-1 gene expression by histone acetylation and factor recruitment at the LTR promoter*. Embo. J., 2003. **22**(24): p. 6550–6561.
101. Ho, T.H., et al., *A screen for epstein-barr virus proteins that inhibit the DNA damage response reveals a novel histone binding protein*. J. Virol., 2018. **92**(14): 00262-18.
102. Udenwobele, D.I., et al., *Myristoylation: An important protein modification in the immune response*. Front. Immunol., 2017. **8**: p. 751.
103. Guan, X. and Fierke, C.A., *Understanding protein palmitoylation: biological significance and enzymology*. Sci. China Chem., 2011. **54**(12): p. 1888–1897.
104. Pal, R., et al., *Myristoylation of gag proteins of HIV-1 plays an important role in virus assembly*. AIDS Res. Hum. Retroviruses, 1990. **6**(6): p. 721–730.
105. Bryant, M.L., et al., *Replication of human immunodeficiency virus 1 and Moloney murine leukemia virus is inhibited by different heteroatom-containing analogs of myristic acid*. Proc. Natl. Acad. Sci. U. S. A., 1989. **86**(22): p. 8655–8659.
106. Panjwani, A., et al., *Capsid protein VP4 of human rhinovirus induces membrane permeability by the formation of a size-selective multimeric pore*. PLoS Pathog., 2014. **10**(8): e1004294.
107. Majeau, N., et al., *Palmitoylation of hepatitis C virus core protein is important for virion production*. J. Biol. Chem., 2009. **284**(49): p. 33915–33925.
108. Chlanda, P., et al., *Palmitoylation contributes to membrane curvature in influenza A virus assembly and hemagglutinin-mediated membrane fusion*. J. Virol., 2017. **91**(21): 00947-17.
109. Rousso, I., et al., *Palmitoylation of the HIV-1 envelope glycoprotein is critical for viral infectivity*. Proc. Natl. Acad. Sci. U. S. A., 2000. **97**(25): p. 13523–13525.
110. Oliver, S.L., *The structures and functions of VZV glycoproteins*. Curr. Top. Microbiol. Immunol., 2023. **438**: p. 25–58.
111. Homman-Loudiyi, M., et al., *Envelopment of human cytomegalovirus occurs by budding into Golgi-derived vacuole compartments positive for gB, Rab 3, trans-golgi network 46, and mannosidase II*. J. Virol., 2003. **77**(5): p. 3191–3203.
112. Pool, M.R., *Targeting of proteins for translocation at the endoplasmic reticulum*. Int. J. Mol. Sci., 2022. **23**(7): p. 3773.
113. Pauwels, E., Schülein, R., and Vermeire, K., *Inhibitors of the Sec61 complex and novel high throughput screening strategies to target the protein translocation pathway*. Int. J. Mol. Sci., 2021. **22**(21): 12007.

114. Gazina, E.V., et al., *Differential requirements for COPI coats in formation of replication complexes among three genera of Picornaviridae*. J. Virol., 2002. **76**(21): p. 11113–11122.
115. Crawford, S.E., et al., *COPII vesicle transport is required for rotavirus NSP4 interaction with the autophagy protein LC3 II and trafficking to viroplasms*. J. Virol., 2019. **94**(1): e01341-19.
116. Zhang, J., et al., *SARS-CoV-2 remodels the Golgi apparatus to facilitate viral assembly and secretion*. bioRxiv, 2024.
117. Romero-Brey, I. and Bartenschlager, R., *Endoplasmic reticulum: the favorite intracellular niche for viral replication and assembly*. Viruses, 2016. **8**(6): p. 160.
118. Chatel-Chaix, L. and Bartenschlager, R., *Dengue virus- and hepatitis C virus-induced replication and assembly compartments: the enemy inside--caught in the web*. J. Virol., 2014. **88**(11): p. 5907–5911.
119. Gong, Y., et al., *The glycosylation in SARS-CoV-2 and its receptor ACE2*. Signal Transduct. Target. Ther., 2021. **6**(1): p. 396.
120. Guo, S., et al., *Structure of the HIV immature lattice allows for essential lattice remodeling within budded virions*. eLife, 2023. **12**: e84881.
121. Grohman Jacob, K., et al., *An immature retroviral RNA genome resembles a kinetically trapped intermediate state*. J. Virol., 2014. **88**(11): p. 6061–6068.
122. Wagner, R.N., Reed, J.C., and Chanda, S.K., *HIV-1 protease cleaves the serine-threonine kinases RIPK1 and RIPK2*. Retrovirology, 2015. **12**(1): p. 74.
123. Jiang, H., et al., *Human adenovirus type 5 induces cell lysis through autophagy and autophagy-triggered caspase activity*. J. Virol., 2011. **85**(10): p. 4720–4729.
124. Rheinemann, L. and Sundquist, W.I., *Virus budding*, in *Encyclopedia of Virology*. 2021. p. 519–528. doi: 10.1016/B978-0-12-814515-9.00023-0 Epub 2021 Mar 1. PMCID: PMC7413661. https://pmc.ncbi.nlm.nih.gov/articles/PMC7413661/.
125. Bigalke, J.M. and Heldwein, E.E., *The great (nuclear) escape: new insights into the role of the nuclear egress complex of herpesviruses*. J. Virol., 2015. **89**(18): p. 9150–9153.
126. Laliberte, J.P. and Moss, B., *Lipid membranes in poxvirus replication*. Viruses, 2010. **2**(4): p. 972–986.
127. Aguilar, H.C., et al., *Paramyxovirus glycoproteins and the membrane fusion process*. Curr. Clin. Microbiol. Rep., 2016. **3**(3): p. 142–154.
128. Lomonosova, E., Subramanian, T., and Chinnadurai, G., *Requirement of BAX for efficient adenovirus-induced apoptosis*. J. Virol., 2002. **76**(22): p. 11283–11290.
129. Yang, J.E., et al., *Complexity and ultrastructure of infectious extracellular vesicles from cells infected by non-enveloped virus*. Sci. Rep., 2020. **10**(1): p. 7939.
130. Mardi, N., et al., *Exosomal transmission of viruses, a two-edged biological sword*. Cell Commun. Signal., 2023. **21**(1): p. 19.
131. Taylor, M.P., et al., *Role of microtubules in extracellular release of poliovirus*. J. Virol., 2009. **83**(13): p. 6599–6609.

3 Principles of Cell Biology During Virus Infections

3.1 Cell Cycle Progression During Virus Replication

The relationship between the eukaryotic cell cycle and viral replication is dynamic. It is based on the ability of viruses to manipulate and exploit the regulatory and cellular machinery inherent in the host cell's division cycle.

The eukaryotic cell cycle is a precisely regulated process. It is central to the growth and development of multicellular organisms. It comprises distinct phases, each governed by different regulatory networks and checkpoints. Despite this tightly controlled process, viruses have evolved strategies to effectively synchronize their replication with specific stages of the host cell cycle, optimizing the conditions for their propagation. For example, cell cycle regulators, such as cyclins and cyclin-dependent kinases (CDKs), are frequently targeted by viruses to ensure favorable conditions for replication [1]. Also, some DNA viruses often rely on the host cell's DNA synthesis during the S-phase to support their DNA replication. Conversely, the host cell must employ defense mechanisms against virus infection (see Table 3.1).

3.1.1 Cell Cycle in Eukaryotes

The cell cycle is a series of events that a cell undergoes to grow and divide. It ensures that cells replicate accurately and divide at the right time. It can be divided into two main phases: (i) Interphase, consisting of the G1 phase, S-phase, and G2 phase, and (ii) mitosis, which consists of the M phase (see Figure 3.1).

3.1.1.1 Interphase

G1 Phase (Gap 1) The cell grows and carries out normal functions. Here, the cell grows physically (increase in size), copies of organelles increase, and it makes the molecular building blocks it will need in later steps.

Checkpoint: At the end of G1, the cell checks for DNA damage and ensures it is ready for DNA synthesis.

S Phase (Synthesis) Function: The cell synthesizes a complete DNA replicate of itself in its nucleus, resulting in two copies of each chromosome. It also copies its

Table 3.1 Key cell cycle regulators and examples of viruses that exploit these regulators.

Cell cycle regulator	Function	Viruses implicated	Mechanism of exploitation
Tumor protein p53 (p53)	Tumor suppressor regulates G1/S checkpoint	Human papillomavirus (HPV), adenovirus	HPV E6 protein promotes the degradation of p53, allowing uncontrolled cell division
Rb (Retinoblastoma protein)	Tumor suppressor regulates G1/S transition	HPV, adenovirus	HPV E7 protein binds and inactivates Rb, leading to uncontrolled cell cycle progression
Cyclin D	Promotes transition from G1 to S-phase	Hepatitis B virus (HBV), human T-cell leukemia virus (HTLV-1)	HBV X protein can upregulate cyclin D expression, promoting cell cycle progression
Cyclin E	Promotes G1/S transition	Human cytomegalovirus (HCMV)	HCMV can upregulate cyclin E, driving the cell into S-phase to support viral replication
Cyclin A	Promotes S-phase progression and entry into mitosis	Kaposi's sarcoma-associated herpesvirus (KSHV)	KSHV latency-associated nuclear antigen (LANA) upregulates cyclin A, aiding viral replication
Cyclin B	Promotes transition from G2 to M phase	Epstein–Barr virus (EBV)	EBV LMP1 protein can upregulate cyclin B, facilitating cell division and viral propagation
CDK (cyclin-dependent kinases)	Key regulators of cell cycle progression	Human immunodeficiency virus, HPV	HIV Tat protein can interact with CDKs to modulate the cell cycle; HPV proteins can activate CDKs indirectly
p21 (Cip1/Waf1)	CDK inhibitor regulates G1/S checkpoint	Hepatitis C virus (HCV)	HCV core protein can suppress p21, leading to unchecked cell cycle progression
p27 (Kip1)	CDK inhibitor regulates G1/S transition	HCMV	HCMV immediate-early proteins can downregulate p27, promoting cell cycle progression
E2F transcription factors	Regulates genes required for S-phase	HPV, adenovirus	HPV E7 and Adenovirus E1A proteins can release E2F from Rb, leading to the activation of S-phase genes

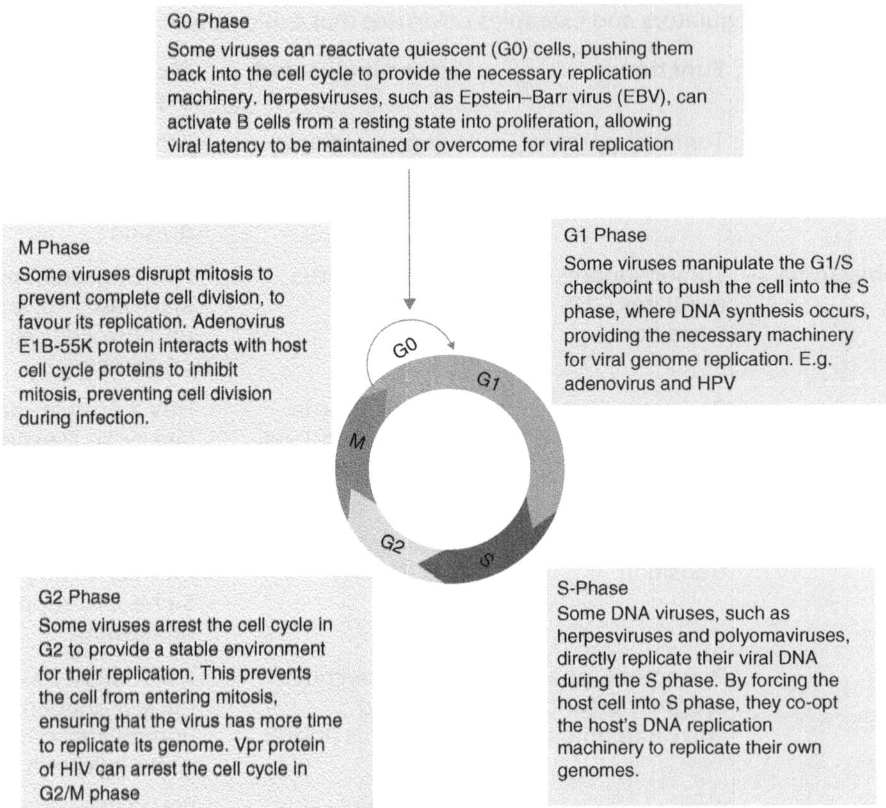

Figure 3.1 Overview of virus preference for different phases of the cell cycle.

microtubule-organizing structure, called the centrosome.

Checkpoint: The cell ensures that DNA replication is complete and accurate.

G2 Phase (Gap 2) Function: The cell continues to grow and prepares for mitosis. It produces proteins and organelles that start to reorganize in preparation for needed division.

Checkpoint: The cell checks for DNA damage and ensures all DNA is replicated properly.

3.1.1.2 Mitosis

M Phase (Mitosis) Function: The cell divides its replicated DNA and cytoplasmic contents to form two new daughter cells. Mitosis is further divided into sub-phases: prophase, metaphase, anaphase, telophase, and cytokinesis.

Checkpoint: The cell ensures that all chromosomes are properly aligned and attached to the spindle fibers before division.

Each phase of the cell cycle serves specific purposes, and the cellular environment and resources vary as the cell transitions from one phase to the next. It is within this highly regulated cell cycle that viruses have found opportunities to optimize their replication strategies. The ability of viruses to replicate efficiently often relies on their synchronization with the specific phases of the cell cycle. One striking example is the affinity of certain DNA viruses for the S-phase of the cell cycle, during which the host cell's DNA synthesis machinery is in full function.

DNA viruses, such as herpesviruses, adenoviruses, and papillomaviruses, have evolved to seize the advantages offered by the S-phase. This phase is characterized by the replication of the host cell's genomic DNA, which necessitates an active DNA synthesis machinery, including DNA polymerases and various factors involved in DNA replication. By replicating their genetic material during the S-phase, these DNA viruses can exploit the host cell's DNA synthesis machinery. The alignment with the host cell's activities not only provides an environment conducive to viral replication but also ensures that the necessary components for DNA replication are readily available. Beyond DNA viruses that replicate during the S-phase, several other viruses have evolved to replicate efficiently during distinct cell cycle phases.

3.1.2 Virus Manipulations of the Cell Cycle Phases

3.1.2.1 Virus Preference for the G1 Phase

Hepatitis B virus (HBV) demonstrates a preference for the G1 phase of the cell cycle. It has evolved mechanisms to promote cell cycle progression by inducing G1 phase arrest, which is favorable for viral replication. In the mouse model of HBV infection, HBx protein encoded by HBV is shown to interact with cell cycle regulators, promoting G1 phase entry [2].

3.1.2.2 Virus Preference for the S and G2 Phases

Papillomaviruses are DNA viruses associated with cervical cancer. Their genome primarily replicates in host cells that are in the S-phase (may extend to G2) [3], as this is when the host cell's DNA synthesis machinery is active. The viral genome replicates in coordination with the host's DNA replication.

3.1.2.3 Virus Preference for G2 and M Phases

Human immunodeficiency virus (HIV) stops the cell cycle during the G2 and M phases of the cell cycle [4]. This occurs because the HIV protein viral protein R (Vpr) actively induces cell cycle arrest at the G2/M transition, creating a favorable environment for viral replication within the host cell [4].

3.1.2.4 Virus Preference for Nondividing Quiescent Cells

Quiescent cells are nondividing cells that have exited the active cell cycle and are in a state of dormancy, often referred to as the G0 phase. These cells are metabolically active but do not proliferate unless stimulated by specific signals.

Herpes simplex virus (HSV): HSV is a DNA virus that employs a more complex strategy. It replicates its DNA during the S-phase, but can also establish latent infections in nondividing cells [5]. During latency, the viral genome remains dormant, blocking the progression of cells into S-phase. When reactivated, it re-enters the S-phase to initiate replication.

3.1.2.5 *Reactivation of the Cell Cycle*

Cells that remain quiescent (nondividing) for their entire life pose a unique challenge to viruses because viruses typically rely on the host cell's machinery, which is more active during cell division. However, some viruses have evolved mechanisms to stimulate these quiescent cells to re-enter the cell cycle and undergo division, thus creating a conducive environment for replication by:

Stimulation of Cell Cycle Entry Some viral mitogenic proteins activate various cellular pathways, including those involved in the G1/S transition of the cell cycle. This stimulation triggers the entry of quiescent cells into the G1 phase, where the cell prepares for DNA synthesis (S-phase) and subsequent division. Examples of viral mitogenic proteins include Epstein–Barr virus (EBV) nuclear antigen 3C (EBNA3C), Kaposi sarcoma-associated herpesvirus (KSHV) viral G protein-coupled receptor (v-GPCR), and HBVx protein [6, 7, 8].

Activation of Cellular Machinery As the cell cycle progresses, the host cell's machinery becomes more active, facilitating DNA replication, transcription, and translation, crucial for the successful replication of the virus. Some viruses encode specific proteins that act as mitogens or growth factors capable of mimicking the activity of cellular proteins. This promotes cell division by activating signaling pathways that drive quiescent cells into the cell cycle.

Notable examples of viruses that can stimulate quiescent cells to re-enter the cell cycle include oncogenic viruses like Papillomaviruses and EBV. These viruses encode specific proteins that activate cell cycle progression, promoting cell division and, in some cases, contributing to the development of cancers.

- **Epstein–Barr virus**
  Infected cells: B lymphocytes and epithelial cells.
  Mitogen-like protein: EBV encodes several proteins that can act as mitogens. One of the most prominent is the EBNA2 [9]. EBNA2 activates cellular genes, including those involved in the G1/S transition, effectively driving quiescent B-cells into the cell cycle [10].
- **Human papillomaviruses**
  Type of infected cells: Epithelial cells.
  Mitogen-like protein: The E6 and E7 oncoproteins of high-risk papillomavirus types are known to act as mitogens [11]. These proteins interfere with the regulation of the cell cycle. Specifically, E7 interacts with the retinoblastoma (Rb)

protein, while E6 promotes the degradation of tumor protein p53 (p53), allowing the host cell to enter the S-phase [12].

- **Herpes simplex virus**
  Type of infected Cells: A variety of cell types, including neurons and epithelial cells.
  Mitogen-like protein: HSV-1 encodes Infected Cell Protein 0 (ICP0), which plays a role in cell cycle control [13]. ICP0 can induce the degradation of certain cellular proteins, including components of the promyelocytic leukemia nuclear bodies, thereby modulating cell cycle regulation.
- **Adenoviruses**
  Type of infected cells: A wide range of cell types.
  Mitogen-like protein: Adenoviruses produce early region 1A (E1A) protein, which acts as a mitogen-like factor [14]. E1A interacts with cellular proteins, including those controlling the G1/S transition, such as Rb [15]. This interaction disrupts the normal cell cycle regulation and stimulates cell division.

3.1.3 Viral Replication and DNA Damage During the Cell Cycle

As mentioned earlier, eukaryotic cells have checkpoint control mechanisms to monitor DNA damage and other abnormalities. Viral replication and its induction of DNA responses represent a fascinating interplay between viruses and host cells that can result in DNA damage [16, 17].

The molecular mechanisms of the host cell's response to DNA damage involve a highly coordinated network of cellular processes designed to detect, repair, and mitigate genetic lesions. Key players in this complex system include the ataxia-telangiectasia mutated (ATM) and ataxia-telangiectasia and Rad3-related (ATR) kinases. ATM and ATR are a family of protein kinases involved in apoptosis and cell cycle regulation during DNA damage [18]. They can serve as sentinel proteins, initiating downstream signaling cascades upon the recognition of DNA damage [19]. These pathways activate cell cycle checkpoints, such as the G1/S and G2/M checkpoints, to temporarily halt cell cycle progression, allowing time for DNA repair.

DNA damage can be repaired by several mechanisms, including nonhomologous end joining (NHEJ), homologous recombination, base excision repair, base excision repair, or direct reversal [20]. In cases of severe or unrepaired damage, cells may trigger programmed cell death (apoptosis) or enter a state of senescence to prevent the propagation of genomic instability.

Types of damage to the DNA include [21]

- **Point mutations:** These involve changes to a single nucleotide base. Inflammation can induce this type of mutation [22]. For example, a base substitution where one base is replaced by another (e.g., A to T).

- **Insertions and deletions:** These are changes in the number of nucleotide bases, which can lead to frameshift mutations and altered protein sequences.
- **Structural aberrations:** These include larger-scale changes in DNA structure, such as chromosomal translocations, inversions, and duplications. These can result in genetic disorders or cancer.
- **Base loss or damage:** Chemical agents or radiation can alter the structure of individual bases, leading to mismatches or incorrect base pairing during DNA replication.
- **DNA strand breaks:** These can be single-strand breaks (SSBs) or double-strand breaks (DSBs) in the DNA backbone. DSBs are particularly harmful, as they can result in the loss of genetic information or chromosomal rearrangements.
- **Cross-links:** These are covalent bonds that form between nucleotides on different DNA strands, preventing the strands from separating and inhibiting DNA replication and transcription.
- **Deamination:** This is the removal of an amino group from a nucleotide base, which can convert cytosine into uracil or adenine into hypoxanthine, leading to incorrect base pairing during replication.

3.1.3.1 Mechanism of Virus-Induced DNA Damage

- **Genome integration:** Some viruses, such as retroviruses (e.g., HIV's Vpr), directly induce DNA damage [23]. The repair of these breaks can inadvertently create mutations and genomic instability.
- **Replication stress:** Inflammation, and the rapid and extensive replication of viral genomes, often leads to replication stress in host cells [24, 25]. The increased demand for nucleotides, DNA polymerases, and other factors can strain the host's DNA replication machinery, resulting in the stalling and collapse of replication forks and the generation of SSBs.

3.1.3.2 Cellular Responses to Virus-Induced DNA Damage

- **Activation of ATM and ATR kinases:** Virally induced DNA damage activates cellular kinases, such as ATM and ATR [26, 27]. These kinases are key players in the DNA damage response pathway. They phosphorylate downstream targets, triggering cell cycle checkpoints and DNA repair processes.
- **Checkpoint activation:** The activation of ATM and ATR initiates cell cycle checkpoints, such as the G1/S and G2/M checkpoints. These checkpoints halt the cell cycle, allowing time for DNA repair and minimizing the transmission of damaged DNA to daughter cells.
- **Activation of DNA repair pathways:** The DNA damage response promotes the recruitment of repair proteins to the sites of damage. The choice of repair pathway depends on the type of DNA damage,

including NHEJ and homologous recombination.

- **p53 and apoptosis:** In response to severe DNA damage, p53, a tumor suppressor protein, can be activated [28]. It plays a pivotal role in either allowing DNA repair or triggering apoptosis to eliminate cells with irreparable damage.

3.1.3.3 *Viral Strategies and Countermeasures*

- **Evasion of the DNA damage response:** Some viruses encode proteins that inhibit DNA damage responses to promote their replication. For instance, the adenovirus early region 4 open reading frame 6 protein/early region 1B 55-kilodalton protein (E4orf6/E1B55K) complex targets and degrades p53, suppressing apoptosis [29].
- **Modulation of repair pathways:** Viruses may manipulate the host cell's DNA repair pathways for their benefit. For example, EBV enhances nonhomologous end-joining to facilitate the circularization of its genome [30].

3.2 Cellular Response to Virus-Induced Stress

How do cells respond to the stress induced by viral infections?

Viral infections pose a significant challenge to host cells. In return, the cells trigger a variety of stress responses aimed at counteracting the detrimental effects of the invading pathogen. This is critical for maintaining homeostasis and ensuring survival. Some of the few cellular responses to alleviate virus-induced stress are examined below.

3.2.1 Activation of the Unfolded Protein Response (UPR)

The UPR is an essential cellular stress response that becomes activated when unfolded or misfolded proteins build up in the ER [31]. During viral replication, particularly during the synthesis and assembly of viral proteins, the ER may recognize these viral proteins as unfolded or misfolded, or the replication process itself may place stress on the ER. This leads to the production of misfolded host proteins and stress on the ER, triggering the UPR.

Many viruses, such as HBV, hepatitis C virus (HCV), and certain flaviviruses, synthesize large amounts of proteins in the host cell. This increased load can overwhelm the protein-folding capacity of the ER, leading to ER stress and activation of the UPR [32, 33].

A. **Effector pathways and mechanisms of UPR**
   1. **PKR-like Endoplasmic Reticulum Kinase (PERK) pathway:** Reduces overall protein synthesis to decrease the load on the ER and promotes the translation of ATF4, which activates genes involved in oxidative stress response (OSR) and amino acid metabolism.
   2. **Inositol-Requiring Enzyme 1 (IRE1) pathway:** Splices X-box binding protein 1 (XBP1) messenger

RNA (mRNA), leading to the production of a potent transcription factor that enhances the expression of chaperones and other proteins that aid in protein folding and ER-associated degradation (ERAD).

3. **Activating Transcription Factor 6 (ATF6) pathway:** In response to ER stress, ATF6 translocates to the Golgi apparatus, where it is cleaved by proteases. The resulting cytosolic fragment then moves to the nucleus to activate the transcription of UPR target genes.

B. **Functions and outcomes of UPR**

1. **Restoration of protein homeostasis:** By increasing the expression of chaperones such as binding immunoglobulin protein (BiP) and other folding enzymes, the UPR enhances the cell's ability to fold proteins correctly.
2. **Degradation of misfolded proteins:** The UPR promotes the degradation of misfolded proteins via the ERAD, preventing their accumulation and maintaining ER function [34].
3. **Apoptosis:** If the stress is severe and persistent, the UPR can trigger apoptosis to eliminate damaged cells, preventing the propagation of dysfunctional proteins.

C. **Viral manipulation of UPR pathways to enhance replication**

Even though these responses are upregulated with the intent to alleviate the pathologic effects of virus replication on the cellular machinery, viruses have developed ways of manipulating these effects by:

1. **Hijacking UPR pathways:** By hijacking UPR pathways, some viruses can create a more favorable environment for their replication. The upregulation of chaperones and folding enzymes helps ensure the proper folding of viral proteins, while the inhibition of apoptosis prevents premature cell death.
2. **Immune evasion:** Some viruses use the UPR to evade the host immune response. For instance, UPR activation can lead to the suppression of interferon responses, reducing the host's ability to mount an effective antiviral defense [35].
3. **Host cell damage:** Prolonged UPR activation due to chronic viral infection can lead to cellular dysfunction and contribute to disease pathogenesis [36]. For example, chronic hepatitis virus infections are associated with liver diseases, including fibrosis and hepatocellular carcinoma, partly due to sustained ER stress and UPR activation [37].

3.2.2 Activation of Oxidative Stress Response

The OSR is a crucial cellular defense mechanism for maintaining homeostasis, which addresses the damage caused by reactive oxygen species (ROS). ROS are highly reactive molecules generated as byproducts of normal cellular metabolism, particularly in the mitochondria, but their levels can increase significantly under certain stress conditions. For example, virus infections can significantly influence the OSR within host cells. This response can affect both viral replication and host cell survival.

A. **Effector pathways and mechanisms of oxidative stress response**
   1. **Activation of antioxidant defense systems:** Cells employ a variety of antioxidants to neutralize ROS, such as:
      - Production of enzymatic antioxidants such as superoxide dismutases (SOD) capable of converting superoxide anion to hydrogen peroxide [38]; catalase capable of converting hydrogen peroxide to water and oxygen; glutathione peroxidases (GPx) capable of reducing hydrogen peroxide and lipid peroxides using glutathione [39].
      - Production of nonenzymatic antioxidants such as glutathione – a tripeptide that directly scavenges ROS and is a substrate for GPx; vitamin C (ascorbic acid) and vitamin E (tocopherol) capable of scavenging free radicals and regenerating other antioxidants [40, 41].
   2. **Transcriptional regulation:** The OSR is tightly regulated at the genetic level to ensure an appropriate response:
      - **Nuclear factor erythroid 2-related factor 2 (Nrf2):** A key transcription factor that, upon activation, translocates to the nucleus and binds to antioxidant response elements (ARE) in the promoters of antioxidant and detoxification genes [42, 43].
      - **Activator protein-1(AP-1) and nuclear factor kappa B (NF-κB):** Also involved in regulating the expression of genes that manage oxidative stress [44, 45].
   3. **Damage repair and removal:** Response to oxidative stress can also direct repair or degradation of damaged molecules, such as upregulating DNA repair enzymes to fix oxidative DNA damage, proteasomes and lysosomes to degrade oxidatively damaged proteins, or lipid repair mechanisms to repair or remove oxidized lipids.

B. **Viral manipulation of OSR pathways**

Some viruses manipulate the host's antioxidant defenses to create a favorable environment for replication. For example, HIV infection increases ROS production, leading to oxidative stress and damage to immune cells [46]. Oxidative stress to the $CD4^+$ cells may contribute to the depletion or dysfunction of $CD4^+$ T-cells and the progression of acquired immunodeficiency syndrome (AIDS). Also, HBV and HCV induce oxidative stress in liver cells [47], contributing to liver inflammation, fibrosis, and an increased risk of hepatocellular carcinoma. Influenza virus infection causes oxidative stress in respiratory epithelial cells, contributing to inflammation. Severe acute respiratory syndrome coronavirus 2 (SARS-CoV-2) also induces oxidative stress, leading to lung tissue damage and exacerbating the severity of the disease [48]. Elevated ROS levels are associated with the cytokine storm seen in severe cases.

Mechanisms of virus manipulation include

- **Upregulating antioxidant genes:** Certain viruses activate the Nrf2 pathway to upregulate antioxidant genes, reducing ROS levels and preventing premature cell death. For instance, HCV can induce Nrf2 activation to protect infected cells from oxidative damage [49].
- **Suppression of antioxidant enzymes:** Other viruses may downregulate antioxidant enzymes to maintain a level of oxidative stress that promotes viral replication without killing the host cell.
- **Induction of apoptosis:** High levels of ROS can trigger apoptosis, which some viruses exploit to facilitate the release of viral particles. For example, HIV-induced oxidative stress can lead to apoptosis of infected T-cells, contributing to immune system depletion [50].

3.2.3 Heat Shock Response

The heat shock response (HSR) is a fundamental cellular defense mechanism activated by exposure to elevated temperatures, but it is also activated in the presence of other stressors that can damage proteins. This response helps cells survive by enhancing their ability to refold damaged proteins and prevent protein aggregation.

During viral infections, HSR plays a significant role in the cellular response to virus-induced stress. HSR also aids in the degradation of misfolded viral proteins, helping to limit viral spread [51] (see Table 3.2). By enhancing protein stability and refolding, the HSR contributes to the cell's ability to manage the stress imposed by viral infections, ultimately influencing the outcome of the infection and the host's immune response.

A. **Effector pathways and mechanisms**

The primary effectors of HSR are a group of proteins known as heat shock proteins (HSPs), which function as molecular chaperones. Examples of HSPs include HSP70, which binds to and refolds misfolded proteins, preventing aggregation [52]; HSP90, which assists in the proper folding of proteins, particularly signaling proteins and receptors; and HSP60, a chaperone protein in the mitochondria. Small HSPs, e.g., HSP27, assist in stabilizing unfolded proteins and protect cells from stress-induced apoptosis [53].

B. **Functions and outcomes**

- **Cell survival:** The HSR enhances cell survival under stress by preventing the accumulation of toxic protein aggregates and maintaining protein homeostasis. Specifically, HSPs interact with

Table 3.2 Examples of heat shock proteins (HSPs) and other chaperones involved during viral infections.

Chaperone	Function	Role in viral infections
HSP70	Protein folding, assembly, and degradation	Essential for viral protein synthesis, assembly, and trafficking. It acts as a molecular chaperone, facilitating the proper folding and maturation of viral proteins
HSP90	Protein folding and stability	Interacts with viral oncoproteins and regulatory proteins, e.g., HCV NS5A
HSP40	Co-chaperone with HSP70	Assist in the nuclear import of viral ribonucleoproteins, e.g., influenza A virus
Small HSPs (e.g., HSP27, αB-crystallin)	Protein aggregation and stress response	Involved in viral protein stability and stress response
Hsp60	Mitochondrial chaperonin	Often acting as a positive regulator for virus progression
Bag proteins	Regulate HSP70 and HSP90 activity	Involved in protein folding, degradation, and signaling

nascent and misfolded proteins to facilitate proper folding, prevent aggregation, and assist in the refolding or degradation of damaged proteins. Some HSPs, like HSP70 and HSP90, also participate in the degradation of irreparably damaged proteins via the proteasome pathway [54].

- **Protection against apoptosis:** HSPs can inhibit stress-induced apoptosis by stabilizing mitochondrial membranes and interacting with key apoptotic regulators [55, 56].

C. **Viral manipulation of HSR**

- **Enhanced replication:** Some viruses exploit HSPs to facilitate their replication and assembly. For example, HSP70 and HSP90 can assist in the folding and stabilization of viral proteins [51].
- **Immune evasion:** By upregulating HSPs, viruses may protect infected cells from stress-induced apoptosis, allowing for prolonged viral replication.

3.2.4 Formation of Stress Granules

Stress granules are dynamic aggregates of proteins and RNAs that are formed in the cytoplasm of cells in response to various types of stress, such as oxidative stress, heat shock, and viral infection. These granules play a crucial role in cellular stress responses by halting the translation of most mRNAs and sequestering them into small aggregates called granules [57]. This process prevents their degradation and allows the cell to conserve resources and energy until conditions improve. When the stress is alleviated, stress granules may disassemble, allowing the translation of mRNAs to resume and the cell to return to normal functioning.

A. **Function**
- **Translational control:** Stress granules play a role in regulating translation by sequestering untranslated mRNAs, and some key proteins, such as T-cell-restricted intracellular antigen-1 (TIA-1) and GTPase-activating protein (SH3 domain)-binding protein (G3BP), act as nucleating factors to initiate the assembly of stress granules [57].
- **Protection of mRNA:** They help protect mRNAs from degradation during stress conditions.
- **Cellular signaling:** Stress granules are involved in various signaling pathways that help the cell respond to and recover from stress.

B. **Viral manipulation of stress granule formation**

Viruses often interact with and manipulate these processes to their advantage. Some viruses can inhibit stress granule assembly, while others may co-opt components of stress granules for their benefit. Stress granules can be induced by a variety of viruses, such as RNA viruses like HIV, Zika virus, and influenza virus [58].

a. **Inhibition of stress granule formation**
- **Inhibition of key proteins:** Some viruses inhibit proteins essential for stress granule assembly, such as G3BP1 and TIA-1. By doing so, they prevent the formation of stress granules, allowing uninterrupted viral replication. For example, West Nile virus can inhibit stress granule formation at some point in its replication [59].
- **Cleavage of stress granule components:** Viral proteases can cleave key stress granule components, preventing their assembly. For example, some enterovirus enzymes, proteases, cleave G3BP1, thereby disrupting stress granule formation. This helps in efficient viral replication [60].

b. **Modulation and utilization of stress granules**
- **Hijacking stress granule components:** Certain viruses hijack stress granule components for their replication. For example, some RNA viruses use stress granule proteins to stabilize their RNA and enhance replication [61].
- **Redistribution of stress granule proteins:** Viruses can cause the redistribution of stress granule proteins to viral replication complexes. For example, flaviviruses [62].

3.2.5 Formation of RNA Processing Bodies (P-Bodies)

RNA P-bodies are key cytoplasmic structures involved in the regulation and processing of RNA [63]. They play crucial roles in controlling gene expression, managing RNA stability, and facilitating the degradation of RNA molecules.

RNA P-bodies are distinct from stress granules, although both are involved in RNA regulation and can share some components. Stress granules are like emergency response centers – they handle storage. They are formed when a cell is under stress (like when it is sick or overloaded). The cell puts certain messages (in mRNA) and proteins into these granules to protect them and keep them safe until the stress is over. In contrast, P-bodies are like recycling centers – they handle breakdown. Their main function is to manage and clean up old or unneeded materials. They help break down and recycle mRNA messages that the cell no longer needs or that are faulty.

A. **Effector pathways and mechanisms**

P-bodies form through the aggregation of RNA and protein components. This aggregation is believed to be driven by interactions between RNA-binding proteins and specific RNAs. They can also interact with other cellular structures, such as stress granules, especially under stress conditions [64].

B. **Functions of P-bodies**

- **mRNA decay:** P-bodies are central to the decay process of mRNA through the process of mRNA decapping and degradation [65]. They contain enzymes that remove the 7-methyl guanylate cap from mRNA (decapping) and degrade the RNA body. This is an essential mechanism for regulating gene expression by eliminating unwanted or faulty mRNAs, like viral mRNA.
- **RNA storage and regulation:** P-bodies can store mRNAs that are not immediately needed for translation [66]. These mRNAs can be stored in P-bodies and later released for translation when needed, preventing virus access to cellular RNA.
- **Regulation of translation:** P-bodies help regulate the translation of mRNAs by sequestering them in a repressed state. They contain factors that inhibit translation and can control the availability of mRNAs for protein synthesis. Specifically, cytoplasmic viral RNA can be sequestered in P-bodies, preventing its translation or decay [67].

Virus manipulation of P-bodies

Viruses can manipulate processing bodies (P-bodies) to enhance their replication, evade the host's immune system, and achieve other goals that benefit their lifecycle. Examples are stated below.

- **Inhibition of P-body functions:** Some viruses produce proteins that interfere with the normal functions of P-bodies, such as the decapping and degradation of mRNA.

By inhibiting these processes, viruses can prevent the degradation of viral mRNA, allowing more viral proteins to be produced. Examples include HCV, which inhibits P-body functions to stabilize its mRNA and ensure efficient replication [68].

- **Disruption of P-body assembly**: Some viruses can prevent the formation of P-bodies or disrupt their structure. This can interfere with the cell's ability to regulate and degrade mRNAs, benefiting the virus by allowing the accumulation of viral mRNA and proteins. The poliovirus can disrupt P-body formation through cleavage of key cellular proteins such as eIF4G and G3BP to evade host defenses and enhance viral replication [69].
- **Hijacking P-body components:** Viruses can hijack components of P-bodies for their benefit. For instance, some viral proteins can bind to P-body components, redirecting them to support viral replication or to avoid detection by the immune system. Influenza virus uses components of P-bodies to enhance its replication process and to evade host immune responses [70].
- **Manipulating mRNA storage and enhancing viral mRNA stability**: By interacting with P-body components, some viruses can enhance the stability of their mRNA while reducing the stability of host mRNAs. This allows the virus to produce more proteins and replicate more efficiently.

3.2.6 Autophagy

Autophagy is a cellular process involved in the degradation and recycling of cellular components. Autophagy, which means "self-eating" in Greek, is a natural cellular housekeeping mechanism where cells degrade and recycle their components [71]. This process helps to remove damaged organelles, misfolded proteins, and other cellular debris, ensuring the cell's survival and proper functioning. Unlike apoptosis (programmed cell death), where the whole cell is killed, autophagy only destroys or recycles a few of the cell components or organelles [57]. Think of it as the cell's way of taking out the trash and reusing valuable parts. When a cell is stressed, damaged, or lacking nutrients, it can break down its old or unnecessary parts into smaller pieces. These pieces can then be used to build new cell components or generate energy. This helps the cell stay healthy and function properly.

Depending on the context, it can be both a pro-survival and a pro-death mechanism. Autophagy involves the formation of autophagosomes, double-membraned structures that engulf cytoplasmic cargo, including damaged organelles, and deliver them to lysosomes for degradation [72] (see Figure 3.2). Autophagy plays a crucial role in the host cell's antiviral defense mechanisms. It can selectively target and degrade viruses, making it a part of the intrinsic immune response.

A. **Effector pathways and mechanisms**
   - **Initiation and formation of the autophagosome:** When a cell is under stress, infected, damaged,

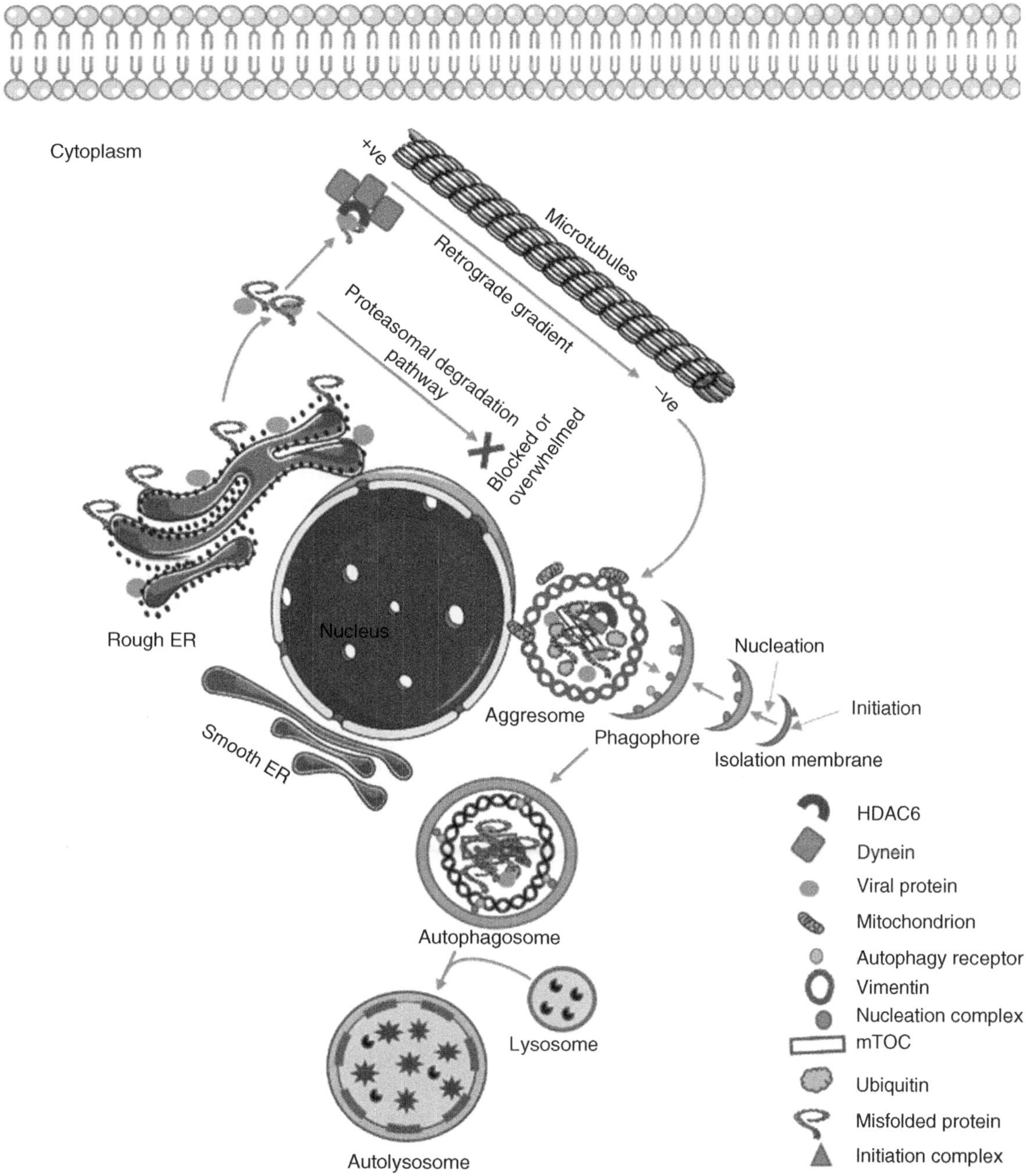

Figure 3.2 Systematic illustration of virus-induced autophagy formation. HDAC6: Histone Deacetylase 6, and mTOC: Microtubule-Organizing Center.

Source: Ref. [57]/MDPI/CC BY 4.0.

or deprived of nutrients, autophagy can be triggered. A membrane begins to form around the cellular material to be degraded. The membrane expands and engulfs the targeted cellular components, forming a structure called an autophagosome.

- **Fusion with lysosomes:** The autophagosome then fuses with a lysosome, an organelle containing digestive enzymes. These fusion forms are autolysosomes.
- **Degradation and recycling:** The contents of the autolysosome are broken down by the lysosomal enzymes. The resulting molecules, such as amino acids and fatty acids, are then released back into the cytoplasm for reuse by the cell [72].

Functions of autophagy

- **Nutrient recycling:** During periods of virus infection, viruses may shut down host protein synthesis, leading to cell starvation. Autophagy provides an internal source of nutrients, helping the cell to survive [73].
- **Cellular quality control:** Autophagy can also help in quality control of the cell processes by removing damaged organelles and proteins. Autophagy helps to maintain cellular health and prevent the accumulation of toxic materials like viral proteins [74].
- **Defense against infection:** Autophagy can help to eliminate infected cell organelles by degrading them.

Virus manipulation of autophagy

Viruses have evolved various strategies to manipulate, either by upregulating or preventing autophagosome formation, to benefit their replication and survival within host cells. By influencing autophagy, viruses can create a more favorable environment for their lifecycle, helping them to evade host immune responses and ensure the production of viral components.

A. **Enhancement of autophagy for viral replication**
   - **Promoting autophagosome formation:** Some viruses enhance the formation of autophagosomes to create membrane structures that can serve as sites for viral replication. For example, the poliovirus and DENV can induce the formation of double-membrane vesicles, which resemble autophagosomes, to facilitate viral RNA replication or evade immune response [75].
   - **Utilizing autophagic machinery:** Viruses like the HCV exploit autophagic machinery to support their replication. HCV uses autophagy-related proteins to enhance the formation of membranous replication complexes and membrane trafficking to aid the efficient production of viral RNA [76].

B. **Inhibition of autophagy to evade immune responses**
   - **Blocking autophagosome–lysosome fusion:** Some viruses can inhibit the fusion of autophagosomes with lysosomes,

preventing the degradation of viral components. This allows viruses to avoid destruction and continue replicating. For instance, the HSV expresses a neurovirulence factor called infected cell protein 34.5 (ICP34.5) that interferes with autophagosome–lysosome fusion [77].

- **Modulating autophagic flux:** By modulating the flow of autophagic degradation, some viruses can create a balance that favors their survival. Influenza A virus, for example, can temporarily block autophagic degradation to prevent the destruction of viral proteins while still using autophagic machinery for its benefit [78].

HCV and autophagy

a. **Induction of autophagy:** HCV infection induces autophagy in liver cells [79]. This autophagic response aids virus replication by providing a membranous environment for viral RNA replication complexes.
b. **Formation of autophagosomes:** During HCV infection, autophagosomes form and surround HCV replication complexes [80]. These structures create a protected environment for viral RNA to replicate and evade host cell defenses.
c. **Viral assembly and release:** Importantly, HCV takes advantage of the autophagy process for viral particle assembly and maturation. Newly synthesized HCV genomic RNA and structural proteins, including core and envelope proteins, are trafficked to autophagosomes [81].
d. **Viral particle maturation:** Within the autophagosomes, HCV particles undergo maturation, not to gain infectivity but for assembly [81, 82]. Lipid droplets and host cell membranes may fuse with autophagosomes, leading to the envelopment of HCV particles in host-derived membranes. This process is crucial for the formation of infectious virions.
e. **Release of virions:** Upon maturation, HCV particles are trafficked by autophagic mechanisms from the infected hepatocyte [83].

Other examples of virus manipulation of autophagy

a. **Herpesviruses:** HSV can inhibit autophagy in neuronal cells by expressing the ICP34.5 protein [84, 85], which binds to the autophagy protein Beclin-1, preventing autophagosome formation to enhance virus replication.
b. **Flaviviruses:** Dengue virus and Zika virus can induce autophagy to create membrane structures that support viral replication [86, 87]. These viruses manipulate the autophagy pathway to enhance the production of viral particles.
c. **Coronaviruses:** SARS-CoV-2 interacts with autophagy-related pathways. It induces autophagy to facilitate its replication and may also inhibit autophagic degradation to evade immune responses [88].

3.2.7 Apoptosis

Apoptosis, which is programmed cell death, plays a pivotal role fundamental role in the host defense mechanism against viral infections. When a cell senses a viral invasion, it can initiate apoptosis to limit viral replication and prevent the spread of the infection to neighboring cells (see Figure 3.3). As a way of defense, this process can help eliminate infected cells before they release newly formed viral particles.

A. **Effector pathways and mechanisms**

Apoptosis can be initiated through two main pathways: the intrinsic (mitochondrial) pathway and the extrinsic (death receptor) pathway. Both pathways converge on a common set of molecular players that execute the death program [89, 90].

- **The intrinsic pathway (mitochondrial pathway)** is often initiated by internal stress signals, such as DNA damage, oxidative stress, or other cellular stressors like virus infection. The main event of this pathway involves the release of cytochrome c from mitochondria into the cytoplasm, activating caspase-9, which then activates other caspases (such as caspase-3), leading to the dismantling of the cell [91].
- **The extrinsic pathway (death receptor pathway)** is often activated by external signals, such as the binding of death ligands (e.g., Fas ligand or tissue necrotic factor (TNF)-related apoptosis-inducing ligand) to their respective death receptors on the cell surface. The main event of this pathway involves the recruitment of adaptor proteins (like Fas-associated protein with death domain) and activation of caspase-8, which then activates downstream effector caspases (e.g., caspase-3) [92].

Key steps in apoptosis

- **Signal initiation:** Signals are received either from within the cell (intrinsic pathway) or from outside the cell (extrinsic pathway).
- **Caspase activation:** Activation of caspases, a family of protease enzymes, which cleave specific cellular proteins, leading to the morphological and

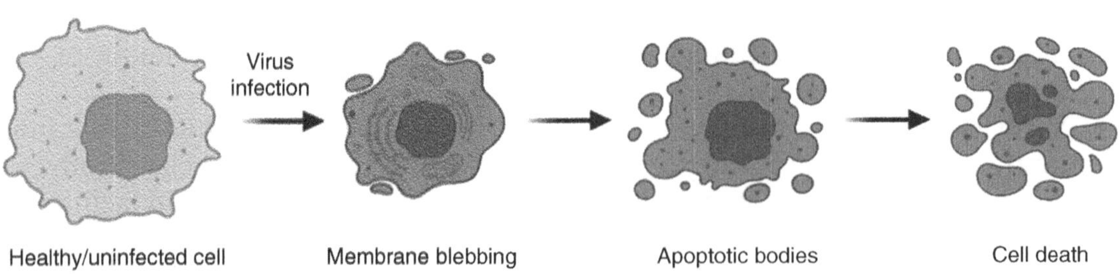

Figure 3.3 Overview of virus-induced apoptosis.

biochemical changes associated with apoptosis.

- **Cellular changes:** Such as chromatin condensation leading to DNA fragmentation and chromatin condensation; membrane blebbing, where the cell membrane forms blebs or bulges; and formation of apoptotic bodies, that is, the cell breaks into smaller, membrane-bound fragments.
- **Phagocytosis:** The apoptotic bodies are recognized and engulfed by phagocytic cells, such as macrophages, without inducing an inflammatory response.

B. **Significance of apoptosis during viral infection**

- **Limiting viral replication:** By triggering apoptosis early in the infection, host cells can suppress the replication of viruses [93]. When an infected cell undergoes apoptosis, it prevents the virus from using the cellular machinery to produce more viral particles.
- **Preventing virus spread:** Apoptosis helps to isolate and contain the infection, preventing the spread of the virus to neighboring cells. The apoptotic cell is engulfed and digested by phagocytes, which helps to remove the viral particles within it.
- **Activating immune responses:** Apoptotic cells release signals that can activate immune responses, including the recruitment of immune cells like macrophages and dendritic cells [94]. These cells can then present viral antigens to T-cells, enhancing the adaptive immune response.

C. **Virus modulation of apoptosis**

Like other mechanisms, viruses activate various strategies to manipulate apoptosis to enhance their survival and replication. Some viruses may encode specific anti-apoptotic proteins to counteract the host cell's apoptotic response. This manipulation can involve either inhibiting or inducing apoptosis, depending on the stage of infection and the virus's strategy.

- **Inhibition of apoptosis:** Many viruses inhibit apoptosis during the early stages of infection to allow sufficient time for replication. Examples are HSV and adenoviruses. Adenovirus early region 1B 19 kilodalton protein (E1B-19 K) and E1A inhibit apoptotic pathways, allowing the virus to replicate before the host cell dies [95]. Some viruses also interfere with the release of cytochrome c, a critical step in the intrinsic apoptotic pathway. For instance, the HBV can activate the Phosphoinositide 3-Kinase/Protein Kinase B (PI3K/Akt) pathway, which has anti-apoptotic effects and prevents p53-mediated apoptosis [96]. Also, herpesviruses produce proteins like Bcl-2 homologs that inhibit the activation of pro-apoptotic factors.
- **Induction of apoptosis:** At later stages of infection, some viruses may induce apoptosis to facilitate the release and spread of progeny virions. Examples are

HIV and influenza virus. HIV can induce apoptosis in infected $CD4^+$ T cells, leading to immune system depletion and facilitating virus spread [97]. Also, influenza A virus can induce apoptosis in infected epithelial cells, promoting the release of viral particles and facilitating viral spread. In some cases, apoptosis can help viruses escape the host's immune response.

3.3 Cell Trafficking Machinery and Cytoskeleton During Viral Infection

Cell trafficking refers to the processes by which molecules and vesicles are transported within cells. This trafficking is essential for maintaining cellular functions. Key components of cell trafficking machinery are the cytoskeleton (such as microtubules, actin filaments, and intermediate filaments) and the motor proteins (such as kinesins, dynein, and myosin).

Types of trafficking pathways

- **Endocytic trafficking pathway:** The process by which cells internalize large molecules and particles from their environment is known as endocytosis [98]. Endocytosis can be divided into receptor-mediated endocytosis (also known as clathrin-mediated endocytosis), caveolae, pinocytosis, and phagocytosis.
- **Exocytic trafficking pathway:** The process by which a cell transports large molecules and particles from inside the cell to the exterior through vesicular trafficking is known as exocytosis. Viral and cellular proteins are trafficked during assembly and egress using this pathway [99, 100].
- **Retrograde transport:** Retrograde transport is the process by which molecules and vesicles are transported from one compartment of the cell to another. It is a specific type of intracellular transport that moves materials (toxins, cellular or viral proteins) from a later compartment in the cell back to an earlier one [101, 102]. This is crucial for recycling components, correcting errors in protein sorting, and ensuring the proper functioning of cellular organelles. Common examples are the movement of particles from the Golgi to the ER for correction or degradation and from the endosome to the Golgi for recycling, further processing, or modification.

Cell cytoskeleton

The cytoskeleton is a complex interlinking of protein filaments and tubules that extend throughout the cytoplasm of a cell. This provides structural support, maintains cell shape, enables cellular movement, and plays essential roles in intracellular transport, cellular division, and signaling. The main components of the cytoskeleton are microfilaments (actin filaments), microtubules, and intermediate filaments, each with unique properties and functions.

a. **Microfilaments (actin filaments):** Microfilaments are thin, flexible fibers composed of actin, a globular protein. They are about 7 nm in diameter and are most concentrated just beneath the plasma membrane, forming a dense network known as the cell cortex [103]. Key roles of microfilaments include providing mechanical support, determining and maintaining the shape of the cell, amoeboid movement, cytokinesis (the division of a cell's cytoplasm), and muscle contraction. Specifically, actin filaments can facilitate the movement of organelles, vesicles, and proteins within the cell [104].
b. **Microtubules:** Microtubules are hollow cylinders made of tubulin proteins (alpha and beta tubulin) [105]. They are about 25 nm in diameter and are the largest components of the cytoskeleton. The functions of microtubules include maintaining the cell's shape by resisting compressive forces. Microtubules also form tracks for the movement of organelles and vesicles, facilitated by motor proteins like kinesin and dynein. They form the structural basis of cilia and flagella, enabling cell motility and the movement of extracellular fluids [106]. Microtubules form the mitotic spindle, which is essential for chromosome separation during mitosis and meiosis.

Viruses exploit these cellular structures and processes to facilitate their entry, replication, assembly, and release.

3.3.1 Trafficking Machinery During Viral Entry

Use of actin filaments: Many viruses use the actin cytoskeleton to enter host cells. For instance, enveloped viruses such as influenza and HIV induce actin polymerization to facilitate membrane fusion or endocytosis [107, 108].

- **Use of receptor-mediated endocytosis:** The binding of viral surface proteins to host cell receptors often triggers actin remodeling, leading to the formation of endocytic vesicles that internalize the virus.
- **Use of macropinocytosis:** Some viruses, like Ebola, exploit actin-driven micropinocytosis-like endocytosis to enter cells, depending on the target cells [109]. The micropinocytosis forms large vesicles that engulf the virus.
- **Use of microtubules:** After entry, many viruses utilize microtubules for intracellular transport.
- **Use of dynein and kinesin motor proteins:** Viruses often use dynein and kinesin motor proteins, which move along microtubules to transport the viral particles to specific locations within the cell, such as the nucleus or replication factories [110, 111].

3.3.2 Trafficking Machinery During Viral Replication and Assembly

- **Replication complexes:** For many RNA viruses, replication occurs in specialized structures called replication complexes, which are often associated with the cytoskeleton [112]. These complexes can be anchored to actin filaments or microtubules to stabilize the replication machinery and facilitate efficient replication.

- **Viral factories:** DNA viruses like herpesviruses and poxviruses form viral factories that rely on microtubule networks for the assembly and maturation of new virions [113, 114].
- **Intermediate filaments:** These structures provide mechanical support and help maintain the integrity of viral replication sites.

3.3.3 Trafficking Machinery During Viral Protein Modification

Some viruses encode proteins that directly interact with cytoskeletal elements to manipulate the host cell's architecture and trafficking machinery.

- **Viral actin-binding proteins**: Proteins like HIV Negative Factor (Nef) can bind to and modulate actin dynamics to promote viral replication and egress [115].
- **Microtubule-associated proteins:** HSV-1 VP22 protein stabilizes and hyperacetylates microtubules to enhance viral transport within the cell [116].

3.3.4 Trafficking Machinery During Viral Egress and Spread

- **Actin filaments:** During egress, viruses can induce actin polymerization to propel themselves out of the cell or to spread to neighboring cells.
- **Actin tails:** Viruses such as vaccinia virus form actin tails that push the virus out of the cell, enhancing cell-to-cell spread [117].
- **Syncytium formation:** Some viruses, like measles, cause infected cells to fuse with neighboring cells, forming multinucleated syncytia, which facilitate viral spread without exposure to the extracellular environment [118].
- **Microtubules:** Microtubule networks are involved in the transport of viral particles to the cell membrane for release.
- **Exocytosis:** Enveloped viruses, such as coronaviruses, use the microtubule network to transport viral particles to the Golgi apparatus and then to the plasma membrane for exocytosis [119].

3.4 Cell Transformation and Oncogenesis

Cell transformation refers to the process by which normal cells acquire characteristics of cancer cells. This transformation involves genetic and epigenetic changes that lead to uncontrolled cell growth, resistance to apoptosis (programmed cell death), and the ability to invade other tissues. This transformation can occur through the action of viral genes that disrupt normal cellular processes.

Mechanisms of virus-induced cell transformation and oncology are described below:

3.4.1 Integration of Viral Genome

Some viruses integrate their DNA into the host genome, disrupting normal cell functions. This integration can activate oncogenes or deactivate tumor suppressor genes. Examples include human papillomavirus (HPV), which integrates its DNA into the host genome, particularly in cervical cells, affecting genes such as p53 and Rb [120].

3.4.2 Expression of Viral Oncogenes

Oncogenes are genes that encode proteins that interfere with the host cell's control of cell division and apoptosis, leading to uncontrolled cell proliferation and tumor formation. Viruses encode oncogenes that can directly induce cell transformation. These viral genes can interfere with normal cellular control mechanisms, such as those regulating cell cycle progression. Examples include EBV-expressed latent membrane protein 1 (LMP1), which mimics a constitutively active receptor, to enhance cell survival and proliferation [121] (see Table 3.3).

3.4.3 Inactivation of Tumor Suppressor Genes

Tumor suppressor genes are a class of genes that play a crucial role in regulating cell growth and maintaining the stability of the genome. They act as the "brakes" of the cell cycle, preventing cells from undergoing uncontrolled division and growth, which can lead to cancer.

Table 3.3 Example of some notable virus-induced oncogenes.

Oncogenes	Viruses	Mechanism of action
E6	Human papillomavirus (HPV)	Inactivates p53, leading to reduced apoptosis and increased cell proliferation
E7	Human papillomavirus (HPV)	Inactivates Rb, causing uncontrolled cell cycle progression
Latent membrane protein 1 (LMP1)	Epstein–Barr virus (EBV)	Mimics a constitutively active receptor, promoting cell survival and proliferation
Tax	Human T-cell leukemia virus Type 1 (HTLV-1)	Activates nuclear factor kappa-light-chain-enhancer of activated B cells (NF-κB) pathway; inhibits tumor suppressor functions
HBx	Hepatitis B virus (HBV)	Interferes with p53 function; promotes cell survival and proliferation
Core, non-structural protein 3 (NS3), NS5A, and NS5B	Hepatitis C virus (HCV)	Activates signaling pathways like Wnt/β-catenin, leading to uncontrolled cell growth

When these genes are mutated or inactivated, the cell loses its ability to control cell division, leading to tumor formation. Some viruses produce proteins that inhibit the function of tumor suppressor genes, such as p53 and Rb, which normally prevent uncontrolled cell division (see Table 3.4).

3.4.4 Alteration of Signaling Pathways

Viruses can hijack signaling pathways that control cell growth, survival, and differentiation, leading to the deregulation of these processes. Examples include HBV and HCV, which can activate signaling pathways like the Wnt/β-catenin pathway, promoting cell proliferation [122].

3.4.5 Induction of Genetic Instability and Mutation

Viral infections can also increase the mutation rate within the host cell genome, leading to genetic instability and the accumulation of mutations that can drive carcinogenesis [123]. Integration of viral DNA in the host genome can also cause chromosomal rearrangements and mutations, contributing to cancer development.

Table 3.4 Examples of viral tumor suppressor protein targets.

Viruses	Viral proteins	Tumor suppressor targets	Associated cancers	Mechanism of actions
Human papillomavirus (HPV)	E6	Tumor protein 53 (p53)	Cervical cancer, anogenital cancers, oropharyngeal cancer	E6 binds to p53, leading to its ubiquitination and proteasomal degradation
HPV	E7	Retinoblastoma protein (Rb)	Cervical cancer, anogenital cancers, oropharyngeal cancer	E7 binds to Rb, disrupting its ability to regulate the cell cycle
Epstein–Barr virus (EBV)	LMP1	p53, Rb (indirectly)	Burkitt's lymphoma, Hodgkin's lymphoma, nasopharyngeal carcinoma	LMP1 activates pathways that can inhibit p53 and Rb functions indirectly
Human T-cell leukemia virus Type 1 (HTLV-1)	Tax	p53	Adult T-cell leukemia/lymphoma	Tax interferes with p53's transcriptional activity and promotes its degradation.

Table 3.4 (Continued)

Viruses	**Viral proteins**	**Tumor suppressor targets**	**Associated cancers**	**Mechanism of actions**
Hepatitis B virus (HBV)	HBx	p53	Hepatocellular carcinoma	HBx binds to p53, inhibiting its transcriptional activity (alters p53 binding site) and promoting degradation.
Adenovirus	E1A	Rb	Various cancers (in experimental models)	E1A binds to Rb, preventing it from controlling cell cycle progression
Adenovirus	E1B 55 K, E1B 19 K	p53	Various cancers (in experimental models)	E1B 55 K binds to p53, inhibiting its transcriptional activity
Simian virus 40 (SV40)	Large T antigen	p53, Rb	Mesothelioma, brain tumors (in experimental models)	Large T antigen binds and inactivates both p53 and Rb
Merkel cell polyomavirus (MCV)	Large T antigen	Rb	Merkel cell carcinoma	Large T antigen binds to Rb, leading to uncontrolled cell cycle progression.
Kaposi's sarcoma-associated herpesvirus (KSHV)	Latency-associated nuclear antigen (LANA)	p53, Rb	Kaposi's sarcoma, primary effusion lymphoma	LANA binds to and inactivates p53 and Rb

3.5 Study Questions

Case Study

3.1 A 60-year-old patient with a history of chronic HBV infection is diagnosed with hepatocellular carcinoma. The tumor shows overexpression of HBV-encoded oncoproteins. What roles do HBV-encoded oncoproteins play in the development and progression of hepatocellular carcinoma, and how can targeting these viral oncogenes alter treatment strategies for this patient?

3.2 A 35-year-old female with cervical cancer has been found to have high levels of human papillomavirus (HPV) E6 and E7 oncogene expression in her tumor tissue. How does the expression of HPV E6 and E7 oncogenes affect the clinical management?

3.3 What potential complications might arise from dysregulated apoptosis in viral infections, and how can these complications impact the clinical management of patients?

3.4 In a patient with a persistent viral infection (e.g., HPV or EBV), how does the virus influence the host cell cycle to promote its replication? How does this interaction contribute to the development of associated malignancies, such as cervical cancer or lymphomas?

Objectives

3.1 **Which of the following best describes the role of viral oncoproteins in the host cell cycle?**
(a) They inhibit cellular DNA replication to prevent viral replication.
(b) They modulate cell cycle progression to facilitate viral replication.
(c) They repair damaged DNA to enhance cell survival.
(d) They induce apoptosis to limit viral spread.

3.2 **During which phase of the cell cycle do many viruses, such as herpesviruses, primarily replicate?**
(a) G1 phase.
(b) S-phase.
(c) G2 phase.
(d) M phase.

3.3 **Which viral protein is commonly involved in manipulating host cell cycle regulation to favor viral oncogenesis and replication?**
(a) Capsid protein.
(b) Matrix protein.
(c) E6/E7 protein.
(d) Reverse transcriptase.

3.4 **What is the primary cellular target for viral entry in many enveloped viruses?**
(a) Mitochondria.
(b) Nucleus.
(c) Cell membrane receptors.
(d) Endoplasmic reticulum.

3.5 **In the context of viral infections, what does the term "latency" refer to?**
(a) A period when the virus actively replicates and causes symptoms.
(b) A period when the virus is present in the host but does not actively replicate.
(c) A stage where the virus causes acute disease.
(d) The process of viral assembly and release from the host cell.

3.6 **Which cellular structure is crucial for the assembly and budding of many enveloped viruses?**
(a) Nucleus.
(b) Golgi apparatus.
(c) Ribosome.
(d) Lysosome.

3.7 **How do viruses typically enter host cells?**
(a) By direct penetration through the cell membrane.
(b) By endocytosis or receptor-mediated entry.
(c) By using cellular transport proteins.
(d) By inducing cellular apoptosis.

3.8 What is the primary purpose of viral genome integration into the host cell genome in retroviruses?

(a) To enhance the host cell's metabolism.

(b) To evade the immune response and ensure long-term persistence.

(c) To directly cause cellular apoptosis.

(d) To disrupt cellular protein synthesis.

3.9 Which of the following best describes the lytic cycle of viral replication?

(a) The virus integrates its genome into the host DNA and remains dormant.

(b) The virus replicates and eventually causes cell death through lysis.

(c) The virus exits the host cell without causing damage.

(d) The virus forms a latent infection without replicating.

References

1. Bagga, S. and Bouchard, M.J., *Cell cycle regulation during viral infection*. Methods Mol. Biol., 2014. **1170**: p. 165–227.
2. Schollmeier, A., Glitscher, M., and Hildt, E., *Relevance of HBx for hepatitis B virus-associated pathogenesis*. Int. J. Mol. Sci., 2023. **24**(5): p. 4964.
3. Reinson, T., et al., *The cell cycle timing of human papillomavirus DNA replication*. PLoS One, 2015. **10**(7): e0131675.
4. Li, G., et al., *Cell cycle G2/M arrest through an S phase-dependent mechanism by HIV-1 viral protein R*. Retrovirology, 2010. **7**(1): p. 59.
5. Serquiña, A.K. and Ziegelbauer, J.M., *How herpesviruses pass on their genomes*. J. Cell Biol., 2017. **216**(9): p. 2611–2613.
6. Gearhart, T.L. and Bouchard, M.J., *The hepatitis B virus X protein modulates hepatocyte proliferation pathways to stimulate viral replication*. J. Virol., 2010. **84**(6): p. 2675–2686.
7. Knight, J.S. and Robertson, E.S., *Epstein-Barr virus nuclear antigen 3C regulates cyclin A/p27 complexes and enhances cyclin A-dependent kinase activity*. J. Virol., 2004. **78**(4): p. 1981–1991.
8. Bottero, V., et al., *Kaposi sarcoma-associated herpes virus (KSHV) G protein-coupled receptor (vGPCR) activates the ORF50 lytic switch promoter: a potential positive feedback loop for sustained ORF50 gene expression*. Virology, 2009. **392**(1): p. 34–51.
9. Zhao, B., et al., *RNAs induced by Epstein-Barr virus nuclear antigen 2 in lymphoblastoid cell lines*. PNAS, 2006. **103**(6): p. 1900–1905.
10. Beer, S., et al., *EBNA2-EBF1 complexes promote MYC expression and metabolic processes driving S-phase progression of Epstein-Barr virus-infected B cells*. PNAS, 2022. **119**(30): e2200512119.
11. Dust, K., et al., *Human papillomavirus 16 E6 and E7 oncoproteins alter the abundance of proteins associated with DNA damage response, immune signaling and epidermal differentiation*. Viruses, 2022. **14**(8): p. 1764.
12. Yim, E.K. and Park, J.S., *The role of HPV E6 and E7 oncoproteins in HPV-associated cervical carcinogenesis*. Cancer Res. Treat., 2005. **37**(6): p. 319–324.
13. Rodríguez, M.C., et al., *The HSV-1 ubiquitin ligase ICP0: Modifying the cellular proteome to promote infection*. Virus Res., 2020. **285**: 198015.
14. Stacey, D.W., et al., *The adenovirus E1A protein overrides the requirement for cellular ras in initiating DNA synthesis*. EMBO J., 1994. **13**(24): p. 6107–6114.

15. Zamanian, M. and La Thangue, N.B., *Adenovirus E1a prevents the retinoblastoma gene product from repressing the activity of a cellular transcription factor*. EMBO J., 1992. **11**(7): p. 2603–2610.
16. Weitzman, M.D. and Fradet-Turcotte, A., *Virus DNA replication and the host DNA damage response*. Annu. Rev. Virol., 2018. **5**(1): p. 141–164.
17. Lopez, A., et al., *Viral modulation of the DNA damage response and innate immunity: two sides of the same coin*. J. Mol. Biol., 2022. **434**(6): 167327.
18. Maréchal, A. and Zou, L., *DNA damage sensing by the ATM and ATR kinases*. Cold Spring Harb. Perspect. Biol., 2013. **5**(9): a012716.
19. Burdak-Rothkamm, S., Rothkamm, K., and Prise, K.M., *ATM acts downstream of ATR in the DNA damage response signaling of bystander cells*. Cancer Res., 2008. **68**(17): p. 7059–7065.
20. Alhmoud, J.F., et al., *DNA damage/repair management in cancers*. Cancers (Basel), 2020. **12**(4): p. 1050.
21. Chatterjee, N. and Walker, G.C., *Mechanisms of DNA damage, repair, and mutagenesis*. Environ. Mol. Mutagen., 2017. **58**(5): p. 235–263.
22. Kay, J., et al., *Inflammation-induced DNA damage, mutations and cancer*. DNA Repair (Amst), 2019. **83**: 102673.
23. Li, D., et al., *HIV Vpr modulates the host DNA damage response at two independent steps to damage DNA and repress double-strand DNA break repair*. mBio, 2020. **11**(4): e00940-20.
24. Ragu, S., Matos-Rodrigues, G., and Lopez, B.S., *Replication stress, DNA damage, inflammatory cytokines and innate immune response*. Genes (Basel), 2020. **11**(4): p. 409.
25. Szewczyk-Roszczenko, O., et al., *Genotoxic consequences of viral infections*. Npj Viruses, 2025. **3**(1): p. 5.
26. Sowd, G.A., et al., *SV40 utilizes ATM kinase activity to prevent non-homologous end joining of broken viral DNA replication products*. PLoS Pathog., 2014. **10**(12): e1004536.
27. Li, P., et al., *Single-stranded RNA viruses activate and hijack host apical DNA damage response kinases for efficient viral replication*. Genome Instab. Dis., 2022. **3**(2): p. 83–87.
28. Abuetabh, Y., et al., *DNA damage response revisited: the p53 family and its regulators provide endless cancer therapy opportunities*. Exp. Mol. Med., 2022. **54**(10): p. 1658–1669.
29. Querido, E., et al., *Degradation of p53 by adenovirus E4orf6 and E1B55K proteins occurs via a novel mechanism involving a Cullin-containing complex*. Genes Dev., 2001. **15**(23): p. 3104–3117.
30. Kudoh, A., et al., *Homologous recombinational repair factors are recruited and loaded onto the viral DNA genome in Epstein-Barr virus replication compartments*. J. Virol., 2009. **83**(13): p. 6641–6651.
31. Read, A. and Schröder, M., *The unfolded protein response: an overview*. Biology (Basel), 2021. **10**(5): p. 384.
32. Choi, J.A. and Song, C.H., *Insights into the role of endoplasmic reticulum stress in infectious diseases*. Front. Immunol., 2019. **10**: p. 3147.
33. Zhang, L. and Wang, A., *Virus-induced ER stress and the unfolded protein response*. Front. Plant Sci., 2012. **3**: p. 293.
34. Hwang, J. and Qi, L., *Quality control in the endoplasmic reticulum: crosstalk between ERAD and UPR pathways*. Trends Biochem. Sci, 2018. **43**(8): p. 593–605.
35. Zhu, Z., et al., *Induction of UPR promotes interferon response to inhibit PRRSV replication via PKR and NF-κB pathway*. Front. Microbiol., 2021. **12**: 757690.
36. Lin, J.H., Walter, P., and Yen, T.S., *Endoplasmic reticulum stress in disease pathogenesis*. Annu. Rev. Pathol., 2008. **3**: p. 399–425.
37. Pavlović, N. and Heindryckx, F., *Exploring the role of endoplasmic reticulum stress in*

hepatocellular carcinoma through mining of the human protein atlas. Biology (Basel), 2021. **10**(7): p. 640.

38. Sakamoto, T. and Imai, H., *Hydrogen peroxide produced by superoxide dismutase SOD-2 activates sperm in Caenorhabditis elegans*. J. Biol. Chem., 2017. **292**(36): p. 14804–14813.
39. Lubos, E., Loscalzo, J., and Handy, D.E., *Glutathione peroxidase-1 in health and disease: from molecular mechanisms to therapeutic opportunities*. Antioxid. Redox Signaling, 2011. **15**(7): p. 1957–1997.
40. Traber, M.G. and Stevens, J.F., *Vitamins C and E: beneficial effects from a mechanistic perspective*. Free Radic. Biol. Med., 2011. **51**(5): p. 1000–1013.
41. Hemilä, H., *Vitamin C and infections*. Nutrients, 2017. **9**(4): p. 339.
42. Herengt, A., Thyrsted, J., and Holm, C.K., *NRF2 in viral infection*. Antioxid.(Basel), 2021. **10**(9): p. 1491.
43. Liu, X., et al., *Nuclear factor erythroid 2-related factor 2 potentiates the generation of inflammatory cytokines by intestinal epithelial cells during hyperoxia by inducing the expression of interleukin 17D*. Toxicology, 2021. **457**: 152820.
44. Hsu, T.C., et al., *Activator protein 1 (AP-1)- and nuclear factor kappaB (NF-kappaB)-dependent transcriptional events in carcinogenesis*. Free Radic. Biol. Med., 2000. **28**(9): p. 1338–1348.
45. Catani, M.V., et al., *Nuclear factor kappaB and activating protein 1 are involved in differentiation-related resistance to oxidative stress in skeletal muscle cells*. Free Radic. Biol. Med., 2004. **37**(7): p. 1024–1036.
46. Ivanov, A.V., et al., *Oxidative stress during HIV infection: mechanisms and consequences*. Oxid. Med. Cell. Longevity, 2016. **2016**: p. 8910396.
47. Ivanov, A.V., et al., *Oxidative stress, a trigger of hepatitis C and B virus-induced liver carcinogenesis*. Oncotarget, 2017. **8**(3): p. 3895–3932.
48. Greenberger, J.S., et al., *SARS-CoV-2 spike protein induces oxidative stress and senescence in mouse and human lung*. In Vivo, 2024. **38**(4): p. 1546–1556.
49. Ivanov, A.V., et al., *Hepatitis C virus proteins activate NRF2/ARE pathway by distinct ROS-dependent and independent mechanisms in HUH7 cells*. PLoS One, 2011. **6**(9): e24957.
50. Gil, L., et al., *Contribution to characterization of oxidative stress in HIV/AIDS patients*. Pharmacol. Res., 2003. **47**(3): p. 217–224.
51. Zhang, X. and Yu, W., *Heat shock proteins and viral infection*. Front. Immunol., 2022. **13**: 947789.
52. Mayer, M.P. and Bukau, B., *Hsp70 chaperones: cellular functions and molecular mechanism*. Cell. Mol. Life Sci., 2005. **62**(6): p. 670–684.
53. Sharp, F.R., Zhan, X., and Liu, D.Z., *Heat shock proteins in the brain: role of Hsp70, Hsp 27, and HO-1 (Hsp32) and their therapeutic potential*. Transl. Stroke Res., 2013. **4**(6): p. 685–692.
54. Sharma, K., et al., *Quantitative proteomics reveals that Hsp90 inhibition preferentially targets kinases and the DNA damage response*. Mol. Cell. Proteomics, 2012. **11**(3): M111.014654.
55. Lanneau, D., et al., *Heat shock proteins: essential proteins for apoptosis regulation*. J. Cell. Mol. Med., 2008. **12**(3): p. 743–761.
56. De Maio, A. and Hightower, L.E., *Heat shock proteins and the biogenesis of cellular membranes*. Cell Stress Chaperones, 2021. **26**(1): p. 15–18.
57. Olasunkanmi, O.I., et al., *Virus-induced cytoplasmic aggregates and inclusions are critical cellular regulatory and antiviral factors*. Viruses, 2020. **12**(4): p. 399.
58. Li, W. and Wang, Y., *Stress granules: potential therapeutic targets for infectious and*

inflammatory diseases. Front. Immunol., 2023. **14**: 1145346.
59. Courtney, S.C., et al., *West Nile virus infections suppress early viral RNA synthesis and avoid inducing the cell stress granule response*. J. Virol., 2012. **86**(7): p. 3647–3657.
60. Visser, L.J., et al., *Essential role of enterovirus 2A protease in counteracting stress granule formation and the induction of type I interferon*. J. Virol., 2019. **93**(10): e00222-19.
61. Reineke, L.C. and Lloyd, R.E., *Diversion of stress granules and P-bodies during viral infection*. Virology, 2013. **436**(2): p. 255–267.
62. Lloyd, R.E., *Regulation of stress granules and P-bodies during RNA virus infection*. Wiley Interdiscip. Rev.: RNA, 2013. **4**(3): p. 317–331.
63. Luo, Y., Na, Z., and Slavoff, S.A., *P-bodies: composition, properties, and functions*. Biochemistry, 2018. **57**(17): p. 2424–2431.
64. Decker, C.J. and Parker, R., *P-bodies and stress granules: possible roles in the control of translation and mRNA degradation*. Cold Spring Harbor Perspect. Biol., 2012. **4**(9): a012286.
65. Brothers, W.R., et al., *P-bodies directly regulate MARF1-mediated mRNA decay in human cells*. Nucleic Acids Res., 2022. **50**(13): p. 7623–7636.
66. Wang, C., et al., *Context-dependent deposition and regulation of mRNAs in P-bodies*. Elife, 2018. **7**: e29815.
67. Khong, A. and Jan, E., *Modulation of stress granules and P bodies during dicistrovirus. Infection*., 2011. **85**(4): p. 1439–1451.
68. Pérez-Vilaró, G., et al., *Hepatitis C virus infection inhibits P-body granule formation in human livers*. J. Hepatol., 2015. **62**(4): p. 785–790.
69. Dougherty, J.D., White, J.P., and Lloyd, R.E., *Poliovirus-mediated disruption of cytoplasmic processing bodies*. J. Virol., 2011. **85**(1): p. 64–75.
70. Mok, B.W.-Y., et al., *The NS1 protein of influenza A virus interacts with cellular processing bodies and stress granules through RNA-associated protein 55 (RAP55) during virus infection*. J. Virol., 2012. **86**(23): p. 12695–12707.
71. Glick, D., Barth, S., and Macleod, K.F., *Autophagy: cellular and molecular mechanisms*. J. Pathol., 2010. **221**(1): p. 3–12.
72. Florey, O. and Overholtzer, M., *Autophagy proteins in macroendocytic engulfment*. Trends Cell Biol., 2012. **22**(7): p. 374–380.
73. Liang, S., et al., *Autophagy in viral infection and pathogenesis*. Front. Cell Dev. Biol., 2021. **9**: 766142.
74. White, E., et al., *Role of autophagy in suppression of inflammation and cancer*. Curr. Opin. Cell Biol., 2010. **22**(2): p. 212–217.
75. Choi, Y., Bowman, J.W., and Jung, J.U., *Autophagy during viral infection – a double-edged sword*. Nat. Rev. Microbiol., 2018. **16**(6): p. 341–354.
76. Dreux, M., et al., *The autophagy machinery is required to initiate hepatitis C virus replication*. PNAS, 2009. **106**(33): p. 14046–14051.
77. Ripa, I., et al., *Interplay between autophagy and herpes simplex virus type 1: ICP34.5, one of the main actors*. Int. J. Mol. Sci., 2022. **23**(21): p. 13643.
78. Zhang, R., et al., *The regulation of autophagy by influenza A virus*. Biomed Res. Int., 2014. **2014**: 498083.
79. Shrivastava, S., et al., *Hepatitis C virus upregulates beclin1 for induction of autophagy and activates mTOR signaling*. J. Virol., 2012. **86**(16): p. 8705–8712.
80. Dash, S., et al., *Hepatitis C virus infection induces autophagy as a prosurvival mechanism to alleviate hepatic ER-stress response*. Viruses, 2016. **8**(5): p. 150.
81. Chu, J.Y.K. and Ou, J.J., *Autophagy in HCV replication and protein trafficking*. Int. J. Mol. Sci., 2021. **22**(3): p. 1089.

82. Wang, L. and Ou, J.J., *Regulation of autophagy by hepatitis C virus for its replication*. DNA Cell Biol., 2018. **37**(4): p. 287–290.
83. Shrivastava, S., et al., *Knockdown of autophagy inhibits infectious hepatitis C virus release by the exosomal pathway*. J. Virol., 2016. **90**(3): p. 1387–1396.
84. Lussignol, M., et al., *The herpes simplex virus 1 Us11 protein inhibits autophagy through its interaction with the protein kinase PKR*. J. Virol., 2013. **87**(2): p. 859–871.
85. Orvedahl, A., et al., *HSV-1 ICP34.5 confers neurovirulence by targeting the Beclin 1 autophagy protein*. Cell Host Microbe, 2007. **1**(1): p. 23–35.
86. Chiramel, A.I. and Best, S.M., *Role of autophagy in Zika virus infection and pathogenesis*. Virus Res., 2018. **254**: p. 34–40.
87. Acharya, B., et al., *Significance of autophagy in dengue virus infection: a brief review*. Am. J. Trop. Med. Hyg., 2019. **100**(4): p. 783–790.
88. Zhou, H., Hu, Z., and Castro-Gonzalez, S., *Bidirectional interplay between SARS-CoV-2 and autophagy*. mBio, 2023. **14**(4): e0102023.
89. Wang, C. and Youle, R.J., *The role of mitochondria in apoptosis*. Annu. Rev. Genet., 2009. **43**: p. 95–118.
90. Cavalcante, G.C., et al., *A cell's fate: an overview of the molecular biology and genetics of apoptosis*. Int. J. Mol. Sci., 2019. **20**(17): p. 4133.
91. Mustafa, M., et al., *Apoptosis: a comprehensive overview of signaling pathways, morphological changes, and physiological significance and therapeutic implications*. Cells, 2024. **13**(22): p. 1838.
92. Yanumula, A.C.J., *Biochemistry, Extrinsic Pathway of Apoptosis*. 2023, StatPearls Publishing.
93. Orzalli, M.H. and Kagan, J.C., *Apoptosis and necroptosis as host defense strategies to prevent viral infection*. Trends Cell Biol., 2017. **27**(11): p. 800–809.
94. Trahtemberg, U. and Mevorach, D., *Apoptotic cells induced signaling for immune homeostasis in macrophages and dendritic cells*. Front. Immunol., 2017. **8**: p. 1356.
95. Tarakanova, V.L. and Wold, W.S., *Adenovirus E1A and E1B-19K proteins protect human hepatoma cells from transforming growth factor beta1-induced apoptosis*. Virus Res., 2010. **147**(1): p. 67–76.
96. Lin, S. and Zhang, Y.J., *Interference of apoptosis by hepatitis B virus*. Viruses, 2017. **9**(8): p. 230.
97. Jansen, J.M., et al., *Influenza virus-specific CD4+ and CD8+ T cell-mediated immunity induced by infection and vaccination*. J. Clin. Virol., 2019. **119**: p. 44–52.
98. Gillespie, E.J., et al., *Selective inhibitor of endosomal trafficking pathways exploited by multiple toxins and viruses*. PNAS, 2013. **110**(50): p. E4904–E4912.
99. Dogrammatzis, C., et al., *Tracing the STING exocytosis pathway during herpes viruses infection*. mBio, 2024. **15**(4): e0037324.
100. Hogue, I.B., Scherer, J., and Enquist, L.W., *Exocytosis of alphaherpesvirus virions, light particles, and glycoproteins uses constitutive secretory mechanisms*. mBio, 2016. **7**(3). doi: 10.1128/mbio.00820-16.
101. Spooner, R.A., et al., *Retrograde transport pathways utilised by viruses and protein toxins*. Virol. J., 2006. **3**: p. 26.
102. Bender, D. and Hildt, E., *Back to the roots: noncanonical retrograde trafficking of the HBV nucleocapsids*. Cell. Mol. Gastroenterol. Hepatol., 2023. **15**(3): p. 797–798.
103. Approach., C.G.T.C.A.M., *Structure and Organization of Actin Filaments*. 2nd Edn. 2000, Sinauer Associates: Sunderland (MA).
104. Pollard, T.D. and Cooper, J.A., *Actin, a central player in cell shape and movement*. Science, 2009. **326**(5957): p. 1208–1212.

105. Bera, A. and Gupta, M.L. Jr., *Microtubules in microorganisms: how tubulin isotypes contribute to diverse cytoskeletal functions*. Front. Cell Dev. Biol., 2022. **10**: 913809.
106. Satarić, M.V., et al., *Calcium ions tune the beats of cilia and flagella*. Biosystems, 2020. **196**: 104172.
107. Jiang, M., et al., *Adducin-1 facilitates influenza virus endosomal trafficking and uncoating by regulating branched actin dynamics and myosin IIB activity*. Adv. Sci. (Weinh), 2025. **12**(28): e2417318.
108. Anand, A.R., et al., *N-terminal Slit2 inhibits HIV-1 replication by regulating the actin cytoskeleton*. Retrovirology, 2013. **10**: p. 2.
109. Aleksandrowicz, P., et al., *Ebola virus enters host cells by macropinocytosis and clathrin-mediated endocytosis*. J. Infect. Dis., 2011. **204 Suppl 3**(Suppl 3): p. S957–S967.
110. Dodding, M.P. and Way, M., *Coupling viruses to dynein and kinesin-1*. EMBO J., 2011. **30**(17): p. 3527–3539.
111. Hsieh, M.J., White, P.J., and Pouton, C.W., *Interaction of viruses with host cell molecular motors*. Curr. Opin. Biotechnol., 2010. **21**(5): p. 633–639.
112. Li, M., et al., *The host cytoskeleton functions as a pleiotropic scaffold: orchestrating regulation of the viral life cycle and mediating host antiviral innate immune responses*. Viruses, 2023. **15**(6): p. 1354.
113. White, S., et al., *Herpes simplex virus organizes cytoplasmic membranes to form a viral assembly center in neuronal cells*. J. Virol., 2020. **94**(19).
114. Seo, D., et al., *Poxvirus A51R proteins regulate microtubule stability and antagonize a cell-intrinsic antiviral response*. Cell Rep., 2024. **43**(3): 113882.
115. Nobile, C., et al., *HIV-1 nef inhibits ruffles, induces filopodia, and modulates migration of infected lymphocytes*. J. Virol., 2010. **84**(5): p. 2282–2293.
116. Elliott, G. and O'Hare, P., *Herpes simplex virus type 1 tegument protein VP22 induces the stabilization and hyperacetylation of microtubules*. J. Virol., 1998. **72**(8): p. 6448–6455.
117. Hollinshead, M., et al., *Vaccinia virus utilizes microtubules for movement to the cell surface*. J. Cell Biol., 2001. **154**(2): p. 389–402.
118. Cathomen, T., Naim, H.Y., and Cattaneo, R., *Measles viruses with altered envelope protein cytoplasmic tails gain cell fusion competence*. J. Virol., 1998. **72**(2): p. 1224–1234.
119. Aminpour, M., Hameroff, S., and Tuszynski, J.A., *How COVID-19 hijacks the cytoskeleton: therapeutic implications*. Life (Basel), 2022. **12**(6): p. 814.
120. Wrede, D., et al., *Expression of RB and p53 proteins in HPV-positive and HPV-negative cervical carcinoma cell lines*. Mol. Carcinog., 1991. **4**(3): p. 171–175.
121. Kilger, E., et al., *Epstein-Barr virus-mediated B-cell proliferation is dependent upon latent membrane protein 1, which simulates an activated CD40 receptor*. EMBO J., 1998. **17**(6): p. 1700–1709.
122. Wang, W., et al., *Action and function of Wnt/β-catenin signaling in the progression from chronic hepatitis C to hepatocellular carcinoma*. J. Gastroenterol., 2017. **52**(4): p. 419–431.
123. Chen, Y., et al., *Viral carcinogenesis: factors inducing DNA damage and virus integration*. Cancers (Basel), 2014. **6**(4): p. 2155–2186.

4 Overview of Viral Genetics and Evolution in Tropical Virology

Viruses display remarkable genetic diversity and evolve rapidly, driven by high mutation rates, short replication cycles, and large population sizes. While DNA viruses tend to maintain greater genetic stability, RNA viruses are especially prone to variation due to the lack of proofreading activity in their RNA-dependent RNA polymerases, resulting in frequent replication errors [1, 2]. Over time, these errors, combined with evolutionary pressures, contribute to extensive structural and functional changes in viral genomes, enhancing their adaptability and replication efficiency. This diversity arises through various mechanisms, including mutation, recombination, and reassortment, each playing a role in shaping viral evolution and driving significant changes in replication dynamics and host adaptability.

4.1 Overview of Viral Genome Structure

Viral genomes are compact and highly optimized to carry the information necessary for replication, assembly, and host interaction. Despite their limited genome sizes, viruses have developed highly efficient strategies to encode and regulate essential functions. Viral genomes are typically divided into two main regions.

4.1.1 Coding Regions

This carries the information that codes for specific proteins, structural proteins (capsid, envelope), and nonstructural proteins (polymerases, proteases, and regulatory proteins). In many viruses, such as flaviviruses (e.g., dengue and Zika), the entire genome consists of a single long open reading frame (ORF) that is translated into a large polyprotein, which is subsequently cleaved during a later stage of virus replication into individual functional proteins. In contrast, other viruses, such as retroviruses (for example, human immunodeficiency virus [HIV-1]) and DNA viruses (e.g., hepatitis B virus [HBV]), contain multiple coding regions that may overlap or be arranged in complex patterns to maximize the limited genomic space [3, 4]. Coding regions are highly efficient, often using various gene expression strategies (canonical and noncanonical) (Table 4.1), such as overlapping genes, leaky scanning, ribosomal frameshifting, and stop codon readthrough, to generate multiple proteins from the same nucleotide sequence, maximize replication, and ensure precise control of protein production.

4.1.2 Noncoding Regions

The noncoding regions (NCRs) of viral genomes, though not translated into

Table 4.1 Canonical and noncanonical viral gene expression strategies of the coding region.

Strategy	Description	Examples
Canonical viral gene expression strategies		
Monocistronic mRNA translation (cap-dependent initiation)	Viral mRNAs mimic host mRNAs with 5′ caps and 3′ poly(A) tails, allowing efficient ribosome recognition.	Herpesviruses (HSV) [5]
Temporal regulation of gene expression	Viral genes are expressed in an ordered sequence (early, intermediate, late) depending on the infection stage.	Adenoviruses, poxviruses, herpesviruses [6–8]
Use of host-like promoters and enhancers	Viral genomes utilize promoters and enhancers resembling those of the host to recruit RNA polymerase efficiently.	Human papillomavirus (HPV) [9]
Canonical RNA splicing	Viruses exploit host splicing machinery to process pre-mRNAs into mature transcripts.	Adenoviruses (extensive alternative splicing), retroviruses [10]
Polyadenylation of mRNAs	Viral mRNAs gain poly(A) tails to increase stability and translation efficiency, mirroring host processes.	Influenza virus [11]
Noncanonical viral gene expression		
Overlapping genes	Two or more open reading frames share nucleotide sequences, allowing multiple proteins to be encoded from the same genomic region.	Hepatitis B virus (HBV)
Leaky scanning	Ribosomes bypass the first AUG start codon if its context is suboptimal, initiating translation downstream to produce alternative proteins.	Flaviviruses (dengue, Zika), paramyxoviruses (measles virus) [12, 13]

Table 4.1 (Continued)

Strategy	Description	Examples
Ribosomal frameshifting	Ribosomes shift reading frames (commonly −1 or + 1) during translation to produce distinct fusion proteins, controlling protein ratios.	HIV-1 (Gag-Pol), SARS-CoV, SARS-CoV-2 [14, 15]
Stop codon readthrough	Ribosomes ignore a stop codon, extending the translation product to create an elongated protein isoform.	Chikungunya virus (nsP3/nsP4) [16]
RNA editing by polymerases (polymerase stuttering)	Viral RNA polymerases insert or delete nucleotides during transcription, generating multiple protein variants from a single template.	Paramyxoviruses (Measles virus, Sendai virus) [17]
Cap-snatching	Viruses steal capped primers from host mRNAs to initiate viral mRNA synthesis.	Influenza virus, bunyaviruses [18, 19]
Internal ribosome entry site (IRES)-mediated translation	Ribosomes initiate translation internally on viral RNA, bypassing the 5′ cap structure, often under host translation shutdown.	Poliovirus, hepatitis C virus (HCV), enteroviruses

proteins, play essential regulatory roles in the viral life cycle (Table 4.2).

a. In RNA viruses, these regions are often located at the 5′ and 3′ ends of the genome, commonly referred to as untranslated regions (UTRs). They contain conserved RNA structures and sequence motifs that regulate key processes such as replication, transcription, translation, and genome packaging. For example, in enteroviruses such as poliovirus and enterovirus A71, the 5′ UTR harbors an internal ribosome entry site (IRES) that enables cap-independent translation of viral proteins [20], allowing efficient synthesis even when host cap-dependent translation is shut down. In flaviviruses (dengue, Zika, and yellow fever), the NCRs form stem-loop structures that facilitate genome cyclization [12], an essential step for replication by the viral RNA-dependent RNA polymerase. Similarly, in bunyaviruses and arenaviruses, short intergenic NCRs function as transcription termination signals and help regulate the production of subgenomic RNAs [21].

Table 4.2 Noncoding genomic elements of viral genome.

Noncoding genomic structures	Function	Example	Remarks
5′ Untranslated region (5′ UTR)	Signals for ribosome binding, translation initiation, and RNA stability.	Dengue virus 5′ UTR stem-loops	May contain secondary structures like stem-loops critical for viral replication
3′ Untranslated region (3′ UTR)	Signals for RNA stability, replication termination, and polyadenylation.	Poliovirus 3′ UTR	Often interacts with viral and host proteins for replication complex formation
Regulatory motifs and enhancers	Facilitate transcription and translation regulation.	Hepatitis B virus epsilon stem-loop	Can modulate the expression of viral genes depending on host cell conditions
Packaging signals	Direct genome incorporation into virions.	Retrovirus Psi (Ψ) packaging signal	Often includes specific secondary or tertiary RNA structures recognized by viral capsid proteins
Conserved RNA structures	Scaffold for viral or host protein binding, ribosome recruitment.	IRES in picornaviruses	Enables cap-independent translation; critical in viruses lacking 5′ cap
Riboswitches and ribozymes	Riboswitches are RNA sequences that regulate gene expression without being translated into protein. They typically occur in the 5′ UTR or other untranslated regions. Ribozymes are RNA molecules with catalytic activity, capable of self-cleavage or processing other RNAs.	Hepatitis delta virus ribozyme	Self-cleaving or regulatory RNAs that influence viral genome maturation
Replication origins (Ori) and promoters	In most viral genomes, they are considered noncoding regions. They are regulatory sequences, not translated into protein, that initiate RNA or DNA replication	Adenovirus origin of replication	Often contains conserved sequences recognized by viral polymerases or host factors

Table 4.2 (Continued)

Noncoding genomic structures	Function	Example	Remarks
Terminal structures (caps, poly(A) tails)	Mimic host mRNA, promote translation, protect RNA from degradation. 3′ Poly(A) tail is a stretch of adenine nucleotides at the 3′ end of RNA, which forms part of the UTR.	Many positive-sense single-stranded RNA viruses	Essential for efficient translation and genome stability
Repetitive and inverted sequences	Repetitive sequences are repeated nucleotide motifs that often play structural or regulatory roles, such as helping genome circularization, recombination, or replication initiation. Inverted sequences are functional RNA elements with sequences that can form secondary structures like hairpins or stem-loops, which are important for replication, packaging, or transcription regulation.	Coronaviruses stem-loop structures in the 5′ UTR	Can act as packaging signals or replication enhancers

b. In DNA viruses, NCRs often contain promoters, enhancers, and origin of replication sequences essential for controlling gene expression and viral DNA replication. For example, the HBV genome includes regulatory elements such as the basal core promoter (core, preS1, preS2, and X) and enhancer regions (enhancers I and II) that govern transcription of viral RNAs [22]. Similarly, in papillomaviruses, the long control region (LCR) is a major noncoding segment containing the origin of replication and transcriptional regulatory sites [23].

4.1.3 Special Features of Viral Genomic Structure

Virus genomes often contain structural or sequence features beyond standard coding and NCRs, which enhance replication, translation, and adaptability. Examples include:

- **Terminal structures:** Some RNA genomes possess 5′ caps or 3′ poly(A) tails (e.g., many positive-sense single-stranded RNA viruses) that mimic host mRNA, facilitating ribosome recognition and translation.

- **Repeats and inverted sequences:** Direct or inverted repeats contribute to genome circularization, replication initiation, and serve as packaging signals for virion assembly.
- **Quasispecies:** Especially prominent in RNA viruses, high mutation rates generate a dynamic population of closely related genome variants.

Mutations or deletions within UTRs have been linked to altered replication efficiency, viral resistance, and pathogenicity [24], as seen in attenuated strains of dengue virus used for vaccine development (3′-UTR chimeric virus, rDEN3–3’D4Δ30) [25]. Thus, while they do not encode proteins, NCRs act as molecular "switchboards," fine-tuning viral gene expression and ensuring the precise regulation of the replication cycle.

4.2 Genetic Basis of Emergence and Cross-Species Transmission in Tropical Virology

The genetic makeup of viruses determines their ability to emerge in new populations and cross species barriers. Ecological diversity and human–animal interactions promote adaptation, emergence, and cross-species infections from animals to humans. Viral emergence remains a recurring global public health threat. In tropical regions, specific cases such as Zika virus, Ebola virus, Nipah virus, and Lassa fever virus illustrate this, facilitated by various viral genetic adaptations. Mechanisms of viral adaptability and emergence of cross-species transmission include:

1. **High mutation rates and genetic drift:** Most emerging viruses in the tropics are RNA viruses, which replicate using error-prone polymerases that lack proofreading capacity. This results in high mutation rates, producing numerous variants during each replication cycle.
   - **Genetic drift:** These random mutations can accumulate over time, altering surface glycoproteins or internal proteins critical for replication. Even small changes may enable recognition of new host receptors or evasion of immune responses.
   - **Example:** Lassa fever virus (an arenavirus) accumulates point mutations in its glycoprotein gene, which may have enhanced its ability to infect human cells compared to its rodent reservoir [26, 27].
2. **Quasispecies dynamics:** Viral populations do not exist as single uniform genomes, but rather as a swarm of genetically related variants, known as a quasispecies. This swarm increases the probability that one variant will be suited for survival in a new host, enabling cross-species transmission.
   - **Adaptive potential:** Within-host diversity ensures that viruses can rapidly respond to selective pressures, such as crossing into a new host species.
   - **Example:** Simian immunodeficiency viruses (SIVs) circulating in

primates likely lead to HIV-1 and HIV-2 in humans through variants that were genetically preadapted for replication in human cells [28].

3. **Recombination and reassortment:** Genetic exchange between viral genomes is another powerful cause of emergence. Recombination occurs when two related viruses co-infect a single host cell, leading to the exchange of genomic segments. This can create hybrid viruses with novel characteristics. In reassortment, particularly in segmented viruses (e.g., orthomyxoviruses, bunyaviruses, reoviruses), entire gene segments can be swapped, leading to abrupt changes in viral phenotype.
   - **Example:** Reassortment among arboviruses in mosquito vectors can give rise to variants with altered host tropism [29]. Influenza reassortment in avian or swine reservoirs occasionally yields pandemic strains [30].
4. **Adaptation in viral surface glycoproteins:** A central barrier to cross-species transmission is the ability of a virus to recognize and bind host cell receptors. Mutations in viral attachment proteins (such as glycoproteins or spikes) are often the key genetic adaptation required for emergence.
   - **Example:** Ebola virus glycoprotein acquired changes that enhanced binding to Niemann-Pick C1 (NPC1) receptor, a critical determinant of its zoonotic potential. Nipah virus glycoproteins are adapted to bind ephrin-B2 and ephrin-B3 receptors in humans, facilitating outbreaks in tropical Asia [31, 32].
5. **Antagonism of host restriction factors:** For successful cross-species infection, viruses must also overcome the innate immune defenses of the new host. Many viruses carry accessory genes or evolve mutations that antagonize host restriction factors.
6. **Genome plasticity and structural features:** Some viruses possess genetic elements such as terminal repeats, RNA secondary structures, or accessory ORFs that enhance adaptability. These features facilitate recombination, allow alternative splicing, or provide flexibility in gene expression, further supporting emergence.

Case studies of cross-species emergence of viral infection in the tropics:

1. **Ebola virus:** Cross-species infection from bat reservoirs to humans
   Ebola virus outbreaks, first recognized in Central Africa, have been linked to contact with fruit bats and other wildlife [33].
   - **Genetic adaptation:** Mutations in the Ebola glycoprotein enhanced binding to the NPC1 receptor in human cells, a critical step for cross-species transmission [34, 35].
   - **Quasispecies advantage:** The virus exhibits rapid within-host evolution, allowing it to adapt to human immune pressures during outbreaks.
   - **Outcome:** Though often self-limiting due to high case fatality and restricted transmission chains,

Ebola illustrates how small genetic changes in viral glycoproteins can enable zoonotic spillover in tropical environments.

2. **Nipah virus:** Transmission from fruit bats via intermediate hosts

   Nipah virus, first identified in Malaysia in the late 1990s, emerged from Pteropus fruit bats with spillover into domestic pigs before infecting humans. Subsequent outbreaks in Bangladesh and India involved direct bat-to-human transmission.

   - **Genetic adaptation:** Viral envelope proteins evolved to bind ephrin-B2 and ephrin-B3 receptors in human endothelial and neuronal tissues, explaining the severe encephalitis seen in infections [31].
   - **Recombination potential**: The virus maintains genetic diversity within bat populations, increasing the likelihood of variants capable of infecting humans.
   - **Outcome:** Repeated outbreaks in South and Southeast Asia highlight the ongoing risk of cross-species transmission in tropical agricultural systems.

3. **Zika virus:** Emergence and adaptation to mosquito vectors

   Zika virus, first isolated in Uganda in 1947, remained obscure until its explosive spread in the Americas in 2015–2016.

   - **Genetic adaptation:** Mutations in the viral envelope protein enhanced infectivity in *Aedes aegypti* mosquitoes, the predominant tropical vector [36].
   - **Evolution of pathogenicity:** Genetic changes in Zika have also been linked to increased neurotropism [37], explaining its association with congenital microcephaly during recent epidemics.
   - **Outcome:** The ability of virus to adapt to both human hosts and efficient mosquito vectors demonstrates how small genetic changes can transform a relatively benign virus into a major public health threat.

4.3 Overview of Virus Mutation

Mutation is a process by which the genetic sequence (RNA or DNA) of an organism's genome is changed. They can occur in various forms and can have different effects on the virus or even have no obvious effect. Mutations are a fundamental mechanism of genetic variation and evolution, but they can also play a significant role in the development of diseases.

4.3.1 Forms of Mutations

Mutations during virus infection are typically categorized into two main types: spontaneous and induced mutations.

4.3.1.1 Spontaneous Mutations

Spontaneous mutations occur naturally due to errors in nucleic acid replication or repair processes. These mutations usually occur without any significant external influence and are a result of the inherent chemical instability of nucleotides or errors during viral genome replication. Essentially, they can be caused by:

- **Errors in DNA/RNA replication:** During replication, the viral polymerase may incorporate incorrect nucleotides. Although many of these errors are corrected by proofreading mechanisms, some can persist and result in mutations. For RNA viruses, the error rate during replication is typically higher due to the lack of proofreading activity in RNA polymerases.
- **Chemical instability:** Nucleotides can undergo spontaneous chemical changes, such as deamination or oxidation, which lead to base substitutions or structural alterations in the DNA or RNA.
- **Internal cellular processes:** Host cell processes, such as oxidative stress or metabolic byproducts, can cause DNA damage and lead to mutations in the viral genome.

4.3.1.2 Induced Mutations

Induced mutations are caused by external factors or agents that alter the viral genome. These mutations are typically the result of exposure to mutagens or environmental stresses [38]. This can be caused by:

- **Chemical mutagenic agents:** Nucleotide analogs, that is, chemicals that mimic nucleotides, can be incorporated into the viral genome, leading to mispairing during replication. For example, certain antiviral drugs act as nucleotide analogs and can induce mutations. Alkylating agents, chemicals that add alkyl groups to nucleotides, can cause mutations by altering normal base pairing or inducing cross-linking between DNA strands.
- **Physical mutagenic agent:** Radiation, such as ultraviolet (UV) radiation, can cause the formation of pyrimidine dimers in DNA [39], leading to replication errors and mutations. Ionizing radiation (X-rays and gamma rays) can induce double-strand breaks in DNA, which can result in various types of mutations if not properly repaired.
- **Biological mutagenic agents:** Some virus infections can induce mutations in their genome or the host's genome by triggering cellular processes that generate reactive oxygen species (ROS) or through integration into the host genome [40]. For example, human papillomavirus (HPV) induces oxidative stress and DNA damage in host genes, capable of contributing to tumorigenesis [41, 42].

4.3.2 Types of Mutation During Virus Infection

4.3.2.1 Point Mutations

In this type of mutation, a single nucleotide is replaced by another, also known as substitutions (see Figure 4.1). A point mutation can lead to:

- **Silent mutations**: No change in the amino acid sequence of the viral protein. Often has minimal impact on virus function.

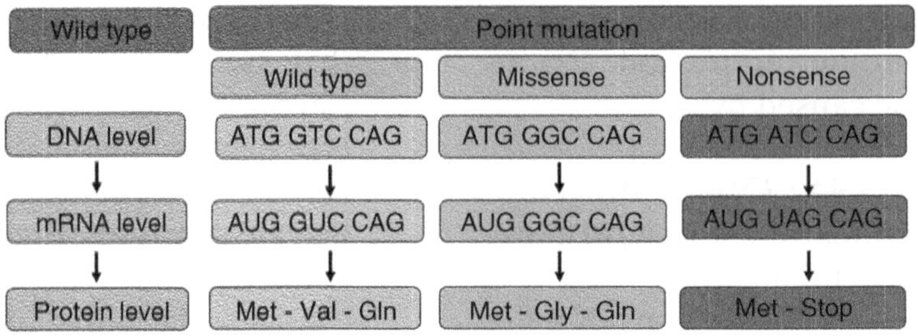

Figure 4.1 Representation of various types of point mutations.

- **Missense mutations**: Change in a single amino acid in a viral protein, potentially altering its function or antigenicity. For example, mutations in the spike protein of severe acute respiratory syndrome coronavirus 2 (SARS-CoV-2) can affect its binding to the angiotensin-converting enzyme 2 (ACE2) receptor [43].
- **Nonsense mutations**: Introduction of a premature stop codon in the viral protein, leading to a truncated protein that may be nonfunctional. This can severely impact viral replication or virulence.

4.3.2.2 Insertion and Deletion

- **Insertion** is the addition of one or more nucleotides into the viral genome that alters the reading frame. This can also lead to frameshift mutations, a situation where there is an alteration of the reading frame of the mRNA, resulting in a completely different amino acid sequence downstream of the mutation, which can produce nonfunctional or dysfunctional proteins.
- **Deletion:** It is the removal of one or more nucleotides. This can also lead to frameshift mutations or loss of essential viral proteins.

4.3.2.3 Duplication

Gene duplication is the repetition of a segment of the viral genome, but it does not shift the frame. This can lead to increased expression of certain viral proteins or the creation of fusion proteins with altered functions. In some cases, duplications can enhance viral replication or pathogenesis [44] (see Figure 4.2).

4.3.3 Medical Importance of Mutations in Virus Infections Includes

4.3.3.1 Role in Viral Evolution and Adaptation

- **Genetic variability:** Mutations can introduce genetic diversity into viral populations. This variability can help viruses adapt to changing environments, such as immune responses or antiviral treatments.

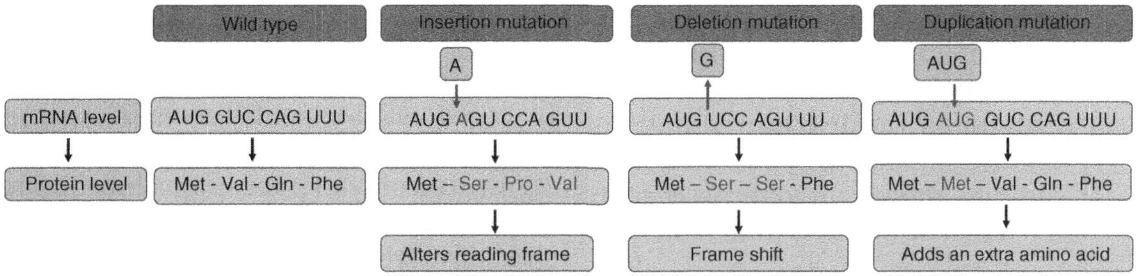

Figure 4.2 Representation of various types of mutations.

- **Virulence:** Mutations can enhance the ability of a virus to infect host cells, replicate more efficiently, or cause more severe disease.

4.3.3.2 Role in Viral Pathogenicity

- **Host switching, altered or broad tropism:** Mutations can change the affinity of virus for specific cell types or tissues, affecting its pathogenicity. When viruses jump between species (e.g., from animals to humans, known as zoonosis infection), they may undergo rapid adaptation to survive and replicate in a new host environment, accelerating evolution.
- **Disease severity:** Variants of viruses with specific mutations may lead to more severe disease manifestations or different clinical outcomes. Variants of concern (VOCs) for SARS-CoV-2, such as the Delta and Omicron variants [45], have shown differences in transmissibility and disease severity compared to earlier strains.
- **Oncogenic potential:** In some cases, viral mutations can affect the integration of viral DNA into the host genome, leading to the activation of oncogenes or the inactivation of tumor suppressor genes. For instance, mutations in viral genes can alter the function of viral oncogenes, contributing to cancer development.

4.3.3.3 Role in Immune Evasion

- **Antigenic drift:** Small, gradual mutations in viral surface proteins can alter the antigenic properties of a virus, helping it evade recognition by the host's immune system. This is common in influenza viruses and can necessitate annual updates to flu vaccines.
- **Antigenic shift:** Large-scale genetic reassortment events, such as those occurring in influenza viruses, can lead to major changes in surface antigens, potentially resulting in new viral strains that the immune system has never encountered.

4.3.3.4 Role in Drug Resistance

- **Mutations in drug targets:** Mutations can alter the structure of viral proteins that are targeted by antiviral drugs, leading to reduced drug efficacy. For example, mutations in the reverse transcriptase enzyme of HIV can confer resistance to antiretroviral medications [46].

- **Enhanced replication:** Mutations can also enhance the virus replication capacity, allowing it to thrive even in the presence of antiviral treatments.

4.4 Genetic Recombination in Virus Infection

Genetic recombination involves the exchange of genetic material between different viral strains or between virus and host genomes. This can lead to the creation of hybrid viruses with new properties. For instance, recombination events in polioviruses have produced new variants with altered transmissibility or immune evasion characteristics [47, 48].

4.4.1 Mechanisms of Genetic Recombination in Viruses

4.4.1.1 Homologous Recombination

This occurs when two similar or identical sequences of DNA or RNA exchange genetic material. In viruses, this often happens during the replication of their genomes. This is common in DNA viruses like herpesviruses and in some RNA viruses like coronaviruses.

4.4.1.2 Nonhomologous Recombination

This occurs when genetic exchange occurs between sequences that are not necessarily similar. This can result in the creation of entirely new genomic structures.

4.4.1.3 Segmented Genome Reassortment

In viruses with segmented genomes, such as influenza viruses, reassortment occurs when segments from different strains mix during the co-infection of a single host cell. This is a major mechanism behind the generation of new influenza virus strains.

4.4.2 Importance of Genetic Recombination in Virus Replication

4.4.2.1 Increased Genetic Diversity

Once genetic recombination has occurred during virus infection, it may lead to genetic diversity if the circulating virus type could enhance the adaptability of the virus to its host or cause a survival advantage.

- **Enhanced adaptability:** Genetic recombination introduces new genetic variations, enabling viruses to adapt to different environments, hosts, and selective pressures such as immune responses or antiviral treatments.
- **Survival advantage:** Greater genetic diversity can increase a virus population's chances of surviving adverse conditions, such as changes in host species or the introduction of vaccines.

4.4.2.2 Emergence of New Viral Strains

Genetic diversity can cause the emergence of a new strain of viruses that has different or more complicated pathogenesis and causes severe or less severe effects on health, for example:

- **Novel pathogens:** Recombination can create diverse new viral strains with unique genetic combinations, potentially leading to new outbreaks or epidemics [49].
- **Public health challenges:** These new strains may exhibit altered virulence, transmissibility, and antigenic properties, complicating disease control and prevention efforts.

4.4.2.3 Antigenic Shift and Drift

- **Antigenic shift:** In viruses with segmented genomes, such as influenza, reassortment of genome segments can lead to major antigenic changes (antigenic shift) [50]. This process can produce novel viral subtypes against which the population has little or no preexisting immunity, potentially leading to epidemics.
- **Antigenic drift:** Even in non-segmented viruses, recombination can contribute to gradual changes in antigenic sites, helping the virus evade host immune responses over time.

4.4.2.4 Drug Resistance

- **Resistance genes:** Recombination can facilitate the acquisition and spread of drug-resistance genes within viral populations. For example, if a virus acquires a resistance gene from another strain through recombination [51], it can become resistant to previously effective treatments.
- **Therapeutic challenges:** The emergence of recombinant drug-resistant viruses poses significant challenges for the development and implementation of antiviral therapies.

4.4.2.5 Altered Pathogenicity

- **Virulence factors:** Recombination can result in the transfer of virulence factors between viruses, potentially increasing the pathogenicity of the recombinant virus.
- **Host range:** Changes in the viral genome through recombination can also affect the host range, enabling viruses to infect new species or tissues within the same host [49].

4.4.2.6 Epidemiological Implications

- **Outbreak dynamics:** Recombinant viruses can have different transmission dynamics, affecting the spread and control of viral infections. Public health measures need to be adapted to address these new dynamics.
- **Surveillance:** Continuous monitoring and surveillance are essential to detect and respond to recombinant viruses promptly. Molecular epidemiology plays a crucial role in tracking these changes.

Clinical significance

Wild-type viruses: These are the original or naturally occurring forms of a virus found in nature, representing the standard genetic makeup of the virus before any induced or natural mutations occur.

Variant type: Variants are strains of a virus that have acquired genetic changes

Table 4.3 Classification of viral mutants and variants.

Type of variant/ mutant	Definition	Example
Antibody escape variants	Variants with mutations that reduce recognition by neutralizing antibodies.	SARS-CoV-2 Omicron variant.
Vaccine escape variants	Mutants capable of infecting vaccinated individuals due to immune evasion.	Hepatitis B virus surface antigen mutants.
Drug-resistant variants	Variants carrying mutations that confer resistance to antiviral drugs.	HIV reverse transcriptase-resistant strains.
Temperature-sensitive (ts) mutants	Mutants that replicate at permissive temperatures but fail at restrictive ones.	Influenza live-attenuated vaccine strains.
Attenuated variants	Variants with reduced pathogenicity, often used as vaccines.	Sabin strains (oral polio vaccine).
Reversion mutants	Variants where prior mutations revert, restoring the original virulence or phenotype.	Circulating vaccine-derived polioviruses (cVDPVs).
Recombinant variants	Variants are formed through recombination between distinct strains.	SARS-CoV-2 XE recombinant variant.
Reassortant variants	Variants are formed by exchanging genome segments between segmented viruses.	Influenza A 2009 H1N1 pandemic strain.
Tropism-shift variants	Variants that acquire the ability to infect new cell types or tissues.	Feline coronavirus (FCoV). Mutations that enable coronavirus to infect macrophages.
Replication-defective mutants	Variants are unable to replicate efficiently without helper functions.	Adenovirus vectors are used in vaccines.

compared to the wild type. Examples include (see Table 4.3):

- **Mutation-based variants**: These result from point mutations, insertions, deletions, or small duplications in the viral genome. A good example of a circulating mutation-based variant of a virus of medical importance is the SARS-CoV-2 Delta variant (B.1.617.2) [52]. It emerged due to the accumulation of point mutations, especially in the spike protein gene, which enhanced its transmissibility and partially reduced neutralization by antibodies. This variant spread globally and caused significant surges in coronavirus disease 2019 (COVID-19) cases, representing a mutation-driven adaptation. Other examples include the Omicron variant of SARS-CoV-2 (B.1.1.529) [53].

HBV vaccine escape mutants [54]. These are viral strains that have developed mutations in the Hepatitis B surface antigen (HBsAg), allowing them to replicate and infect individuals despite being vaccinated against HBV.

- **Recombination-based variants:** Formed when two related viruses exchange genetic material during co-infection, creating hybrid genomes. This is common in RNA viruses like coronaviruses and enteroviruses. HIV-1 also exhibits extensive recombination, contributing to the generation of circulating recombinant forms (CRFs) [55]. CRFs of HIV-1 are hybrid viruses formed when different subtypes of HIV-1 recombine [56]. These recombinants are significant because they can become established circulating strains in the HIV epidemic. Influenza C virus, while recombination is less common than reassortment in influenza A and B, recombination events have also been reported in influenza C, contributing to viral diversity (this is highly debated among scientists). Also, circulating vaccine-derived polioviruses (cVDPVs) evolve through recombination and mutation between attenuated vaccine strains and non-polio enteroviruses, leading to new neurovirulent variants.
- **Reassortment-based variants:** Specific to segmented viruses (like influenza). When two viruses co-infect the same host cell, genome segments can mix, generating a variant with segments from both parent strains. For example, Influenza A reassortment can lead to new pandemic strains, such as the 1957 Asian flu (H2N2), 1968 Hong Kong flu (H3N2), and the 2009 H1N1 "swine flu" pandemic [57]. Rotavirus reassortment contributes to the emergence of new rotavirus strains with altered antigenic properties and epidemiology [58]. Crimean–Congo hemorrhagic fever virus (CCHFV) has a tripartite segmented genome [59], shows evidence of reassortment, contributing to its genetic diversity and adaptation.

4.5 Molecular Evolution of Viruses

Viral evolution refers to the changes that occur in viral populations over time, leading to the emergence of new strains, altered virulence, and sometimes the ability to infect new hosts. Viral evolution is a fundamental aspect of virology because it explains why viruses adapt rapidly to environmental pressures, including host immunity, antiviral drugs, and ecological changes.

Importantly, viruses have contributed to host evolution [60]. By mediating gene transfer and leaving enduring genetic imprints, viruses can actively shape host evolution [61]. They introduce genetic novelty, accelerate adaptation to environmental pressures, and can even drive speciation events. Beyond coding sequences, viral integration can modulate regulatory networks, immune responses, and developmental pathways, leaving a lasting evolutionary footprint. This dual role, as both agents of horizontal gene transfer (HGT) and as embedded genomic elements, reveals the significance of viruses as evolutionary catalysts [62].

4.5.1 Forms of Evolutionary Mechanisms in Viruses

4.5.1.1 *Adaptive Evolution in Viruses*

Adaptive evolution occurs when mutations confer a selective advantage, increasing viral fitness [63]. Genetic changes (mutations, recombination, or reassortment) provide a selective advantage to a virus ("less fit gene to a more fit gene" [63]). Unlike neutral evolution, adaptive evolution is shaped by natural selection, favoring variants that enhance viral fitness. Examples include mutations (Table 4.4) that enhance binding to host receptors and changes in viral epitopes that allow escape from neutralizing antibodies. It is often observed in RNA viruses due to high mutation rates and rapid replication.

Outcomes of Adaptive Evolution Adaptive evolution in viruses can lead to several outcomes that enhance their survival and spread. Increased fitness allows viruses to replicate more efficiently, enter host cells more effectively, or transmit more readily between hosts. Immune escape occurs when alterations in antigenic sites enable the virus to evade recognition by antibodies or T cells [68]. Drug resistance arises when mutations in viral enzymes targeted by therapy reduce the effectiveness of antiviral treatments.

Adaptive changes can also expand host range or tropism, enabling viruses to infect new host species or additional tissue types. In some cases, mutations may enhance pathogenicity or virulence, thereby

Table 4.4 Examples of adaptive evolution (mutation).

Virus	Adaptive mutation	Effect
Dengue virus	Envelope protein mutations (V324I, V351L, and V380I)	Altered antigenicity, immune escape [64]
Zika virus	Mutations in E-V763M and NS1 genes	Increased replication and neurovirulence [65]
Chikungunya virus	E1-A226V mutation	Enhanced vector competence in *Aedes albopictus* [66]
SARS-CoV-2	Spike protein RBD mutations (e.g., N501Y)	Increased transmissibility and immune escape [67]

increasing disease severity; however, this is not always the case if it negatively impacts transmission.

4.5.1.2 Neutral Evolution in Viruses

This is a kind of mutation that neither benefits nor harms the virus ("null effect on fitness") [69]. Usually, it neither increases nor decreases the ability of virus to replicate, infect, or transmit, but it can create opportunities for the emergence of new viral phenotypes [69]. This is often caused by genetic drift rather than gene selection in the regions of the genome that do not affect protein function (e.g., synonymous to silent mutations in coding regions or some NCRs). These are nucleotide changes that do not change the amino acid sequence of the encoded protein.

4.5.2 Importance of Viral Evolutionary Events

4.5.2.1 Molecular Clock and Viral Evolutionary Rates

The molecular clock is based on the observation that neutral mutations accumulate at an approximately constant rate over time. By measuring these mutations across viral genomes, scientists can estimate the timing of evolutionary events, such as the divergence of strains or the emergence of a new outbreak. It enables the estimation of divergence times between viral strains or lineages. It also helps in estimating the time to the most recent common ancestor (tMRCA) for viral lineages [70].

4.5.2.2 Viral Evolutionary Rates

Viral evolutionary rate refers to the speed at which viral genomes accumulate genetic changes over time. It is usually measured in substitutions per nucleotide site per year and reflects how quickly a virus can adapt to its environment, hosts, or selective pressures.

- **RNA viruses:** $\sim 10^{-3}$–10^{-5} substitutions per site per year (high rate due to error-prone polymerases) [71].
- **DNA viruses:** $\sim 10^{-8}$–10^{-6} substitutions per site per year (lower, due to proofreading) [72].

High evolutionary rates facilitate rapid adaptation, especially in RNA viruses like dengue, chikungunya, and SARS-CoV-2.

Applications:

- Tracking outbreak origins and spread.
- Estimating the time to the most recent common ancestor (tMRCA) of viral lineages.
- Predicting potential emergence of new variants.

4.5.2.3 Predicting Viral Evolution and Pandemic Preparedness

Forecasting viral evolution is essential for proactive public health strategies. Rapid genomic sequencing, combined with epidemiological and immunological data, enables comprehensive modeling of viral emergence and adaptation. Platforms such as GISAID (`https://gisaid.org`) have demonstrated the value of global data sharing in tracking viral spread and variant emergence.

Computational and machine learning models are increasingly applied to predict mutations that confer fitness advantages, such as enhanced transmissibility or immune escape [73]. Identifying mutations that may reduce susceptibility to preexisting immunity allows the design of next-generation vaccines and informs containment strategies. These anticipatory approaches move virology from reactive measures to proactive pandemic preparedness.

4.5.2.4 Virus-Mediated Horizontal Gene Transfer

Viruses are not only agents of disease but also important mediators of genetic exchange across species [74, 75]. Through mechanisms collectively termed HGT, viruses can shuttle genetic material between organisms, influencing genome evolution, adaptation, and diversity. This capacity positions viruses as both drivers and recorders of evolutionary processes.

Viruses can facilitate gene transfer between unrelated species by packaging host genetic material during replication and delivering it to new hosts. This process is particularly evident in bacteriophages, which can transduce bacterial genes, including those encoding antibiotic resistance or metabolic functions, between bacterial populations. In eukaryotes, viral-mediated gene transfer contributes to the introduction of novel sequences into animal, plant, and fungal genomes, occasionally resulting in new traits or adaptive advantages.

Endogenous viral elements (EVEs) are viral sequences that have become integrated into the germline of host organisms and are inherited across generations [76, 77]. It can influence host gene regulation, contribute coding sequences, or provide regulatory elements that shape gene expression. For example, certain placental development genes, known as syncytins, in mammals are derived from viral envelope proteins, illustrating how EVEs can be co-opted for critical biological functions [78].

4.6 Future Directions in Viral Genetics and Evolution

The field of viral genetics and evolution is at a transformative crossroads, driven by technological advances and the urgent need to address emerging infectious threats. Modern approaches combine synthetic biology, high-throughput experimentation, genomic surveillance, and computational modeling to understand viral behavior, predict evolutionary trajectories, and develop novel interventions. These strategies are reshaping our capacity to anticipate, prevent, and respond to viral outbreaks.

4.6.1 Implications of Mutation and Evolution for Viral Genetics in Vaccines, Diagnostics, and Therapeutics

The insights from viral genetics and evolution have profound implications for public health.

- **Vaccines:** The ability to predict viral evolution is essential for vaccine design. For example, for influenza, this information is used to select the strains for the annual vaccine. In the future, this predictive capability could lead to "preemptive" vaccines for pathogens like coronaviruses or dengue virus, targeting variants before they become widespread. Furthermore, synthetic virology is enabling the development of new vaccine platforms, such as self-amplifying RNA vaccines, which can be rapidly designed and manufactured in response to an emerging threat.

- **Diagnostics:** As viruses evolve, diagnostic tests must keep up. A mutation in a viral gene targeted by a diagnostic test (e.g., a PCR primer) could lead to a false negative result. By tracking viral evolution, diagnostic companies can ensure their tests remain accurate and effective. New technologies, such as clustered regularly interspaced short palindromic repeats (CRISPR)-based diagnostics, are also being developed that are designed to be more robust against viral evolution.
- **Therapeutics:** Viral evolution is a primary driver of antiviral drug resistance. A classic example is the evolution of drug resistance in HIV, which requires patients to take a combination of drugs like highly active antiretroviral therapy (HAART) to prevent the virus from evolving a way to escape treatment. By understanding the genetic pathways to resistance, researchers can design new drugs that are less likely to be evaded or develop combination therapies that make it more difficult for the virus to evolve. Additionally, engineered viruses (oncolytic viruses) are being used as a therapeutic tool to specifically target and destroy cancer cells.

4.6.2 CRISPR-Based Gene Editing, Diagnostics, and Therapeutics in Viral Infections

The advent of CRISPR technology has revolutionized the field of molecular biology, offering precise tools for genome editing, detection, and therapeutic intervention. In the context of viral infections, CRISPR-based approaches are emerging as powerful platforms for rapid diagnostics and novel antiviral therapies [79, 80]. This genome-editing system was derived from the adaptive immune defense of bacteria and archaea. Its core components include a guide RNA (gRNA) that recognizes a specific DNA or RNA sequence and a Cas nuclease (e.g., Cas9 for DNA or Cas13 for RNA) that introduces precise cleavage at the target site (Figure 4.3) [81]. This system allows highly specific modifications of genetic material, enabling deletion, insertion, or correction of genes.

1. **The components:** CRISPR gene editing requires two main components:
   - **Guide RNA (gRNA):** A short RNA sequence (~20 nucleotides) designed to match the target DNA sequence. It guides the Cas nuclease to the exact spot in the genome.
   - **Cas nuclease:** A protein (like Cas9 for DNA or Cas13 for RNA) that can cut genetic material. Cas9 is the most commonly used for editing DNA.

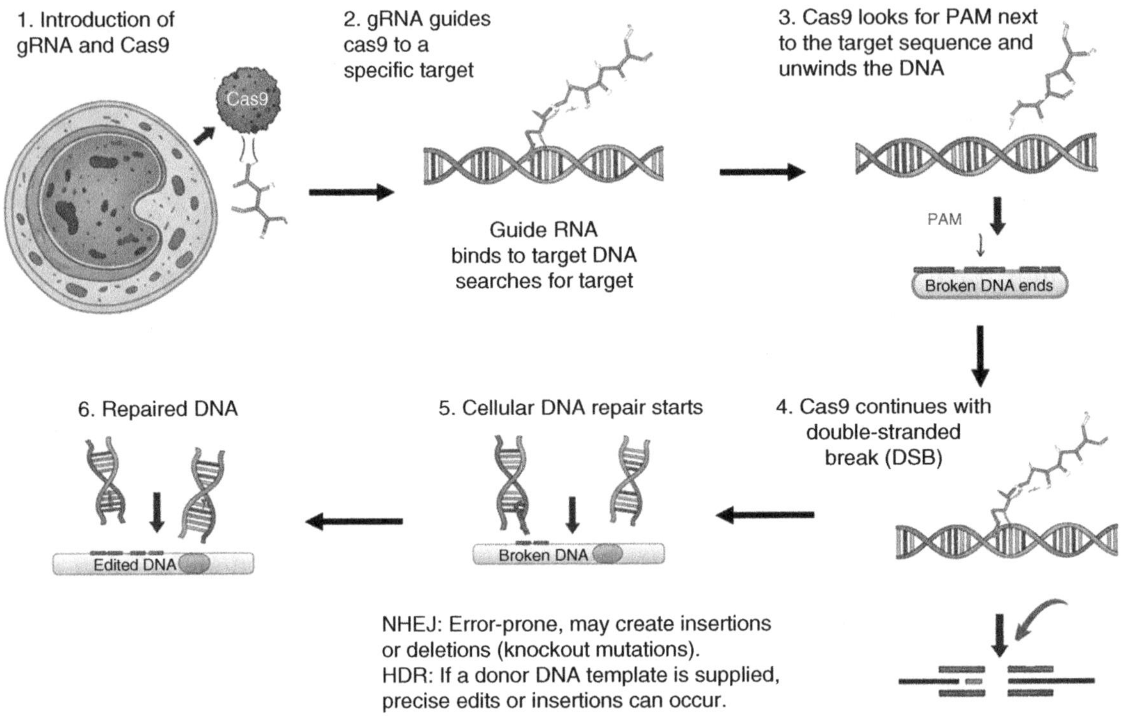

Figure 4.3 Overview of the CRISPR Gene editing process.

Some systems also include a repair template, which is used to introduce a new sequence at the cut site.

2. **Target recognition:** The gRNA binds to the complementary DNA sequence in the genome, allowing Cas9 to recognize a short DNA motif called the protospacer adjacent motif (PAM), which is required for the nuclease to bind and cut.
3. **DNA cleavage:** Once the gRNA and Cas9 locate the target, Cas9 introduces a double-strand break (DSB) at that precise site in the DNA [82].
4. **DNA repair and editing:** After the break, the cell repairs the DNA using one of two pathways:
   - **Nonhomologous end joining (NHEJ):** The cell ligates the broken DNA ends back together. This repair is error-prone and often introduces small insertions or deletions (indels) that can disrupt a gene, effectively "knocking it out" [83].
   - **Homology-directed repair (HDR):** If a repair template is provided, the cell can use it to insert a new sequence at the cut site precisely. This allows for gene correction, insertion, or replacement [84].
5. **Outcome:** The gene can be knocked out, corrected, or modified depending on the design.

CRISPR is highly versatile because it can be programmed to target virtually any DNA sequence in the genome.

4.6.2.1 CRISPR-Based Diagnostics

CRISPR-based diagnostics leverage the sequence-specific targeting capability of CRISPR-associated (Cas) proteins, particularly Cas12 and Cas13, to detect viral nucleic acids with high sensitivity and specificity (Table 4.5). In these systems, a gRNA directs the Cas protein to a viral RNA or DNA sequence. Upon target recognition, Cas proteins exhibit collateral cleavage activity, which can be coupled to fluorescent or colorimetric reporter molecules, generating a detectable signal. These diagnostics can be completed in under an hour, require minimal equipment, and are adaptable to point-of-care testing, making them invaluable in outbreak response and surveillance.

4.6.2.2 CRISPR-Based Therapeutics

CRISPR technology also holds promise as a direct antiviral therapeutic strategy. Cas proteins, guided by specific RNA sequences, can target and cleave viral genomes, inhibiting replication. For RNA viruses, Cas13 can be programmed to degrade viral RNA, while for DNA viruses, Cas9-mediated genome editing can disrupt essential viral genes.

Therapeutic applications include:

1. **Direct antiviral therapy:** Targeting viral genomes to reduce viral load in infected cells.
2. **Prophylactic gene editing:** Engineering resistance in susceptible host cells by introducing protective genetic modifications.
3. **Immunotherapy enhancement:** Modifying host immune genes to boost antiviral responses.

Table 4.5 Examples and mechanisms of CRISPR-based point-of-care diagnostic tools for virus infection in low-resource settings.

Diagnostic	CRISPR system	Target type	Principle	Examples
SHERLOCK	Cas13	RNA	Cas13 binds target RNA → collateral cleavage of reporter RNA → fluorescent/colorimetric signal	Zika virus [85], SARS-CoV-2 (STOPCovid) [86]
DNA endonuclease-targeted CRISPR trans reporter (DETECTR)	Cas12a	DNA	Cas12a binds target DNA → nonspecific cleavage of ssDNA reporter → detectable signal	HPV detection [87–89]
FELUDA	FnCas9	DNA	FnCas9 binds target DNA → lateral flow assay readout	SARS-CoV-2 [90]

While these approaches are promising, challenges remain, including the development of safe and efficient delivery systems, minimizing off-target effects, and preventing potential viral escape mutations. Nonetheless, CRISPR-based diagnostics and therapeutics represent a transformative frontier in virology, offering tools that are rapid, precise, and adaptable to emerging viral threats.

4.6.3 Synthetic Virology and Directed Evolution

Synthetic virology allows researchers to design and construct viral genomes or protein (e.g. capsid) in the laboratory, moving beyond the study of naturally occurring viruses. This enables systematic investigation of gene function, virus–host interactions, replication mechanisms, and immune evasion strategies. A critical application is the development of vaccines, where viruses can be engineered to retain immunogenicity while reducing pathogenicity [91].

Directed evolution, on the other hand, applies laboratory-based selection to generate viral variants with desired traits. By iteratively inducing mutations and selecting for specific phenotypes, researchers can optimize viral vectors for gene delivery, enhance protein interactions, or identify molecular determinants of pathogenicity [92]. These approaches provide powerful tools to probe viral biology, develop therapeutic agents, and innovate vaccine design.

4.7 MicroRNAs in Viral Genomes and Evolution

Traditionally, studies of viral replication and pathogenesis have focused on DNA and RNA, as well as mRNAs that encode structural and nonstructural proteins. However, aside from these conventional nucleic acids, there are small RNAs (sRNAs). These sRNAs are classes of noncoding, single-stranded RNA about 18–25 nucleotides long [93]. They fine-tune the suppression of complementary sequence, leading to endonucleolytic cleavage of the target mRNA.

Viruses can also produce regulatory small noncoding RNAs, particularly microRNAs (miRNAs), that help control host–virus interactions. Viral miRNAs are short noncoding RNAs generated from hairpin precursors, similar to host miRNAs. They typically regulate gene expression through posttranscriptional silencing, either by degrading complementary mRNA or inhibiting its translation.

4.7.1 Classification of Small RNA

The two major classes of sRNAs that have been implicated in virus replication are miRNA and short-interference RNA (siRNA). Fundamentally, these two sRNAs, siRNA (generated from long double-stranded RNA [dsRNA] precursors, usually exogenous), and miRNA (derived from primary miRNA transcripts [pri-miRNAs]), are similar in terms of molecular characteristics, effector function, and biogenesis; current differences between the two seem to be arbitrary. Unlike other sRNA, piwi-interacting RNAs (piRNAs) are larger (24–31 nucleotides) and primarily found in germ cells, where they protect the genome from transposon activity. Other examples sRNA and their function in virology are listed in Table 4.6.

Table 4.6 Types of small RNA.

Type of small RNA	Size (nt)	Origin/biogenesis	Function	Virus–host relevance
MicroRNA (miRNA)	~20–24	From pri-miRNA → pre-miRNA (Drosha/DGCR8 → Dicer)	Posttranscriptional repression (translation inhibition or mRNA degradation)	Host and viral miRNAs regulate infection, latency, and immune evasion
Small-interfering RNA (siRNA)	~20–23	From long dsRNA or viral RNA (Dicer cleavage)	RNA silencing via Argonaute → target RNA cleavage	Antiviral immunity in plants/invertebrates; therapeutic RNA interference (RNAi)
PIWI-interacting RNA (piRNA)	~24–31	From long single-stranded RNA precursors, Dicer-independent	Silencing of transposons in germline	Protects against retroelement-driven viral evolution
shRNA (short hairpin RNA)	~19–29	Artificial/engineered hairpin transcripts → Dicer processing	Gene knockdown, therapeutic RNAi	Antiviral research and therapy
Virus-derived siRNA (vsiRNA)	~21–24	From viral dsRNA via Dicer during infection	Silences viral transcripts (antiviral defense)	Key antiviral mechanism in plants/invertebrates
Endogenous siRNA (endo-siRNA)	~21	Derived from pseudogenes, repetitive elements, or structured RNAs	Regulate transposons, gene expression	Modulates host response during viral infection
snoRNA-derived RNAs (sdRNAs)	~17–30	Processed from small nucleolar RNAs	Regulate splicing, mRNA stability	Stress and viral replication influence
snRNA-derived fragments	~18–30	From spliceosomal snRNAs	Gene regulation, RNA processing	Can be altered during viral infection
tRNA-derived fragments (tRFs/tsRNAs)	~14–35	Cleaved from tRNAs by angiogenin or nucleases	Regulate translation, gene silencing	Viral infections induce tRFs (HIV, RSV) to modulate host translation
rRNA-derived small RNAs (rsRNAs/rRFs)	~18–30	Cleaved from rRNA precursors	Stress response, ribosome regulation	Found altered in viral infection and cancers

(*Continued*)

Table 4.6 (Continued)

Type of small RNA	Size (nt)	Origin/biogenesis	Function	Virus–host relevance
lncRNA-derived small RNAs	Variable (~18–30)	Processed from long noncoding RNAs	Gene silencing, chromatin regulation	Can regulate viral replication indirectly
YRNA-derived fragments (YsRNAs)	~22–33	Derived from Y RNAs	Stress signaling, apoptosis regulation	Induced during viral infection and stress
Vault RNA-derived small RNAs (svRNAs)	~23	From vault RNAs processed by Dicer	Regulate drug resistance, gene expression	May contribute to host–virus interactions
Dicer-related siRNA (DR-siRNA)	~22	From structured regions of transcripts	RNA interference	Some viral RNAs trigger their production
tRNA halves/ stress-induced tRNA fragments (tiRNA)	~29–50	Cleaved in anticodon loop of tRNA under stress	Inhibit translation initiation, stress response	Induced by viral infection, regulate host shutoff
Centromere small RNAs (cenRNAs)	~20–30	Derived from centromeric repeats	Chromatin regulation	Indirect evolutionary role
circRNA-derived small RNAs	~18–30	From circular RNAs	miRNA sponges, regulatory roles	Viral circRNAs (e.g., hepatitis delta virus) may produce them
Secondary piRNA-like RNAs (siwiRNA)	~27–30	Amplified from piRNA targets (ping-pong cycle)	Reinforce transposon silencing	Limit retrovirus mobilization
exogenous siRNA (exo-siRNA)	~21–23	From exogenous dsRNA	RNA interference	Experimentally used against viruses
Centromere repeat-associated small (RNAcrasiRNA)	~35–42	From centromeric repeats	Centromere function, genome stability	Evolutionary influence on integration of viral elements

Source: Adapted from Ref. [95].

The canonical biogenesis of sRNA (e.g., miRNA) involves two major processes: the transcription of nascent pre-miRNA (about 53-70 nt precursor) from the gene (intron or exon), followed by the cleavage of the pre-miRNA into a mature sRNA. These two processes involve RNase III enzymes – Drosha and Dicer, respectively [94]. While the canonical biogenesis of the host and some virus-derived sRNA is initiated in the nucleus, rare noncanonical sRNA biosynthesis has been described for RNA viruses of cytoplasmic origin. An initial characteristic of noncanonical sRNA biosynthesis is that transcripts of cytoplasmic-derived sRNA are processed in a manner that is independent of Drosha but still requires Dicer.

Hundreds of human virus-derived sRNAs have been identified. Previously identified virus-derived sRNA was thought to be exclusively produced from dsDNA viruses, mainly of herpesvirus and adenovirus families. However, retrovirus and some extensively debated cytoplasmic RNA viruses, such as West Nile virus have been shown to produce virus-derived sRNA. The Kunjin strain encodes viral miRNA (vmiRNA) [95]. Other examples are listed in Table 4.7.

Table 4.7 Examples of viruses that encode small RNAs.

Virus family/ virus	Type of small RNA encoded	Examples/names	Functional roles
Herpesviridae (DNA viruses)	miRNAs	EBV (miR-BARTs, miR-BHRF1), Kaposi's sarcoma-associated herpesvirus (KSHV) (miR-K12 series), CMV (miR-UL112)	Regulate latency/lytic switch, evade host immunity, suppress apoptosis
Polyomaviridae	miRNAs	SV40 miRNA, BK virus miRNAs, JC virus miRNAs	Downregulate viral antigen expression, evade immune detection
Adenoviridae	Viral-associated RNAs (VA-RNAs, processed into miRNA-like small RNAs, called mivaRNAs)	Adenovirus VA RNA I and II	Modulate RNAi machinery, regulate viral gene expression
Retroviridae (Retroviruses)	miRNA-like RNAs	HIV-1 (TAR-derived miRNA, Nef-derived miRNA)	Control viral replication, suppress apoptosis, modulate host gene expression

(Continued)

Table 4.7 (Continued)

Virus family/ virus	Type of small RNA encoded	Examples/names	Functional roles
Flaviviridae (RNA viruses)	Viral small RNAs (vsRNAs)	West Nile virus vsRNAs, dengue virus-derived small RNAs	Regulate replication, influence host response
Rhabdoviridae	Virus-derived siRNAs (vsiRNAs)	Vesicular stomatitis virus (VSV)-derived siRNAs	Host antiviral RNAi interaction
Orthomyxoviridae (Influenza A virus)	miRNA-like RNAs	Influenza A virus (miRNA-like RNAs from segment 7)	Modulate host immune response
Papillomaviridae (HPV)	No canonical miRNAs, but reports of viral sRNA fragments	HPV-derived small RNAs (not classical miRNA)	Potential regulation of viral and host genes (under investigation)
Poxviridae	Small RNAs (vsiRNAs, miRNA-like RNAs)	Vaccinia virus-derived sRNAs	Modulate immune response, regulate host–virus interactions
Other RNA viruses	vsiRNAs (virus-derived siRNAs via host RNAi machinery)	Plant RNA viruses, insect RNA viruses	Trigger or evade host RNA interference pathways

4.7.2 Role of Small RNA in Modulating Host Antiviral and Immune Response

About a decade ago, miRNA response to viral infection was described as an archaic immune system. However, recent events and evidence have shown that it is an active part of the innate immune response. It can play an antiviral role as an independent entity or augment other established innate immune responses [96]. Small-interfering RNA responds to exogenous vRNA as one of the endogenous immune defense mechanisms in mammalian cells. Mutation of the siRNA pathway in mammalian cells results in highly compromised resistance to viral infection.

Independently, miRNA and siRNA could target invading viruses that harbor complementary nucleotide sequences in their genome, thereby interrupting their replicative process. Reports on EV-71 (Enterovirus-71) infection show that EV-71 VP1 (viral protein 1) RNA coding region can be targeted directly by miR-296-5p and miR-23b [97]. In addition, the RNA interference (RNAi) mechanism could mediate the antiviral state within the host cell by suppressing host or viral factors through targeting of mRNA that encodes viral or proviral protein. Examples of this have been observed in human cells infected with Dengue virus serotype 2 (DENV-2). miR-223 downregulates the production of microtubule-destabilizing protein stathmin 1, which consequently downregulates viral infection [98]. Also, miR-342-5p exerts a broad antiviral response against several unrelated viruses, such as cytomegalovirus (CMV) and influenza virus, by targeting the steroid pathway requirement for the replication of these viruses [99]. Likewise, evidence suggests that cellular miRNA mediates antiviral defense against two types of influenza virus infections [100].

Furthermore, miRNA modulates interferon (IFN) activities, a vital component of antiviral innate immunity, by binding to a perfect or near-perfect complementarity recognition element, a seed region of 6–8 nucleotides at the 3′ end of IFN mRNA [101]. Also, evidence is accumulating in support of the hypothesis that mammalian cells, through the IFN system, use RNAi mechanisms to fight viral infection. IFNβ rapidly modulates the expression of several cellular miRNAs called IFN-stimulated miRNA (ISM) as a defense mechanism against HCV in the human hepatoma cell line. Using sequence complimentary analysis, eight IFNβ-induced miRNAs (miR-1, miR-30, miR-128, miR-196, miR-448, miR-431, miR-296, and miR-351) were identified [101]. Remarkably, these miRNAs have a complementary seed sequence to HCV. The introduction of synthetic miRNA that mimics these eight miRNAs enhances the antiviral state of IFNβ, while neutralization of these miRNAs decreases IFNβ antiviral effect, hence enhancing HCV replication.

4.7.3 Viral Evasion of the RNAi Antiviral Mechanism

Like any other immune response, the viral resistance to RNAi antiviral effect is inevitable. Viruses tend to develop a defensive niche and evade this defense mechanism to replicate successfully. Viruses could evade RNAi defense mechanisms by interfering with RNAi function and overwhelming its contribution to the host immune response [102]. Viruses achieve this evasion by protecting vRNA from RNAi. Also, viruses can directly destroy siRNA, for example, during EV-71 infection, 3A protein has been shown to suppress RNAi [103]. In some situations, the virus can produce complementary anti-siRNA that competitively attenuates RNAi antiviral response, as observed in adenovirus infection [104].

4.8 Study Questions

Case Study/ Applied Questions

4.1 A patient living in a tropical region is infected with Dengue virus. Sequencing reveals several nucleotide changes in the envelope protein gene compared to the reference strain. Discuss the potential implications of these mutations.

4.2 During a chikungunya outbreak, sequencing reveals several recombinant viruses with genes from different lineages. Discuss how recombination contributes to viral evolution and outbreak dynamics.

4.3 A laboratory develops a CRISPR-based diagnostic for Zika virus. Explain the principle of detection and how knowledge of viral genome variation is critical for assay accuracy.

4.4 An Ebola virus outbreak shows several variants circulating simultaneously. Discuss how mutations and selection pressures could influence the outbreak's trajectory and control measures.

4.5 Explain how viral quasispecies affect pathogenesis and treatment outcomes.

Objectives

4.1 **Genetic reassortment in viruses occurs when**
- (a) Two viruses exchange segments of their genome during co-infection
- (b) A virus mutates spontaneously in a host
- (c) A virus integrates into the host genome
- (d) Viral proteins fold incorrectly

4.2 **Which type of viral mutation results in a single nucleotide change?**
- (a) Silent mutation
- (b) Point mutation
- (c) Frameshift mutation
- (d) Reassortment

4.3 **RNA viruses typically have higher mutation rates than DNA viruses because**
- (a) They have smaller genomes
- (b) Their polymerases lack proofreading ability
- (c) They have segmented genomes
- (d) They replicate more slowly

4.4 **Mutations in viral genomes can**
- (a) Lead to antiviral resistance
- (b) Alter viral pathogenicity
- (c) Allow immune escape
- (d) All of the above

4.5 **Viral recombination contributes to**
- (a) Evolution of new viral strains
- (b) Maintenance of genome stability
- (c) Only structural protein changes
- (d) Inactivation of the virus

4.6 **Which of the following describes a viral variant?**
- (a) A virus that is identical to the reference strain
- (b) A virus with genetic changes leading to different biological properties
- (c) A virus that cannot replicate
- (d) Only viruses with noncoding mutations

4.7 **Viral mutants can be selected for by**
- (a) Host immune pressure
- (b) Antiviral drugs
- (c) Environmental factors
- (d) All of the above

4.8 **Endogenous viral elements (EVEs) are examples of**
- (a) Mutations in coding regions
- (b) Viral sequences integrated into host genomes through horizontal transfer

(c) CRISPR-based edits
(d) Viral reassortments

4.9 **CRISPR-Cas systems in viral infections can be used to:**
(a) Edit viral genomes
(b) Detect viral RNA or DNA
(c) Develop antiviral therapies
(d) All of the above

4.10 **A CRISPR-based diagnostic tool works by:**
(a) Editing the host genome
(b) Detecting specific viral sequences via Cas enzyme activity
(c) Reassorting viral genomes
(d) Only targeting viral proteins

References

1. Domingo, E., *Molecular basis of genetic variation of viruses: error-prone replication*, in *Virus as Populations*. 2020. p. 35–71. doi: 10.1016/B978-0-12-816331-3.00002-7 Epub 2019 Nov 8. PMCID: PMC7153327. https://pmc.ncbi.nlm.nih.gov/articles/PMC7153327/.
2. Sanjuán, R. and Domingo-Calap, P., *Genetic diversity and evolution of viral populations*, in *Encyclopedia of Virology*. 2021. p. 53–61. doi: 10.1016/B978-0-12-809633-8.20958-8 Epub 2021 Mar 1. PMCID: PMC7157443. https://pmc.ncbi.nlm.nih.gov/articles/PMC7157443/.
3. Fernandes, J.D., et al., *Functional segregation of overlapping genes in HIV*. Cell, 2016. **167**(7): p. 1762–1773.e12.
4. Li, S., et al., *Adaptive evolution of proteins in hepatitis B virus during divergence of genotypes*. Sci. Rep., 2017. **7**(1): p. 1990.
5. Saffran, H.A., Read, G.S., and Smiley, J.R., *Evidence for translational regulation by the herpes simplex virus virion host shutoff protein*. J. Virol., 2010. **84**(12): p. 6041–6049.
6. Fessler, S.P. and Young, C.S., *Control of adenovirus early gene expression during the late phase of infection*. J. Virol., 1998. **72**(5): p. 4049–4056.
7. Moss, B., *Poxvirus cell entry: how many proteins does it take?* Viruses, 2012. **4**(5): p. 688–707.
8. Gruffat, H., Marchione, R., and Manet, E., *Herpesvirus late gene expression: a viral-specific pre-initiation complex is key*. Front. Microbiol., 2016. **7**: p. 869.
9. Tian, R., et al., *HPV integration generates a cellular super-enhancer which functions as ecDNA to regulate genome-wide transcription*. Nucleic Acids Res., 2023. **51**(9): p. 4237–4251.
10. Biasiotto, R. and Akusjärvi, G., *Regulation of human adenovirus alternative RNA splicing by the adenoviral L4-33K and L4-22K proteins*. Int. J. Mol. Sci., 2015. **16**(2): p. 2893–2912.
11. Poon, L.L., et al., *Direct evidence that the poly(A) tail of influenza A virus mRNA is synthesized by reiterative copying of a U track in the virion RNA template*. J. Virol., 1999. **73**(4): p. 3473–3476.
12. Sanford, T.J., et al., *Circularization of flavivirus genomic RNA inhibits de novo translation initiation*. Nucleic Acids Res., 2019. **47**(18): p. 9789–9802.
13. Griffin, D.E., Lin, W.-H., and Pan, C.-H., *Measles virus, immune control, and persistence*. FEMS Microbiol. Rev., 2012. **36**(3): p. 649–662.
14. Roman, C., et al., *The SARS-CoV-2 programmed −1 ribosomal frameshifting element crystal structure solved to 2.09 Å using chaperone-assisted RNA crystallography*. ACS Chem. Biol., 2021. **16**(8): p. 1469–1481.

15. Biswas, P., et al., *The human immunodeficiency virus type 1 ribosomal frameshifting site is an invariant sequence determinant and an important target for antiviral therapy*. J. Virol., 2004. **78**(4): p. 2082–2087.
16. Li, R., et al., *A structural and functional analysis of opal stop codon translational readthrough during chikungunya virus replication*. J. Gen. Virol., 2023. **104**(10): 001909.
17. Vidal, S., Curran, J., and Kolakofsky, D., *A stuttering model for paramyxovirus P mRNA editing*. Embo J., 1990. **9**(6): p. 2017–2022.
18. Loutan, A.J., et al., *Bunyaviral cap-snatching endonuclease activity and inhibition with Baloxavir-like inhibitors in the context of full-length L proteins*. Viruses, 2025. **17**(3): p. 420.
19. De Vlugt, C., Sikora, D., and Pelchat, M., *Insight into influenza: a virus cap-snatching*. Viruses, 2018. **10**(11): p. 461.
20. Lai, M.C., et al., *Translation control of enterovirus A71 gene expression*. J. Biomed. Sci., 2020. **27**(1): p. 22.
21. Ferron, F., et al., *Transcription and replication mechanisms of Bunyaviridae and Arenaviridae L proteins*. Virus Res., 2017. **234**: p. 118–134.
22. Mohamadkhani, A., Montazeri, G., and Poustchi, H., *The importance of hepatitis B virus genome diversity in Basal core promoter region*. Middle East J. Dig. Dis., 2011. **3**(1): p. 13–19.
23. Alimohammadi, M., et al., *Long noncoding RNAs and HPV-related cervical cancer: uncovering molecular mechanisms and clinical applications*. Transl. Oncol., 2025. **55**: 102363.
24. Xu, Z., et al., *Statistical analysis supports UTR (untranslated region) deletion theory in SARS-CoV-2*. Virulence, 2022. **13**(1): p. 1772–1789.
25. Blaney, J.E. Jr., et al., *Dengue virus type 3 vaccine candidates generated by introduction of deletions in the 3′ untranslated region (3'-UTR) or by exchange of the DENV-3 3'-UTR with that of DENV-4*. Vaccine, 2008. **26**(6): p. 817–828.
26. Emonet, S.F., et al., *Arenavirus genetic diversity and its biological implications*. Infect., Genet. Evol., 2009. **9**(4): p. 417–429.
27. Carr, C.R., et al., *Deep mutational scanning reveals functional constraints and antibody-escape potential of Lassa virus glycoprotein complex*. Immunity, 2024. **57**(9): p. 2061–2076.e11.
28. Sharp, P.M. and Hahn, B.H., *Origins of HIV and the AIDS pandemic*. Cold Spring Harb Perspect Med., 2011. **1**(1): a006841.
29. Reese, S.M., et al., *Potential for La Crosse virus segment reassortment in nature*. Virol. J., 2008. **5**: p. 164.
30. Khiabanian, H., Trifonov, V., and Rabadan, R., *Reassortment patterns in Swine influenza viruses*. PLoS One, 2009. **4**(10): e7366.
31. Xu, K., Broder, C.C., and Nikolov, D.B., *Ephrin-B2 and ephrin-B3 as functional henipavirus receptors*. Semin. Cell Dev. Biol., 2012. **23**(1): p. 116–123.
32. Wong, J.J., et al., *EphrinB2 clustering by Nipah virus G is required to activate and trap F intermediates at supported lipid bilayer-cell interfaces*. Sci. Adv., 2021. **7**(5): eabe1235.
33. Koch, L.K., et al., *Bats as putative Zaire ebolavirus reservoir hosts and their habitat suitability in Africa*. Sci. Rep., 2020. **10**(1): p. 14268.
34. Diehl, W.E., et al., *Ebola virus glycoprotein with increased infectivity dominated the 2013–2016 epidemic*. Cell, 2016. **167**(4): p. 1088–1098.e6.
35. Bornholdt, Z.A., et al., *Host-primed Ebola virus GP exposes a hydrophobic NPC1 receptor-binding pocket, revealing a target for broadly neutralizing antibodies*. MBio, 2016. **7**(1). doi: 10.1128/mbio.02154-15.
36. Regla-Nava, J.A., et al., *A Zika virus mutation enhances transmission potential and*

confers escape from protective dengue virus immunity. Cell Rep., 2022. **39**(2): 110655.
37. Gaburro, J., et al., *Neurotropism and behavioral changes associated with Zika infection in the vector Aedes aegypti*. Emerg. Microbes Infect., 2018. **7**(1): p. 68.
38. Elena, S.F. and de Visser, J.A., *Environmental stress and the effects of mutation*. J. Biol., 2003. **2**(2): p. 12.
39. Douki, T., von Koschembahr, A., and Cadet, J., *Insight in DNA repair of UV-induced pyrimidine dimers by chromatographic methods*. Photochem. Photobiol., 2017. **93**(1): p. 207–215.
40. Wieczfinska, J., Kleniewska, P., and Pawliczak, R., *Oxidative stress-related mechanisms in SARS-CoV-2 infections*. Oxid. Med. Cell. Longevity, 2022. **2022**: 5589089.
41. Williams, V.M., et al., *Human papillomavirus type 16 E6* induces oxidative stress and DNA damage*. J. Virol., 2014. **88**(12): p. 6751–6761.
42. Letafati, A., et al., *Emerging paradigms: unmasking the role of oxidative stress in HPV-induced carcinogenesis*. Infect. Agent Cancer, 2024. **19**(1): p. 30.
43. Jafary, F., et al., *Different aspects in explaining how mutations could affect the binding mechanism of receptor binding domain of SARS-CoV-2 spike protein in interaction with ACE2*. PLoS One, 2023. **18**(9): e0291210.
44. Jacquet, S., et al., *Adaptive duplication and genetic diversification of protein kinase R contribute to the specificity of bat-virus interactions*. Sci. Adv., 2022. **8**(47): eadd7540.
45. Andre, M., et al., *From alpha to omicron: how different variants of concern of the SARS-Coronavirus-2 impacted the world*. Biology (Basel), 2023. **12**(9): p. 1267.
46. Das, K. and Arnold, E., *HIV-1 reverse transcriptase and antiviral drug resistance. Part 1*. Curr. Opin. Virol., 2013. **3**(2): p. 111–118.
47. Duggal, R., et al., *Genetic recombination of poliovirus in a cell-free system*. PNAS, 1997. **94**(25): p. 13786–13791.
48. Guo, Q., et al., *Genetic characterization and molecular evolution of type 3 vaccine-derived polioviruses from an immunodeficient patient in China*. Virus. Res., 2023. **334**: 199177.
49. Pérez-Losada, M., et al., *Recombination in viruses: mechanisms, methods of study, and evolutionary consequences*. Infect., Genet. Evol., 2015. **30**: p. 296–307.
50. Kim, H., Webster, R.G., and Webby, R.J., *Influenza virus: dealing with a drifting and shifting pathogen*. Viral Immunol., 2018. **31**(2): p. 174–183.
51. Miras, M., et al., *Interfamilial recombination between viruses led to acquisition of a novel translation-enhancing RNA element that allows resistance breaking*. New Phytol., 2014. **202**(1): p. 233–246.
52. Zhan, Y., Yin, H., and Yin, J.Y., *B.1.617.2 (Delta) Variant of SARS-CoV-2: features, transmission and potential strategies*. Int. J. Biol. Sci., 2022. **18**(5): p. 1844–1851.
53. Sah, R., et al., *Omicron (B.1.1.529) variant and its subvariants and lineages may lead to another COVID-19 wave in the world? -an overview of current evidence and counteracting strategies*. Int. J. Surg. Open, 2023. **55**: 100625.
54. Washizaki, A., et al., *Neutralization of hepatitis B virus with vaccine-escape mutations by hepatitis B vaccine with large-HBs antigen*. Nat. Commun., 2022. **13**(1): p. 5207.
55. Olabode, A.S., et al., *Revisiting the recombinant history of HIV-1 group M with dynamic network community detection*. PNAS, 2022. **119**(19): e2108815119.
56. Chin, M.P.S., et al., *Molecular determinants of HIV-1 intersubtype recombination potential*. Virology, 2007. **363**(2): p. 437–446.
57. Lycett, S.J., Duchatel, F., and Digard, P., *A brief history of bird flu*. Philos. Trans. R. Soc. Lond. B. Biol. Sci., 2019. **374**(1775): 20180257.

58. Hakim, M.S., et al., *Driving forces of continuing evolution of rotaviruses*. World J. Virol., 2024. **13**(2): p. 93774.
59. Zivcec, M., et al., *Molecular insights into Crimean-Congo hemorrhagic fever virus*. Viruses, 2016. **8**(4): p. 106.
60. Broecker, F. and Moelling, K., *What viruses tell us about evolution and immunity: beyond Darwin?* Ann. N. Y. Acad. Sci., 2019. **1447**(1): p. 53–68.
61. Irwin, N.A.T., et al., *Systematic evaluation of horizontal gene transfer between eukaryotes and viruses*. Nat. Microbiol., 2022. **7**(2): p. 327–336.
62. Nicioli, T. *Ancient viral DNA may play a key role in early human development, new study suggests*. 2025; Available from: `https://www.cnn.com/2025/08/09/science/ancient-viruses-junk-dna-primate-genome`.
63. Olson-Manning, C.F., Wagner, M.R., and Mitchell-Olds, T., *Adaptive evolution: evaluating empirical support for theoretical predictions*. Nat. Rev. Genet., 2012. **13**(12): p. 867–877.
64. Jiang, T., et al., *Envelope domain III E(324), E(351), and E(380) mutations lever adaptive evolution of DENV-1 genotype I*. J. Virol., 2024. **98**(10): e0118324.
65. Hung, S.J. and Huang, S.W., *Contributions of genetic evolution to Zika virus emergence*. Front. Microbiol., 2021. **12**: 655065.
66. Fortuna, C., et al., *Vector competence of Aedes albopictus for the Indian Ocean lineage (IOL) chikungunya viruses of the 2007 and 2017 outbreaks in Italy: a comparison between strains with and without the E1:A226V mutation*. Euro. Surveill., 2018. **23**(22): 1800246.
67. Liu, Y., et al., *The N501Y spike substitution enhances SARS-CoV-2 infection and transmission*. Nature, 2022. **602**(7896): p. 294–299.
68. Carabelli, A.M., et al., *SARS-CoV-2 variant biology: immune escape, transmission and fitness*. Nat. Rev. Microbiol., 2023. **21**(3): p. 162–177.
69. Tenaillon, O. and Matic, I., *The impact of neutral mutations on genome Evolvability*. Curr. Biol., 2020. **30**(10): p. R527–R534.
70. Hu, X., et al., *Classification of a massive number of viral genomes and estimation of time of Most recent common ancestor (tMRCA) of SARS-CoV-2 using Phylodynamic analysis*. Bio. Protoc., 2024. **14**(6): e4955.
71. Holmes, E.C., *Molecular clocks and the puzzle of RNA virus origins*. J. Virol., 2003. **77**(7): p. 3893–3897.
72. Sanjuán, R., et al., *Viral mutation rates*. J. Virol., 2010. **84**(19): p. 9733–9748.
73. Choi, W.J., et al., *A prediction of mutations in infectious viruses using artificial intelligence*. Genomics Inform., 2024. **22**(1): p. 15.
74. de la Casa-Esperón, E., *Horizontal transfer and the evolution of host-pathogen interactions*. Int. J. Evol. Biol., 2012. **2012**: 679045.
75. Gozashti, L. and Corbett-Detig, R., *Double-stranded DNA viruses may serve as vectors for horizontal transfer of intron-generating transposons*. bioRxiv, 2025.
76. Nino Barreat, J.G. and Katzourakis, A., *Deep mining reveals the diversity of endogenous viral elements in vertebrate genomes*. Nat. Microbiol., 2024. **9**(11): p. 3013–3024.
77. Frank, J.A. and Feschotte, C., *Co-option of endogenous viral sequences for host cell function*. Curr. Opin. Virol., 2017. **25**: p. 81–89.
78. Cornelis, G., et al., *Retroviral envelope gene captures and syncytin exaptation for placentation in marsupials*. PNAS, 2015. **112**(5): p. E487–E496.
79. Binnie, A., et al., *CRISPR-based strategies in infectious disease diagnosis and therapy*. Infection, 2021. **49**(3): p. 377–385.
80. Kaminski, M.M., et al., *CRISPR-based diagnostics*. Nat. Biomed. Eng., 2021. **5**(7): p. 643–656.

81. Gostimskaya, I., *CRISPR-Cas9: a history of its discovery and ethical considerations of its use in genome editing*. Biochemistry (Mosc), 2022. **87**(8): p. 777–788.
82. Xue, C. and Greene, E.C., *DNA repair pathway choices in CRISPR-Cas9-mediated genome editing*. Trends Genet., 2021. **37**(7): p. 639–656.
83. Geisinger, J.M., et al., *In vivo blunt-end cloning through CRISPR/Cas9-facilitated non-homologous end-joining*. Nucleic Acids Res., 2016. **44**(8): p. e76–e76.
84. Liao, H., et al., *CRISPR-Cas9-mediated homology-directed repair for precise gene editing*. Mol. Ther. Nucleic Acids, 2024. **35**(4): 102344.
85. Myhrvold, C., et al., *Field-deployable viral diagnostics using CRISPR-Cas13*. Science, 2018. **360**(6387): p. 444–448.
86. Joung, J., et al., *Point-of-care testing for COVID-19 using SHERLOCK diagnostics*. medRxiv, 2020.
87. Gong, J., et al., *A simple and rapid diagnostic method for 13 types of high-risk human papillomavirus (HR-HPV) detection using CRISPR-Cas12a technology*. Sci. Rep., 2021. **11**(1): p. 12800.
88. He, C., et al., *Application of CRISPR-Cas system in human papillomavirus detection using biosensor devices and point-of-care technologies*. BME Front., 2025. **6**: p. 0114.
89. Qian, X., et al., *CRISPR for companion diagnostics in low-resource settings*. Lab Chip, 2024. **24**(20): p. 4717–4740.
90. Rajan, A., et al., *CRISPR-Cas system: from diagnostic tool to potential antiviral treatment*. Appl. Microbiol. Biotechnol., 2022. **106**(18): p. 5863–5877.
91. White, R.A. 3rd, *The future of virology is synthetic*. mSystems, 2021. **6**(4): 101128msystems0077021.
92. Guenther, C.M., et al., *Synthetic virology: engineering viruses for gene delivery*. Wiley Interdiscip. Rev. Nanomed. Nanobiotechnol., 2014. **6**(6): p. 548–558.
93. Ratti, M., et al., *MicroRNAs (miRNAs) and long non-coding RNAs (lncRNAs) as new tools for Cancer therapy: first steps from bench to bedside*. Targeted Oncol., 2020. **15**(3): p. 261–278.
94. Wu, Q., et al., *The RNase III enzyme DROSHA is essential for microRNA production and spermatogenesis*. J. Biol. Chem., 2012. **287**(30): p. 25173–25190.
95. Hussain, M., et al., *West Nile virus encodes a microRNA-like small RNA in the 3′ untranslated region which up-regulates GATA4 mRNA and facilitates virus replication in mosquito cells*. Nucleic Acids Res., 2011. **40**(5): p. 2210–2223.
96. Berkhout, B. and Haasnoot, J., *The interplay between virus infection and the cellular RNA interference machinery*. FEBS Lett., 2006. **580**(12): p. 2896–2902.
97. Zheng, Z., et al., *Human microRNA hsa-miR-296-5p suppresses enterovirus 71 replication by targeting the viral genome*. J. Virol., 2013. **87**(10): p. 5645–5656.
98. Wu, N., et al., *miR-223 inhibits dengue virus replication by negatively regulating the microtubule-destabilizing protein STMN1 in EAhy926 cells*. Microbes Infect., 2014. **16**(11): p. 911–922.
99. Robertson, K.A., et al., *An interferon regulated MicroRNA provides broad cell-intrinsic antiviral immunity through multihit host-directed targeting of the sterol pathway*. PLoS Biol., 2016. **14**(3): e1002364.
100. Lin, X., et al., *Human microRNA-30 inhibits influenza virus infection by suppressing the expression of SOCS1, SOCS3, and NEDD4*. Cell. Microbiol., 2020. **22**(5): e13150.
101. Pedersen, I.M., et al., *Interferon modulation of cellular microRNAs as an antiviral mechanism*. Nature, 2007. **449**(7164): p. 919–922.

102. Zhao, S., Kong, X., and Wu, X., *RNAi-based immunity in insects against baculoviruses and the strategies of baculoviruses involved in siRNA and miRNA pathways to weaken the defense*. Dev. Comp. Immunol., 2021. **122**: 104116.
103. Qiu, Y., et al., *Human virus-derived small RNAs can confer antiviral immunity in mammals*. Immunity, 2017. **46**(6): p. 992–1004.e5.
104. Ketzinel-Gilad, M., Shaul, Y., and Galun, E., *RNA interference for antiviral therapy*. J. Gene. Med., 2006. **8**(8): p. 933–950.

5 Immune Response to Viral Diseases

The immune response to viral diseases is a fascinating and multifaceted area of study that lies at the intersection of virology and immunology. Viral infections have been a significant driver of scientific inquiry, pushing the boundaries of our understanding of how the human body's defenses function to protect us against these microscopic invaders. The understanding of viruses and the immune system has also paved the way for groundbreaking discoveries and led to the development of life-saving vaccines, antiviral drugs, and a deeper comprehension of our body's ability to adapt and remember viral encounters.

From the moment a virus enters the body, our immune system deploys a rapid and precise defense strategy. Innate immune responses act as the initial, nonspecific line of defense. Following this, the immune cells, which are part of the adaptive immune system, then detect the presence of viruses through a series of pattern recognition receptors (PRRs).

The study of immune response to viral diseases also reveals the importance of maintaining the balance between host protection and potential harm. In some cases, the immune system's exuberance in fighting viruses can lead to immunopathology. This is a situation in which inflammation and immune responses cause tissue damage and clinical symptoms. Therefore, understanding this dichotomy of immune responses is not only vital for managing viral infections but has broader implications for understanding autoimmune diseases and the fine-tuning of the immune system. Generally, immune responses can be divided into innate and adaptive immune responses.

5.1 Innate Immunity

Innate immunity provides the first line of defense against viral infections and consists of various defense mechanisms that are "immediately available" and not specific to any pathogen, including viruses. That is, it is present at birth and provides an immediate, but nonspecific, defense against pathogens.

Mechanisms of innate immunity against viral infections

5.1.1 Physical Barriers

5.1.1.1 Skin

The skin acts as a physical barrier, preventing virus entry [1]. The outer layer of the skin, known as the epidermis, forms a tough physical barrier that viruses must breach to enter the body. Healthy skin is generally impermeable to most viruses, preventing their entry into deeper tissues where they could infect cells and replicate [2, 3].

However, viruses can typically enter the body through contact with broken skin or through other mucosal openings such as the mouth, eyes, or nose. For example, herpes simplex virus (HSV), Ebola virus, hepatitis B virus (HBV), and papillomaviruses (strains that cause oral, genital herpes, and cervical cancer) typically enter the body through mucous membranes or breaks in the skin [4–6]. Intact, healthy skin on areas that are not affected by lesions may provide a barrier that reduces the risk of viral entry. Therefore, it is important to wash the skin thoroughly when contact with a virus is suspected and to follow necessary biohazard procedures.

5.1.1.2 Mucous Membranes

The mucous membrane lines the surface of the respiratory, gastrointestinal (GI), and urogenital tracts to trap viruses. These membranes could trap viruses and other pathogens. Cilia, hair-like structures on the surface of cells in the respiratory tract, also help to sweep away trapped viruses, preventing them from reaching susceptible cells [7–9].

However, mucous membranes can also be a route of infection for respiratory viruses such as that primarily infects the respiratory tract through contact with mucous membranes in the nose, mouth, and throat; sexually transmitted infections such as human immunodeficiency virus (HIV) that infect the body through mucous membranes in the genital, anal, and oral areas during sexual contact or through exposure to contaminated blood; HSV that enters the body through mucous membranes of the mouth, lips, and genital areas; and viral hemorrhagic viruses like Marburg virus, Lassa, and Ebola virus that can be transmitted through direct contact with blood, body fluids, or mucous membranes of infected individuals [10].

5.1.2 Chemical Barriers

5.1.2.1 Acidic pH of the Stomach

The normal pH of gastric acid in adults is generally between 1.5 and 3.5 [11]. This level can inhibit virus replication. For example, a study showed that Middle East respiratory syndrome coronavirus (MERS-CoV) treatment with a proton pump inhibitor in an animal model resulted in an exaggerated infection in the small intestine [12]. This suggests that MERS-CoV treatment with a proton pump inhibitor in an animal model resulted in an exaggerated infection in the small intestine [12], playing a critical role in limiting the infection and spread of the virus.

However, the presence of gastric acid pH may not be sufficient to significantly decrease the viral load or prevent the spread of the virus to other organs. Noroviruses and enteric adenoviruses can exploit the compromised acidic barrier to establish infections more effectively [13]. Unlike rhinoviruses, non-rhinovirus species of picornaviruses are acid-stable, able to survive at pHs below 3, and remain infective [14]. Some enteroviruses may interact with sialic acid receptors to stabilize the viral capsid, allowing it to remain intact at higher pH levels than it would without receptor binding. Yet, many viruses are sensitive to the distinctively low pH found in gastric juice.

Patients with hypochlorhydria, a condition characterized by reduced stomach acidity, might be more susceptible to viral infections due to the decreased protective acidic barrier [15]. This lowered acidity could potentially allow viruses to survive longer and penetrate deeper into the GI tract, increasing the risk of infection.

5.1.2.2 Tears and Saliva

Saliva induces both antimicrobial and antiviral activities [16]. It contains a variety of enzymes, antibodies, and proteins that can neutralize pathogens. Specifically, enzymes like lysozyme can inhibit enveloped viruses, while salivary antibodies (such as IgA) can bind to and neutralize some viruses by preventing them from attaching to and entering host cells.

In addition, another significant antiviral component in the saliva is the presence of lactoferrin, defensins, histatins, sialic acid-containing mucin 5, salivary scavenger receptor cysteine-rich glycoprotein-340 (gp-340), alpha-2-microglobulin (A2M) (which functions as a broad-spectrum protease inhibitor), and C-type lectins (surfactant protein D) [17].

Specifically, lectins bind to carbohydrate structures on the surfaces of pathogens, including viruses [18]. This binding can neutralize viruses by preventing their attachment to host cells and enhancing their clearance by immune cells through opsonization. Also, gp-340 and A2M act as classic γ inhibitors that prevent influenza A virus hemagglutinin (HA) from attaching to the target cell [19]. They serve as a sialic acid ligand that binds to the viral HA. Surfactant protein D acts as a classic β inhibitor in influenza A virus infection. Similarly, defensins, small cationic peptides, can penetrate viral envelope (membranes) and potentially neutralize viral infection [20]. Histatins, another class of salivary proteins, possess antiviral properties, although their mechanisms are not as well characterized [21].

5.1.3 Cell-Mediated Innate Immune Responses

Natural killer (NK) cells: NK cells are large granular lymphocytes that are part of the innate immune system. They provide a rapid, nonspecific response to infected or transformed cells. Unlike adaptive (specialized) immune cells that may take days to be fully activated, they act within hours of infection, and their activities peak and decline soon after the appearance of cytotoxic T-cells (3–7 days). NK cells can directly kill virus-infected cells by releasing cytotoxic granules containing perforin and granzymes [22]. Perforin forms pores in the target cell membrane, allowing granzymes to enter and induce apoptosis (programmed cell death).

5.1.3.1 NK Recognition of Virus-Infected Cells

- **Reduced expression of major histocompatibility complex (MHC):** Virus-infected cells often have reduced expression of MHC Class I molecules, which are normally recognized by inhibitory receptors on NK cells [23]. This mechanism allows NK cells to detect and target virus-infected cells that may evade detection by cytotoxic T lymphocyte cells (CTLs) due to their altered MHC Class I expression.
- **Expression of stress-induced ligands:** Infected cells also often upregulate stress-induced molecules or ligands on their surface, such as MHC Class I chain-related proteins A and B (MICA and MICB), and UL16 binding proteins (in humans) [24]. MICA and MICB are considered "stress-induced molecules" or ligands [25], expressed on the surface of cells experiencing stress like viral infection, damage, or transformation. These ligands bind to activating natural killer group 2D (NKG2D) receptors on NK cells, leading to NK cell activation and cytotoxicity.
- **Expression of viral proteins:** Some viruses encode proteins that directly interact with NK cell receptors [26]. For example, certain herpesviruses produce proteins that interact with NKG2D ligands, modulating NK cell recognition [27].
- **Antibody-dependent cellular cytotoxicity (ADCC):** NK cells express Fc receptors (FcγRIIIa) that bind to the Fc portion of antibodies or immunoglobulin (Ig) bound to virus-infected cells [28]. This triggers NK cell activation and targeted killing of the antibody-coated infected cells.
- **Cytokines:** NK cells can sense the cytokine milieu in the microenvironment. Inflammatory cytokines, such as interferons (IFNs), produced during viral infections enhance NK cell cytotoxicity and cytokine secretion, amplifying their antiviral response [29, 30].

Surface markers for laboratory identification of NK cells

NK cells are identified by a distinct set of cell surface markers that distinguish them from other immune cells. These markers include:

- **CD 56:** NK cells are typically characterized by a lack of cluster of differentiation (CD)3 and the expression of CD56 [31], also known as neural cell adhesion molecule (NCAM). There are two isoforms of CD56, which are $CD56^{dim}$ and $CD56^{bright}$, each with different functional properties [32].
- **CD16:** NK cells also express CD16, also known as Fc epsilon receptor I gamma chain (FcεR1γ) or CD3ζ [33]. CD16 is important for mediating ADCC and facilitating the recognition and destruction of antibody-coated target cells.

- **CD57:** CD57 is a marker of NK cell maturation and differentiation [34]. $CD57^+$ NK cells are generally more cytotoxic and exhibit enhanced effector functions compared to $CD57^-$ NK cells.
- **NKp30 (NCR3) and NKp44 (NCR2):** Natural killer cell p30-related protein (NKp30)/natural cytotoxicity receptor 3 (NCR3) are additional natural cytotoxicity receptors expressed by subsets of NK cells, contributing to their recognition and activation [35].
- **NKp46 (NCR1):** NKp46 is a natural cytotoxicity receptor (NCR) expressed by all human NK cells [36]. It plays a crucial role in NK cell-mediated cytotoxicity against infected or transformed cells.
- **DNAx accessory molecule-1 (CD226):** DNAx accessory molecule-1 is a co-stimulatory receptor expressed by NK cells that interacts with its ligands [37], such as poliovirus receptor (CD155) and Nectin-2 (CD112), on target cells to promote NK cell activation and cytotoxicity [38].

5.1.3.2 Macrophages

Macrophages are phagocytic cells that originate from monocytes, a type of white blood cell (WBC) in the peripheral blood [39], which differentiate into macrophages when they migrate into tissues. Macrophages are proficient at recognizing, engulfing, and digesting foreign particles, including virus particles or infected cells, through a process known as phagocytosis [40]. Beyond their phagocytic capabilities, macrophages are also key players in cytokine production, initiating inflammatory response, tissue repair, and antigen presentation, where they display viral antigens on their surface to activate T-cells, thus bridging the gap between innate and adaptive immunity.

Full blood count (FBC) or complete blood count (CBC) during viral infection: A district laboratory may perform an FBC or CBC, which may show several characteristic changes depending on the stage of viral infection. Usually, during the acute phase of viral infection, the body's immune response to the infection may be reflected in FBC results.

Viral infections can predispose individuals to secondary bacterial infections. FBC may also help to detect secondary infections due to existing viral infections. Monitoring the WBC count and differential can help identify increases in neutrophils, which are indicative of bacterial infections.

Interpretation of total white blood cell (WBC) count during infection: A decrease or normal total WBC count can occur, particularly in the early stages of infection [41, 42]. However, an increase can be seen. This increase is primarily due to elevated levels of lymphocyte. Lymphocytes primarily target infection of viruses. Mild elevation in neutrophils, first responders to bacterial infections, can also be seen during viral infections (differential count, but not usually total leucocyte count).

Differential WBC count

Neutrophilia: Elevated levels of neutrophils, which are involved in phagocytosis and the initial response to infection, may be observed, especially with bacterial co-infection.

Lymphopenia: A decrease in lymphocyte count, particularly T-cells and NK cells, which are crucial in the adaptive immune response against viruses. This could be seen in chronic viral infection.

Monocytosis: An increase in monocytes, which are involved in phagocytosis and antigen presentation.

Platelet count: Platelet levels may be normal or decreased (thrombocytopenia) during viral infections [43, 44]. Significant changes may occur in severe cases or specific viral hemorrhagic fevers.

Red blood cell (RBC) indices: Generally, RBC indices (mean corpuscular volume, mean corpuscular hemoglobin, and mean corpuscular hemoglobin concentration) remain largely within normal ranges during viral infections unless there are underlying complications such as anemia due to chronic illness or secondary bacterial infections.

Erythrocyte sedimentation rate (ESR): May be elevated, indicating inflammation, but not as compared to bacterial infection.

5.1.3.3 Process of Phagocytosis During Viral Infection

Phagocytosis is a critical process in the immune response to viral infections, involving several steps that allow macrophages to engulf and destroy virus-infected cells and free virus particles (see Table 5.1 and Figure 5.1). The process of phagocytosis is described below:

Recognition Phagocytic cells recognize virus-infected cells or free virus particles through PRRs on their surface [45]. These receptors identify pathogen-associated molecular patterns (PAMPs) – molecular structures or patterns found on pathogens. Examples of PRR are toll-like receptor (TLR) and retinoic acid inducible gene 1 (RIG-I)-like receptors (RLRs). TLR3 is expressed on the surface of some phagocytotic cells, such as dendritic cells (endosomal compartment) and macrophages (both surface and endosomal compartment). In contrast, RLRs are present in the cytoplasm of cells involved in innate immune responses. Epithelial and endothelial cells, found on the surface of some organs and blood vessels, also express RLRs [46]. This helps to detect viral infection and to prevent the spread of viruses throughout the body. Examples of virus-like structures (i.e., PAMPs) that are recognized by PRRs include:

- **Viral RNA:** Double-stranded RNA (dsRNA), which is typically not found in mammalian cells under normal conditions, serves as a PAMP [47]. It can be recognized by PRRs such as toll-like receptor 3 (TLR3) and RLRs [48]. TLR3 detects dsRNA, whereas RIG-I preferentially recognizes single-stranded RNA (ssRNA) over dsRNA [49].
- **Viral DNA:** Unmethylated CpG motifs in viral DNA are recognized by TLR9 [50]. Once recognized, this can activate innate immune responses. Examples of innate immune cells that express TLR9 are plasmacytoid

Table 5.1 Examples of other cells capable of phagocytosis during infection.

Cell type	Origin	Function
Neutrophils	White blood cells and the most abundant type of granulocytes in the blood	Effective at phagocytosing bacteria, fungi, and other pathogens, including virus-infected cells.
Dendritic cells	Originate from hematopoietic stem cells (HSCs) in the bone marrow.	1. Conventional dendritic cells (cDCs): These cells are further divided into cDC1 and cDC2 subsets. cDC1 cells are primarily responsible for cross-presentation and are important for antiviral responses, whereas cDC2 cells are involved in presenting antigens to helper T-cells. 2. Plasmacytoid dendritic cells (pDCs): They are primarily involved in producing large amounts of type I IFN in response to viral infections. 3. Monocyte-derived dendritic cells (mo-DCs): Inflammatory conditions can drive monocytes to differentiate into dendritic cells, which play roles in both innate and adaptive immune responses.
Eosinophils	Eosinophils are a type of granulocyte in the peripheral blood.	Involved in combating multicellular parasites and certain infections. They can perform phagocytosis, although they are less efficient at it compared to neutrophils and macrophages.
Mast cells	Derived from progenitor cells that leave the bone marrow and circulate in the blood as immature precursors, which then mature in tissues.	Mast cells are like basophils but are found in tissues rather than blood. They can phagocytose pathogens and release granules containing histamine and other inflammatory mediators.

dendritic cells, macrophages, monocytes, and neutrophils.

- **Viral glycoproteins:** Certain glycoproteins on the surface of viruses, such as HA and neuraminidase (NA) on influenza viruses, act as PAMPs [51]. They are recognized by dendritic cell-specific intercellular adhesion molecule-3-grabbing non-integrin (DC-SIGN) and TLRs, triggering immune responses.
- **Capsid and envelope proteins:** Components of the viral capsid can also act as PAMPs. For instance, the capsid proteins of some viruses contain motifs that can be recognized by host cell PRRs, leading to the initiation of antiviral immune responses [52].

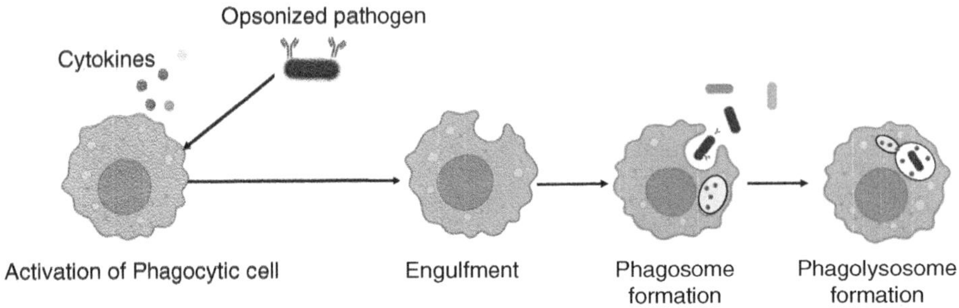

Figure 5.1 Systematic illustration of the progress of phagocytosis.

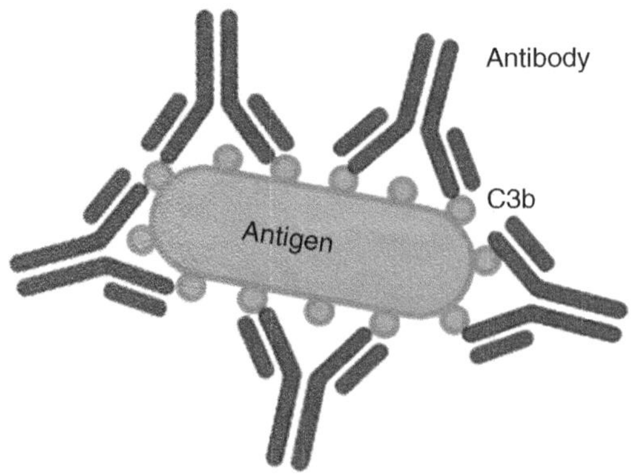

Figure 5.2 Illustration of an opsonized antigen.

Opsonization Opsonization is a process whereby pathogens such as bacteria or viruses are identified and rendered for ingestion and elimination by phagocytic. This process involves the binding of opsonin molecules to the surface of pathogens and enhance the efficiency of phagocytosis by facilitating the recognition and binding of pathogens to phagocytic cells (see Figure 5.2). The main opsonins involved in this process are antibodies (IgG and IgM) and complement proteins (C3b).

Engulfment Once bound to the virus or infected cell, the macrophage extends its plasma membrane around the target, forming pseudopods – temporary protrusions that surround and engulf the target.

Phagosome Formation The phagosome contains pseudopods (long projections). As the pseudopods continue to surround the target, they eventually close in and fuse, forming a membrane-bound compartment called a phagosome. The phagosome now contains the virus or the infected cell.

Phagolysosome Formation The phagosome containing the virus or infected cell fuses with a lysosome, an organelle containing digestive enzymes. This fusion creates a phagolysosome, where the pathogen is exposed to a variety of enzymes, reactive oxygen species (ROS), or reactive nitrogen species (RNS) that degrade and destroy it.

Antigen Presentation After digestion, macrophages process the viral antigens and present them on their surface using MHC Class II molecules (primarily ligand-presenting molecules). This antigen presentation is crucial for the activation of T helper cells, which play a key role in orchestrating the adaptive immune response.

Release of Cytokines Macrophages and other phagocytic cells are capable of releasing cytokines and chemokines in response to viral infection after phagocytosis. This signals other immune cells to the site of infection and enhances the inflammatory response and other immune responses.

5.1.4 Cytokine-Mediated Innate Immune Response

Cytokines are a broad category of small biomolecules, secreted by immune cells and other cells in the body. They act as signaling molecules that regulate immune cell communication and effector functions. Key cytokines involved in viral infections are:

5.1.4.1 Interferons

IFNs are vital antiviral cytokines that inhibit viral replication within infected cells and enhance the expression of proteins involved in viral resistance. They also play roles in activating immune cells and modulating adaptive immune responses. They are classified into three main types based on their interactions with receptor subunits, peptide mapping, and sequencing homology: type I, type II, and type III IFNs.

Classification of IFN

Type I IFN The type I IFN system includes a family of IFN proteins encoded by multiple genes [53]. In humans, this family consists of the following:

- **Thirteen IFNα (IFNA) subtypes:** These are encoded by genes such as *IFNA1, IFNA2, IFNA4, IFNA5, IFNA6, IFNA7, IFNA8, IFNA10, IFNA13, IFNA14, IFNA16, IFNA17,* and *IFNA21* [54]. The IFNα subtypes have a high degree of sequence homology, with 76–96% of their amino acid sequences shared.
- **One IFNβ (IFNB) gene:** Encoded by the *IFNβ* gene.
- One IFNε (epsilon) gene
- One IFNκ (kappa) gene
- One IFNω (omega) gene

These IFNs bind to a common, shared receptor complex composed of two chains (sub-units), IFN alpha receptor 1 (IFNAR1) and IFNAR2. The binding of type I IFNs to this receptor initiates a signaling cascade that activates the expression of numerous IFN-stimulated genes (ISGs), which play crucial roles in antiviral defense mechanisms (see Table 5.2).

Type II IFN Type II IFN consists of a single member, IFNγ (gamma), which is structurally and functionally distinct from type I IFNs. IFNγ is produced mainly by activated T-cells (particularly $CD4^+$, Th cells and $CD8^+$, cytotoxic T-cells) and NK cells [55]. It binds to a different receptor

Table 5.2 Classification of IFN and mechanism of antiviral effects.

Classification	Mechanism of antiviral effect	Examples
Type I interferons (IFNα and IFNβ):	Type I IFNs are produced by a wide variety of cells in response to viral infections. They bind to the IFNα/β receptor on neighboring cells, triggering a signaling cascade that leads to the expression of hundreds of interferon-stimulated genes (ISGs). These ISGs encode proteins that interfere with various stages of the viral life cycle, such as viral entry, replication, assembly, and release.	During an influenza virus infection, epithelial cells and immune cells rapidly produce IFNα and IFNβ, which activate antiviral defenses in both the infected cells and surrounding uninfected cells, limiting viral spread.
Type II interferon (IFNγ)	IFNγ is primarily produced by natural killer (NK) cells and T-cells. It enhances the antiviral activity of macrophages and promotes the adaptive immune response. IFNγ also upregulates the expression of major histocompatibility complex (MHC) molecules, enhancing antigen presentation to T-cells.	In HBV infection, IFNγ produced by NK cells and cytotoxic T lymphocytes (CTLs) helps control viral replication by activating macrophages and promoting the clearance of infected hepatocytes.
Type III IFN (IFNλ)	Type III IFNs, also known as lambda IFNs, bind to a distinct receptor complex expressed primarily on epithelial cells and certain immune cells. They induce a similar set of ISGs as Type I IFN but with more localized effects, particularly at mucosal surfaces.	In respiratory syncytial virus (RSV) infection, IFNλ is produced by infected epithelial cells and acts on neighboring epithelial cells to induce an antiviral state, helping to contain the infection at the site of entry. May also induce an antiviral state in hepatocytes.

complex consisting of IFN gamma receptor 1 (IFNGR1) and IFNGR2. IFNγ plays a critical role in immune regulation and activation, enhancing the antigen-presenting capabilities of macrophages and other cells, and promoting the adaptive immune response against pathogens.

Type III IFN Type III IFNs, also known as IFNλ (lambda), include IFNλ1, IFNλ2, IFNλ3, and IFNλ4 [56]. These cytokines are relatively newly identified and have IFN-like activities. They interact with a receptor complex composed of interleukin-10 receptor subunit 2 (IL10R2) and IFNLR1 complex [57]. Although type III IFNs activate a similar set of ISGs as type I IFNs, they predominantly act on epithelial cells and play a crucial role in mucosal immunity, particularly in the respiratory and GI tracts [58].

Overview of the IFN Signaling Pathway

1. **Interferon production:** When a cell is infected by a virus or other pathogen, PRRs such as TLRs, RIG-I-like receptors (RLRs), or cyclic GMP-AMP synthase - Stimulator of Interferon Genes (cGAS-STING) pathways recognize viral nucleic acids (DNA or RNA).
   These PRRs trigger downstream signaling cascades that activate transcription factors, mainly Interferon Regulatory Factor 3 (IRF3) and Nuclear Factor kappa-light-chain-enhancer of activated B cells (NF-κB), which move into the nucleus to induce the production of IFN [59].
2. **Interferon release:** The IFNs produced by the infected cell are secreted into the extracellular space. These IFNs then bind to their respective receptors on the surface of neighboring cells or the same cell (autocrine signaling) to trigger an antiviral state.
3. **Interferon receptors and signaling:** IFNs bind to the receptor on target cells. IFN binding to its receptor triggers the activation of Janus kinases (JAKs), which phosphorylate the receptors, signal transducers, and activators of transcription (STAT) proteins [60].
   - For Type I and Type III IFNs, JAK1 and TYK2 are activated, leading to the phosphorylation of STAT1 and STAT2 [61].
   - For Type II IFN, JAK1, and JAK2 phosphorylate STAT1.
4. **Formation of transcription complexes:** In type I and III IFNs, phosphorylated STAT1 and STAT2 combine with IRF9 to form the ISG factor 3 complex [62], which translocates to the nucleus. In type II IFN, phosphorylated STAT1 forms homodimers that move into the nucleus to activate gene transcription [63].
5. **Induction of IFN-stimulated genes:** In the nucleus, ISGF3 or STAT1 binds to specific DNA sequences known as interferon-stimulated response elements (ISREs). This activates the transcription of hundreds of ISGs, which produce proteins that:
   - Inhibit viral replication (e.g., protein kinase R, oligoadenylate synthetase [OAS], and Mx proteins)
   - Enhance antigen presentation (e.g., MHC Class I)
   - Activate immune cells (e.g., NK cells)
   - Promote apoptosis in infected cells
6. **Negative regulation and homeostasis:** While signaling is critical for defense, it is tightly regulated to prevent chronic inflammation or autoimmunity.
   - Suppressors of cytokine signaling (SOCS) proteins and protein tyrosine phosphatases (PTPs) help to deactivate the JAK–STAT [64].
   - ISG15 and USP18 also provide feedback inhibition, ensuring the response is balanced.

Specific Antiviral Function of IFN

Inhibition of Viral Protein Synthesis IFNs induce the expression of protein kinase R (PKR), which phosphorylates the eukaryotic translation initiation factor 2 (eIF2α), leading to the inhibition of viral protein synthesis [65], which is crucial for preventing viral replication and spread. This process occurs through a cascade of molecular mechanisms such as:

- **Activation of protein kinase R:** PKR (contains dsRNA-binding domain) is activated upon binding to dsRNA, which is often produced during viral replication [66]. In turn, activated PKR enhances IFNα/β expression.
- **Phosphorylation of eIF2α:** Activated PKR phosphorylates the eIF2α [67]. Phosphorylated eIF2α inhibits the formation of the eIF2-GTP-tRNA complex, which is necessary for the initiation of translation [68].
- **Global translation shutdown:** This phosphorylation results in the shutdown of both viral and host protein synthesis, significantly reducing the production of viral proteins and thus limiting viral replication.

Inhibition of Virus Entry IFN induces the upregulation of interferon-induced transmembrane proteins (IFITMs) [69]. IFITM proteins are capable of inhibiting virus entry and replication by blocking the fusion of viral membranes with cellular membranes [70]. By preventing virus entry, they reduce the number of viruses that can start protein synthesis in the first place.

Degradation of Viral RNA IFNs stimulate the production of endoribonuclease RNase L [71], which degrades viral RNA, preventing viral replication [72].

Mechanisms of activation of RNase L pathway include

- **Upregulation of OAS:** Upon recognition of viral dsRNA, a common feature of many viral genomes or replication intermediates, IFNs induce the expression of OAS enzymes [73, 74]. OAS is an ISG that plays a key role in detecting viral RNA.
- **Production of 2′-5′ oligoadenylates (2-5A):** OAS converts ATP into 2′-5′ linked oligoadenylates (2-5A) in the presence of viral dsRNA [75]. These 2-5A molecules act as activators of RNase L by triggering the enzyme dimerization.
- **Activation of RNase L:** 2-5A binds to and activates RNase L, an endoribonuclease that cleaves both viral and cellular RNA [76]. This RNA degradation halts viral protein synthesis and replication, leading to the inhibition of viral spread.

Indirect RNA Degradation and Upregulation of IFN-Induced RNase Activity Although PKR's primary role is to inhibit translation, the stress response it triggers can lead to the activation of RNase L, contributing to RNA degradation. Moreover, the inhibition of protein synthesis prevents viral RNA from being translated into proteins, indirectly contributing to the degradation and turnover of viral RNA.

Activation of ISG20 Exonuclease IFNs stimulate the expression of ISG20, a 3′-5′ exonuclease that degrades single-stranded RNA. This enzyme directly targets viral RNA for degradation, reducing the amount of viral RNA available for replication and translation [77].

Upregulation of microRNA (miRNA) Activities IFNs modulate the expression of certain host miRNAs that specifically target viral RNA [78]. These miRNAs bind to complementary sequences in the viral RNA, marking them susceptible to degradation by the RNA-induced silencing complex (RISC).

Blocking Viral Assembly and Release IFNs can bind to cell receptors on both infected and neighboring uninfected cells, activating the JAK–STAT signaling pathway [79]. This leads to the transcription of ISGs that encode various antiviral proteins. Also, IFN induces ISGs which modulates the expression of hundreds more IFNs, many of which have antiviral functions that block viral entry, transcription, and replication.

Activates Other Immune Cells IFNs increase the expression of MHC Class I and II molecules, improving the ability of immune cells to recognize and eliminate infected cells [80].

Inducing Apoptosis of Infected Cells IFNs can induce apoptosis in infected cells, thereby preventing the production and release of viral particles. Proteins such as PKR, which are induced by IFNs, can lead to the shutdown of protein synthesis in infected cells, inducing apoptosis.

Clinical Use of IFN

- **Antiviral therapies:** Recombinant IFN (e.g., IFNα) is used in the treatment of chronic viral infections such as hepatitis B and C [81]. Pegylated IFNα is used in combination with antiviral drugs to treat chronic HCV and HBV infections, reducing viral load and improving liver function.
- **Autoimmune diseases:** Dysregulation of IFN signaling is associated with autoimmune diseases like systemic lupus erythematosus (SLE), where excessive Type I IFN production can drive immune-mediated tissue damage [82]. IFNβ is used as a treatment to reduce the frequency and severity of relapses in patients with multiple sclerosis, an autoimmune disease where viral infections are thought to play a role in triggering immune responses against myelin.
- **Cancer therapy:** IFNs are also used in cancer immunotherapy due to their ability to stimulate immune responses against tumors [83, 84].

5.1.4.2 Interleukins

Interleukins are a group of cytokines that act as messengers, signaling between cells, especially WBCs, in the immune system. Interleukins are produced by various cells, including WBCs (leukocytes), fibroblast, and epithelial cells. They also play roles in activating immune cells and regulating adaptive immune responses. Different ILs have diverse functions in viral infections, such as stimulating T-cell proliferation and differentiation (e.g., IL-2, IL-12), promoting B-cell antibody production (e.g., IL-4), and controlling inflammation (e.g., IL-6, IL-10) [85, 86].

Clinical importance

Understanding the roles of different interleukins during viral infections has significant clinical implications, which include:

- **Diagnostic biomarkers:** Elevated levels of certain interleukins can serve as biomarkers for the severity of infection and the inflammatory response.

- **Therapeutic targets:** Drugs that modulate interleukin activity can aid in treating viral infections and related inflammatory conditions. For example, blocking IL-6 and IL-10 has been studied in the treatment of severe coronavirus disease 2019 (COVID-19) [87] and chronic HBV infection [88], respectively. Meanwhile, IL-12 has been examined for its potential to strengthen the immune response against HBV. CTLs and NK cells may be a therapeutic option for treating viral infections in the future [89].

5.1.4.3 Tumor Necrosis Factors (TNFs)

TNFs contribute to inflammation and immune cell activation during viral infections [90]. They can induce apoptosis of infected cells and promote the production of other cytokines and chemokines. TNFα is a key cytokine in the immune response to viral infections. While it plays a crucial role in controlling viral replication and initiating immune responses, its dysregulation can lead to severe inflammation, tissue damage, and disease complications [90, 91]. Understanding the balance of TNFα activity is important for developing therapeutic strategies to manage viral infections and their associated inflammatory responses. Other functions include:

- **Induction of fever and systemic symptoms:** An elevated body temperature can inhibit viral replication and enhance the immune response. TNFα induces fever and other systemic symptoms such as malaise and fatigue, which are common during viral infections.
- **Enhances vascular permeability:** TNFα is a major pro-inflammatory cytokine [92]. It increases the permeability of blood vessels at the intracellular junction, which can lead to edema and hemorrhage [93]. This is particularly significant in infections like dengue virus and Ebola virus, where vascular leakage can be life-threatening [94].
- **Tissue damage:** While TNFα is essential for controlling infections, excessive production can lead to tissue damage and organ dysfunction [95, 96]. This is evident in severe cases of Ebola, yellow fever, and severe influenza, where TNFα contributes to multi-organ failure.
- **Immune modulation:** TNFα modulates the immune response by influencing the activity of various immune cells, including macrophages, T-cells, and NK cells. It helps coordinate the body's defense against viruses but can also contribute to immune system dysregulation in chronic infections, such as HIV.

Balance cytokine interaction

Cytokines can be either pro-inflammatory or anti-inflammatory, which enables them to maintain the balance between immune activation and suppression. Therefore, homeostasis of this important immune response system is essential for preventing excessive tissue damage and chronic inflammation while effectively combating infections.

In the context of viral infection, the body typically mounts a strong pro-inflammatory response to eliminate the virus. However, if this response is too robust or prolonged, it can cause tissue damage and contribute to severe disease outcomes, such as those seen in severe cases of COVID-19. Anti-inflammatory cytokines are then crucial in downregulating the immune response to prevent damage to the host's tissues and to facilitate recovery.

Pro-inflammatory cytokines
Pro-inflammatory cytokines promote inflammation. They help recruit immune cells to sites of infection, enhance the immune cells' ability to kill pathogens, and stimulate the production of other inflammatory mediators. Major pro-inflammatory cytokines are listed in Table 5.3.

Anti-inflammatory cytokines
Anti-inflammatory cytokines counteract the effects of pro-inflammatory cytokines, helping to resolve inflammation and promote tissue repair. They play a crucial role in preventing chronic inflammation and autoimmune diseases. Key examples of anti-inflammatory cytokines are listed in Table 5.4. Some of the main interleukins and their roles during viral infection are listed in Tables 5.5 and 5.6.

Fever
Fever is a common physiological response to infection and inflammation. TNFα triggers prostaglandin production, a significant contributor to fever (an endogenous pyrogen) [97–99], other cytokines, such as IL-1, IL-6, IFNα, and IFNγ, also play vital roles in inducing and regulating fever [100]. The interaction

Table 5.3 Examples of pro-inflammatory cytokines.

Cytokine	Produced by	Functions
Interleukin-1 (IL-1)	Macrophages, other cells	Induces fever, promotes the production of other pro-inflammatory cytokines, and enhances the adhesion of leukocytes to blood vessel walls.
Tumor necrosis factor-alpha (TNFα)	Macrophages, T-cells	Promotes inflammation, activation, and recruitment of immune cells; increases vascular permeability; and induces fever and apoptosis in infected cells.
Interleukin-6 (IL-6)	Various cells (including macrophages and T-cells)	Stimulates the production of acute-phase proteins by the liver; promotes B-cell differentiation and enhances T-cell proliferation.
Interferon-gamma (IFNγ)	T-cells, NK cells	Activates macrophages, enhances antigen presentation, and promotes differentiation of T helper cells toward Th1 phenotype, essential for controlling intracellular pathogens.
Interleukin-12 (IL-12)	Dendritic cells, macrophages	Promotes differentiation of naive T-cells into Th1 cells, crucial for producing IFNγ and activating macrophages.

Table 5.4 Examples of anti-inflammatory cytokines.

Cytokine	Produced by	Functions
Interleukin-10 (IL-10)	Regulatory T-cells, macrophages, and other immune cells	Inhibits the production of pro-inflammatory cytokines, reduces antigen presentation, and suppresses T-cell and macrophage activation.
Transforming growth factor-beta (TGFβ)	Various cells, including regulatory T-cells and macrophages	Promotes tissue repair, inhibits T-cell proliferation and activation, and reduces production of pro-inflammatory cytokines.
Interleukin-4 (IL-4)	Th2 cells	Promotes differentiation of naive T-cells into Th2 cells, suppresses pro-inflammatory responses, and enhances expression of anti-inflammatory cytokines.
Interleukin-13 (IL-13)	Th2 cells	Promotes antibody production, reduces inflammation, and facilitates tissue repair.

Table 5.5 Interleukin-mediated immune response.

Interleukin	Function	Effect during viral infection
Interleukin-1 (IL-1)	Primarily involved in the inflammatory response (pro-inflammatory). Promotes the expression of adhesion molecules on endothelial cells, aiding in the recruitment of immune cells to the site of infection.	Elevated IL-1 levels can enhance the body's ability to respond to viral pathogens by promoting inflammation and recruiting immune cells to the infected tissue.
Interleukin-2 (IL-2)	Critical for the growth, proliferation, and differentiation of T-cells, which are vital for adaptive immunity. It essentially drives the activation and expansion of T-cells necessary to fight infections and other immune challenges	Supports the expansion of virus-specific cytotoxic T-cells that target and destroy infected cells.
Interleukin-6 (IL-6)	Has both pro-inflammatory and anti-inflammatory roles. Promotes the differentiation of B-cells into plasma cells, which produce antibodies.	Often elevated during viral infections and is a key mediator of the acute-phase response, contributing to fever and the production of acute-phase proteins by the liver.
Interleukin-10 (IL-10)	An anti-inflammatory cytokine that helps regulate the immune response to prevent excessive tissue damage.	It can limit the immune response to avoid chronic inflammation and tissue damage, but its overproduction may hinder the clearance of the virus.

Table 5.6 Virus-specific interleukin-mediated immune response.

Virus	Interleukin	Role and therapeutic potential
Hepatitis B virus (HBV)	Interleukin-12 (IL-12)	Enhances the immune response against HBV by boosting the activity of cytotoxic T lymphocytes (CTLs) and natural killer (NK) cells.
	Interleukin-10 (IL-10)	Modulating IL-10 levels may enhance the antiviral immune response by preventing its suppressive effects, which can lead to chronic infection.
Hepatitis C virus (HCV)	Interleukin-2 (IL-2)	Explored for enhancing T-cell responses in chronic HCV infection. Combining with antiviral treatments may improve outcomes.
	Interleukin-12 (IL-12)	Stimulates a stronger immune response against HCV, potentially leading to better viral clearance.
Human immunodeficiency virus (HIV)	Interleukin-2 (IL-2)	Used to expand $CD4^+$ T-cell counts. Showed promise in increasing T-cell counts but did not significantly improve clinical outcomes in large-scale studies.
	Interleukin-7 (IL-7)	Being explored to enhance immune reconstitution, particularly in patients who do not fully recover $CD4^+$ T-cell counts despite antiretroviral therapy.
Influenza virus	Interleukin-22 (IL-22)	Investigated for protecting epithelial cells and enhancing mucosal immunity, reducing lung injury, and promoting repair.
	Interleukin-17 (IL-17)	Targeting IL-17 may help manage excessive inflammation in severe cases, reducing lung damage and improving outcomes.
Respiratory syncytial virus (RSV)	Interleukin-17 (IL-17)	Modulating IL-17 responses may reduce inflammation and lung pathology in severe RSV infections.
	Interleukin-4 (IL-4) and Interleukin-13 (IL-13)	Targeting these interleukins can help control the Th2-mediated immune response, reducing mucus production and airway hyperresponsiveness.
Cytomegalovirus (CMV)	Interleukin-2 (IL-2)	Explored to boost the immune response in CMV-infected individuals, particularly in immunocompromised patients, such as organ transplant recipients. Interleukin-2 (IL-2) enhances cytotoxic lymphocyte activity of CMV-infected cells

between these cytokines and the production of prostaglandins, particularly prostaglandin E2, in the hypothalamus is crucial for the febrile response (see Tables 5.7 and 5.8) [101].

1. **Classification of fever based on duration**
   a. **Acute fever:** Lasts less than 7 days. Commonly associated with infections such as influenza, colds, and gastroenteritis.
   b. **Subacute fever:** Lasts from 7 days to 2 weeks. Often seen in conditions like mononucleosis, typhoid fever, or early stages of chronic infections.
   c. **Chronic or persistent fever:** Lasts more than 2 weeks. This can be due to chronic infections, autoimmune diseases, or malignancies.
2. **Classification based on pattern**
   a. **Continuous fever:** Temperature remains above normal throughout the day and does not fluctuate more than 1 °C (e.g., typhoid fever).

Table 5.7 Cytokine-mediated fever response.

Cytokine	Role in fever
Tumor necrosis factor-alpha (TNFα)	TNFα is a potent pyrogen (fever-inducing substance). It acts on the hypothalamus to raise the body's set point temperature, leading to fever. It is often elevated during infections and inflammatory conditions.
Interleukin-1 (IL-1)	IL-1 is another major pyrogen. IL-1β is known to stimulate the synthesis of PGE2, capable of stimulating the hypothalamus to increase body temperature. It is released by activated macrophages and plays a significant role in the acute-phase response.
Interleukin-6 (IL-6)	IL-6 is a key mediator of fever and the acute-phase response. It acts on the hypothalamus to induce fever and promotes the production of acute-phase proteins by the liver. IL-6 levels are often elevated in infections and inflammatory diseases.
Interferon-alpha (IFNα)	IFNα has pyrogenic effects and can induce fever as part of the body's antiviral response. It enhances the immune response, particularly during viral infections.
Interferon-gamma (IFNγ)	IFNγ can also contribute to fever. It is involved in activating macrophages and promoting the immune response against intracellular pathogens, including viruses. IFNγ can work synergistically with other cytokines to induce fever.
Interleukin-8 (IL-8)	IL-8, while primarily a chemokine involved in the recruitment of neutrophils, can contribute indirectly to the febrile response through its role in inflammation.
Prostaglandins (PGE2)	Although not a cytokine, prostaglandin E2 (PGE2) is a key mediator of fever. Cytokines like IL-1, TNFα, and IL-6 stimulate the production of PGE2 in the hypothalamus, which then acts to raise the body's set point temperature, resulting in fever.

Table 5.8 Fever onset in viral infections.

Viruses	Stage of infection when fever typically appears	Comments
Ebola virus	Incubation period (2–21 days): Fever usually appears suddenly at the onset of symptoms, typically within 4–10 days after exposure.	Fever is often one of the earliest symptoms, along with headache, muscle pain, and fatigue.
COVID-19 (SARS-CoV-2)	Incubation period (2–14 days): Fever usually appears 2–14 days after exposure, commonly around 5–6 days.	Fever is a common initial symptom, often accompanied by cough, fatigue, and loss of taste or smell.
Lassa fever virus	Incubation period (6–21 days): Fever typically appears within 1–3 weeks after exposure.	Fever is a prominent early symptom, along with sore throat, chest pain, and gastrointestinal symptoms.
Rotavirus	Incubation period (1–3 days): Fever usually appears suddenly at the onset of symptoms.	Fever is common in children and infants, often accompanied by vomiting, diarrhea, and dehydration.
Dengue virus	Incubation period (3–10 days): Fever usually appears suddenly at the onset of symptoms.	High fever is a characteristic symptom, often accompanied by severe headache, pain behind the eyes, and joint and muscle pain.
Zika virus	Incubation period (3–14 days): Fever typically appears within a few days of the onset of symptoms.	Mild fever is common, along with rash, conjunctivitis, and joint pain.
Chikungunya virus	Incubation period (3–7 days): Fever usually appears suddenly at the onset of symptoms.	High fever is a hallmark symptom, often accompanied by severe joint pain, muscle pain, and rash.
Yellow fever virus	Incubation period (3–6 days): Fever typically appears within a few days of the onset of symptoms.	High fever is a key symptom, often followed by jaundice, bleeding, and organ failure in severe cases.
Influenza virus	Incubation period (1–4 days): Fever usually appears suddenly at the onset of symptoms.	High fever is common, often accompanied by chills, sore throat, muscle aches, and fatigue.
Human immunodeficiency virus (HIV)	Acute infection (2–6 weeks post-exposure): Fever is a common early symptom during acute HIV infection.	Fever occurs during the acute retroviral syndrome phase, along with sore throat, rash, and swollen lymph nodes.

(*Continued*)

Table 5.8 (Continued)

Viruses	Stage of infection when fever typically appears	Comments
Poliovirus	Incubation period (3–35 days but usually 7–10 days): Fever usually appears early in the symptomatic phase, typically within 6–20 days after exposure.	Fever is a common early symptom, often accompanied by sore throat, headache, nausea, and vomiting. In severe cases, it can progress to paralysis.
Human papillomavirus (HPV)	Fever is not a typical symptom of HPV infection.	HPV infections are often asymptomatic. When symptoms do occur, they are typically warts or lesions rather than systemic symptoms like fever. Persistent infections with high-risk HPV types can lead to cancers.
Measles virus	Incubation period (10–14 days): Fever typically appears as one of the first symptoms, around 10–12 days after exposure.	High fever is common, followed by cough, runny nose, conjunctivitis, and a characteristic rash.
Rubella virus	Incubation period (14–21 days): Fever usually appears early, within 14–17 days after exposure.	Mild fever is common, along with rash, swollen lymph nodes, and joint pain.
Mumps virus	Incubation period (12–25 days): Fever typically appears within 16–18 days after exposure.	Fever is common, along with swelling of the parotid glands, headache, muscle aches, and fatigue.
Cytomegalovirus (CMV)	Primary infection (3–12 weeks): Fever appears during the symptomatic phase of primary infection.	In immunocompetent individuals, fever is often mild and accompanied by fatigue and muscle aches. In immunocompromised patients, fever can be more pronounced and persistent.
Epstein–Barr virus (EBV)	Incubation period (4–6 weeks): Fever typically appears during the symptomatic phase, often referred to as infectious mononucleosis or "mono."	High fever is common, along with sore throat, swollen lymph nodes, and fatigue.
Varicella-zoster virus (VZV) (chickenpox)	Incubation period (10–21 days): Fever usually appears early in the symptomatic phase.	Fever is common, often accompanied by a characteristic itchy rash, fatigue, and headache.
Respiratory syncytial virus (RSV)	Incubation period (2–8 days): Fever appears early, typically within 4–6 days after exposure.	Fever is common in infants and young children, often accompanied by cough, wheezing, and difficulty breathing.

Table 5.8 (Continued)

Viruses	Stage of infection when fever typically appears	Comments
Hepatitis A virus (HAV)	Incubation period (15–50 days): Fever usually appears early in the symptomatic phase.	Mild fever is common, along with jaundice, fatigue, and gastrointestinal symptoms.
Hepatitis B virus (HBV)	Incubation period (30–180 days): Fever can appear in the early symptomatic phase, although it is not always present.	When present, fever is mild and accompanied by jaundice, fatigue, and abdominal pain.
Hepatitis C virus (HCV)	Incubation period (2–26 weeks): Fever is not commonly a prominent feature but can appear in the early symptomatic phase.	Chronic infection is often asymptomatic; the acute phase may include mild fever, fatigue, and jaundice.

   b. **Intermittent fever:** Fever spikes are separated by intervals of normal temperature. Seen in malaria, septicemia, and pyogenic infections.
   c. **Quotidian:** Spikes occur every 24 hours.
   d. **Tertian:** Spikes occur every 48 hours.
   e. **Quartan:** Spikes occur every 72 hours.
   f. **Remittent fever:** Temperature fluctuates more than 1 °C but does not return to normal. Common in bacterial infections, such as infective endocarditis.
   g. **Relapsing fever:** Periods of fever are separated by days or weeks of normal temperature and seen in conditions such as relapsing fever (caused by *Borrelia* species).
3. **Classification based on severity**
   a. **Low-grade fever:** 38 °C (100.4 °F) to 39 °C (102.2 °F). Often seen in mild infections.
   b. **Moderate fever:** 39 °C (102.2 °F) to 40 °C (104 °F). Common in more severe infections.
   c. **Fever:** Above 40 °C (104 °F). It can be dangerous and is seen in severe infections, heatstroke, and some inflammatory conditions.
   d. **Hyperpyrexia:** Above 41.5 °C (106.7 °F). This can lead to medical emergency, often associated with heatstroke, severe infections, or central nervous system hemorrhages.

5.1.4.4 Chemokines

Chemokines are a specialized subset of cytokines that play a crucial role in the immune system to induce cell migration. They are typically 8–12 kDa in size and share a common structural motif that includes conserved cysteine residues. Based on the arrangement of these cysteine residues, chemokines are classified into four main subfamilies, namely CC, CXC, CXC3C, and

XC, which bind to their respective receptors, denoted as CCR, CXCR, CX3CR, and XCR, on susceptible cells.

- **CC chemokines:** The first two cysteines are adjacent (e.g., CCL2, also known as MCP-1).
- **CXC chemokines:** The first two cysteines are separated by one amino acid (e.g., CXCL8, also known as IL-8).
- **CX3C chemokines:** Three amino acids separate the first two cysteines (e.g., CX3CL1, also known as fractalkine).
- **XC chemokines:** Only two cysteines are present, one at the amino terminus and one downstream (e.g., XCL1).

Chemokine-mediated immune response in viral infection:

Chemotaxis: Chemokines modulate a concentration gradient of immune cells following infection and or inflammation, facilitating immune surveillance and response initiation [102]. For example, CC chemokine ligand (CCL) 19 and CCL21 guide T-cells and dendritic cells to and from lymph nodes [103].

Inflammation: During inflammation, chemokines recruit various immune cells, including neutrophils, monocytes, and lymphocytes, to the affected tissues. This is crucial for an effective immune response and tissue repair.

Activation of immune cells and immune homeostasis: Chemokines can also control the release of innate immune cells from the bone marrow during homeostasis, even without infection, as well as in response to infection and inflammation [102].

Clinical significance

Chemokines and their receptors are involved in various pathological conditions

Infections: Pathogens like HIV exploit chemokine receptors (e.g., CCR5 and CXCR4) to enter host cells.

Autoimmune diseases: Uncontrolled chemokine expression can lead to inappropriate immune cell infiltration, contributing to diseases such as rheumatoid arthritis and multiple sclerosis.

Cancer: Tumors can secrete chemokines to attract immune cells that support tumor growth and metastasis [104]. Conversely, some chemokines can recruit immune cells that attack tumors. An overview of virus-specific chemokine-mediated immune responses during viral infections is listed in Table 5.9.

5.1.5 Complement System

The complement system is a system of proteins comprising a series of small proteins that circulate in the blood. They are activated in a cascade manner through a series of interactions to enhance the activity of antibodies and activation of phagocytic cells to clear pathogens, promote inflammation, and directly lyse pathogen cells.

There are over 30 proteins and protein fragments that comprise the complement system, which include (i) complement proteins (C)- (C1-C9), (ii) factor proteins (e.g. Factor B, Factor D, properdin [Factor P]), and (iii) regulatory proteins (e.g., CD55 (decay accelerating factor), CD59 [protectin], and Factor H).

Table 5.9 Examples of chemokine-mediated immune response during viral infection.

Virus	Cytokine	Role and impact
Zika virus	Interleukin-6 (IL-6)	Elevated during infection, IL-6 is associated with inflammation and the acute-phase response. High levels can contribute to severe disease manifestations.
	Interleukin-10 (IL-10)	An anti-inflammatory cytokine that can suppress the immune response. Elevated IL-10 levels may contribute to viral persistence and dampened immune response.
	Tumor necrosis factor-alpha (TNFα)	Pro-inflammatory cytokine is involved in fever and inflammation. High levels are linked to severe disease and complications such as Guillain–Barré syndrome.
Ebola virus	Interleukin-6 (IL-6)	Significantly elevated during infection, IL-6 contributes to the cytokine storm, severe inflammation, and vascular leakage.
	Tumor necrosis factor-alpha (TNFα)	Plays a role in the cytokine storm, contributing to fever, inflammation, and tissue damage. High levels correlate with disease severity.
	Interleukin-10 (IL-10)	Elevated levels can suppress antiviral immune responses, potentially leading to uncontrolled viral replication and disease progression.
	Interferon-gamma (IFNγ)	Critical for antiviral responses, but its dysregulation can contribute to immune pathology and severity of the disease.
Yellow fever virus	Interleukin-6 (IL-6)	Elevated during infection, associated with severe disease and liver damage. High levels can indicate a poor prognosis.
	Tumor necrosis factor-alpha (TNFα)	Contributes to the inflammatory response, fever, and tissue damage. High levels are linked to severe manifestations such as hemorrhage.
	Interferon-alpha (IFNα)	Plays a key role in the early antiviral response. Delayed or insufficient production can lead to uncontrolled viral replication.
Dengue virus	Interleukin-6 (IL-6)	Associated with severe dengue and dengue hemorrhagic fever. High levels contribute to vascular permeability and shock.
	Interleukin-10 (IL-10)	Elevated in severe cases, potentially suppressing effective immune responses and contributing to severe disease.
	Tumor necrosis factor-alpha (TNFα)	Linked to severe disease manifestations, including plasma leakage, shock, and hemorrhage.

(Continued)

Table 5.9 (Continued)

Virus	Cytokine	Role and impact
Chikungunya virus	Interleukin-6 (IL-6)	Associated with joint inflammation and chronic arthritis. High levels correlate with disease severity and chronic symptoms.
	Interleukin-10 (IL-10)	Elevated levels may help control inflammation but can also suppress effective immune responses.
	Tumor necrosis factor-alpha (TNFα)	Contributes to joint inflammation, pain, and swelling. High levels are associated with severe and chronic disease.
Lassa fever virus	Interleukin-6 (IL-6)	Elevated during infection, correlates with disease severity and poor outcomes.
	Tumor necrosis factor-alpha (TNFα)	Plays a role in inflammation, vascular leakage, and shock. High levels are associated with severe disease.
	Interferon-gamma (IFNγ)	Important for antiviral responses, but dysregulation can contribute to immunopathology.

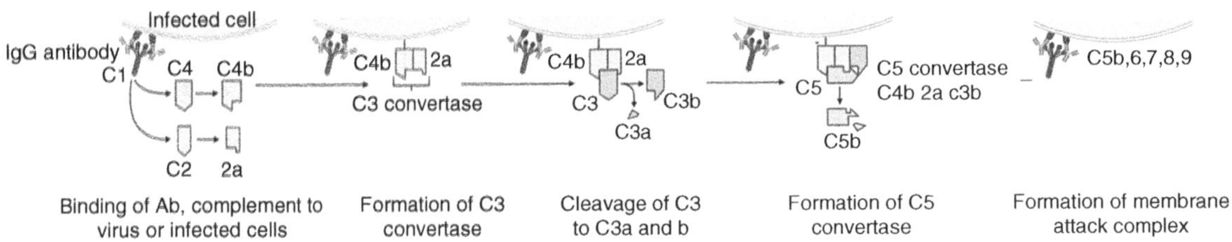

Figure 5.3 Classical pathway.

Activation of the complement pathways The complement system can be activated through three pathways – (i) classical pathway: activated by C1 complex (C1q, C1r, C1s) binding to antibodies (IgM or IgG) on the pathogen surface (IgM- or IgG-antigen complex) (see Figure 5.3); (ii) lectin pathway: activated by mannose-binding lectin on pathogen surfaces; (iii) alternative pathway: initiated by the spontaneous high level of hydrolysis of C3 and its binding to pathogen surfaces such as viruses, bacteria, and fungi (see Figure 5.4). All three pathways converge at a central point – the activation of C3, the central component. C3 convertase cleaves C3 into C3a and C3b, leading to the opsonization of pathogens and the formation of C5 convertase. C5 convertase cleaves C5 into C5a and C5b. C5b initiates the assembly of the membrane attack complex (MAC) (C5b, C6, C7, C8, and C9), which forms pores in the pathogen cell membrane, leading to cell lysis.

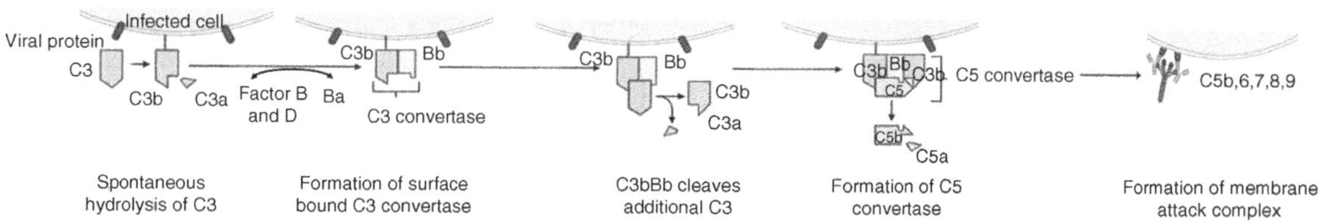

Figure 5.4 Alternate pathway.

5.1.5.1 *Functions of the Complement System and Protein*

- **Opsonization:** C3b, iC3b, C4b, and C1q proteins serve as opsonins [105]. They bind to the surface of pathogens, tagging them for phagocytosis or apoptosis by macrophages and neutrophils.
- **Chemotaxis and inflammation:** C3a and C5a act as anaphylatoxins [106], stimulating mast cells to release histamine and other inflammatory mediators, and attracting phagocytes to the infection site.
- **Cell lysis:** MAC creates pores in the cell membranes of pathogens or pathogen-infected cells, leading to osmotic lysis and death of the pathogen.
- **Clearance of immune complexes:** C1q binds to immune complexes, facilitating their removal by phagocytes [107].

5.1.5.2 *Regulation of the Complement System*

Regulation of the complement system is essential for preventing excessive inflammation, maintaining self-tolerance, and protecting host cells from unintended damage during immune responses. Dysregulation of complement regulation can lead to various diseases, including autoimmune disorders (e.g., systemic lupus erythematosus, rheumatoid arthritis), inflammatory conditions (e.g., glomerulonephritis), and diseases associated with excessive complement activation (e.g., paroxysmal nocturnal hemoglobinuria).

5.1.6 Acute-Phase Response

The acute-phase response (APR) is a rapid and coordinated innate immune reaction that occurs in response to various disturbances in homeostasis, such as infections, tissue injury, trauma, surgery, or inflammation [108]. It is a critical part of the body's defense mechanism aimed at restoring physiological balance and combating potential threats. Although it was initially believed that APR is directed against bacterial infection, studies have shown that acute-phase proteins (APP), such as pentraxins (PTX), a family of highly conserved APP, may target virus particles to facilitate their clearance. In some cases, it may signify disease progression. For instance, high serum levels of short pentraxins such as C-reactive protein (CRP) and serum amyloid proteins have been observed in patients infected with influenza A virus, HCV, HBV, and HIV [109].

5.2 Adaptive Immune System

The adaptive immune system, also known as acquired (specific) immunity, is a remarkable defense mechanism of the body that plays a central role in combating viral infections. It develops over time as the immune system learns to recognize and respond to foreign invaders. The "learning" process starts when a person's immune system encounters non-self-substances (antigens) and mounts a response against them.

In the fight against pathogenic viral invaders, the adaptive immune system mounts a highly specific response, generating tailored antibodies and T-cell receptors (TCRs) that target unique viral antigens. A hallmark of this system is its ability to form memory cells, which are long-lived B and T lymphocytes that "remember" the virus. Upon re-exposure, these memory cells trigger a faster, stronger immune response, often neutralizing the virus before symptoms even arise. This immunological memory forms the basis of the success of vaccines and provides long-term protection against recurrent infections.

The adaptive immune system comprises two main types: the cell-mediated immune response and the humoral immune response. Cell-mediated immunity is primarily mediated by T-cells and is crucial for targeting infected host cells. Humoral immunity, on the other hand, is mediated by B-cells and involves the production of antibodies (Ig) that can neutralize pathogens.

5.2.1 Cell-Mediated Adaptive Immunity

Cell-mediated immunity is an important type of the adaptive immune system responsible for defending the body against intracellular pathogens, such as viruses, as well as abnormal host cells like cancer cells. This arm of the immune response primarily relies on specialized immune cells called T-cells, specifically cytotoxic T-cells (CD^{8+} T-cells) and helper T-cells (CD^{4+} T-cells).

Cell-mediated immunity is characterized by its ability to identify and destroy infected or altered host cells, ensuring that they do not continue to harbor and propagate intracellular pathogens or become malignant.

I. **Cytotoxic T-cells ($CD8^+$ T-cells):** Cytotoxic T-cells are the "effectors" of cell-mediated immunity. Guided by the recognition of antigens on infected cells, they can release toxic substances, such as perforin and granzymes, to induce apoptosis (cell death) in the target cell. This is a precise mechanism to eliminate intracellular pathogens and cancerous cells. Cytotoxic cells, specifically cytotoxic T-cells ($CD8^+$ T-cells), play a crucial role in the immune response against viral infections. While there are various subsets of cytotoxic T-cells with specific functions, the primary cytotoxic cells involved in viral infections are $CD8^+$ T-cells.

II. **Helper T-cells ($CD4^+$ T-cells):** Helper T-cells play a supporting role in cell-mediated immunity. They assist in the activation of cytotoxic T-cells and the coordination of immune responses. Helper T-cells release cytokines that stimulate other immune cells, enhancing the overall immune response.

5.2.1.1 Antigen-Presenting Cells

Antigen presentation is the central aspect of cell-mediated immunity. It involves the recognition, by antigen-presenting cells (APCs), of cells that are dysfunctional or structurally altered [110]. When a cell becomes infected with a virus or transformed due to malignancy, it displays abnormal proteins on its surface, recognized by APCs as antigens [111]. This forms the first step in the recognition of virus-infected cells by cell-mediated adaptive immunity.

APCs, such as dendritic cells, macrophages, and even B-cells [111, 112], are specialized immune cells that capture, process, and present antigens to T-cells. They act as "middlemen" between the innate immune system's detection of pathogens and the adaptive immune system's response. T-cells, $CD8^+$ T-cells, and helper $CD4^+$ T-cells recognize antigens presented on the surface of APCs using MHC molecules [113].

5.2.2 Major Histocompatibility Complex Molecules

MHC molecules, also known as human leukocyte antigen (HLA) molecules in humans, are critical components of the immune system. In the cellular immune response, MHC molecules, particularly MHC Class I and MHC Class II, are central for antigen presentation and activation of cytotoxic T-cells ($CD8^+$) and helper T-cells ($CD4^+$) (see Figure 5.5).

MHC Class I MHC Class I molecules are found on the surface of almost all nucleated cells in the body [114]. They present endogenous antigens, such as viral proteins produced within an infected cell, to cytotoxic T-cells ($CD8^+$ T-cells). This presentation serves to inform $CD8^+$ T-cells about the presence of intracellular pathogens, including viruses. When a $CD8^+$ T-cell recognizes a viral antigen presented by MHC Class I, it becomes activated and kills the infected cell, thereby limiting viral replication.

The molecular structure of MHC Class I consists of a heterodimer composed of

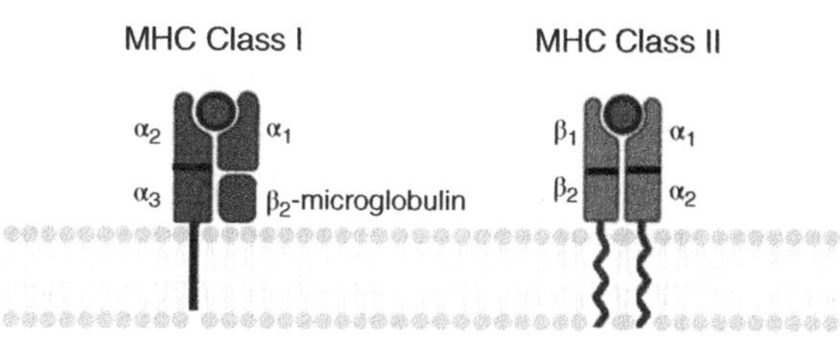

Figure 5.5 Putative structure of MHC Class I and II.

two chains: an alpha chain and a smaller beta-2 microglobulin (β_2m) chain.

Alpha chain (HLA-A, HLA-B, or HLA-C): The alpha chain is the larger of the two chains and is highly polymorphic, that is, it exists in various forms with different amino acid sequences. These polymorphic regions are responsible for the specificity of antigen binding.

α chain: Contains a transmembrane domain and a cytoplasmic tail that anchors the molecule in the cell membrane.

The extracellular portion is divided into three domains:

- **α_1 and α_2 domains:** Together form the peptide-binding groove, which accommodates peptides 8–10 amino acids long. These regions are highly polymorphic and determine which peptides can bind.
- **α_3 domain:** A conserved domain that provides structural stability and serves as the primary binding site for the CD8 co-receptor on cytotoxic T cells.

β_2-microglobulin (β_2m):

- A small, non-polymorphic protein (~100 amino acids).
- Non-covalently associates with the α3 domain.
- Critical for the proper folding, stability, and surface expression of the MHC Class I complex, though it does not participate directly in antigen binding.

Peptide presentation

Each MHC Class I molecule presents one peptide at a time, typically derived from endogenous proteins (e.g., viral, tumor, or self-proteins processed by the proteasome). A single cell expresses many MHC Class I molecules, so it can present a wide variety of peptides simultaneously across its surface.

MHC Class II MHC Class II molecules are primarily expressed on APCs, including dendritic cells, macrophages, and B-cells. They play a key role in presenting exogenous antigens to helper T-cells ($CD4^+$ T-cells) [115]. The molecular structure of MHC Class II molecules also consists of two different chains, known as the alpha and beta chains. Each of these chains consists of two extracellular domains. The distal part of this domain anchors the antigen peptide presented on the APC, while the proximal part is a member of the Ig superfamily.

The molecular structure of MHC Class II consists of the following:

α chain (HLA-DPα, -DQα, -DRα): Length ~ 220–240 amino acids.

- α_1 domain – pairs with β_1 to form the peptide-binding groove.
- α_2 domain – provides structural support and pairs with β_2.

Has a transmembrane and cytoplasmic tail that anchors the molecule in the cell membrane.

β chain (HLA-DPβ, -DQβ, -DRβ): Length ~ 240–260 amino acids.

- β_1 domain – pairs with α1 to form the peptide-binding groove; this domain

is highly polymorphic and influences peptide specificity.
- β_2 domain – interacts with the CD4 co-receptor on helper T cells.

Also contains a transmembrane and cytoplasmic tail for membrane anchoring.

Peptide presentation:
- MHC Class II molecules bind peptides (typically 13–25 amino acids) derived from exogenous antigens that are internalized and processed in APCs (dendritic cells, macrophages, and B cells).
- Each MHC Class II molecule presents one peptide at a time, but a single APC can present many different peptides across its MHC II molecules.
- The β_2 domain provides the binding site for the CD4 co-receptor on helper T cells, which stabilizes the TCR–MHC interaction.

MHC Class II molecules are designed to present a single antigenic peptide simultaneously. This feature is essential for their role in the activation of helper T-cells ($CD4^+$ T-cells), which require diverse signals to coordinate immune responses. APCs, like dendritic cells, macrophages, and B-cells, typically present a range of exogenous antigens, captured from engulfed pathogens, to $CD4^+$ T-cells.

5.2.2.1 T-Cell Recognition of Peptide Antigens

T-cells, also known as T lymphocytes, are a type of WBC. They originate and mature in different parts of the body, with distinct processes occurring at each stage. T-cells originate from hematopoietic stem cells (HSCs) in the bone marrow. These stem cells are multipotent, meaning they can differentiate into various types of blood cells, including all types of lymphocytes (B-cells, T-cells, and NK cells). Early T-cell progenitors (also called thymocytes) leave the bone marrow and migrate to the thymus via the bloodstream. The thymus is a specialized primary lymphoid organ located in the upper anterior part of the chest. Mature, naive T-cells that have successfully undergone positive and negative selection exit the thymus and enter the peripheral circulation; they are referred to as "recent thymic emigrants" (RTEs), ready to participate in immune responses throughout the body.

Stages of T-cell maturation
- **Double-negative (DN) stage:** Thymocytes enter the thymus as double-negative cells lacking surface markers (both CD4 and CD8) for specific identification of antigens. They progress through four substages (DN1, DN2, DN3, and DN4) characterized by the expression of different surface markers and the rearrangement of the TCR genes.
- **Double-positive (DP) stage:** After successful expression of TCR β-chain rearrangement, thymocytes express both CD4 and CD8 surface markers and are referred to as double-positive cells. They undergo TCR α-chain rearrangement and express a complete TCR.
- **Positive selection:** Thymocytes with functional TCRs that can moderately bind self-MHC molecules presented

by thymic epithelial cells in the cortex are positively selected. This ensures that the T-cells can recognize antigens presented by self-MHC molecules.

- **Single-positive (SP) stage:** Positively selected thymocytes differentiate into either $CD4^+$ (helper) or $CD8^+$ (cytotoxic) single-positive cells, depending on which class of MHC molecule their TCR recognizes (MHC Class II for $CD4^+$ T-cells and MHC Class I for $CD8^+$ T-cells).
- **Negative selection:** Thymocytes that bind too strongly to self-antigens presented by MHC molecules on thymic dendritic cells and medullary epithelial cells are negatively selected (deleted) to prevent autoimmunity. This process occurs primarily in the thymic medulla. Following this, the T-cells undergo maturation, egress, and become functional cells.

T-cell recognition of peptide antigens is a highly specific and complex process crucial for the adaptive immune response [116]. It centers around the TCR, a complex protein on the surface of T-cells. The TCR consists of two chains, alpha and beta, that together form a heterodimer [117]. Each TCR chain contains variable (V), diversity (D), C (constant), and joining (J) regions, like the Ig superfamily of proteins.

The V regions of the TCR, particularly the complementarity-determining regions (CDRs), are crucial for recognizing peptide antigens presented by MHC molecules on APCs [118]. These CDRs are responsible for the specificity of antigen recognition by the TCR. The C region, located at the constant end of each TCR chain, provides structural stability and is involved in interactions with signaling molecules during T-cell activation. On the other hand, the diversity (D) and joining (J) regions in the TCR are primarily involved in the V(D)J recombination process during T-cell development. This process generates diversity in the TCR repertoire, allowing T-cells to recognize a wide range of potential antigens. The combination of V and C regions, along with the diversity introduced by the V(D)J recombination process, ensures both the specificity and diversity of T-cell responses, contributing to the adaptive immune system's ability to respond to a variety of pathogens and antigens [119].

Activated T-cells may differentiate into cytotoxic T-cells, helper T-cells, or regulatory T-cells, depending on the specific signals they receive. Once a particular T-cell develops its effector function, it expresses a specific TCR forever; however, further recombination is possible in certain scenarios.

Cytotoxic T-Cell-Mediated Immunity

Cytotoxic T-cells are mostly $CD8^+$ T-cells [120]; the minority of $CD4^+$ are cytotoxic T-cells [121]. They play a pivotal role in the immune response to viral infections. These specialized immune cells are essential for identifying and eliminating virus-infected host cells (see Figure 5.6).

Steps involved in cytotoxic T-cell-mediated immunity

- **Antigen recognition:** Cytotoxic T-cells recognize viral infections by recognizing specific antigens displayed on and presented by APC. These antigens are typically viral

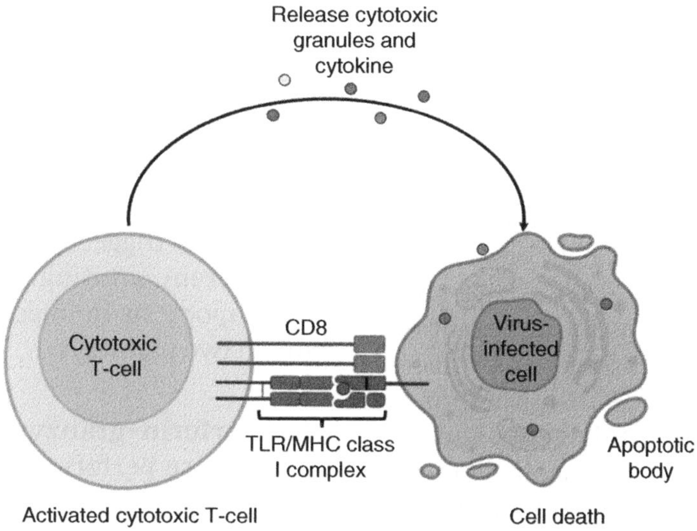

Figure 5.6 Overview of cytotoxic T-cell interaction with virus-infected cells.

proteins, often derived from viral replication.

- **Activation:** When a cytotoxic T-cell encounters an APC displaying the appropriate viral antigen on its MHC-I molecules, the T-cell becomes activated. This interaction involves the TCR on the cytotoxic T-cell binding to the viral antigen–MHC-I complex.
- **Clonal expansion:** Upon activation, cytotoxic T-cells undergo clonal expansion, leading to the proliferation of a large population of effector $CD8^+$ T-cells. This is a critical step for amplifying the immune response and ensuring enough cytotoxic T-cells to combat the infection. Clonally expanded immune cells differentiate into effector cells with specialized functions tailored to combat the antigen.
- **Effector function:** Activated cytotoxic T-cells are equipped with cytotoxic granules containing proteins like perforin and granzymes. These granules are released, and perforin creates pores in the membrane of the infected cell. Granzymes then enter the target cell through these pores, inducing apoptosis (programmed cell death) in the infected cell.
- **Elimination of infected cells:** The induction of apoptosis in infected cells by cytotoxic T-cells effectively eliminates the host cells harboring the virus. This helps control viral replication and limits the spread of the virus within the body.
- **Memory formation:** Cytotoxic T-cells not only eliminate infected cells during the current infection but also give rise to memory $CD8^+$ T-cells. These memory cells persist in the body, providing long-term immunity. In case of future encounters with the

same virus, memory cytotoxic T-cells can respond more rapidly and effectively, preventing a severe or prolonged infection.

Virus-Specific Cytotoxic T-Cells Virus-specific CTLs are a critical component of the adaptive immune system, specifically designed to identify and eliminate virus-infected cells. They play an important role in controlling and clearing virus-infected cells and could provide insights into the development of antiviral agents and vaccines. This signifies that vaccine-induced protection is mediated not only by humoral immune response but also by robust cellular immune response (see Table 5.10).

Direct Cytotoxicity Direct cytotoxicity is a mechanism by which immune cells, such as CTLs and NK cells, recognize and eliminate virus-infected cells through direct contact [122]. They achieve this by releasing cytotoxic molecules like perforin and granzymes, which trigger apoptosis in the infected cells, thereby limiting viral spread. CTLs induce apoptosis in infected cells primarily through two pathways:

- **Perforin–granzyme pathway:** CTLs release perforin, which forms pores in the target cell membrane, allowing granzymes (proteolytic enzymes) to enter the cell and trigger apoptosis.

Table 5.10 Examples of virus-specific CTL-mediated immune response during viral infections.

Viral infection	Role of CTLs
HIV (Human immunodeficiency virus)	CTLs target and destroy HIV-infected cells, helping to control viral replication.
HBV (Hepatitis B virus)	CTLs contribute to the elimination of HBV-infected hepatocytes, limiting liver damage.
HCV (Hepatitis C virus)	CTLs play a role in controlling viral load by killing infected liver cells
EBV (Epstein–Barr virus)	CTLs help control EBV by killing virus-infected cells, especially during latent infection.
CMV (cytomegalovirus)	CTLs are essential for controlling CMV, especially in immunocompromised individuals.
HSV (Herpes simplex virus)	CTLs help clear infected cells during both primary and recurrent HSV infections.
SARS-CoV-2 (COVID-19)	CTLs target infected cells, aiding in the reduction of viral replication and symptom severity.
VZV (Varicella-zoster virus)	CTLs assist in controlling VZV infections, especially during reactivation (shingles).
HPV (Human papillomavirus)	CTLs targeting E6 and E7 should theoretically be capable of eliminating even HPV-infected basal cells, where only the initial stage of viral replication occurs.

- **First apoptosis signal (Fas) and first apoptosis signal receptor–ligand (Fas–FasL) pathway:** CTLs express FasL on their surface, which binds to the Fas receptor on the target cell, initiating a cascade that leads to apoptosis.

Indirect Cytotoxicity Indirect cytotoxicity is a mechanism in which immune cells eliminate virus-infected cells through the release of soluble factors, such as cytokines or antibodies, rather than direct cell-to-cell contact. These intermediaries can trigger apoptosis, enhance phagocytosis, or activate other immune components to clear infected cells. For example, CTLs produce and secrete cytokines such as IFNγ and TNFα, which have antiviral effects and can enhance the immune response by recruiting and activating other immune cells.

T Helper Cell-Mediated Immunity T helper cells, also known as $CD4^+$ T-cells, are a crucial component of the adaptive immune system responsible for the coordination of most immune responses against pathogens, including viruses. These cells play a central role in recognizing antigens presented by APCs and regulating immune responses tailored to eliminate specific pathogens [110, 123]. One of the key functions of T helper cells is to provide help and guidance to other immune cells, including CTLs, B-cells, and macrophages. They do this through the release of cytokines, which are signaling molecules that modulate the behavior and function of other immune cells. For instance, T helper cells secrete cytokines including IL-2, IL-4, IL-5, IL-10, and IFNγ, among others [124].

Specific roles of T helper cells in promoting immune responses include:

1. **Activation of B-cells:** T helper cells assist B-cells in differentiating into plasma cells that produce antibodies specific to the invading virus [125, 126]. This process, known as antibody-mediated immunity, helps neutralize viruses by binding to viral particles and marking them for destruction by other immune cells.
2. **Activation of CTLs:** T helper cells stimulate CTLs to recognize and kill virus-infected cells [127]. They do this by releasing cytokines that promote CTL activation, proliferation, and recruitment to sites of infection. CTLs play a crucial role in directly killing cells harboring intracellular viruses, thus preventing viral replication and spread.
3. **Regulation of inflammation:** T helper cells also regulate the inflammatory response during viral infections. Some subsets of T helper cells, such as Th1 cells, produce IFNγ, which activates macrophages and enhances their ability to phagocytose and kill viruses [127, 128]. Th1 cells also stimulate the production of complement proteins and opsonins that facilitate the clearance of viral particles.
4. **Memory response:** Upon encountering a pathogen (virus), T helper cells differentiate into memory T-cells that persist in the body for years or even decades [129]. These memory T-cells provide rapid and robust responses upon re-exposure to the same virus, leading to faster clearance and enhanced protection against subsequent infections.
5. **Modulation of immune tolerance:** T helper cells also play a role in

maintaining immune tolerance [130], preventing the immune system from attacking healthy tissues and cells. This balance is crucial for preventing autoimmune diseases while mounting effective responses against viral infections.

Virus-Specific T Helper Cells T helper cells play a vital role in orchestrating the immune response against various viral infections. There are several subsets of $CD4^+$ T. The primary types of Th cells include:

- Th1 cells – produce cytokines such as IFNγ and TNF, which activate macrophages and enhance the killing of intracellular pathogens like viruses and certain bacteria [131].
- Th2 cells – involved in humoral immunity and the defense against extracellular parasites [132].
- Th17 cells – associated with the defense against extracellular bacteria and fungi [133].
- Th9 cells – participate in immune response against parasitic infections and have roles in allergic inflammation and autoimmunity [134].
- Treg cells (regulatory T cells) – play a crucial role in maintaining immune tolerance and preventing autoimmune diseases in virus-infected cells.

Among all the T helper cells, Th1 cells are specifically geared toward orchestrating effective antiviral responses by enhancing cellular immunity and cytokine-mediated antiviral mechanisms against viruses such as:

T helper cells response during HIV infection

During early HIV infection, a strong T helper cells response, characterized by IFNγ and IL2 production, supports cell-mediated immunity by activating CTLs and macrophages, helping to initially control viral replication. However, HIV targets and infects $CD4^+$ T helper cells [135]. This can lead to a decrease in the number of $CD4^+$ cells in the whole blood. A significant decrease over time can eventually lead to immunodeficiency if untreated.

T helper cells response during influenza virus

Influenza infection triggers a robust immune response, where T helper cells are key players [136]. $CD4^+$ T-cells employ several effector mechanisms to defend and promote recovery from influenza virus infection. Activated T helper cells release cytokines such as IL-2, IL-6, and IFNγ, enhancing the antiviral response and activating CTLs to destroy infected epithelial cells in the respiratory tract. Also, $CD4^+$ T-cells promote the differentiation of B-cells into plasma cells that secrete neutralizing antibodies targeting the influenza virus, enhancing virus neutralization and clearance.

T helper cells response during hepatitis B virus infection

T helper cells are essential in the immune response to HBV infection [137]. They also play a vital role in the activation of CTL in HBV infection. T helper cells produce cytokines like IL-2, which help in the proliferation and activation of HBV-specific CTLs that can recognize and destroy infected liver cells. Similarly, they help B-cells produce antibodies against HBV surface antigens, which are critical for neutralizing the virus and preventing its spread.

T helper cells response during human papillomavirus (HPV) infection

In HPV infections, T helper cells contribute significantly [138]. T helper cells recognize viral antigens presented by MHC Class II molecules on APCs such as dendritic cells. Upon activation, $CD4^+$ T-cells differentiate into various subsets, including T helper 1 and T helper 2 cells, which secrete different cytokines. Th1 cells produce IFNγ, which enhances the cytotoxic activity of CTLs and macrophages, aiding in the clearance of infected cells. Th2 cells, on the other hand, assist in the activation and proliferation of B-cells, promoting antibody production against HPV.

T helper cells response during poliovirus infection

Upon encountering APCs that display poliovirus antigen peptides via MHC Class II molecules, T helper cells become activated. This activation leads to their proliferation and differentiation into various subtypes, including T helper 1 and T helper 2 cells, which secrete different cytokines that shape the immune response [139].

T helper cells response during Ebola virus infection

The T-cell response is known to target Ebola virus glycoprotein, which is a major antigenic target. They contribute to Ebola virus glycoprotein-specific T-cell memory and are capable of producing IFNγ, TNFα, and IL-2. However, Ebola virus has evasion mechanisms that can directly interfere with the T-helper cells' antiviral mechanism. For example, the Ebola glycoprotein can block TCRs from interacting with HLA class-1, which makes it difficult for T-cells to attack [140].

Clinical significance of $CD4^+$ T helper cells in HIV infection

$CD4^+$ T helper cells play a very important role in HIV infection, which influences the diagnosis, management, and prognosis of acquired immunodeficiency syndrome (AIDS). The normal value of $CD4^+$ T-cells in a healthy adult (who is not infected with HIV) typically ranges between 500 and 1500 cells per cubic millimeter (cells/mm^3) of blood. This range can vary slightly depending on the laboratory and the specific population being tested.

A. **Diagnostic marker**
   1. **Indicator for monitoring disease**
      - CD4 count: The number of $CD4^+$ T (generally known as CD4 counts) cells in the whole blood is a primary marker used to monitor the progression of HIV infection. A declining CD4 count indicates worsening immunodeficiency and progression toward AIDS. Hence, a need for a standard hospital or laboratory to easily access the facility.
      - Viral load: Alongside CD4 counts, measuring the amount of HIV RNA in the blood provides a comprehensive picture of disease status. High viral loads are usually associated with rapid $CD4^+$ T-cell decline.
   2. **Indicator of disease progression Stages of HIV infection**
      - **Stage 1**: Primary HIV infection. Patients are often asymptomatic or with only mild symptoms.

May include flu-like symptoms such as fever, headache, rash, and swollen lymph nodes. CD4 count is typically normal or only slightly reduced.

- **Stage 2**: Clinical latency (asymptomatic or mild symptoms). The virus is active but reproduces at very low levels. Individuals may remain in this stage for several years with little to no symptoms. Persistent generalized lymphadenopathy (PGL) may occur. CD4 count begins to decline but remains above 350 cells/mm^3.
- **Stage 3:** Symptomatic HIV infection. More severe symptoms and opportunistic infections begin to emerge. Conditions may include unexplained chronic diarrhea, persistent fever, severe bacterial infections, and oral candidiasis (thrush). Weight loss of more than 10% of body weight. CD4 count drops below 350 cells/mm^3 but remains above 200 cells/mm^3.
- **Stage 4:** AIDS. Marked by the presence of one or more severe opportunistic infections or HIV-related cancers. Conditions include pneumocystis pneumonia (PCP), Kaposi's sarcoma, tuberculosis (TB), and severe wasting syndrome. CD4 count falls below 200 cells/mm^3. Severe immune system damage occurs, leading to life-threatening infections.

3. **Indicator for treatment decisions Initiation of antiretroviral therapy (ART):**

Guidelines: Current WHO guidelines recommend initiating ART in people with HIV regardless of CD4 and WHO clinical stage, but historically, ART was started based on $CD4^+$ T-cell thresholds to prevent opportunistic infections and delay progression to AIDS. However, in some places where there can be limited ART coverage, priority should be given to those with severe or advanced HIV disease (WHO clinical stages 3 or 4) or a $CD4^+$ cell count at or below 350 cells/mm^3. This guideline varies in different countries based on availability, accessibility, affordability, efficacy, and ease of administration of antiretroviral, among others.

Examples

Nigeria's 2014 ART clinical guidelines recommend that all people living with HIV with a $CD4^+$ cell count less than 500 cells/mm^3 should be given ART irrespective of clinical symptoms as soon as necessary laboratory and clinical assessments have been carried out and the person is deemed ready to receive ART. It, however, puts priority on individuals with a $CD4^+$ cell count <350 cells/mm^3.

South Africa's 2019 ART clinical guidelines recommended that people living with HIV (PLHIV) should start ART regardless of age, $CD4^+$ cell count, or clinical stage if they do not have any contraindications.

Ghana and Kenya's guidelines for ART 2011 recommend that for patients in WHO stages 1 and 2 with a $CD4^+$ cell count of 350 cells/mm^3 or below, ART should be initiated. For those in the same stage with a $CD4^+$ cell count above 350 cells/mm^3, ART can be deferred. For patients in WHO stages 3 and 4, ART should be started regardless of the $CD4^+$ cell count.

Generally, patients with TB disease, regardless of their $CD4^+$ cell count, are placed on ART. Similarly, individuals with HIV and HBV co-infection showing evidence of active or chronic liver disease, HIV-associated nephropathy, characterized by proteinuria, and impaired kidney function with or without peripheral edema are placed on ART regardless of their $CD4^+$ cell count. Laboratory results do not need to be available to start clients on ART on the same day.

4. **Monitoring response:** $CD4^+$ T-cell counts are used to monitor the effectiveness of ART. An increase in $CD4^+$ T-cells and a decrease in viral load indicate successful treatment.
5. **Management of opportunistic infections**
   - **Prophylaxis and treatment:** Patients with low $CD4^+$ T-cell counts are at increased risk for opportunistic infections like Pneumocystis jirovecii pneumonia (PCP), Mycobacterium avium complex (MAC), and fungal infections. Prophylactic treatments are often initiated based on $CD4^+$ T-cell counts.
   - **Immune reconstitution inflammatory syndrome (IRIS):** Some patients may develop IRIS when starting ART, as the recovering immune system begins to respond to preexisting infections. $CD4^+$ T-cell monitoring helps manage and anticipate this condition.

Differences between T helper cells, cytotoxic T lymphocytes, and natural killer cells:

Cytotoxic T lymphocytes (CTLs or $CD8^+$ T-cells):

- **Function:** CTLs directly kill infected or cancerous cells. They do this by inducing apoptosis (programmed cell death) in their target cells.
- **Activation:** They are activated by recognizing antigens presented by MHC Class I molecules, which are found on almost all nucleated cells.
- **Mechanism:** Once activated, CTLs release cytotoxic granules containing perforin and granzymes that penetrate the target cell membrane and trigger apoptosis. They can also kill cells through Fas–FasL interactions.
- **Markers:** They express the CD8 molecule on their surface, distinguishing them from $CD4^+$ T-cells.

T helper cells ($CD4^+$ T-cells):

- **Function:** T helper cells are primarily responsible for coordinating the immune response. They do not directly kill infected cells but instead help activate and direct other immune cells by releasing cytokines [141].
- **Activation:** They are activated by recognizing antigens presented by MHC Class II molecules on the

surface of APCs like dendritic cells, macrophages, and B-cells.

- **Mechanism:** Once activated, T helper cells can differentiate into various subsets (e.g., Th1, Th2, Th17, and Treg) that perform different functions, such as aiding B-cells to produce antibodies, activating macrophages, and recruiting other immune cells to the site of infection.
- **Markers:** They express the CD4 molecule on their surface, which helps in their identification.

Natural killer (NK) cells:

- **Function:** NK cells are part of the innate immune system and can kill virus-infected cells and tumor cells without prior sensitization. They play a crucial role in early defense against infections and tumors.
- **Activation**: Unlike T-cells, NK cells do not require antigen presentation by MHC molecules. They are activated by a lack of "self" signals (absence or reduced levels of MHC Class I molecules) and the presence of stress-induced ligands on target cells.
- **Mechanism:** NK cells kill target cells by releasing cytotoxic granules containing perforin and granzymes, like CTLs. They also produce cytokines like IFNγ to activate other immune cells.
- **Markers:** NK cells express a variety of surface markers, including CD16 (FcγRIII) and CD56 (a NCAM), but lack the CD3 marker that is present on T-cells.

5.2.3 Antibody-Mediated Adaptive Immunity (Humoral Immunity)

Antibody-mediated adaptive immunity, also known as humoral immunity, is a crucial component of the adaptive immune system. It protects against extracellular pathogens, such as bacteria and viruses, as well as their toxins, through the production of antibodies by B-cells.

B-cells, also known as B lymphocytes, are responsible for producing antibodies, presenting antigens, and forming memory cells that provide long-term immunity. They originate from HSCs in the bone marrow. After production in the HSCs, B-cells undergo a series of developmental stages, including the pro-B cell, pre-B cell, immature B-cell, and mature B-cell stages in the bone marrow [142]. During this process, they rearrange their Ig genes to create a unique B-cell receptor (BCR) for binding to antigens.

When a pathogen enters the body, APCs present fragments of the pathogen (antigens) to helper T-cells, which then activate B-cells. Also, APCs can directly present antigens to B-cells; in fact, B-cells themselves are considered a type of APC. The activated B-cells differentiate into plasma cells that secrete antibodies specific to the antigen. These antibodies neutralize pathogens by binding to them, marking them for destruction by phagocytes, or activating the complement system

5.2.3.1 *Antibody-Mediated Cytotoxicity*

Antibody-mediated cytotoxicity, also known as ADCC [143], is a critical immune mechanism through which antibodies, produced by B-cells, facilitate the destruction of infected or abnormal cells. In ADCC, antibodies bind to antigens on the surface of target cells, marking them for destruction. This binding forms a bridge between the target cell and immune effector cells, such as NK cells, macrophages, and neutrophils.

Fc receptors on the surface of NK cells and other effector cells recognize the Fc region of the bound antibodies [144]. Upon binding, NK cells are activated and release cytotoxic molecules like perforin and granzymes. Perforin creates pores in the membrane of the target cell, allowing granzymes to enter and induce apoptosis, which leads to cell death. This process is crucial for eliminating cells infected with viruses, as well as tumor cells and cells coated with specific antibodies.

5.2.3.2 Subsets of Mature B-Cells

- **Naive B-cells:** These are mature B-cells that have not yet encountered their specific antigen.
- **Activated B-cells:** These are B-cells that have encountered specific antigens and received necessary signals for activation.
- **Plasma cells:** These are differentiated B-cells that produce and secrete large amounts of antibodies.
- **Memory B-cells:** These are long-lived B-cells that provide rapid responses upon re-exposure to their specific antigen.
- **Regulatory B-cells (Breg):** These are a subset of B-cells that produce anti-inflammatory cytokines like IL-10, contributing to immune regulation and tolerance (see Table 5.10).

5.2.3.3 Process of Activation and Differentiation of B-Cells [145]

1. **Antigen encounter:** Circulating mature B-cells in the blood and lymphoid organs, such as the spleen and lymph nodes, may encounter different types of antigens. They bind to the antigen via the BCR.
2. **Helper T-cell assistance:** For most antigens, B-cells require help from $CD4^+$ Th cells (see Figure 5.7). With the help of Th cells, B-cells internalize the antigen, process it, and present peptide fragments on their surface via MHC Class II molecules. The cells again recognize these fragments and provide additional signals through cytokines and costimulatory molecules (e.g., CD40-CD40L interaction), which fully activate the B-cell. Activated B-cells undergo proliferation and differentiation.
3. **Antigen presentation:** B-cells express both MHC Class I and MHC Class II molecules, allowing other immune cells to present antigens to them. Rather than directly attacking antigens alone, B-cells can recognize and process antigens presented by APC, contributing to a more robust activation and coordination of the immune response. This dual expression enhances B-cells' ability to interact with both cytotoxic T-cells (via MHC I) and helper T-cells (via MHC II).
4. **Clonal expansion:** Activated B-cells proliferate, creating a clone of cells specific to the encountered antigen.

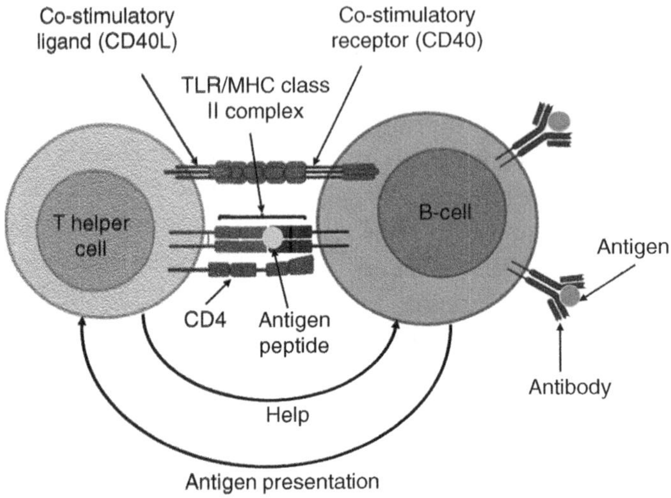

Figure 5.7 Overview of B-cell and T-cell interactions.

5. **Germinal center maturation:** Following clonal expansion, activated B-cells migrate into germinal centers, where they undergo somatic hypermutation and affinity maturation to improve antibody specificity. During this process, they may also undergo class-switch recombination (CSR), enabling them to produce different antibody isotypes. Only B-cells with high-affinity receptors are selected to continue toward differentiation into plasma cells or memory B-cells (Table 5.11).
6. **Differentiation:** These activated B-cells differentiate into two main types of cells – plasma cells and memory cells. During this differentiation process, the B-cell can undergo CSR, which makes it produce specific antibodies (see Table 5.12).
   - **Plasma cells:** These cells are specialized for producing and secreting large amounts of antibodies. Plasma cells typically reside in the bone marrow or lymphoid tissues.
   - **Memory B-cells:** These are long-lived cells that persist in the body with the "immunological memory" of the antigen and provide rapid and robust responses upon re-exposure to the same antigen.

Class switch recombination (CSR) in B-cells

CSR is a biological mechanism that changes the B-cell's production of Ig from one class to another (such as from IgM to IgG, IgA, or IgE) during differentiation, without altering the antigen specificity. This process involves the recombination of the heavy chain constant (C) region gene segments in the DNA. The specific cytokines released by the helper T-cells influence which antibody class the B-cell will produce. For example, IL-4 promotes switching to IgE, while TGF-β and IL-5 promote switching to IgA [146, 147].

Table 5.11 Stages of maturation of B-cells.

Stage	Location	Key characteristics	Gene rearrangement	Surface markers	Importance
Pro-B-cell	Bone marrow	Early stage; initiation of heavy chain rearrangement	Rearrangement of the immunoglobulin (Ig) heavy chain (IgH) genes begins. D-J rearrangement; V-DJ rearrangement	CD19, CD10 c-Kit/CD117	Sets the foundation for the BCR by beginning heavy chain gene rearrangement.
Pre-B cell	Bone marrow	Formation of pre-BCR; initiation of light chain rearrangement	Successful rearrangement of the IgH gene leads to the expression of the μ heavy chain. The μ chain pairs with surrogate light chains (VpreB and λ5) to form the pre-BCR. Light chain rearrangement: Initiation of light chain (IgL) gene rearrangement, starting with the κ (kappa) locus. If unsuccessful, the λ (lambda) locus undergoes rearrangement.	CD19, CD10, pre-BCR (μ heavy chain with surrogate light chains)	Signaling through pre-BCR promotes survival, proliferation, and further differentiation.

(*Continued*)

Table 5.11 (Continued)

Stage	Location	Key characteristics	Gene rearrangement	Surface markers	Importance
Immature B-cell	Bone marrow	Expression of complete BCR (IgM); undergoes selection for self-tolerance	Successful light chain gene rearrangement results in the expression of a complete BCR (IgM).	IgM, CD19, CD20	Undergoes positive and negative selection to ensure functionality and self-tolerance. Immature B-cells are tested for self-reactivity in the bone marrow. Those that strongly bind self-antigens are either deleted (clonal deletion), rendered nonfunctional (energy), or undergo receptor editing to change their specificity, thereby preventing autoimmunity.
Mature B-cell	Bone marrow (initially), then peripheral lymphoid organs	Expression of IgM and IgD; fully capable of responding to antigens	These cells are now fully capable of responding to antigens and participating in immune responses. Alternative mRNA splicing may occur.	IgM, IgD, CD19, CD20	Naive mature B-cells circulate in blood and lymphoid organs, ready to respond to antigens. Upon activation, they differentiate into plasma or memory B-cells.

Table 5.12 Classification of antibodies.

Type	Structure	Characteristics	Functions	General location
IgM	Pentamer (in serum) or monomer (on B-cell surface)	Largest antibody; 5 units linked by J chain (in serum). The first antibody produced in response to infection or antigen exposure	Effective at agglutination and complement activation; initial immune response	Blood, lymph, B-cell surface
IgG	Monomer	The most abundant antibody in blood and extracellular fluid; can cross the placenta (providing passive immunity to the fetus); long-lived	Neutralization, opsonization, complement activation, and antibody-dependent cellular cytotoxicity (ADCC)	Blood, lymph, and tissues cross the placenta
IgA	Monomer (in serum) or dimer (in mucosal secretions like saliva, tears, milk)	Found in mucosal areas (saliva, tears, breast milk) Prevents pathogen adherence to mucosal surfaces	Mucosal immunity, neutralization, prevents colonization by pathogens	Mucosal surfaces, blood, lymph, and secretions
IgE	Monomer	Involved in allergic reactions and defense against parasitic infections. Binds to Fc receptors on mast cells and basophils, triggering the release of histamine and other mediators	Allergic responses, defense against parasites	Blood, tissues; bind to mast cells and basophils
IgD	Monomer	Found in small amounts in the blood (primarily on the surface of B lymphocytes) and tissues. The exact function is not fully understood, but is likely involved in the initiation of immune responses	May regulate B-cell activation and antibody production	Blood, tissues; mainly on the B-cell surface

5.3 Virus Persistence and Immune Evasion

5.3.1 Virus Persistence

Virus persistence refers to the ability of viruses to remain in the host organism for extended periods, weeks, and sometimes for a lifetime, without being completely cleared by the immune system [148]. There are two main mechanisms of virus persistence: (a) Chronic infection – continuous, long-term presence of the virus in the host, usually with ongoing virus replication and periodic shedding in the presence of immune response. Examples include HBV and HCV infection. (b) Latent infection – a dormant state in which the virus is present within host cells but does not actively replicate. However, reactivation can occur under certain conditions. Examples include HSV and varicella-zoster virus (VZV) [149]. Similarly, some viruses can integrate their genome into the host's DNA, becoming part of the host cell's genetic material, where it can remain dormant and be reactivated later. Also, HIV integrates into the host's genome and can persist in a latent state within T-cells [150].

5.3.2 Virus Immune Evasion

Virus immune evasion refers to the various strategies that viruses use to avoid detection and destruction by the host's immune system [151, 152]. These strategies enhance their ability to infect and replicate within the host. For example, viruses can evade immune response by (i) changing surface proteins to avoid recognition by antibodies in a process termed genetic and antigenic variability (e.g., Influenza virus, HIV) [153, 154]. (ii) Inhibiting antigen presentation by blocking the presentation of viral antigens on MHC molecules to prevent T-cell recognition (e.g., cytomegalovirus [CMV], Epstein–Barr virus (EBV)) [155, 156]. (iii) Interfering with immune signaling. For example, producing proteins that inhibit IFN signaling or other immune pathways (e.g., influenza virus, HCV). (iv) Mimicry and decoys. Viruses can produce viral proteins that mimic host proteins to interfere with immune responses (e.g., poxviruses) [157]. (v) Replication in immune-privileged sites. Some viruses can avoid being targeted and eliminated by the immune system by hiding in body regions where immune surveillance is limited, such as the central nervous system. Examples include the rabies virus and CMV.

5.3.2.1 Viral Strategies for Evading Innate Immunity

Interfering with IFN Signaling Viruses can interfere with IFN signaling in several ways, such as blocking the production of IFN, preventing IFN from binding to its receptor, or inhibiting the effects of IFN on infected cells.

Examples of virus invasion of the immune response:

Influenza A virus: The nonstructural protein 1 (NS1) of influenza A virus can inhibit the activation of IFN regulatory factors (IRFs) and other transcription factors necessary for IFN production [158].

HCV: The HCV NS3/4A protease cleaves the adaptor protein mitochondrial antiviral signaling protein (MAVS) (also known as IPS-1, VISA, or Cardif) on mitochondria, which is essential for signaling to IRF3 and NF-κB [159].

Ebola virus: Ebola virus glycoprotein (GP) can be shed from the viral surface and act as a decoy receptor for IFNs [160]. Additionally, secreted GP (sGP) may bind to cellular IFN receptors.

HIV: The Vpu protein downregulates the expression of tetherin, a protein induced by IFN that prevents the release of viral particles from the cell surface [161].

Vaccinia virus: Vaccinia virus encodes the protein E3L, which binds dsRNA and prevents the activation of PKR, a key IFN-stimulated antiviral protein [162].

HSV: HSV ICP0 protein degrades basal levels of IFI16, which are responsible for mediating the antiviral effects of IFN [163].

SARS-CoV-2: Several SARS-CoV-2 proteins, including ORF6, ORF8, and NSP1, can interfere with various steps in the type 1 IFN signaling pathway. For example, ORF6 blocks the nuclear translocation of STAT1 and STAT2, critical transcription factors in the IFN pathway [164].

Adenoviruses: The E1A protein of adenovirus can bind to and inhibit the function of IRF3 [165], a key transcription factor in the IFN production pathway.

Zika virus: Zika virus NS5 protein targets and degrades STAT2, a crucial component of the IFN signaling pathway.

NOTE: The JAK–STAT pathway is a critical signaling mechanism used by a variety of growth factors and IFNs to regulate immune responses, cell growth, differentiation, and survival [166]. This pathway is highly conserved across many species and is essential for the proper functioning of the immune system.

Disguising Viral PAMPs

Under normal conditions, PRRs on immune cells can detect PAMPs and trigger an immune response. PAMPs are molecules that are unique to viruses and other pathogens. However, viruses have evolved mechanisms to hide their PAMPs from immune recognition. They do so either by modifying their viral genomes or proteins or by hiding them within the host cell.

Modifying Viral Genomes

1. **Capping viral RNA:** Many viruses modify the 5′ end of their RNA with a cap structure like that found in host mRNA. This cap structure helps in evading detection by PRRs such as RIG-I. Example: Influenza virus adds a 5′ cap to its RNA through mechanisms like "cap snatching," where it "steals" caps from host mRNAs [167].
2. **RNA methylation:** Methylation of viral RNA can help it mimic host RNA, making it less recognizable by PRRs. Example: DENV methylates its RNA cap, reducing recognition by melanoma differentiation protein 5 (MDA5) and RIG-I [168].
3. **Protein glycosylation:** Viruses can add sugar molecules to their proteins, a process known as glycosylation, to mimic host cell proteins and avoid immune detection.

Hiding PAMPs Within Host Cells

1. **Replication in membrane-bound compartments:** Some viruses replicate within membrane-bound compartments derived from the host cell, effectively hiding their replication intermediates from PRRs. Example: HCV replicates in membrane-associated replication complexes within the cytoplasm, which helps shield its RNA from PRRs like RIG-I and MDA5 [169].
2. **Sequestration in the nucleus:** DNA viruses can replicate within the nucleus, where detection by cytoplasmic PRRs is avoided. Example: HSV replicates its DNA in the nucleus, limiting exposure to cytosolic DNA sensors [170].

Producing Viral Proteins That Interfere with Immune Components Such as PRRs

1. **NS1 Protein of influenza virus:** The NS1 protein of influenza virus binds to viral RNA and hides it from detection by RIG-I [171]. Additionally, NS1 can inhibit the processing of viral RNA to prevent recognition.
2. **Viral proteins that degrade host PRRs:** Some viral proteins can degrade or inhibit PRRs directly. Example: Kaposi's sarcoma-associated herpesvirus (KSHV) lytic proteins K3 and K5 have E3 ubiquitin ligase activities that degrade MHC Class 1, intercellular adhesion molecule 1 (ICAM-1), and B7–2 (CD86) to reduce their surface expression [172].

Inhibiting Apoptosis Apoptosis is programmed cell death that helps to eliminate virus-infected cells. Viruses can prevent apoptosis in several ways, including by expressing proteins that inhibit apoptosis or interfering with the apoptotic signaling pathway such as

Expression of Anti-apoptotic Proteins

1. **Mimic anti-apoptotic proteins:** Some viruses encode proteins that mimic the function of cellular Bcl-2 [173], which is a key regulatory protein in the process of apoptosis, particularly in the mitochondrial (intrinsic) pathway. Example: EBV expresses Bam HI fragment H rightward open reading frame 1 (BHRF1) and Bam HI fragment H leftward open reading frame 1 (BALF1), which are homologs of Bcl-2 and can bind to and inhibit pro-apoptotic proteins, preventing the initiation of apoptosis [174].
2. **Inhibition of pro-apoptotic factors:** Viral proteins can directly bind and inhibit pro-apoptotic factors [175]. Example: CMV encodes the protein viral mitochondrial inhibitor of apoptosis (vMIA), which binds to and inhibits the pro-apoptotic protein Bax, preventing mitochondrial outer membrane permeabilization (MOMP) and cytochrome c release [176].

Interference with the Death Receptor Pathway The death receptor pathway, also known as the extrinsic pathway of apoptosis, involves the modulation of death receptors on the cell surface by specific ligands, leading to the activation of downstream signaling cascades that result in apoptosis. Viruses can interfere with this pathway by:

1. **Blocking death receptors:** Viruses can express proteins that interfere

with death receptor signaling. Example: Cowpox virus infection causes the expression of cytokine response modifier A (CrmA), a serpin that inhibits caspase-8 [177]. Caspase-8 is a key initiator of caspases in the extrinsic apoptotic pathway triggered by death receptors like Fas and TNF receptors.

2. **Producing decoy receptors:** Some viruses produce soluble decoy receptors that bind to and neutralize death ligands. Example: Myxoma virus (mouse virus) produces a soluble factor called MT-2 (a TNF receptor homolog) [178], preventing it from triggering apoptosis through its receptor. Due to the ability of myxoma virus to infect and kill rabbit cells, researchers are investigating the potential of using myxoma virus as an oncolytic virotherapy for treating certain human cancers.

Inhibition of Mitochondrial Pathway

1. **Targeting mitochondrial proteins:** Viral proteins can target mitochondrial proteins to inhibit apoptosis. Example: Adenovirus E1B 19 K protein interacts with and inhibits pro-apoptotic proteins like Bax and Bak, preventing mitochondrial outer membrane permeabilization [179].
2. **Stabilizing mitochondrial membrane:** Viral proteins can stabilize the mitochondrial membrane to prevent cytochrome c release. Example: HBV X protein can interact with mitochondrial proteins, especially voltage-dependent anion channels (VDACs), to stabilize the membrane and inhibit apoptosis [180].

Inhibition of Caspases

i. **Direct caspase inhibition:** Some viral proteins bind to and inhibit already active (executioner) caspases to stop the apoptotic process. Example: Poxviruses, such as vaccinia virus, express serpins like Serpin-2 (SPI-2) or B13R that inhibit caspases 3 and 7 [181, 182], preventing the execution phase of apoptosis.

ii. **Preventing caspase activation:** Other viruses act earlier in the pathway by preventing the activation of initiator caspases, stopping apoptosis before it fully begins. Example: HSV ICP27 protein inhibits apoptosis [183] and allows the virus to replicate more efficiently.

NOTES: Caspases are a family of cysteine proteases that cleave substrates after aspartic acid residues. They exist in the cell as inactive proenzymes (zymogens) and are activated in a proteolytic cascade during apoptosis. Upon activation, caspases cleave specific target proteins, including structural and regulatory molecules, which dismantle critical cellular components. This leads to the systematic disassembly of the cell, resulting in controlled and noninflammatory cell death.

Caspases are categorized into initiator caspases (e.g., caspase-8, caspase-9), which trigger the apoptotic cascade, and effector caspases (e.g., caspase-3, caspase-7), which execute the destruction of cellular components.

Modulation of Cellular Signaling Pathways

1. **Activation of survival pathways:** Viruses can activate cellular survival pathways to counteract apoptotic

signals. Example: HPV E6 protein activates the PI3K/Akt pathway, promoting cell survival and inhibiting apoptosis that drives carcinogenesis [184].

2. **Inhibition of pro-apoptotic signaling:** Viral proteins can inhibit signaling pathways that lead to apoptosis. Example: HIV Nef protein inhibits the activation of pro-apoptotic, caspase-mediated pathways in infected cells [185].

5.3.2.2 *Viral Strategies for Evading Adaptive Immunity*

Antigenic Variation This process involves changes in viral antigens – the molecules on the surface of the virus that are recognized by B-cells and T-cells. By altering these antigens, viruses can evade the immune responses that would otherwise neutralize them. The underlined mechanism is involved in antigenic variation are:

a. **Genetic mutation:** Small changes (point mutations) in the viral genome can alter amino acids in viral proteins, including surface antigens. These mutations can occur rapidly, especially in viruses with high mutation rates, such as RNA viruses. Example: The influenza virus undergoes frequent point mutations in its HA and NA proteins, a process known as antigenic drift [186]. This results in the emergence of new strains that can evade preexisting immunity in the population, necessitating an annual dose for influenza vaccines.

b. **Reassortment:** In viruses with segmented genomes, segments can be exchanged between different strains during the co-infection of a single host cell. This reassortment can lead to significant changes in viral antigens. Example: Influenza A virus can undergo an antigenic shift, where reassortment between human, avian, and swine influenza viruses creates novel HA or NA subtypes [187–189]. These new subtypes can potentially cause pandemics, as seen with the 2009 H1N1 pandemic.

c. **Recombination:** Some viruses can undergo recombination, where segments of genetic material are exchanged between different viral genomes. This can lead to new antigenic profiles. Example: Coronaviruses, including SARS-CoV-2, have shown evidence of recombination events that can alter spike protein epitopes, contributing to immune escape, known as "variant of concern" [190, 191]. Influenza virus and enteroviruses (e.g., poliovirus, coxsackievirus) also undergo genetic recombination. Recombination observed in poliovirus and non-polio enteroviruses, called recombinant circulating vaccine-derived polioviruses, has contributed to outbreaks in Nigeria and other countries, complicating eradication efforts [192].

d. **Glycosylation:** The addition of sugar molecules to viral proteins (glycosylation) can mask antigenic sites from recognition by antibodies. For example, HIV heavily glycosylates its envelope glycoprotein gp120, creating a "glycan shield" that hides conserved epitopes from neutralizing antibodies [193, 194].

Immunosuppression Viruses can suppress the immune system, making it less effective at fighting infection. Specifically, it can express proteins that directly inhibit the function of immune cells, such as T-cells, B-cells, and NK cells. For example, HIV infects and depletes $CD4^{+}$ T helper cells, which are crucial for coordinating the immune response. This depletion leads to a weakened immune system, making the host susceptible to opportunistic infections. Likewise, CMV encodes proteins like UL18, which mimic MHC Class I molecules to engage inhibitory receptors on NK cells, preventing their activation and killing of infected cells [195, 196].

Immune Checkpoint Inhibition Viruses can express molecules that bind to immune checkpoint receptors on T-cells. This can prevent T-cells from becoming activated and killing virus-infected cells. Viral manipulation of immune checkpoints includes:

1. **Human immunodeficiency virus:** HIV infection leads to upregulation of programmed cell death protein 1 (PD-1), an inhibitory receptor expressed on $CD8^{+}$ T-cells, which contributes to T-cell exhaustion and reduced antiviral activity. Additionally, HIV upregulates cytotoxic T-lymphocyte antigen 4 (CTLA-4) on T-cells, further contributing to T-cell dysfunction. CTLA-4 is an inhibitory receptor that competes with the stimulatory receptor CD28 for binding to B7 molecules (CD80/CD86) on APCs [197].
2. **Hepatitis B virus:** HBV-infected individuals exhibit increased expression of PD-1 on HBV-specific T-cells, which is associated with T-cell exhaustion and impaired antiviral response [198]. Elevated levels of programmed cell death-ligand 1 (PD-L1) in hepatocytes and other liver cells contribute to the persistence of HBV by inhibiting effective T-cell responses [199]. CTLA-4 expression is also increased in T-cells in chronic HBV infection [200], contributing to immune suppression and chronicity of the infection.
3. **Hepatitis C virus:** Also, chronic HCV infection is characterized by high levels of PD-1 expression on HCV-specific $CD8^{+}$ T-cells, leading to T-cell exhaustion and impaired viral clearance [201]. Increased PD-L1 expression on liver cells and immune cells within the liver microenvironment facilitates the virus's evasion of immune detection and elimination [202].
4. **EBV:** EBV infection is associated with the upregulation of PD-1 on T-cells [203], contributing to immune evasion during both primary infection and reactivation. EBV-infected cells can also express PD-L1, which helps them evade immune surveillance.

5.4 Virus-Induced Autoimmune Diseases

Virus-induced autoimmune diseases occur when a viral infection triggers an immune response that mistakenly targets the body's tissues. This phenomenon can lead to chronic conditions where the immune system continues to attack self-antigens even after the virus is cleared. The mechanisms behind virus-induced autoimmunity involve genetic predisposition, molecular mimicry, bystander activation, epitope spreading, and persistent viral infection (see Table 5.13).

5.4.1 Mechanisms of Virus-Induced Autoimmunity

5.4.1.1 Molecular Mimicry

This occurs when viral proteins share structural similarities with host proteins [204]. The immune system's attempt to eliminate the virus results in cross-reactivity with self-antigens, leading to autoimmune disease. For example, the EBV has been linked to multiple sclerosis due to shared epitopes between viral and myelin proteins [205, 206].

Table 5.13 Overview of the mechanism of virus-induced autoimmune diseases.

Autoimmune disease	Trigger/ organisms	Mechanism	Pathology
Guillain–Barré syndrome (GBS)	CMV, EBV, Zika virus and *Campylobacter jejuni*	Molecular mimicry	Demyelination of peripheral nerves leads to muscle weakness and paralysis.
Type 1 diabetes mellitus (T1DM)	Coxsackievirus B	Bystander activation, molecular mimicry	Autoimmune destruction of pancreatic beta cells leads to insulin deficiency.
Multiple sclerosis	EBV, HHV-6	Epitope spreading, molecular mimicry	Autoimmune attack on the myelin sheath of CNS neurons, leading to neurological dysfunction.
Systemic lupus erythematosus (SLE)	EBV	Persistent infection, continuous B-cell activation	Production of autoantibodies against nuclear and cytoplasmic antigens leads to systemic inflammation and multi-organ damage.
Autoimmune hepatitis	HCV, HBV	Molecular mimicry, chronic inflammation	Autoimmune destruction of liver cells leads to liver inflammation and fibrosis.

5.4.1.2 *Bystander Activation*

Viral infections can cause widespread inflammation, immune activation, and autoimmune disease [207, 208]. This environment can lead to the activation of autoreactive T-cells that were previously dormant. Cytokines and other inflammatory mediators released during viral infection can contribute to this process.

5.4.1.3 *Epitope Spreading*

During an immune response to a viral infection, the initial immune attack on viral antigens can expand to include a broader range of host antigens. This can happen due to the release of self-antigens from damaged tissues or due to the processing of viral and self-antigens together by APCs. This can either be intramolecular spreading, where the immune response spreads from one epitope to another within the same protein, or intermolecular spreading, where the immune response spreads from epitopes on one antigen to epitopes on different antigens.

5.5 Study Question

Case Study Questions

5.1 A 32-year-old patient presents with symptoms of joint pain, fatigue, and skin rashes. The patient reports a recent clinical case of EBV that was treated successfully. Laboratory tests reveal the presence of autoantibodies, and the patient is diagnosed with SLE.

(a) Analyze how the recent viral infection might have contributed to the development of autoimmune disease in this patient. What mechanisms are proposed to explain the link between viral infections and the onset of autoimmunity?

(b) Discuss the role of molecular mimicry and how it might be involved in the pathogenesis of SLE following a viral infection. Provide examples of how viral proteins may resemble self-antigens.

(c) Evaluate the implications of viral-induced autoimmunity for clinical management. What considerations should be considered when treating patients with a history of viral infections who present with autoimmune symptoms?

(d) How can understanding the relationship between viral infections and autoimmune diseases improve preventive strategies or therapeutic approaches for such conditions?

5.2 A 45-year-old patient presents with chronic HBV infection. Despite antiviral therapy, the virus remains detectable in the patient's blood over several years, and the patient experiences intermittent episodes of liver inflammation. Analyze the case in terms of viral persistence mechanisms.

(a) What factors contribute to the persistence of HBV in this patient despite antiviral treatment? Discuss the role of viral and host factors in maintaining chronic infection.

(b) How does the HBV lifecycle, including its ability to establish a stable reservoir in hepatocytes, contribute to the difficulty of eradicating the virus?

(c) What are the potential impacts of viral persistence on the patient's liver health, and how might this influence long-term management and treatment strategies?

5.3 A 50-year-old patient with a history of HIV infection presents with an opportunistic infection and a recent decline in her CD4 count. Additionally, her blood test shows an increase in the number of cytotoxic T-cells.

(a) How does the decline in CD4 count affect the patient's overall immune system function and susceptibility to opportunistic infections? Provide a detailed explanation of the role $CD4^+$ T-cells play in the immune response and how their loss impacts the patient's health.

(b) Interpret the significance of the elevated cytotoxic T-cell count in the context of this patient's HIV infection. How can increased cytotoxic T-cell activity relate to the patient's current clinical symptoms and progression of the disease?

Objectives

5.1 Which of the following cells is primarily responsible for the direct killing of virus-infected cells?

(a) B-cells
(b) Helper T-cells
(c) Cytotoxic T-cells
(d) Macrophages

5.2 Which antibody class is most associated with the primary immune response to viral infections?

(a) IgG
(b) IgD
(c) IgM
(d) IgE

5.3 What is the main role of IFN in the immune response to viral infections?

(a) To enhance the production of antibodies
(b) To directly destroy virus particles
(c) To inhibit viral replication in infected cells
(d) To facilitate the activation of T-cells

5.4 Which of the following best describes the role of MHC Class I molecules in viral infections?

(a) They present viral antigens to cytotoxic T-cells.
(b) They present viral antigens to helper T-cells.
(c) They directly neutralize viral particles.
(d) They enhance the activity of B-cells.

5.5 What is the primary function of helper T-cells ($CD4^+$ T-cells) in the immune response to viral infections?

(a) To kill infected cells directly
(b) To produce antibodies
(c) To assist in the activation of B-cells and cytotoxic T-cells
(d) To destroy viral particles

5.6 Which immune response is characterized by the production of memory cells that provide long-term protection against re-infection by the same virus?

(a) Acute inflammatory response
(b) Primary immune response

(c) Adaptive immune response
(d) Innate immune response

5.7 In which part of the immune response do natural killer (NK) cells play a crucial role?
(a) In recognizing and destroying virus-infected cells without prior sensitization
(b) In producing antibodies specific to viral antigens
(c) In presenting viral antigens to helper T-cells
(d) In secreting cytokines to enhance the immune response

5.8 Which type of immune response is typically enhanced by vaccination against viral diseases?
(a) Innate immune response
(b) Primary immune response
(c) Adaptive immune response
(d) Acute inflammatory response

5.9 Which of the following cytokines is most associated with the activation of the adaptive immune response in viral infections?
(a) Interleukin-1 (IL-1)
(b) Tumor necrosis factor-alpha (TNFα)
(c) Interferon-gamma (IFNγ)
(d) Interleukin-6 (IL-6)

5.10 Which immune component is responsible for the neutralization of extracellular viral particles?
(a) Cytotoxic T-cells
(b) Helper T-cells
(c) Antibodies (immunoglobulins)
(d) Natural killer (NK) cells

References

1. Coates, M., Blanchard, S., and MacLeod, A.S., *Innate antimicrobial immunity in the skin: a protective barrier against bacteria, viruses, and fungi*. PLoS Pathog., 2018. **14**(12): e1007353.
2. Nguyen, A.V. and Soulika, A.M., *The dynamics of the skin's immune system*. Int. J. Mol. Sci., 2019. **20**(8): p. 1811.
3. Trompette, A. and Ubags, N.D., *Skin barrier immunology from early life to adulthood*. Mucosal Immunol., 2023. **16**(2): p. 194–207.
4. Messingham, K.N., et al., *Multiple cell types support productive infection and dynamic translocation of infectious Ebola virus to the surface of human skin*. Sci. Adv., 2025. **11**(1): eadr6140.
5. Kappus, M.R. and Sterling, R.K., *Extrahepatic manifestations of acute hepatitis B virus infection*. Gastroenterol. Hepatol. (N. Y), 2013. **9**(2): p. 123–126.
6. Cozzani, E., et al., *Cutaneous manifestations of HAV, HBV. HCV*. Ital. J. Dermatol. Venerol., 2021. **156**(1): p. 5–12.
7. Eccles, R., *Respiratory mucus and persistence of virus on surfaces*. J. Hosp. Infect., 2020. **105**(2): p. 350.
8. Wallace, L.E., et al., *Respiratory mucus as a virus-host range determinant*. Trends Microbiol., 2021. **29**(11): p. 983–992.
9. Schaefer, A. and Lai, S.K., *The biophysical principles underpinning muco-trapping functions of antibodies*. Hum. Vaccines Immunother., 2022. **18**(2): 1939605.
10. Brainard, J., et al., *Risk factors for transmission of Ebola or Marburg virus disease: a systematic review and meta-analysis*. Int. J. Epidemiol., 2016. **45**(1): p. 102–116.
11. Price, E., *Could the severity of COVID-19 be increased by low gastric acidity?* Crit. Care, 2020. **24**(1): p. 456.

12. Zhou, J., et al., *Human intestinal tract serves as an alternative infection route for Middle East respiratory syndrome coronavirus*. Sci. Adv., 2017. **3**(11): eaao4966.
13. Glass, R.I., Parashar, U.D., and Estes, M.K., *Norovirus gastroenteritis*. N. Engl. J. Med., 2009. **361**(18): p. 1776–1785.
14. Andino, R., et al., *The Picornaviridae family: knowledge gaps, animal models, countermeasures, and prototype pathogens*. J. Infect. Dis., 2023. **228**(Supplement_6): p. S427–S445.
15. Tennant, S.M., et al., *Influence of gastric acid on susceptibility to infection with ingested bacterial pathogens*. Infect. Immun., 2008. **76**(2): p. 639–645.
16. Malamud, D., et al., *Antiviral activities in human saliva*. Adv. Dent. Res., 2011. **23**(1): p. 34–37.
17. White, M.R., et al., *Multiple components contribute to ability of saliva to inhibit influenza viruses*. Oral Microbiol. Immunol., 2009. **24**(1): p. 18–24.
18. Mason, C.P. and Tarr, A.W., *Human lectins and their roles in viral infections*. Molecules, 2015. **20**(2): p. 2229–2271.
19. Kobayashi, K., et al., *Protein-bound sialic acid in saliva contributes directly to salivary anti-influenza virus activity*. Sci. Rep., 2022. **12**(1): p. 6636.
20. Gwyer Findlay, E., Currie, S.M., and Davidson, D.J., *Cationic host defence peptides: potential as antiviral therapeutics*. BioDrugs, 2013. **27**(5): p. 479–493.
21. Groot, F., et al., *Histatin 5-derived peptide with improved fungicidal properties enhances human immunodeficiency virus type 1 replication by promoting viral entry*. J. Virol., 2006. **80**(18): p. 9236–9243.
22. Ramírez-Labrada, A., et al., *All about (NK cell-mediated) death in two acts and an unexpected encore: initiation, execution and activation of adaptive immunity*. Front. Immunol., 2022. **13**: 896228.
23. Zhang, F., et al., *Inhibition of major histocompatibility complex-I antigen presentation by sarbecovirus ORF7a proteins*. PNAS, 2022. **119**(41): e2209042119.
24. Chan, C.J., Smyth, M.J., and Martinet, L., *Molecular mechanisms of natural killer cell activation in response to cellular stress*. Cell Death Differ., 2014. **21**(1): p. 5–14.
25. Weiss-Steider, B., et al., *Expression of MICA, MICB and NKG2D in human leukemic myelomonocytic and cervical cancer cells*. J. Exp. Clin. Cancer Res., 2011. **30**(1): p. 37.
26. Orange, J.S., et al., *Viral evasion of natural killer cells*. Nat. Immunol., 2002. **3**(11): p. 1006–1012.
27. Greene, T.T., et al., *A Herpesviral induction of RAE-1 NKG2D ligand expression occurs through release of HDAC mediated repression*. Elife, 2016. **5**: e14749.
28. Wang, W., et al., *NK cell-mediated antibody-dependent cellular cytotoxicity in Cancer immunotherapy*. Front. Immunol., 2015. **6**: p. 368.
29. Kwaa, A.K.R., Talana, C.A.G., and Blankson, J.N., *Interferon alpha enhances NK cell function and the suppressive capacity of HIV-specific CD8(+) T cells*. J. Virol., 2019. **93**(3): e01541-18.
30. Madera, S., et al., *Type I IFN promotes NK cell expansion during viral infection by protecting NK cells against fratricide*. J. Exp. Med., 2016. **213**(2): p. 225–233.
31. Van Acker, H.H., et al., *CD56 in the immune system: more than a marker for cytotoxicity?* Front. Immunol., 2017. **8**: 892.
32. Poli, A., et al., *CD56bright natural killer (NK) cells: an important NK cell subset*. Immunology, 2009. **126**(4): p. 458–465.
33. Aguilar, O.A., et al., *The CD16 and CD32b Fc-gamma receptors regulate antibody-mediated responses in mouse natural killer cells*. J. Leukocyte Biol., 2023. **113**(1): p. 27–40.

34. Lopez-Vergès, S., et al., *CD57 defines a functionally distinct population of mature NK cells in the human CD56dimCD16+ NK-cell subset*. Blood, 2010. **116**(19): p. 3865–3874.
35. Pazina, T., et al., *Regulation of the functions of natural cytotoxicity Receptors by interactions with diverse ligands and alterations in splice variant expression*. Front. Immunol., 2017. **8**: p. 369.
36. Sivori, S., et al., *NKp46 is the major triggering receptor involved in the natural cytotoxicity of fresh or cultured human NK cells. Correlation between surface density of NKp46 and natural cytotoxicity against autologous, allogeneic or xenogeneic target cells*. Eur. J. Immunol., 1999. **29**(5): p. 1656–1666.
37. Shibuya, A., et al., *DNAM-1, a novel adhesion molecule involved in the cytolytic function of T lymphocytes*. Immunity, 1996. **4**(6): p. 573–581.
38. Bottino, C., et al., *Identification of PVR (CD155) and Nectin-2 (CD112) as cell surface ligands for the human DNAM-1 (CD226) activating molecule*. J. Exp. Med., 2003. **198**(4): p. 557–567.
39. Hirayama, D., Iida, T., and Nakase, H., *The phagocytic function of macrophage-enforcing innate immunity and tissue homeostasis*. Int. J. Mol. Sci., 2017. **19**(1): p. 92.
40. Chen, S., et al., *Macrophages in immunoregulation and therapeutics*. Signal Transduct. Target. Ther., 2023. **8**(1): p. 207.
41. Palladino, M., *Complete blood count alterations in COVID-19 patients: a narrative review*. Biochem. Med. (Zagreb), 2021. **31**(3): 030501.
42. Guo, Z., et al., *Lymphopenia caused by virus infections and the mechanisms beyond*. Viruses, 2021. **13**(9): p. 1876.
43. Raadsen, M., et al., *Thrombocytopenia in virus infections*. J. Clin. Med., 2021. **10**(4): p. 877.
44. Zheng, S.-Y., et al., *Association between secondary thrombocytosis and viral respiratory tract infections in children*. Sci. Rep., 2016. **6**(1): p. 22964.
45. Thompson, M.R., et al., *Pattern recognition receptors and the innate immune response to viral infection*. Viruses, 2011. **3**(6): p. 920–940.
46. Wicherska-Pawłowska, K., Wróbel, T., and Rybka, J., *Toll-like receptors (TLRs), NOD-like Receptors (NLRs), and RIG-I-like Receptors (RLRs) in innate immunity. TLRs, NLRs, and RLRs ligands as immunotherapeutic agents for hematopoietic diseases*. Int. J. Mol. Sci., 2021. **22**(24): 13397.
47. Wang, W., et al., *Activation of innate antiviral immune response via double-stranded RNA-dependent RLR receptor-mediated necroptosis*. Sci. Rep., 2016. **6**(1): p. 22550.
48. Yu, M. and Levine, S.J., *Toll-like receptor, RIG-I-like receptors and the NLRP3 inflammasome: key modulators of innate immune responses to double-stranded RNA viruses*. Cytokine Growth Factor Rev., 2011. **22**(2): p. 63–72.
49. Saito, T. and Gale, M. Jr., *Differential recognition of double-stranded RNA by RIG-I-like receptors in antiviral immunity*. J. Exp. Med., 2008. **205**(7): p. 1523–1527.
50. Kumagai, Y., Takeuchi, O., and Akira, S., *TLR9 as a key receptor for the recognition of DNA*. Adv. Drug Delivery Rev., 2008. **60**(7): p. 795–804.
51. Eisfeld, H.S., et al., *Viral glycoproteins induce NLRP3 Inflammasome activation and Pyroptosis in macrophages*. Viruses, 2021. **13**(10): 2076.
52. Shepardson, K.M., et al., *Induction of antiviral immune response through recognition of the repeating subunit pattern of viral capsids is toll-like receptor 2 dependent*. mBio, 2017. **8**(6): 01356-17.
53. López de Padilla, C.M. and Niewold, T.B., *The type I interferons: basic concepts and clinical relevance in immune-mediated*

inflammatory diseases. Gene, 2016. **576**(1 Pt 1): p. 14–21.
54. Moreau, T.R.J., et al., *Heterogeneity and functions of the 13 IFN-α subtypes - lucky for some?* Eur. J. Immunol., 2023. **53**(8): e2250307.
55. Mah, A.Y. and Cooper, M.A., *Metabolic regulation of natural killer cell IFN-γ production*. Crit. Rev. Immunol., 2016. **36**(2): p. 131–147.
56. Yu, D., et al., *Design and evaluation of novel interferon lambda analogs with enhanced antiviral activity and improved drug attributes*. Drug Des. Devel. Ther., 2016. **10**: p. 163–182.
57. Andreakos, E., Zanoni, I., and Galani, I.E., *Lambda interferons come to light: dual function cytokines mediating antiviral immunity and damage control*. Curr. Opin. Immunol., 2019. **56**: p. 67–75.
58. Walker, F.C., Sridhar, P.R., and Baldridge, M.T., *Differential roles of interferons in innate responses to mucosal viral infections*. Trends Immunol., 2021. **42**(11): p. 1009–1023.
59. Muhammad, I., et al., *Chasing virus replication and infection: PAMP-PRR interaction drives type I interferon production, which in turn activates ISG expression and ISGylation*. Viruses, 2025. **17**(4): p. 528.
60. Nan, Y., Wu, C., and Zhang, Y.J., *Interplay between Janus kinase/signal transducer and activator of transcription signaling activated by type I interferons and viral antagonism*. Front. Immunol., 2017. **8**: p. 1758.
61. de Weerd, N.A., et al., *Structure–function of type I and III interferons*. Curr. Opin. Immunol., 2024. **86**: 102413.
62. Wang, Y., et al., *Negative regulation of type I IFN signaling by phosphorylation of STAT2 on T387*. Embo J., 2017. **36**(2): p. 202–212.
63. Sadzak, I., et al., *Recruitment of Stat1 to chromatin is required for interferon-induced serine phosphorylation of Stat1 transactivation domain*. PNAS, 2008. **105**(26): p. 8944–8949.
64. Tamiya, T., et al., *Suppressors of cytokine signaling (SOCS) proteins and JAK/STAT pathways: regulation of T-cell inflammation by SOCS1 and SOCS3*. Arterioscler., Thromb., Vasc. Biol., 2011. **31**(5): p. 980–985.
65. Bonnet, M.C., et al., *PKR stimulates NF-kappaB irrespective of its kinase function by interacting with the IkappaB kinase complex*. Mol. Cell. Biol., 2000. **20**(13): p. 4532–4542.
66. Puthenveetil, S., et al., *Controlling activation of the RNA-dependent protein kinase by siRNAs using site-specific chemical modification*. Nucleic Acids Res., 2006. **34**(17): p. 4900–4911.
67. Ventoso, I., et al., *Translational resistance of late alphavirus mRNA to eIF2alpha phosphorylation: a strategy to overcome the antiviral effect of protein kinase PKR*. Genes Dev., 2006. **20**(1): p. 87–100.
68. Rios-Fuller, T.J., et al., *Translation regulation by eIF2α phosphorylation and mTORC1 signaling pathways in non-communicable diseases (NCDs)*. Int. J. Mol. Sci., 2020. **21**(15): 5301.
69. Gómez-Herranz, M., Taylor, J., and Sloan, R.D., *IFITM proteins: understanding their diverse roles in viral infection, cancer, and immunity*. J. Biol. Chem., 2023. **299**(1): 102741.
70. Smith, S., et al., *IFITM proteins-cellular inhibitors of viral entry*. Curr. Opin. Virol., 2014. **4**: p. 71–77.
71. Manivannan, P., Siddiqui, M.A., and Malathi, K., *RNase L amplifies interferon signaling by inducing protein kinase R-mediated antiviral stress granules*. J. Virol., 2020. **94**(13): e00205-20.
72. Gusho, E., Baskar, D., and Banerjee, S., *New advances in our understanding of the "unique" RNase L in host pathogen interaction and immune signaling*. Cytokine, 2020. **133**: 153847.
73. Schwartz, S.L. and Conn, G.L., *RNA regulation of the antiviral protein 2′-5′-oligoadenylate*

synthetase. Wiley Interdiscip. Rev. RNA, 2019. **10**(4): e1534.

74. Zhang, J., et al., *Characterization, expression pattern and antiviral activities of oligoadenylate synthetase in Chinese Giant Salamander, Andrias davidianus*. Dev. Comp. Immunol., 2022. **129**: 104347.
75. Calderon, B.M. and Conn, G.L., *A human cellular noncoding RNA activates the antiviral protein 2′-5′-oligoadenylate synthetase 1*. J. Biol. Chem., 2018. **293**(41): p. 16115–16124.
76. Huang, H., et al., *Dimeric structure of pseudokinase RNase L bound to 2-5A reveals a basis for interferon-induced antiviral activity*. Mol. Cell, 2014. **53**(2): p. 221–234.
77. Weiss, C.M., et al., *The interferon-induced exonuclease ISG20 exerts antiviral activity through upregulation of type I interferon response proteins*. Msphere, 2018. **3**(5). doi: 10.1128/msphere.00209-18.
78. Han, Y., et al., *Critical role of cellular microRNAs in virus infection: decades of progress*. Anim. Zoonoses, 2025. **1**(4): p. 385–393.
79. Fleming, S.B., *Viral inhibition of the IFN-induced JAK/STAT Signalling pathway: development of live attenuated vaccines by mutation of viral-encoded IFN-antagonists*. Vaccines (Basel), 2016. **4**(3): p. 23.
80. Seliger, B., Ruiz-Cabello, F., and Garrido, F., *IFN inducibility of major histocompatibility antigens in tumors*. Adv. Cancer Res., 2008. **101**: p. 249–276.
81. Donnelly, R.P., Dickensheets, H., and O'Brien, T.R., *Interferon-lambda and therapy for chronic hepatitis C virus infection*. Trends Immunol., 2011. **32**(9): p. 443–450.
82. Wang, X., et al., *Recent advances of type I interferon on the regulation of immune cells and the treatment of systemic lupus erythematosus*. J. Inflammation Res., 2025. **18**: p. 4533–4549.
83. Aricò, E., et al., *Type I interferons and Cancer: an evolving story demanding novel clinical applications*. Cancers (Basel), 2019. **11**(12): p. 1943.
84. Zhang, X., et al., *Double-edged effects of interferons on the regulation of cancer-immunity cycle*. Oncoimmunology, 2021. **10**(1): 1929005.
85. Verbist, K.C. and Klonowski, K.D., *Functions of IL-15 in anti-viral immunity: multiplicity and variety*. Cytokine, 2012. **59**(3): p. 467–478.
86. Ma, W.T., et al., *The protective and pathogenic roles of IL-17 in viral infections: friend or foe?* Open Biol., 2019. **9**(7): 190109.
87. Azaiz, M.B., et al., *Deciphering the balance of IL-6/IL-10 cytokines in severe to critical COVID-19 patients*. Immunobiology, 2022. **227**(4): 152236.
88. Dimitriadis, K., et al., *The role of interleukins in HBV infection: a narrative review*. J. Pers. Med., 2023. **13**(12): p. 1675.
89. Laeremans, T., Janssens, A., and Aerts, J.L., *From natural defenders to therapeutic warriors: NK cells in HIV immunotherapy*. Immunotherapy, 2025. **17**(2): p. 133–145.
90. Chaudhary, R., et al., *Interplay of host and viral factors in inflammatory pathway mediated cytokine storm during RNA virus infection*. Curr. Res. Immunol., 2023. **4**: 100062.
91. Sergerie, Y., Rivest, S., and Boivin, G., *Tumor necrosis factor-α and interleukin-1β play a critical role in the resistance against lethal herpes simplex virus encephalitis*. J. Infect. Dis., 2007. **196**(6): p. 853–860.
92. Jang, D.I., et al., *The role of tumor necrosis factor alpha (TNF-α) in autoimmune disease and current TNF-α inhibitors in therapeutics*. Int. J. Mol. Sci., 2021. **22**(5): 2719.
93. Kunimura, K., et al., *S-1-propenylcysteine improves TNF-α-induced vascular endothelial barrier dysfunction by suppressing the GEF-H1/RhoA/Rac pathway*. Cell Commun. Signal., 2021. **19**(1): p. 17.
94. Spiropoulou, C.F. and Srikiatkhachorn, A., *The role of endothelial activation in dengue*

hemorrhagic fever and hantavirus pulmonary syndrome. Virulence, 2013. **4**(6): p. 525–536.
95. Mohd Zawawi, Z., et al., *Prospective roles of tumor necrosis factor-alpha (TNF-α) in COVID-19: prognosis, therapeutic and management*. Int. J. Mol. Sci., 2023. **24**(7): p. 6142.
96. Riyaz Tramboo, S., et al., *The critical impacts of cytokine storms in respiratory disorders*. Heliyon, 2024. **10**(9): e29769.
97. Stefferl, A., et al., *The role of TNF-alpha in fever: opposing actions of human and murine TNF-alpha and interactions with IL-beta in the rat*. Br. J. Pharmacol., 1996. **118**(8): p. 1919–1924.
98. Leis, A.A., et al., *Tumor necrosis factor-alpha signaling may contribute to chronic West Nile virus post-infectious Proinflammatory state*. Front. Med. (Lausanne), 2020. **7**: p. 164.
99. Silva, L.B., et al., *The role of TNF-α as a Proinflammatory cytokine in pathological processes*. Open Dent. J., 2019. **13**: p. 332–338.
100. Traynor, T.R., et al., *Sleep and body temperature responses in an acute viral infection model are altered in interferon type I receptor-deficient mice*. Brain, Behav., and Immun., 2006. **20**(3): p. 290–299.
101. Zampronio, A.R., Soares, D.M., and Souza, G.E., *Central mediators involved in the febrile response: effects of antipyretic drugs*. Temperature (Austin), 2015. **2**(4): p. 506–521.
102. Sokol, C.L. and Luster, A.D., *The chemokine system in innate immunity*. Cold Spring Harbor Perspect. Biol., 2015. **7**(5): a016303.
103. Shannon, L.A., et al., *CCR7/CCL19 controls expression of EDG-1 in T cells*. J. Biol. Chem., 2012. **287**(15): p. 11656–11664.
104. Chow, M.T. and Luster, A.D., *Chemokines in cancer*. Cancer Immunol. Res., 2014. **2**(12): p. 1125–1131.
105. Dunkelberger, J.R. and Song, W.-C., *Complement and its role in innate and adaptive immune responses*. Cell Res., 2010. **20**(1): p. 34–50.
106. Klos, A., et al., *The role of the anaphylatoxins in health and disease*. Mol. Immunol., 2009. **46**(14): p. 2753–2766.
107. Kouser, L., et al., *Emerging and novel functions of complement protein C1q*. Front. Immunol., 2015. **6**: p. 317.
108. Chen, L., et al., *Inflammatory responses and inflammation-associated diseases in organs*. Oncotarget, 2018. **9**(6): p. 7204–7218.
109. Perez, L., *Acute phase protein response to viral infection and vaccination*. Arch. Biochem. Biophys., 2019. **671**: p. 196–202.
110. Chaplin, D.D., *Overview of the immune response*. J. Allergy Clin. Immunol., 2010. **125**(2 Suppl 2): p. S3–S23.
111. Rastogi, I., et al., *Role of B cells as antigen presenting cells*. Front. Immunol., 2022. **13**: 954936.
112. Popi, A.F., Longo-Maugéri, I.M., and Mariano, M., *An overview of B-1 cells as antigen-presenting cells*. Front. Immunol., 2016. **7**: p. 138.
113. Wieczorek, M., et al., *Major histocompatibility complex (MHC) class I and MHC class II proteins: conformational plasticity in antigen presentation*. Front. Immunol., 2017. **8**: p. 292.
114. Hewitt, E.W., *The MHC class I antigen presentation pathway: strategies for viral immune evasion*. Immunology, 2003. **110**(2): p. 163–169.
115. Roche, P.A. and Furuta, K., *The ins and outs of MHC class II-mediated antigen processing and presentation*. Nat. Rev. Immunol., 2015. **15**(4): p. 203–216.
116. Huang, J., Meyer, C., and Zhu, C., *T cell antigen recognition at the cell membrane*. Mol. Immunol., 2012. **52**(3–4): p. 155–164.

117. Mariuzza, R.A., Agnihotri, P., and Orban, J., *The structural basis of T-cell receptor (TCR) activation: an enduring enigma*. J. Biol. Chem., 2020. **295**(4): p. 914–925.
118. Petrova, G., Ferrante, A., and Gorski, J., *Cross-reactivity of T cells and its role in the immune system*. Crit. Rev. Immunol., 2012. **32**(4): p. 349–372.
119. Xu, X., Li, H., and Xu, C., *Structural understanding of T cell receptor triggering*. Cell. Mol. Immunol., 2020. **17**(3): p. 193–202.
120. Liu, Q., et al., *Circulating tumor cells counts are associated with CD8+ T cell levels in programmed death-ligand 1-negative non-small cell lung cancer patients after radio therapy: a retrospective study*. Medicine (Baltimore), 2021. **100**(29): e26674.
121. Xue, D., et al., *Role of regulatory T cells and CD8(+) T lymphocytes in the dissemination of circulating tumor cells in primary invasive breast cancer*. Oncol. Lett., 2018. **16**(3): p. 3045–3053.
122. Andersen, M.H., et al., *Cytotoxic T cells*. J. Invest. Dermatol., 2006. **126**(1): p. 32–41.
123. Campos Canesso, M.C., et al., *Identification of antigen-presenting cell–T cell interactions driving immune responses to food*. Science, 2025. **387**(6739): eado5088.
124. Zhu, X. and Zhu, J., *CD4 T helper cell subsets and related human immunological disorders*. Int. J. Mol. Sci., 2020. **21**(21): p. 8011.
125. McHeyzer-Williams, M., et al., *Helper T-cell-regulated B-cell immunity*. Microbes Infect., 2003. **5**(3): p. 205–212.
126. Szodoray, P., et al., *Integration of T helper and BCR signals governs enhanced plasma cell differentiation of memory B cells by regulation of CD45 phosphatase activity*. Cell Rep., 2021. **36**(6): 109525.
127. Alberts, B.J.A., Lewis, J., et al., *Molecular Biology of the Cell*. 2002, Garland Science.
128. Sun, L., et al., *T cells in health and disease*. Signal Transduc. Target. Ther., 2023. **8**(1): p. 235.
129. Kumar, B.V., Connors, T.J., and Farber, D.L., *Human T cell development, localization, and function throughout life*. Immunity, 2018. **48**(2): p. 202–213.
130. Oparaugo, N.C., et al., *Human regulatory T cells: understanding the role of Tregs in select autoimmune skin diseases and post-transplant nonmelanoma skin cancers*. Int. J. Mol. Sci., 2023. **24**(2): p. 1527.
131. Huang, L., et al., *Induction and regulation of Th1-inducing cytokines by bacterial DNA, lipopolysaccharide, and heat-inactivated bacteria*. Infect. Immun., 1999. **67**(12): p. 6257–6263.
132. Kumar, S., et al., *Dendritic cell-mediated Th2 immunity and immune disorders*. Int. J. Mol. Sci., 2019. **20**(9): p. 2159.
133. Curtis, M.M. and Way, S.S., *Interleukin-17 in host defence against bacterial, mycobacterial and fungal pathogens*. Immunology, 2009. **126**(2): p. 177–185.
134. Licona-Limón, P., et al., *Th9 cells drive host immunity against gastrointestinal worm infection*. Immunity, 2013. **39**(4): p. 744–757.
135. Walker, B. and McMichael, A., *The T-cell response to HIV*. Cold Spring Harb. Perspect. Med., 2012. **2**(11): a007054.
136. Chen, X., et al., *Host immune response to influenza a virus infection*. Front. Immunol., 2018. **9**: p. 320.
137. Ye, B., et al., *T-cell exhaustion in chronic hepatitis B infection: current knowledge and clinical significance*. Cell Death Dis., 2015. **6**(3): p. e1694.
138. Rosales, R. and Rosales, C., *Immune therapy for human papillomaviruses-related cancers*. World J. Clin. Oncol., 2014. **5**(5): p. 1002–1019.

139. Wahid, R., Cannon, M.J., and Chow, M., *Virus-specific CD4+ and CD8+ cytotoxic T-cell responses and long-term T-cell memory in individuals vaccinated against polio*. J. Virol., 2005. **79**(10): p. 5988–5995.
140. Edri, A., et al., *The Ebola-Glycoprotein Modulates the Function of Natural Killer Cells*. Front Immunol., 2018. **9**: 1428.
141. Swain, S.L., McKinstry, K.K., and Strutt, T.M., *Expanding roles for CD4+ T cells in immunity to viruses*. Nat. Rev. Immunol., 2012. **12**(2): p. 136–148.
142. Kim, D.R., *Proteomic changes during the B cell development*. J. Chromatogr. B, 2005. **815**(1): p. 295–303.
143. Gómez Román, V.R., Murray, J.C., and Weiner, L.M., *Chapter 1 – antibody-dependent cellular cytotoxicity (ADCC)*, in *Antibody Fc*, M.E. Ackerman and F. Nimmerjahn, Editors. 2014, Academic Press: Boston. p. 1–27.
144. Dixon, K.J., Wu, J., and Walcheck, B., *Engineering anti-tumor monoclonal antibodies and fc Receptors to enhance ADCC by human NK cells*. Cancers (Basel), 2021. **13**(2): p. 312.
145. Janeway, C.A. Jr., et al., *B-Cell Activation by Armed Helper T Cells*. 2001, Garland Science: New York.
146. Stavnezer, J., Guikema, J.E., and Schrader, C.E., *Mechanism and regulation of class switch recombination*. Annu. Rev. Immunol., 2008. **26**: p. 261–292.
147. Dauba, A., et al., *The immunoglobulin heavy chain super enhancer controls class switch recombination in developing B cells*. Sci. Rep., 2024. **14**(1): p. 7370.
148. Oldstone, M.B.A., *Viral persistence: parameters, mechanisms and future predictions*. Virology, 2006. **344**(1): p. 111–118.
149. Costa-Silva, M., et al., *Concurrent reactivation of varicella zoster virus and herpes simplex virus in an immunocompetent elderly male*. An. Bras. Dermatol., 2019. **94**(6): p. 762–763.
150. Chen, J., et al., *The reservoir of latent HIV*. Front. Cell. Infect. Microbiol., 2022. **12**: 945956.
151. Alcami, A. and Koszinowski, U.H., *Viral mechanisms of immune evasion*. Trends Microbiol., 2000. **8**(9): p. 410–418.
152. Nelemans, T. and Kikkert, M., *Viral innate immune evasion and the pathogenesis of emerging RNA virus infections*. Viruses, 2019. **11**(10): p. 961.
153. van de Sandt, C.E., Kreijtz, J.H., and Rimmelzwaan, G.F., *Evasion of influenza a viruses from innate and adaptive immune responses*. Viruses, 2012. **4**(9): p. 1438–1476.
154. Fang, Z., et al., *Targeting HIV-1 immune escape mechanisms: key advances and challenges in HIV-1 vaccine design*. Microbiol. Res., 2025. **299**: 128229.
155. Silva, J.M., Alves, C.E.C., and Pontes, G.S., *Epstein-Barr virus: the mastermind of immune chaos*. Front. Immunol., 2024. **15**: 1297994.
156. Peters, V.B. and Sperber, K.E., *The effect of viruses on the ability to present antigens via the major histocompatibility complex*. Microbes Infect., 1999. **1**(4): p. 335–345.
157. Johnston, J.B. and McFadden, G., *Poxvirus immunomodulatory strategies: current perspectives*. J. Virol., 2003. **77**(11): p. 6093–6100.
158. Tisoncik, J.R., et al., *The NS1 protein of influenza a virus suppresses interferon-regulated activation of antigen-presentation and immune-proteasome pathways*. J. Gen. Virol., 2011. **92**(Pt 9): p. 2093–2104.
159. Li, X.-D., et al., *Hepatitis C virus protease NS3/4A cleaves mitochondrial antiviral signaling protein off the mitochondria to evade innate immunity*, 2005. **102**(49): p. 17717–17722.

160. Dolnik, O., et al., *Ectodomain shedding of the glycoprotein GP of Ebola virus*. Embo J., 2004. **23**(10): p. 2175–2184.
161. Douglas, J.L., et al., *Vpu directs the degradation of the human immunodeficiency virus restriction factor BST-2/Tetherin via a {beta}TrCP-dependent mechanism*. J. Virol., 2009. **83**(16): p. 7931–7947.
162. Romano, P.R., et al., *Inhibition of double-stranded RNA-dependent protein kinase PKR by vaccinia virus E3: role of complex formation and the E3 N-terminal domain*. Mol. Cell. Biol., 1998. **18**(12): p. 7304–7316.
163. Sodroski, C.N. and Knipe, D.M., *Nuclear interferon-stimulated gene product maintains heterochromatin on the herpes simplex viral genome to limit lytic infection*. PNAS, 2023. **120**(45): e2310996120.
164. Miorin, L., et al., *SARS-CoV-2 Orf6 hijacks Nup98 to block STAT nuclear import and antagonize interferon signaling*. PNAS, 2020. **117**(45): p. 28344–28354.
165. Zemke, N.R., et al., *Adenovirus E1A binding to DCAF10 targets proteasomal degradation of RUVBL1/2 AAA+ ATPases required for quaternary assembly of multiprotein machines, innate immunity, and responses to metabolic stress*. J. Virol., 2023. **97**(12): e0099323.
166. Seif, F., et al., *The role of JAK-STAT signaling pathway and its regulators in the fate of T helper cells*. Cell Commun. Signal., 2017. **15**(1): p. 23.
167. Gu, W., et al., *Influenza a virus preferentially snatches noncoding RNA caps*. Rna, 2015. **21**(12): p. 2067–2075.
168. Chang, D.C., et al., *Evasion of early innate immune response by 2'-O-methylation of dengue genomic RNA*. Virology, 2016. **499**: p. 259–266.
169. Neufeldt, C.J., et al., *The hepatitis C virus-induced membranous web and associated nuclear transport machinery limit access of pattern recognition Receptors to viral replication sites*. PLoS Pathog., 2016. **12**(2): e1005428.
170. Zheng, C., *Evasion of cytosolic DNA-stimulated innate immune responses by herpes simplex virus 1*. J. Virol., 2018. **92**(6): e00099-17.
171. Jureka, A.S., et al., *The influenza NS1 protein modulates RIG-I activation via a strain-specific direct interaction with the second CARD of RIG-I*. J. Biol. Chem., 2020. **295**(4): p. 1153–1164.
172. Brulois, K., et al., *Kaposi's sarcoma-associated herpesvirus K3 and K5 ubiquitin E3 ligases have stage-specific immune evasion roles during lytic replication*. J. Virol., 2014. **88**(16): p. 9335–9349.
173. Kvansakul, M., Caria, S., and Hinds, M.G., *The Bcl-2 family in host-virus interactions*. Viruses, 2017. **9**(10): p. 290.
174. Fitzsimmons, L. and Kelly, G.L., *EBV and apoptosis: the viral master regulator of cell fate?* Viruses, 2017. **9**(11): p. 339.
175. Clarke, P. and Tyler, K.L., *Apoptosis in animal models of virus-induced disease*. Nat. Rev. Microbiol., 2009. **7**(2): p. 144–155.
176. Deng, Y., Águeda-Pinto, A., and Brune, W., *No time to die: how cytomegaloviruses suppress apoptosis, necroptosis, and Pyroptosis*. Viruses, 2024. **16**(8): p. 1272.
177. Renatus, M., et al., *Crystal structure of the apoptotic suppressor CrmA in its cleaved form*. Structure, 2000. **8**(7): p. 789–797.
178. Rahman, M.M. and McFadden, G., *Myxoma virus-encoded host range protein M029: a multifunctional antagonist targeting multiple host antiviral and innate immune pathways*. Vaccines (Basel), 2020. **8**(2): p. 244.
179. Tarakanova, V.L. and Wold, W.S., *Adenovirus E1A and E1B-19K proteins protect human hepatoma cells from transforming growth factor beta1-induced apoptosis*. Virus Res., 2010. **147**(1): p. 67–76.

180. Rahmani, Z., et al., *Hepatitis B virus X protein colocalizes to mitochondria with a human voltage-dependent anion channel, HVDAC3, and alters its transmembrane potential.* J. Virol., 2000. **74**(6): p. 2840–2846.
181. Koehler, H.S. and Jacobs, B.L., *Subversion of programed cell death by poxviruses.* Curr. Top Microbiol. Immunol., 2023. **442**: p. 105–131.
182. Chea, L.S., et al., *Novel modified vaccinia virus Ankara vector expressing anti-apoptotic gene B13R delays apoptosis and enhances humoral responses.* J. Virol., 2019. **93**(5). doi: 10.1128/jvi.01648-18.
183. Yu, X. and He, S., *The interplay between human herpes simplex virus infection and the apoptosis and necroptosis cell death pathways.* Virol. j., 2016. **13**(1): p. 77.
184. Zhang, L., et al., *The role of the PI3K/Akt/mTOR signalling pathway in human cancers induced by infection with human papillomaviruses.* Mol. Cancer, 2015. **14**: p. 87.
185. Segura, J., et al., *HIV-1 release requires Nef-induced caspase activation.* PLoS One, 2023. **18**(2): e0281087.
186. Perofsky, A.C., et al., *Antigenic drift and subtype interference shape A(H3N2) epidemic dynamics in the United States.* medRxiv, 2024.
187. Petrova, V.N. and Russell, C.A., *The evolution of seasonal influenza viruses.* Nat. Rev. Microbiol., 2018. **16**(1): p. 47–60.
188. Ganti, K., et al., *Influenza A virus reassortment in mammals gives rise to genetically distinct within-host subpopulations.* Nat. Commun., 2022. **13**(1): p. 6846.
189. Markin, A., et al., *Revealing reassortment in influenza A viruses with TreeSort.* Mol. Biol. Evol., 2025. **42**(8): msaf133.
190. Carabelli, A.M., et al., *SARS-CoV-2 variant biology: immune escape, transmission and fitness.* Nat. Rev. Microbiol., 2023. **21**(3): p. 162–177.
191. Magazine, N., et al., *Immune epitopes of SARS-CoV-2 spike protein and considerations for universal vaccine development.* bioRxiv, 2023.
192. Castro, C.J., et al., *Complete genome sequences of nine double recombinant vaccine-derived novel oral poliovirus type 2 genomes from Nigeria 2023–2024.* Microbiol. Resour. Announc., 2024. **13**(12): e00881–24.
193. Rathore, U., et al., *Glycosylation of the core of the HIV-1 envelope subunit protein gp120 is not required for native trimer formation or viral infectivity.* J. Biol. Chem., 2017. **292**(24): p. 10197–10219.
194. Behrens, A.-J. and Crispin, M., *Structural principles controlling HIV envelope glycosylation.* Curr. Opin. Struct. Biol., 2017. **44**: p. 125–133.
195. Revilleza, M.J., et al., *How the virus outsmarts the host: function and structure of cytomegalovirus MHC-I-like molecules in the evasion of natural killer cell surveillance.* J. Biomed. Biotechnol., 2011. **2011**: 724607.
196. Beck, S. and Barrell, B.G., *Human cytomegalovirus encodes a glycoprotein homologous to MHC class-I antigens.* Nature, 1988. **331**(6153): p. 269–272.
197. Larsson, M., et al., *Molecular signatures of T-cell inhibition in HIV-1 infection.* Retrovirology, 2013. **10**: p. 31.
198. Li, M., et al., *HBcAg induces PD-1 upregulation on CD4+T cells through activation of JNK, ERK and PI3K/AKT pathways in chronic hepatitis-B-infected patients.* Lab. Invest., 2012. **92**(2): p. 295–304.
199. Zhang, Z., et al., *Dynamic programmed death 1 expression by virus-specific CD8 T cells correlates with the outcome of acute hepatitis B.* Gastroenterology, 2008. **134**(7): p. 1938–1949.e3.
200. Apol, Á.D., et al., *The role of CTLA-4 in T cell exhaustion in chronic hepatitis B virus infection.* Viruses, 2023. **15**(5): p. 1141.

201. Nakamoto, N., et al., *Functional restoration of HCV-specific CD8 T cells by PD-1 blockade is defined by PD-1 expression and compartmentalization*. Gastroenterology, 2008. **134**(7): p. 1927–1937.
202. Pipitone, R.M., et al., *The PD-1/PD-L1 axis in the biology of MASLD*. Int. J. Mol. Sci., 2024. **25**(7): p. 3671.
203. Wang, H., et al., *The PD-1/PD-L1 pathway and Epstein-Barr virus*. Eur. J. Med. Res., 2025. **30**(1): p. 486.
204. Sundaresan, B., et al., *The role of viral infections in the onset of autoimmune diseases*. Viruses, 2023. **15**(3): p. 782.
205. Thomas, O.G., Rickinson, A., and Palendira, U., *Epstein-Barr virus and multiple sclerosis: moving from questions of association to questions of mechanism*. Clin. Transl. Immunology, 2023. **12**(5): e1451.
206. Robinson, W.H. and Steinman, L., *Epstein-Barr virus and multiple sclerosis*. Science, 2022. **375**(6578): p. 264–265.
207. Shim, C.H., et al., *Emerging role of bystander T cell activation in autoimmune diseases*. BMB Rep., 2022. **55**(2): p. 57–64.
208. Fujinami, R.S., et al., *Molecular mimicry, bystander activation, or viral persistence: infections and autoimmune disease*. Clin. Microbiol. Rev., 2006. **19**(1): p. 80–94.

6 Antivirals and Vaccines

6.1 Antivirals

Antivirals are a class of medications designed for the treatment of viral infections. They act by inhibiting different replication stages of the viral life cycle, playing a crucial role in both treating acute viral infections and managing chronic viral conditions. Unlike antibiotics, which target a broad range of bacterial infections, antivirals are usually tailored to specific viruses. However, scientists are working on developing broad-spectrum antiviral agents.

6.1.1 History and Development of Antiviral Agents

The need for the development of antivirals began in the late 20th century, primarily to create antiviral agents that combat life-threatening viruses such as human immunodeficiency virus (HIV), influenza, and herpes simplex virus (HSV). However, the discovery of the first antiviral drug, idoxuridine (IDU), was in the early 1960s [1]. IDU was approved in 1963 for the topical treatment of HSV [1]. It was initially synthesized in 1959 by Prusoff as a potential anticancer drug [2]; however, Herrmann demonstrated in 1961 that it also exhibited activity against HSV and vaccinia virus [3]. Following this, advancements in molecular biology and virology have significantly accelerated the development of more effective and targeted antiviral therapies (see Table 6.1).

Most targets of antiviral agents include viral enzymes such as reverse transcriptase and protease, viral entry and fusion mechanisms, and viral genome replication processes. However, there are several drawbacks to these therapies, which include the development of viral resistance, potential side effects, limited effectiveness against latent or dormant viruses, and the challenge of targeting viruses that rapidly mutate.

6.1.2 Mechanism of Antiviral Agents

Antiviral agents are a critical component in the fight against viral infections. Unlike bacteria, which can be targeted by antibiotics

Table 6.1 List of antiviral agents and their mechanism of action.

Viruses indicated for	Chemical name	Brand name(s)	Mechanism of action	Common side effects
Herpes simplex virus (HSV), varicella-zoster virus (VZV)	Acyclovir	Zovirax	Inhibits viral DNA polymerase	Nausea, diarrhea, headache, dizziness, and fatigue
	Valacyclovir	Valtrex	Prodrug of acyclovir; inhibits viral DNA polymerase	Nausea, headache, dizziness, and abdominal pain
	Famciclovir	Famvir	Prodrug of penciclovir; inhibits viral DNA polymerase	Headache, nausea, diarrhea, and fatigue
Influenza A and B	Oseltamivir	Tamiflu	Inhibits neuraminidase, blocking viral release	Nausea, vomiting, headache, and fatigue
	Zanamivir	Relenza	Inhibits neuraminidase, blocking viral release	Cough, throat irritation, headache, and dizziness
Influenza A	Amantadine	Symmetrel	Inhibits viral M2 protein, preventing uncoating	Nausea, dizziness, insomnia, and nervousness
	Rimantadine	Flumadine	Inhibits viral M2 protein, preventing uncoating	Nausea, dizziness, insomnia, and headache
Human immunodeficiency virus (HIV), hepatitis B virus (HBV)	Tenofovir disoproxil fumarate	Viread	Nucleotide reverse transcriptase inhibitor (NRTI)	Nausea, diarrhea, headache, and renal toxicity
	Lamivudine	Epivir	NRTI	Nausea, diarrhea, headache, and fatigue
HIV	Zidovudine	Retrovir	NRTI	Anemia, neutropenia, nausea, headache, and muscle pain
	Emtricitabine	Emtriva	NRTI	Nausea, diarrhea, headache, hyperpigmentation of palms/soles
	Abacavir	Ziagen	NRTI	Hypersensitivity reaction, nausea, headache, and fatigue

(*Continued*)

Table 6.1 (Continued)

Viruses indicated for	Chemical name	Brand name(s)	Mechanism of action	Common side effects
	Efavirenz	Sustiva	Non-nucleoside reverse transcriptase inhibitor (NNRTI)	Rash, dizziness, insomnia, and abnormal dreams
	Dolutegravir	Tivicay	Integrase strand transfer inhibitor (INSTI)	Insomnia, headache, weight gain, and hypersensitivity reaction
	Raltegravir	Isentress	Integrase strand transfer inhibitor (INSTI)	Nausea, headache, diarrhea, and fever
	Darunavir	Prezista	Protease inhibitor	Diarrhea, nausea, rash, increased cholesterol/ triglycerides
	Atazanavir	Reyataz	Protease inhibitor	Jaundice, nausea, headache, and abdominal pain
	Lopinavir/ ritonavir	Kaletra	Protease inhibitor	Diarrhea, nausea, and increased cholesterol/ triglycerides
Hepatitis C virus (HCV)	Sofosbuvir	Sovaldi	Inhibits NS5B RNA-dependent RNA polymerase	Fatigue, headache, nausea, and insomnia
	Ledipasvir/ sofosbuvir	Harvoni	Inhibits NS5A (Ledipasvir) and NS5B (Sofosbuvir)	Fatigue, headache, and nausea
	Velpatasvir/ sofosbuvir	Epclusa	Inhibits NS5A (Velpatasvir) and NS5B (Sofosbuvir)	Fatigue, headache, and nausea
	Glecaprevir/ pibrentasvir	Mavyret	Inhibits NS3/4A protease (Glecaprevir) and NS5A (pibrentasvir)	Headache, fatigue, and nausea
HCV	Peginterferon alfa-2b	PegIntron	Enhances immune response	Flu-like symptoms, fatigue, depression, and hematologic toxicity
Hepatitis B virus (HBV)	Entecavir	Baraclude	Inhibits viral DNA polymerase	Headache, fatigue, dizziness, and nausea

Table 6.1 (Continued)

Viruses indicated for	Chemical name	Brand name(s)	Mechanism of action	Common side effects
HCV and HBV	Peginterferon alfa-2a	Pegasys	Enhances immune response	Flu-like symptoms, fatigue, depression, and hematologic toxicity
	Valganciclovir	Valcyte	Prodrug of ganciclovir; inhibits viral DNA polymerase	Neutropenia, thrombocytopenia, anemia, renal toxicity
	Cidofovir	Vistide	Inhibits viral DNA polymerase	Nephrotoxicity, neutropenia, and ocular toxicity
Cytomegalovirus (CMV), HSV, and varicella-zoster virus (VZV)	Foscarnet	Foscavir	Inhibits viral DNA polymerase	Nephrotoxicity, electrolyte imbalances, and seizures
Ebola virus	Remdesivir (Veklury)	Veklury	Inhibits viral RNA-dependent RNA polymerase	Nausea, headache, increased liver enzymes, and allergic reactions
Lassa virus	Ribavirin		Inhibits RNA synthesis and mRNA capping	Anemia, fatigue, headache, nausea
Crimean–Congo Hemorrhagic fever virus	Ribavirin		Inhibits RNA synthesis and mRNA capping	Anemia, fatigue, headache, and nausea

due to their unique cellular machinery, viruses are obligate intracellular pathogens. This means they replicate within the host's living cells, making it challenging to target the virus without also harming the host cell. The mechanisms by which antiviral agents work to inhibit viral replication and manage infections are examined in this session (see Figure 6.1).

6.1.2.1 Antiviral That Targets Viral Entry

Some antiviral agents prevent viruses from entering host cells. They achieve this by blocking the receptors on the host cell surface or interfering with the viral proteins necessary for attachment and fusion.

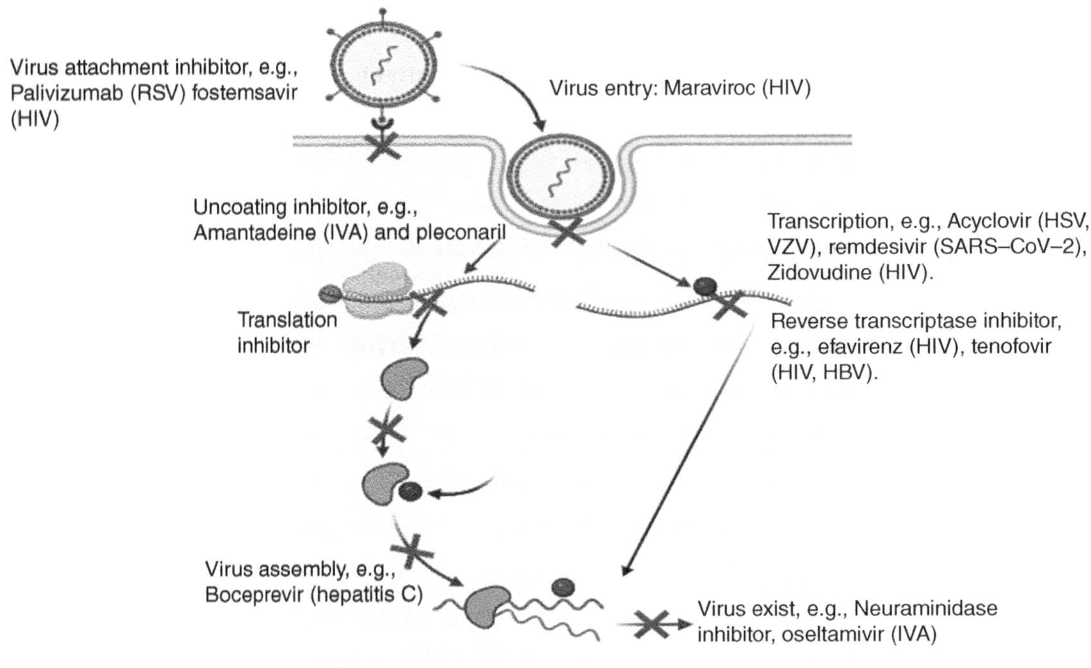

Figure 6.1 Overview of the mechanism of action of some antiviral agents.

Examples of mechanism of antiviral agents in this category include:

Blocking receptors on the host cell surface: Certain antiviral agents prevent viruses from binding to and entering host cells by targeting the receptors that viruses use to attach themselves to the host cell surface. This approach involves either direct antagonism of the receptor or modification of the receptor so that the virus can no longer recognize and bind to it.

- **Receptor antagonism:** These are antiviral agents that can act as antagonists to specific host cell receptors. For instance, maraviroc binds to the transmembrane pocket of the C—C chemokine receptor type 5 (CCR5) receptor on the surface of T-cells [4], preventing HIV from recognizing and binding to this receptor. This blockade stops the virus from entering the cell, thus preventing infection.
- **Receptor modification:** Some agents can modify the host cell receptor, making it unrecognizable or inaccessible to the virus. This modification can be achieved through chemical alteration or by using decoy molecules that mimic the receptor structure, thereby trapping the virus.
- **Interfering with viral proteins:** Viruses have specific proteins that facilitate their attachment to host cells and subsequent fusion with

the cell membrane. Antiviral agents can inhibit these viral proteins, thereby blocking the virus's ability to attach to and fuse with the host cell membrane.

- **Attachment inhibitors:** Certain antiviral agents can bind to viral attachment proteins, such as glycoproteins on the viral envelope. For example, soluble heparin binds to glycoproteins on HSV, preventing the virus from attaching to its actual receptor on the host cell [5].
- **Fusion inhibition:** Some antiviral drugs target viral fusion proteins, preventing the necessary conformational changes for membrane fusion. Enfuvirtide, for instance, binds to the HIV gp41 protein, inhibiting the fusion of the viral envelope with the host cell membrane [6].

Clinical significance of antivirals that block virus entry

Blocking receptors on the host cell surface or preventing viral entry is particularly effective in the early stages of viral infection. By blocking the initial entry of the virus, these antiviral agents can significantly reduce viral load, limit the spread of infection within the host, and prevent the establishment of disease. This approach is also beneficial in prophylactic settings, where individuals are at high risk of exposure.

However, the development of resistance remains a challenge. Viruses can mutate their attachment and fusion proteins, rendering some antiviral agents less effective. Therefore, ongoing research and the development of combination therapies are essential to maintain the efficacy of these treatments.

6.1.2.2 Antiviral That Targets Viral Uncoating

The uncoating of a virus is a critical step in its replication cycle. Inhibiting viral uncoating can effectively halt the replication process, making it a potent target for antiviral therapies. Mechanisms of antiviral agents in this category include:

- **Targeting viral proteins involved in uncoating:** Some antiviral drugs target specific viral proteins that are essential for the uncoating process. By inhibiting these proteins, the drugs prevent the virus from successfully shedding its capsid, thereby trapping the viral genome inside and stopping the replication cycle. For example, pleconaril binds (active against enteroviruses and rhinoviruses) to a hydrophobic pocket within the viral capsid, stabilizing it and preventing the uncoating process [7]. This stabilization inhibits the release of the viral RNA into the host cell, thus blocking replication. Pleconaril has been investigated for the treatment of viral respiratory infections, though its approval and use vary by country. The drug caused a rise in the activity of cytochrome P3A enzymes, which can lead to dangerous interactions with other medications [8]. It has not been approved by the US Food and Drug Administration (FDA).

- **Disrupting acidification of endosomes:** After entry, many viruses rely on the acidification of endosomes (cellular compartments) to trigger uncoating. Antiviral agents can disrupt this acidification process, preventing the necessary microenvironmental changes required for uncoating to occur. Examples of drugs in this category are amantadine and rimantadine for the treatment of influenza A virus [9]. They target M2 protein, an ion channel that facilitates the acidification of the endosome, which is necessary for the uncoating of the viral genome [10]. Amantadine and rimantadine block this ion channel, preventing the influx of hydrogen ions into the endosome. This disruption in acidification prevents the viral capsid from releasing the RNA genome, thus inhibiting viral replication.

6.1.2.3 Antiviral That Targets Nucleic Acid Synthesis

The inhibition of nucleic acid synthesis is a primary mechanism by which antiviral agents prevent the replication and propagation of viruses. This process targets the viral nucleic acid machinery or enzymes responsible for replicating the viral genome, which can be either DNA or RNA. By interfering with these processes, antiviral drugs can effectively stop the production of new viral particles, thereby controlling the infection. Mechanisms of antiviral agents in this category include:

Nucleoside and Nucleotide Analogs These drugs mimic the natural building blocks- nucleoside or nucleotide of DNA or RNA, respectively. When incorporated into the viral genome during replication, they act as chain terminators because they lack the necessary chemical group to allow further elongation of the nucleic acid chain. Examples include acyclovir (for the treatment of HSV and varicella-zoster virus [VZV]) and Sofosbuvir [11, 12]. Acyclovir is a guanosine analog that is phosphorylated by viral thymidine kinase and subsequently incorporated into viral DNA by viral DNA polymerase, causing premature chain termination in HSV and VZV [13]. Sofosbuvir is a nucleotide analog inhibitor of the hepatitis C virus (HCV) RNA-dependent RNA polymerase, nonstructural protein 5B (NS5B), which leads to chain termination during RNA replication [14].

Non-nucleoside Polymerase Inhibitors These drugs bind to viral polymerases at sites other than the active site, causing conformational changes that inhibit the enzyme's activity. Examples include Foscarnet use for the treatment of cytomegalovirus (CMV), HSV, and VZV infections in AIDS and immunocompromised patients [15]. It directly inhibits viral DNA polymerase and reverse transcriptase by binding to the pyrophosphate binding site, preventing the cleavage of pyrophosphate from deoxynucleotide triphosphates. Remdesivir (use for the treatment of viruses like Ebola virus and SARS-CoV-2) inhibits the RNA-dependent RNA polymerase of these viruses [16].

Reverse Transcriptase Inhibitors These inhibitors target the reverse transcriptase enzyme that converts viral RNA into complementary DNA (cDNA). They are used primarily for retroviruses like HIV. Examples include zidovudine (AZT). Zidovudine is a thymidine analog that inhibits reverse transcriptase by causing premature termination of the DNA chain [17].

Non-nucleoside Reverse Transcriptase Inhibitors (NNRTIs) These bind to the reverse transcriptase enzyme in a site distinct from the active site, causing conformational changes that inhibit the enzyme's activity. Examples include Efavirenz. It binds to the HIV-1 reverse transcriptase and inhibits RNA- and DNA-dependent polymerase activities [18].

Advantages and disadvantages of NNRTIs

Advantages

High efficacy: Nucleic acid synthesis inhibitors are highly effective at reducing viral load and controlling infections.

Specificity: Many of these drugs are designed to target viral enzymes specifically, minimizing their impact on host cellular processes.

Disadvantages

Resistance: Viral mutations can lead to the development of drug-resistant strains. For instance, mutations in the HIV reverse transcriptase gene can render NNRTIs and NRTIs ineffective.

Toxicity: Some drugs can have toxic effects on host cells, particularly those that incorporate into host DNA or RNA. For example, AZT can cause bone marrow suppression.

Inhibition of Viral Genome Integration (Integrase Inhibitor) Some viruses, particularly retroviruses like HIV, integrate their genetic material (cDNA) into the host cell's DNA as a critical step in their replication cycle. Inhibiting this process prevents the virus from establishing a permanent infection and disrupting the normal cellular processes of the host. Antiviral drugs targeting this stage are known as integrase inhibitors.

Steps of integration and points of inhibition:

- **Viral DNA processing:** Integrase processes the viral DNA ends by cleaving a dinucleotide from each 3′ end of the viral DNA, creating reactive 3′ hydroxyl groups. This prepares them for integration. Integrase inhibitors can prevent the proper processing of viral DNA ends.
- **Transport to the nucleus:** The processed viral DNA, along with integrase and other viral and cellular proteins, forms a pre-integration complex (PIC) that is transported into the nucleus of the host cell.
- **Strand transfer**: After processing, integrase enzymes also facilitate the insertion of viral DNA into the host cell's DNA. Integrase inhibitors can block the strand transfer reaction, where the viral DNA is covalently linked to the host DNA.

Examples of integrase inhibitors include raltegravir (Isentress), elvitegravir (Vitekta), and dolutegravir (Tivicay) [19]. The three inhibit the strand transfer activity of HIV integrase, preventing the integration of viral DNA into the host genome. Dolutegravir is known for having a lower resistance rate compared to earlier integrase inhibitors.

Clinical implications

Integrase inhibitors have revolutionized the treatment of HIV by providing a potent means to block the virus from establishing a long-term reservoir within the host's genome. By preventing integration, these drugs help maintain lower viral loads, reduce the risk of disease progression, and improve patient outcomes.

Advantages and disadvantages of integrase inhibitors

Advantages

Potency: Integrase inhibitors are highly effective in reducing viral load quickly.

Resistance barrier: Newer integrase inhibitors, like dolutegravir, have a higher barrier to resistance, making them valuable in long-term HIV management.

Disadvantages

Resistance: While newer integrase inhibitors have a high barrier to resistance, earlier drugs like raltegravir and elvitegravir can still encounter resistance, particularly if not used in combination with other antiretrovirals.

Side effects: Though generally well tolerated, some patients may experience side effects, which need to be managed.

Inhibition of Viral RNA Translation Viral translation is the process by which viral messenger RNA (mRNA) is translated into viral proteins by the host cell's ribosomes. Inhibiting this process stops the production of viral proteins required for replication and assembly. Translation inhibitors directly interfere with the host cell's ribosomal machinery to prevent the translation of viral mRNA. These inhibitors can effectively reduce viral replication and control infections. This strategy offers specificity and broad-spectrum potential. Although not commonly used due to toxicity and specificity, as this does not discriminate, some research explores inhibitors that target eIF4E, a eukaryotic initiation factor crucial for the translation of certain viral mRNAs.

Advantages and disadvantages

Advantages

High specificity: By targeting viral enzymes or processes that are distinct from the host's normal cellular functions, these inhibitors can provide high specificity.

Broad spectrum: Some inhibitors, particularly those targeting aminoacyl-tRNA synthetase, can be effective against a range of viruses.

Disadvantages

Resistance: Viruses can mutate, leading to resistance against transcription and translation inhibitors. Combination therapies are often used to mitigate this issue.

Toxicity: Because these inhibitors often interact with the host's cellular machinery, there can be significant side effects and toxicity, particularly with translation inhibitors that target aminoacyl-tRNA synthetase.

Selective targeting: Ensuring that the inhibitors selectively target viral processes without adversely affecting host cell functions remains a challenge.

6.1.2.4 Inhibition of Viral Assembly and Release

Inhibition of viral assembly and release is a fundamental strategy in antiviral therapy aimed at disrupting the final stages of the viral life cycle. By targeting these processes, antiviral agents can effectively prevent the production of infectious viral particles, thereby controlling replication and reducing the spread of infection. Without proteolytic cleavage, the viral polyproteins cannot be processed into functional proteins, resulting in immature, noninfectious viral particles.

Mechanisms of inhibition include

- **Protease inhibitors:** Protease inhibitors play a crucial role in the treatment of viral infections, particularly HIV and HCV, by targeting and inhibiting the viral protease enzyme necessary for the maturation of viral particles. These inhibitors are highly effective when used as part of a combination therapy, helping to achieve and maintain viral suppression.
- **Inhibition of viral polyprotein processing:** Many viruses, including HIV and HCV, produce large polyproteins that need to be cleaved into functional individual proteins by viral proteases. These proteins are necessary for the assembly of new viral particles and for the virus to be infectious. Protease inhibitors can bind to the active site of the viral protease enzyme, blocking its ability to cleave polyproteins.

Examples of protease inhibitors include:

- **HIV protease inhibitors**
  - **Ritonavir (Norvir):** Often used as a pharmacokinetic enhancer due to its ability to inhibit cytochrome P450 enzymes, increasing the levels of other protease inhibitors [20].
  - **Lopinavir (Kaletra, used in combination with ritonavir):** Inhibits the HIV-1 protease enzyme, preventing the maturation of viral particles [21].
  - **Darunavir (Prezista):** A second-generation protease inhibitor that has a high barrier to resistance and is effective against many resistant HIV strains [22].
- **HCV protease inhibitors**
  - **Simeprevir (Olysio):** Inhibits the HCV nonstructural protein 3/4A (NS3/4A) protease, essential for viral replication [23].
  - **Glecaprevir (in combination with pibrentasvir in Mavyret):** Inhibits the HCV NS3/4A protease, used in combination therapy for chronic HCV infection [24].

Benefits and challenges

- **Advantages**
  - **High potency:** Protease inhibitors are highly effective in reducing viral load, often leading to rapid decreases in viral replication.
  - **Combination therapy:** When used in combination with other antiretrovirals, protease inhibitors help to achieve sustained viral suppression and reduce the risk of resistance.
- **Disadvantages**
  - **Resistance:** Although newer protease inhibitors have higher barriers to resistance, the emergence of resistant viral strains is still a concern.
  - **Side effects:** Common side effects can include gastrointestinal issues,

liver enzyme elevations, and metabolic disturbances such as dyslipidemia and insulin resistance.
Drug interactions: Protease inhibitors, particularly those like ritonavir that inhibit cytochrome P450 enzymes, can interact with other medications, requiring careful management.

6.1.3 Combination Therapy in Viral Infections

Combination therapy in viral infections involves using two or more antiviral drugs simultaneously to improve treatment outcomes. This approach can enhance efficacy, prevent resistance, and provide a more robust response against viral infections. Below is an overview of the principles and benefits of combination therapy:

6.1.3.1 Principles of Combination Therapy

Synergistic Effect Combination therapy leverages drugs with different mechanisms of action to achieve a synergistic effect. This means that the combined effect of the drugs is greater than the sum of their individual effects.

Example: In HIV treatment, using a combination of reverse transcriptase inhibitors (e.g., tenofovir, emtricitabine) and integrase inhibitors (e.g., dolutegravir) targets different stages of the viral life cycle, leading to more effective suppression of viral replication [25].

Reduction of Viral Load By targeting different viral processes or pathways, combination therapy can more effectively reduce the viral load compared to monotherapy.

Example: In hepatitis C, the use of direct-acting antivirals (DAAs) that target different nonstructural proteins (e.g., NS3/4A protease inhibitors, NS5A inhibitors) leads to higher cure rates and quicker viral suppression [26].

Prevention of Resistance Combining drugs with different mechanisms of action reduces the likelihood of the virus developing resistance. Resistance mutations are less likely to occur simultaneously in multiple targets.

Broad Coverage Combination therapy can cover a broader range of viral strains or types, especially if different drugs target different viral subtypes or mutations.

Example: In influenza, using a combination of neuraminidase inhibitors and other antiviral agents may help in managing different strains or variants of the virus [27].

Reduced Toxicity Lower doses of each drug can be used in combination therapy, which may reduce the risk of adverse effects compared to using high doses of a single drug.

Example: In HIV treatment, using a combination of drugs allows for lower doses of each, potentially reducing side effects compared to higher doses of a single drug.

6.1.3.2 Advantages and Disadvantages of Combination Therapy

Combination therapy offers both benefits and challenges in medical treatments. Refer to Table 6.2.

Table 6.2 Advantages and disadvantages of combination therapy.

Advantages	Disadvantages
1. Enhanced efficacy Combining drugs with different mechanisms of action can result in a more effective reduction of viral load and improved treatment outcomes.	**1. Increased complexity** Managing multiple medications can be complex for patients, requiring careful adherence to dosing schedules.
2. Prevention of resistance Reduces the likelihood of viral resistance by targeting different stages of the viral life cycle or different viral targets, making it harder for the virus to develop resistance.	**2. Potential for drug interactions** Combining drugs may lead to interactions that affect the efficacy of the drugs or cause adverse effects.
3. Broader coverage This can cover a wider range of viral strains or mutations by using drugs that target different viral subtypes or mutations.	**3. Increased side effects** Each drug in the combination may have its side effects, potentially leading to a higher overall incidence of adverse effects.
4. Reduced dosage requirements Allows for lower doses of each drug to be used, which may reduce the risk of toxicity compared to higher doses of a single drug.	**4. Higher cost** Combination therapy can be more expensive due to the cost of multiple drugs, which may impact accessibility for some patients.
5. Improved clinical outcomes This may lead to better clinical outcomes, such as higher cure rates and reduced disease progression.	**5. Adherence challenges** The complexity of taking multiple medications can lead to difficulties in adherence and an increased risk of missed doses.
6. Potential for reduced viral load Lower viral loads can be achieved more effectively, which can lead to improved health and reduced transmission risk.	**6. Need for monitoring** Requires regular monitoring to manage side effects, check drug interactions, and ensure therapeutic efficacy.

6.1.4 Long-Acting Antiviral Drug: Overview

Long-acting drugs are medications designed to be administered less frequently than daily oral drugs. These drugs are typically formulated for extended-release or depot injection, allowing for sustained therapeutic levels in the body.

The primary goal of long-acting antiretroviral drugs is to inhibit HIV replication through mechanisms like traditional antiretrovirals. They target various stages of the HIV life cycle, including reverse transcriptase inhibitors, integrase inhibitors, and fusion inhibitors.

Examples of Long-Acting Antiretroviral Drugs

i. Cabotegravir [28]
   - **Formulation:** Intramuscular injection
   - **Mechanism:** Integrase strand transfer inhibitor
   - **Dosing schedule:** Every 1–2 months

ii. Rilpivirine (Janssen's Edurant)
   - **Formulation:** Intramuscular injection
   - **Mechanism:** NNRTI
   - **Dosing schedule:** Every 1 month

iii. Cabotegravir-rilpivirine (Cabenuva) [29, 30]
   - **Formulation:** Fixed-dose combination of cabotegravir and rilpivirine for intramuscular injection
   - **Mechanism:** Combines integrase inhibitor (cabotegravir) and NNRTI (rilpivirine)
   - **Dosing schedule:** Every 1 month or every 2 months, depending on the patient's response and tolerability

Advantages of Long-Acting Antiretroviral Drugs Long-acting antiretroviral drugs provide significant benefits, such as improved adherence and reduced dosing frequency, but they also present challenges like potential drug resistance and side effects. For a detailed overview, refer to Tables 6.3 and 6.4.

Table 6.3 Advantages of long-acting antiretroviral drugs.

Advantage	Explanation
Improved adherence	Reduces the frequency of dosing, which can help improve adherence to the treatment regimen.
Simplified treatment regimen	Less frequent administration simplifies the treatment regimen compared to daily oral medications.
Enhanced convenience	Reduces the burden of daily medication, improving patients' quality of life.
Better privacy and discretion	Long-acting formulations can be administered discreetly, which may reduce the stigma associated with HIV treatment.
Potential for improved virological outcomes	Sustained drug levels may lead to more consistent viral suppression and potentially better virological outcomes.

Table 6.4 Disadvantages of long-acting antiretroviral drugs.

Disadvantage	Explanation
Risk of injection site reactions	Injectable formulations may cause pain or reactions at the injection site, such as swelling or redness.
Limited formulation options	Fewer long-acting options are available compared to oral medications, which may limit treatment choices.
Potential for drug interactions	Long-acting antiretrovirals may interact with other medications, requiring careful management.
Adherence to the injection schedule	While dosing is less frequent, patients must still adhere to scheduled appointments for injections, which can be a barrier.
Limited data on long-term safety	Long-term safety data for some long-acting formulations may be limited, necessitating ongoing monitoring.

6.1.5 Immunotherapy for Viral Diseases

Immunotherapy for viral diseases is an emerging and promising approach that aims to enhance the body's natural immune response to fight viral infections. Unlike traditional antiviral drugs, which directly target the virus, immunotherapy focuses on modulating the immune system to recognize and eliminate viruses (see Table 6.5). Some key strategies in viral immunotherapy include:

1. **Monoclonal antibodies (mAbs):** These are laboratory-produced molecules designed to mimic the immune system's ability to fight pathogens.

Table 6.5 List of monoclonal antibodies (mAbs) that have been approved for use against various viral infections.

Monoclonal antibody	Target virus	Mechanism of action	Indication	Approval year
Palivizumab	Respiratory syncytial virus (RSV)	Binds to RSV F (fusion) protein to prevent viral entry into cells	Prevention of serious RSV infection in high-risk pediatrics	1998
Regdanvimab	SARS-CoV-2	Neutralizes SARS-CoV-2 by binding to the spike protein	COVID-19 treatment (mild to moderate cases)	2021
Sotrovimab	SARS-CoV-2	Binds to the spike protein of SARS-CoV-2, preventing viral entry	Treatment of COVID-19	2021
Casirivimab/imdevimab	SARS-CoV-2	Combination mAbs that bind to different regions of the SARS-CoV-2 spike protein	COVID-19 treatment (mild to moderate cases)	2020
Bamlanivimab/Etesevimab	SARS-CoV-2	Bind to distinct epitopes of SARS-CoV-2 spike protein to block its attachment and entry. For use only as post-exposure prophylaxis for adults and pediatric patients	Treatment of COVID-19	2020
Ebanga (ansuvimab)	Ebola virus (Zaire)	Targets Ebola virus glycoprotein to prevent virus entry	Treatment of Ebola virus disease	2020
Inmazeb (REGN-EB3)	Ebola virus (Zaire)	Cocktail of three mAbs (atoltivimab, maftivimab, and odesivimab) that target different sites of the Ebola virus glycoprotein	Treatment of Ebola virus disease	2020

* Approval date may differ by countries.

Monoclonal antibodies have been used successfully to treat viral diseases like coronavirus disease 2019 (COVID-19), Ebola, and respiratory syncytial virus (RSV) infection [31, 32]. They work by neutralizing the virus or making it more susceptible to destruction by immune cells.

2. **Immune checkpoint inhibitors**: These drugs block proteins that prevent immune cells (T-cells) from attacking other cells. While primarily used in cancer immunotherapy, they are being explored for viral infections like HIV, where the virus can evade the immune system by exploiting checkpoint pathways [33].
3. **Therapeutic vaccines**: Unlike preventive vaccines, therapeutic vaccines are designed to treat individuals already infected by boosting their immune response against the virus. This approach is being studied in chronic viral infections like HIV and hepatitis B [34, 35].
4. **Adoptive T-cell therapy**: This technique involves collecting and expanding virus-specific T-cells from a patient or donor and then infusing them to target and kill virus-infected cells [36]. It has shown promise in treating viral infections in immunocompromised patients, such as those undergoing bone marrow transplants.
5. **Cytokine therapy:** Cytokines, like interferons, are signaling molecules that help regulate the immune response. Administering cytokines can enhance antiviral immunity, as seen in the treatment of chronic hepatitis C with interferon-based therapy. One of the most commonly used therapeutic cytokines is interferon-alpha (IFNα), a type I interferon that plays a key role in inhibiting viral replication [37].

The primary advantage of immunotherapy over traditional antiviral treatments is its potential to provide long-lasting protection by enhancing the memory of the immune system against the virus. Additionally, it may be effective against drug-resistant viruses, making it a critical area of research for diseases with limited treatment options.

Current challenges in viral immunotherapy include minimizing adverse immune reactions (e.g., "cytokine storms") and ensuring the therapy specifically targets virus-infected cells without harming healthy tissues. However, continued advancements in biotechnology are likely to expand its applications in treating viral diseases.

6.2 Traditional Medicine

Traditional medicine, also known as indigenous medicine or folk medicine, is a broad term encompassing the vast array of medical practices, knowledge, and beliefs that have developed over generations within various societies [38]. These practices rely on the combinational or singular use of plant-based remedies, animal products, spiritual beliefs, manipulation techniques, and exercise to diagnose or treat illnesses. Some examples are listed in Table 6.6.

In many regions, particularly in rural and underserved areas of tropical countries, traditional medicine remains a primary source of healthcare. With the growing interest in holistic and alternative health practices,

Table 6.6 Examples of African traditional medicines used as antiviral agents.

Medicinal plant	Virus targeted	Proposed mechanism of action	Country found
Neem (*Azadirachta indica*)	HIV, hepatitis, herpes simplex	Inhibits viral replication and boosts immune response by stimulating the production of $CD4^+$ and $CD8^+$ cells	India, Nigeria, Ghana, Senegal, Cameroon.
Corm of African Potato (*Hypoxis hemerocallidea*)	HIV, hepatitis	Immune system modulation, antiviral properties	South Africa, Botswana, Namibia
Aframomum melegueta K.Schum. (Zingiberaceae)	Measles virus and yellow fever virus		Nigeria
Sutherlandia (*Lessertia frutescens*)	HIV	Immune-boosting, antiviral activity against HIV enzymes	South Africa and some West African
Pelargonium sidoides	Influenza A virus strains (H1N1, H3N2), respiratory syncytial virus, human coronavirus, parainfluenza virus, and coxsackievirus.	Inhibits viral replication, enhances immune response	South Africa, Eswatini, and Lesotho
Acacia sieberiana DC. (Fabaceae)	Coxsackievirus		Rwanda
Artemisia annua (African Wormwood)	Hepatitis B virus, HIV, SARS-CoV-2	Immunological and anti-HIV effects.	Ethiopia, Kenya, Tanzania, Uganda, Madagascar
Adansonia digitata L. (Malvaceae)	Rift Valley fever virus	Antioxidant and anti-inflammatory activity	South Africa
Aphloia theiformis	Zika virus (African and Asian strains) and dengue virus (DENV)	Impaired Zika virus attachment to the host cell surface	East Africa, Madagascar
Rooibos (*Aspalathus linearis*)	HIV, Herpes simplex	Antiviral activity, antioxidant properties	South Africa

there is a significant movement toward scientifically validating traditional remedies and integrating them into modern healthcare systems. This integration seeks to harness the benefits of both traditional knowledge and contemporary medical science, providing a more comprehensive approach to health and wellness.

Key components of traditional medicine include the use of medicinal plants, animal products, and minerals, each tailored to treat specific conditions based on long-standing empirical knowledge. While traditional knowledge serves as a valuable starting point, scientific research is crucial to validate the effectiveness of these herbal remedies. Here is what modern science can do:

- Identifying active components: Researchers can isolate and study the bioactive compounds within these plants to understand their potential antiviral mechanisms.
- Clinical trials: Controlled clinical trials can be conducted to assess the efficacy and safety of specific herbal remedies against viruses.
- Integration with modern medicine: The goal is to integrate effective herbal therapies into mainstream medicine, potentially offering complementary or alternative treatment options for viral infections.

Important considerations

- It is essential to remember that herbal remedies are not a magic bullet. While some show promise, more research is needed for many.
- Consulting a qualified healthcare professional is crucial before using any herbal remedy. They can advise on potential interactions with medications and ensure the chosen herb is appropriate for individual needs.
- There can be variations in the quality and potency of herbal products. Opting for standardized formulations from reputable sources is advisable.

6.3 Screening for Antiviral Drugs

Screening for antiviral drugs is a critical process in the development of effective treatments for viral infections. This process involves identifying compounds that can inhibit viral replication or function without causing significant harm to the host cells. The screening process typically includes several stages, from preclinical investigations, which include the initial high-throughput screening (HTS), target-based screening, cell-based assays, mechanism of action studies, and *in vivo* (animal) studies, to clinical studies or trials.

6.3.1 Preclinical Investigations

The preclinical investigation of drugs is a critical phase in the drug development process, determining whether a drug will show antiviral activity. Preclinical results, which include *in silico* (computer-based for example artificial intelligence [AI] screening), *in vitro* (test tube or culture dish), and *in vivo* (animal) studies, are essential for understanding a drug's potential efficacy and safety. These findings are critical for determining whether an antiviral drug candidate is safe and effective enough to progress to human clinical trials. The goal is to establish a robust profile of the drug's antiviral activity, toxicity, pharmacokinetics, and pharmacodynamics, ensuring that only the most promising candidates are considered in the drug development pipeline.

6.3.1.1 High-Throughput Screening (HTS)

HTS is the initial phase in antiviral drug discovery, where large libraries of chemical compounds, natural extracts, or biological molecules are rapidly tested for antiviral activity. This is done using automated systems that can handle thousands of samples simultaneously. The primary goal is to identify "hits" or compounds that show potential antiviral effects. HTS assays can be designed to target various stages of the viral life cycle, such as entry, replication, assembly, or release.

6.3.1.2 Target-Based Screening

In target-based screening, compounds are tested for their ability to interact with specific viral or host proteins that are crucial for viral replication. This approach requires a detailed understanding of the viral life cycle and the identification of key molecular targets. For example, inhibitors of the viral protease, polymerase, or integrase enzymes are common targets in antiviral drug discovery.

6.3.1.3 Cell-Based Assays

Cell-based assays involve testing compounds in cultured cells infected with the virus of interest. These assays provide a more comprehensive understanding of a compound's antiviral activity and potential cytotoxicity. Cell-based assays can measure various parameters, such as viral load reduction, cytopathic effects, and cell viability. These assays help to confirm the antiviral activity observed in HTS and provide data on the compound's efficacy and safety.

6.3.1.4 Mechanism of Action Studies

Once potential antiviral compounds are identified, it is crucial to understand their mechanism of action. This involves detailed biochemical and molecular studies to elucidate how the compound inhibits viral replication. Understanding the mechanism of action helps in optimizing the compound's efficacy, reducing potential side effects, and predicting resistance development.

6.3.1.5 In Vivo Studies

Promising compounds from *in vitro* studies are further tested in animal models to assess their pharmacokinetics, pharmacodynamics, toxicity, and efficacy in a living organism. These studies provide critical data on the safety profile of the compound and therapeutic potential. Animal models that closely mimic human viral infections are used to evaluate the drug's performance under conditions that replicate human disease (see Figure 6.2).

Steps involve in preclinical antiviral screening:

1. **Antiviral activity tests**
   Effective concentration 50 (EC_{50}): This is the concentration of the drug that achieves 50% of its maximum effect. It is commonly used in the context of agonists (compounds that activate a receptor or biological response) to measure how effective a drug is at eliciting a response. A lower EC_{50} indicates a higher potency of the antiviral drug.

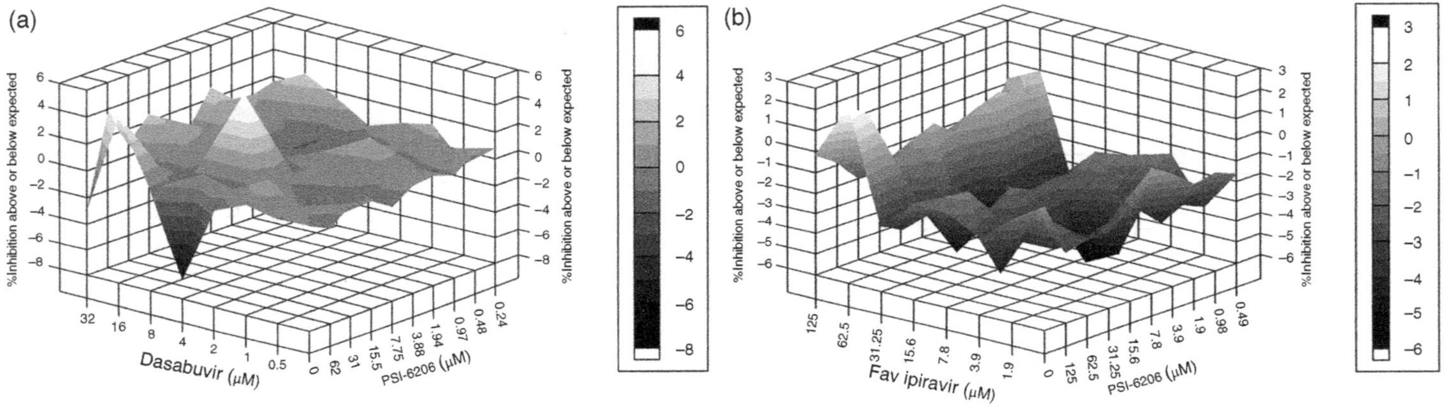

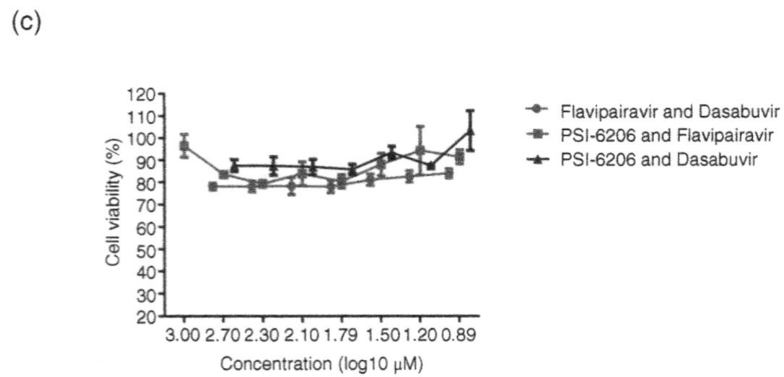

(d)

Drug combination	Synergy (μM²%)*	Antagonism* (μM²%)	Predicted interaction
PSI-6206 + Dasabuvir	+115.08	−59.54	Strong synergism and moderate antagonism
PSI-6206+ Favipiravir	+49.32	−35.77	Minor synergism and minor antagonism

* Mean volumes of synergy or antagonism are presented at 95% confidence levels

Figure 6.2 Example of *in silico* and *in vitro* analysis of a two-drug combination [39]. Analysis of a two-drug combination showing (a and b). Graphic presentation of synergy and antagonism peaks of drug combination graphs plotted using DetalGraph 7. (c) Cytotoxicity of two drugs. (d) Mean volumes of the average of three independent experiments are presented at 95% confidence levels.

Source: Adapted from Ref. [39].

Inhibitory concentration 50 (IC_{50}): The concentration of the drug required to inhibit 50% of viral replication *in vitro*. It is used to measure the effectiveness of a substance in inhibiting a specific biological or biochemical function. IC_{50} is typically used for antagonists or inhibitors to measure how much of a compound is needed to block or reduce a biological process by half. A lower IC_{50} value indicates a more potent inhibitor, as less of the drug is needed to inhibit the target by 50%.

2. **Cytotoxicity tests**
   Cytotoxic concentration 50 (CC_{50}): The concentration of the drug that kills 50% of the host cells. This helps determine the therapeutic index of the drug by comparing CC_{50} with EC_{50} or IC_{50}.
   Selectivity index (SI): The ratio of CC_{50} to EC_{50}. A higher SI indicates a better safety profile, meaning the drug is more effective at inhibiting the virus than harming the host cells.
3. **Mechanism of action studies**
   This involves investigating how the antiviral drug inhibits the virus, such as blocking viral entry, replication, transcription, or assembly.
4. ***In vitro* viral resistance tests**
   Assessing the potential for the virus to develop resistance to the antiviral drug through genetic mutations. This involves passing the virus in the presence of suboptimal drug concentrations and sequencing viral genomes.
5. **Pharmacokinetics (PK) tests**
   Absorption: How well the drug is absorbed into the bloodstream after administration.
   Distribution: How the drug spreads through the body tissues and organs.
   Metabolism: How the drug is broken down by the body, typically by liver enzymes.
   Excretion: How the drug and its metabolites are eliminated from the body, often through urine or feces.
   Half-life: The time it takes for the concentration of the drug in the blood to be reduced by half.
6. **Pharmacodynamics (PD) tests**
   Studying the biochemical and physiological effects of the drug on the body and the relationship between drug concentration and effect.
7. ***In vivo* efficacy studies**
   Testing the antiviral drug in animal models to evaluate its efficacy in a living organism. Common models include mice, ferrets, and non-human primates.
   Viral load reduction: Measuring the decrease in viral RNA or DNA in the animal's tissues or blood.
   Survival rate: Evaluating the impact of the drug on the survival of infected animals.
   Pathology: Assessing the reduction of disease symptoms and damage to tissues or organs.
8. **Toxicology studies**
   Acute toxicity: Determine the immediate toxic effects of a single dose of the drug.
   Chronic toxicity: Assess the toxic effects of long-term administration of the drug.
   Organ toxicity: Identify any toxic effects on specific organs, such as the liver, kidneys, or heart.

Carcinogenicity: Evaluate the potential of the drug to cause cancer over long-term use.
Genotoxicity: Assess the drug's potential to damage genetic information in cells, which could lead to mutations or cancer.

9. **Immunogenicity**
Testing whether the antiviral drug induces an immune response, which could potentially reduce its efficacy or cause adverse reactions.
10. **Combination studies**
Evaluating the effectiveness and safety of the antiviral drug when used in combination with other antiviral agents. This can help identify potential synergistic effects or adverse drug interactions.

6.3.2 Clinical Trials

While preclinical studies are crucial, good results from preclinical investigations do not always translate directly to successful outcomes in human clinical trials. Several factors can lead to discrepancies between preclinical and clinical outcomes. However, compounds that show promise in preclinical studies proceed to clinical trials, which are conducted in multiple phases to ensure the drug's safety and efficacy in humans. Clinical trials are divided into the following major phases:

6.3.2.1 Phase I Clinical Trial

This phase tests safety, dosage, and side effects in a small group of usually healthy human volunteers but may also include patients. This phase can be used to establish the bioavailability, pharmacokinetics, and pharmacodynamics of the test compound.

6.3.2.2 Phase II Clinical Trial

If the result of Phase 1 is promising, a large-scale Phase II clinical trial can be done. This phase evaluates the drug's efficacy and side effects in a larger group of patients.

6.3.2.3 Phase III Clinical Trial

This phase confirms efficacy, monitors side effects, and compares the new drug with standard treatments in a large population.

6.3.2.4 Phase IV Clinical Trial

Conducted after the drug is approved for market, these studies continue to monitor the drug's long-term effectiveness and safety.

6.4 Antiviral Vaccination

Antiviral vaccination is a critical strategy in preventing and controlling viral infections. Vaccines work by stimulating the immune system to subsequently recognize and combat specific viruses, thereby providing immunity without causing the disease. This approach plays an important role in controlling and, in some cases, eradicating several viral diseases (see Table 6.7).

6.4.1 Principles and Theory of Antiviral Vaccination

6.4.1.1 Immune Memory

Vaccines introduce antigens (virus-like particles, inactivated viruses, or attenuated viruses) to the immune system, prompting the body to produce antibodies and memory cells. These memory cells "remember" the

Table 6.7 Different types of vaccines and route of administration.

Vaccine name	Virus targeted	Route of administration	Vaccine type
Measles, mumps, rubella (MMR)	Measles, mumps, rubella	Intramuscular (IM)	Live, attenuated
Varicella (chickenpox)	Varicella-zoster virus (VZV)	Subcutaneous (SC)	Live, attenuated
Influenza (flu shot)	Influenza Virus	Intramuscular (IM)	Inactivated
Polio (inactivated polio vaccine)	Poliovirus	Intramuscular (IM)	Inactivated
Polio (oral polio vaccine)	Poliovirus	Oral	Live, attenuated
Rotavirus	Rotavirus	Oral	Live, attenuated
Hepatitis A (Havrix)	Hepatitis A virus (HAV)	Intramuscular (IM)	Inactivated
Hepatitis B (Engerix-B)	Hepatitis B virus (HBV)	Intramuscular (IM)	Recombinant
Rabies (RabAvert)	Rabies virus	Intramuscular (IM)	Inactivated
Human papillomavirus (HPV)	Human papillomavirus (HPV)	Intramuscular (IM)	Recombinant
Shingles (Shingrix)	Varicella-zoster virus (VZV)	Intramuscular (IM)	Live, attenuated, adjuvanted

virus and can mount a rapid and effective response upon future exposures.

6.4.1.2 Herd Immunity

Widespread vaccination can lead to herd immunity, where enough of the population is immune to the virus, reducing its spread and protecting those who cannot be vaccinated (e.g., immunocompromised individuals).

6.4.2 Types of Antiviral Vaccines

6.4.2.1 Live-Attenuated Vaccines

These vaccines use a weakened form of the virus that cannot cause disease in healthy individuals. Examples include the measles, mumps, and rubella (MMR) vaccine and the varicella (chickenpox) vaccine [40].

Advantages: Strong and long-lasting immune response.
Disadvantages: Not suitable for immunocompromised individuals; requires careful handling and storage to prevent reactivation.

6.4.2.2 Inactivated Vaccines

These vaccines contain viruses that have been killed or inactivated. Examples include the influenza vaccine and the hepatitis A vaccine [41].

Advantages: Safe for immunocompromised individuals.
Disadvantages: Often require booster shots to maintain immunity.

6.4.2.3 *Subunit, Recombinant, and Conjugate Vaccines*

These vaccines use specific pieces of the virus (e.g., proteins or glycoproteins of the virus). Examples include the hepatitis B vaccine and the human papillomavirus (HPV) vaccine [42].

Advantages: Targeted immune response; safe for all populations.
Disadvantages: May require multiple doses.

6.4.2.4 *Messenger RNA Vaccines*

These vaccines use mRNA to instruct cells to produce a viral protein that triggers an immune response. Examples include COVID-19 vaccines produced by Pfizer-BioNTech and Moderna [43].

Advantages: Rapid development and production; strong immune response.
Disadvantages: Requires ultracold storage.

6.4.2.5 *Viral Vector Vaccines*

These vaccines use a different virus as a vector to deliver genes encoding viral antigens. Examples include the Johnson & Johnson COVID-19 vaccine, and Ebola virus that use adenovirus (Ad26) as a vector.

Advantages: Strong and long-lasting immunity; can be developed quickly.
Disadvantages: Preexisting immunity to the vector may reduce efficacy.

6.4.3 Public Health Significance and Impact of Vaccine

a. **Prevention of disease**: Vaccination has been key in preventing viral infections such as smallpox, polio, measles, and influenza, significantly reducing morbidity and mortality rates.
b. **Control of outbreaks**: Vaccines help control and prevent outbreaks of viral diseases, as seen with the rapid development and deployment of COVID-19 vaccines during the pandemic.
c. **Economic benefits**: Preventing viral infections through vaccination reduces healthcare costs associated with treating diseases and managing complications.
d. **Eradication of diseases:** Smallpox is an example of a viral disease eradicated through widespread vaccination. Efforts are ongoing to eradicate polio through global vaccination initiatives.

6.4.4 Challenges in Antiviral Vaccination

a. **Viral mutation and variation:** Some viruses, like influenza and HIV, mutate rapidly, requiring frequent updates to vaccines and complicating the development of effective long-term vaccines.
b. **Mutation and reversion to virulence**: Some viruses, like vaccine-derived poliovirus (VDPV), belong to the category of challenges related to the use of live-attenuated vaccines, specifically the oral polio vaccine

(OPV). VDPV occurs when the weakened virus in the OPV mutates and regains neurovirulence, potentially causing outbreaks in under-immunized populations.

c. **Vaccine hesitancy:** Misinformation and distrust in vaccines can lead to reduced vaccination rates, hindering efforts to achieve herd immunity.
d. **Access and distribution:** Ensuring equitable access to vaccines, particularly in resource-limited settings, remains a significant challenge.
e. **Adverse effects**: While vaccines are generally safe, they can cause side effects in some individuals, ranging from mild (e.g., soreness at the injection site) to severe (e.g., allergic reactions).

HIV vaccine development: Progress and challenges so far

HIV vaccine development has been a critical yet challenging area since the virus was first identified. Despite the complexities, significant progress has been made, however, there is still no US FDA-approved therapeutic HIV vaccine. Various approaches are under study, and some clinical trials are underway. Several approaches can be used, which include the use of:

1. **Nanoparticle-based HIV vaccines:** Nanoparticle HIV vaccines are a promising new approach in the effort to develop an effective vaccine against the virus [44]. These vaccines use nanoparticles to deliver HIV antigens, enhancing immune responses and addressing some of the challenges that are faced by traditional vaccine efforts. Examples of nanoparticle HIV vaccines include:
   - **eOD-GT8 60mer:** This vaccine candidate is a self-assembling nanoparticle that displays engineered versions of the HIV envelope protein [45]. It aims to stimulate the production of broad neutralizing antibodies (bNAbs).
   - **Gag-Env nanoparticles:** Some vaccine designs use nanoparticles that display both the Gag and Env proteins, which are key to viral replication and immune recognition [46].

How nanoparticle HIV vaccines work

a. **Antigen presentation**: Nanoparticles can be engineered to display HIV antigens in a way that mimics the virus [47]. These particles present key viral proteins, such as the envelope glycoprotein (Env), to the immune system, aiming to induce both antibody and T-cell responses.
b. **Improved immune response**: Nanoparticles can enhance the immune response by improving and controlling antigen delivery, stability, and targeting specific cells [48]. They can be designed to target dendritic cells or lymph nodes, which are crucial for initiating strong immune responses.
c. **Broad neutralizing antibodies (bNAbs)**: Some nanoparticle vaccines focus on inducing bNAbs, which can target a wide range of HIV strains [49]. This is critical given the virus's high mutation rate and strain diversity.

   d. **Durability of response**: Nanoparticles can help provide a more durable immune response by slowly releasing antigens over time, simulating a natural infection and thus boosting the body's ability to recognize and respond to HIV.
2. **mRNA-based HIV vaccine:** Leveraging the success of COVID-19 mRNA vaccines, researchers are now using this technology for HIV [44]. The mRNA platform allows for the design of vaccines that can stimulate a strong immune response and be quickly adapted to different HIV variants.
3. **Mosaic-based HIV vaccines:** This approach combined genes from various HIV strains to create a more universal immune response. Mosaic vaccines aimed to address the diversity of HIV and was tested in large-scale clinical trials like the "Mosaico" trial. It did not stimulate sufficient humoral and cellular immune responses, hence discontinued [50].

Study Questions

6.1 A 45-year-old patient with a chronic viral infection (e.g., HIV or hepatitis C) is prescribed a new antiviral medication (combination therapy). The physician aims to optimize treatment outcomes. Identify the rationale for the combination therapy and describe possible mechanisms of action that should be targeted.

6.2 What are the key immunological mechanisms triggered by antiviral vaccines that contribute to protective immunity, and how do these mechanisms vary between live-attenuated and inactivated vaccines?

6.3 A 10-year-old boy is brought to the hospital after being bitten by a stray dog, which is suspected to be rabid. The wound is cleaned, and the boy is immediately given rabies immune globulin and the first dose of the rabies vaccine as post-exposure prophylaxis. The healthcare team plans to administer additional doses of the rabies vaccine over the following weeks. In the context of rabies exposure, how does the rabies post-exposure prophylaxis function as a therapeutic vaccine, and what factors influence its effectiveness in preventing the onset of clinical rabies after potential infection?

6.4 A 50-year-old woman with a history of chronic hepatitis C is considering starting antiviral therapy. She inquired about the availability and mechanism of therapeutic vaccines for hepatitis C, as well as their role in conjunction with standard antiviral treatments. What recommendations would you provide regarding the current state of therapeutic agents for hepatitis C, and how might they impact her treatment plan?

6.5 A 28-year-old female patient presents with symptoms of acute influenza. After diagnosis, she is prescribed oseltamivir, an antiviral agent. Explain the mechanism of action of oseltamivir in inhibiting influenza virus replication. What specific viral processes does it target, and how does this mechanism contribute to its effectiveness in treating influenza?

6.6 A 65-year-old male patient with a history of HIV is being treated with a combination of antiretroviral drugs, including tenofovir and emtricitabine. Describe the mechanisms by which these antiviral agents work to inhibit HIV replication. How do these agents interfere with the viral life cycle, and what implications does this have for the patient's overall treatment strategy?

Objectives

6.1 **Which of the following antiviral agents is primarily used to treat influenza?**
(a) Acyclovir
(b) Oseltamivir
(c) Zidovudine
(d) Ribavirin

6.2 **What is the primary mechanism of action of nucleoside analogs, such as acyclovir?**
(a) Inhibition of viral attachment to host cells
(b) Inhibition of viral RNA synthesis
(c) Competitive inhibition of viral DNA polymerase
(d) Blocking viral entry into the cytoplasm

6.3 **Which type of vaccine is designed to provoke an immune response without causing the disease?**
(a) Live-attenuated vaccine
(b) Inactivated vaccine
(c) Subunit vaccine
(d) All of the above

6.4 **What is the primary purpose of therapeutic vaccines, such as the anti-rabies vaccine?**
(a) To prevent infection before exposure
(b) To treat established infections
(c) To enhance immunity against future infections
(d) To neutralize existing antibodies

6.5 **Which antiviral drug is commonly used as post-exposure prophylaxis for HIV?**
(a) Tenofovir
(b) Amantadine
(c) Nevirapine
(d) Lamivudine

6.6 **What type of immune response is primarily stimulated by vaccines?**
(a) Innate immunity
(b) Humoral immunity
(c) Cellular immunity
(d) All of the above

6.7 **What is the mechanism of action of protease inhibitors in antiviral therapy?**
(a) They inhibit viral RNA synthesis.
(b) They prevent the processing of viral polyproteins.
(c) They block viral entry into host cells.
(d) They enhance the immune response.

6.8 **Which of the following is a live-attenuated vaccine?**
(a) Inactivated polio vaccine (IPV)
(b) Measles, mumps, and rubella (MMR) vaccine
(c) Hepatitis B vaccine
(d) Human papillomavirus (HPV) vaccine

6.9 What is the primary goal of post-exposure prophylaxis (PEP) in the context of viral infections?
(a) To eliminate the virus from the body
(b) To prevent the onset of disease after exposure
(c) To treat existing infections
(d) To boost the immune response

6.10 Which class of antiviral agents is used to inhibit reverse transcriptase in HIV treatment?
(a) Integrase inhibitors
(b) Nucleotide/nucleoside reverse transcriptase inhibitors (NRTIs)
(c) Fusion inhibitors
(d) Entry inhibitors

6.11 What type of vaccine uses a part of the virus (like a protein) to stimulate an immune response?
(a) Live-attenuated vaccine
(b) Inactivated vaccine
(c) Subunit vaccine
(d) mRNA vaccine

6.12 What is the primary advantage of using antiviral combination therapy in HIV treatment?
(a) Increased side effects
(b) Enhanced efficacy and reduced resistance
(c) Lower costs
(d) Simplified dosing regimens

References

1. De Clercq, E. and Li, G., *Approved antiviral drugs over the past 50 years*. Clin. Microbiol. Rev., 2016. **29**(3): p. 695–747.
2. Prusoff, W.H., *Synthesis and biological activities of iododeoxyuridine, an analog of thymidine*. Biochim. Biophys. Acta., 1959. **32**(1): p. 295–296.
3. Herrmann, E.C. Jr., *Plaque inhibition test for detection of specific inhibitors of DNA containing viruses*. Proc. Soc. Exp. Biol. Med., 1961. **107**: p. 142–145 https://pubmed.ncbi.nlm.nih.gov/13713741/.
4. Garcia-Perez, J., et al., *Allosteric model of maraviroc binding to CC chemokine receptor 5 (CCR5)*. J. Biol. Chem., 2011. **286**(38): p. 33409–33421.
5. Schneider, S.M., et al., *Herpes simplex virus 1 glycoprotein C promotes virus penetration from endosomes during entry, independent of interaction with heparan sulfate*. Front. Microbiol., 2025. **16**: 1549349.
6. Greenberg, M.L. and Cammack, N., *Resistance to enfuvirtide, the first HIV fusion inhibitor*. J. Antimicrob. Chemother., 2004. **54**(2): p. 333–340.
7. Egorova, A., et al., *Back to the future: advances in development of broad-spectrum capsid-binding inhibitors of enteroviruses*. Eur. J. Med. Chem., 2019. **178**: p. 606–622.
8. Ma, J.D., et al., *The effect of oral pleconaril on hepatic cytochrome P450 3A activity in healthy adults using intravenous midazolam as a probe*. J. Clin. Pharmacol., 2006. **46**(1): p. 103–108.
9. Jefferson, T., et al., *Amantadine and rimantadine for preventing and treating influenza A in adults*. Cochrane Database Syst. Rev., 2004. **3**: Cd001169.
10. Kumar, G. and Sakharam, K.A., *Tackling influenza A virus by M2 ion channel blockers: latest progress and limitations*. Eur. J. Med. Chem., 2024. **267**: 116172.
11. Andrei, G. and Snoeck, R., *Advances and perspectives in the management of varicella-zoster virus infections*. Molecules, 2021. **26**(4): p. 1132.
12. King, D.H., *History, pharmacokinetics, and pharmacology of acyclovir*. J. Am. Acad. Dermatol., 1988. **18**(1 Pt 2): p. 176–179.
13. Paintsil, E. and Cheng, Y.C., *Antiviral agents*, in *Encyclopedia of Microbiology*. 2019. p. 176–225. 10.1016/B978-0-12-801238-3.02387-4.

Epub 2019 Aug 28. PMCID: PMC7150273. https://pmc.ncbi.nlm.nih.gov/articles/PMC7150273/.
14. Yancey, A., Armbruster, A., and Tackett, S., *Sofosbuvir, a new nucleotide analogue, for the treatment of chronic hepatitis C infection*. J. Pharm. Technol., 2015. **31**(1): p. 29–37.
15. Gérard, L. and Salmon-Céron, D., *Pharmacology and clinical use of foscarnet*. Int. J. Antimicrob. Agents, 1995. **5**(4): p. 209–217.
16. Gordon, C.J., et al., *Remdesivir is a direct-acting antiviral that inhibits RNA-dependent RNA polymerase from severe acute respiratory syndrome coronavirus 2 with high potency*. J. Biol. Chem., 2020. **295**(20): p. 6785–6797.
17. Susan-Resiga, D., et al., *Zidovudine inhibits thymidine phosphorylation in the isolated perfused rat heart*. Antimicrob. Agents Chemother., 2007. **51**(4): p. 1142–1149.
18. Radzio, J. and Sluis-Cremer, N., *Efavirenz accelerates HIV-1 reverse transcriptase ribonuclease H cleavage, leading to diminished zidovudine excision*. Mol. Pharmacol., 2008. **73**(2): p. 601–606.
19. Scarsi, K.K., et al., *HIV-1 integrase inhibitors: a comparative review of efficacy and safety*. Drugs, 2020. **80**(16): p. 1649–1676.
20. Quercia, R., et al., *Ritonavir: 25 years' experience of concomitant medication management. A narrative review*. Infect. Dis. Ther., 2024. **13**(5): p. 1005–1017.
21. Cvetkovic, R.S. and Goa, K.L., *Lopinavir/ritonavir: a review of its use in the management of HIV infection*. Drugs, 2003. **63**(8): p. 769–802.
22. Busse, K.H. and Penzak, S.R., *Darunavir: a second-generation protease inhibitor*. Am. J. Health Syst. Pharm., 2007. **64**(15): p. 1593–1602.
23. Izquierdo, L., et al., *Simeprevir for the treatment of hepatitis C virus infection*. Pharmgenomics Pers. Med., 2014. **7**: p. 241–249.
24. Abutaleb, A., Kottilil, S., and Wilson, E., *Glecaprevir/pibrentasvir expands reach while reducing cost and duration of hepatitis C virus therapy*. Hepatol. Int., 2018. **12**(3): p. 214–222.
25. Kulkarni, R., et al., *The combined anti-HIV-1 activities of emtricitabine and tenofovir plus the integrase inhibitor elvitegravir or raltegravir show high levels of synergy in vitro*. Antimicrob. Agents Chemother., 2014. **58**(10): p. 6145–6150.
26. Li, G. and De Clercq, E., *Current therapy for chronic hepatitis C: the role of direct-acting antivirals*. Antiviral Res., 2017. **142**: p. 83–122.
27. Govorkova, E.A. and Webster, R.G., *Combination chemotherapy for influenza*. Viruses, 2010. **2**(8): p. 1510–1529.
28. WHO. *WHO recommends long-acting cabotegravir for HIV prevention*. 2022; Available from: https://www.who.int/news/item/28-07-2022-who-recommends-long-acting-cabotegravir-for-hiv-prevention.
29. NIH, N.I.o.H.J. *FDA-approved HIV medicines*. 2020.
30. Bernstein, L.J.T.W.P. FDA approves breakthrough injectable HIV medication. 2021.
31. Pantaleo, G., et al., *Antibodies to combat viral infections: development strategies and progress*. Nat. Rev. Drug Discov., 2022. **21**(9): p. 676–696.
32. Esposito, S., et al., *Monoclonal antibodies for prophylaxis and therapy of respiratory syncytial virus, SARS-CoV-2, human immunodeficiency virus, rabies and bacterial infections: an update from the world Association of Infectious Diseases and Immunological Disorders and the Italian Society of Antinfective Therapy*. Front. Immunol., 2023. **14**: 1162342.
33. Cai, H., et al., *Immune checkpoints in viral infections*. Viruses, 2020. **12**(9): p. 1051.
34. Cargill, T. and Barnes, E., *Therapeutic vaccination for treatment of chronic hepatitis B*. Clin. Exp. Immunol., 2021. **205**(2): p. 106–118.

35. Mahmood, F., et al., *HBV vaccines: advances and development*. Vaccines (Basel), 2023. **11**(12): p. 1862.
36. Ottaviano, G., et al., *Adoptive T cell therapy strategies for viral infections in patients receiving Haematopoietic stem cell transplantation*. Cells, 2019. **8**(1): p. 47.
37. Yang, Y.-F., et al., *Long-term efficacy of interferon alpha therapy on hepatitis B viral replication in patients with chronic hepatitis B: a meta-analysis*. Antiviral Res., 2010. **85**(2): p. 361–365.
38. Fokunang, C.N., et al., *Traditional medicine: past, present and future research and development prospects and integration in the National Health System of Cameroon*. Afr. J. Tradit. Complement. Altern. Med., 2011. **8**(3): p. 284–295.
39. Olasunkanmi, O.I. and Zhong, Z.-H., *Combination of dasabuvir and PSI-6206 for the treatment of coxsackievirus B3 infection*. J. Virus Erad., 2022. **8**(2): 100074.
40. Dhillon, S. and Curran, M.P., *Live attenuated measles, mumps, rubella, and varicella zoster virus vaccine (Priorix-tetra)*. Paediatr. Drugs, 2008. **10**(5): p. 337–347.
41. Sanders, B., Koldijk, M., and Schuitemaker, H., *Inactivated viral vaccines*, in *Vaccine Analysis: Strategies, Principles, and Control*. Vol. 28, B.K. Nunnally, R.D. Sitrin, and V.E. Turula, Editors. 2014, Springer. p. 45–80. 10.1007/978-3-662-45024-6_2.
42. Stanley, M., *Tumour virus vaccines: hepatitis B virus and human papillomavirus*. Philos. Trans. R. Soc. Lond. B. Biol. Sci., 2017. **372**(1732): p. 20160268.
43. Gote, V., et al., *A comprehensive review of mRNA vaccines*. Int. J. Mol. Sci., 2023. **24**(3): p. 2700.
44. Willis, J.R., et al., *Vaccination with mRNA-encoded nanoparticles drives early maturation of HIV bnAb precursors in humans*. Science, 2025. **389**(6759): eadr8382.
45. Cohen, K.W., et al., *A first-in-human germline-targeting HIV nanoparticle vaccine induced broad and publicly targeted helper T cell responses*. Sci. Transl. Med., 2023. **15**(697): p. eadf3309.
46. Zhang, P., et al., *A multiclade env–gag VLP mRNA vaccine elicits tier-2 HIV-1-neutralizing antibodies and reduces the risk of heterologous SHIV infection in macaques*. Nat. Med., 2021. **27**(12): p. 2234–2245.
47. Aikins, M.E., Bazzill, J., and Moon, J.J., *Vaccine nanoparticles for protection against HIV infection*. Nanomedicine (Lond), 2017. **12**(6): p. 673–682.
48. Priyanka, et al., *Nanovaccines: a game changing approach in the fight against infectious diseases*. Biomed. Pharmacother., 2023. **167**: 115597.
49. Schiffner, T., et al., *Vaccination induces broadly neutralizing antibody precursors to HIV gp41*. Nat. Immunol., 2024. **25**(6): p. 1073–1082.
50. Julg, B., et al., *Immunogenicity of 2 therapeutic mosaic HIV-1 vaccine strategies in individuals with HIV-1 on antiretroviral therapy*. NPJ Vaccines, 2024. **9**(1): p. 89.

7 Organization and Safety Management in Tropical Virology Laboratory

Effective organization and safety management are fundamental to the operation of virology laboratories, where work involves handling infectious agents that pose significant biohazard risks. Given the high potential for exposure to dangerous pathogens, including those causing viral hemorrhagic fevers, respiratory infections, and emerging viral threats, it is essential to establish strict protocols for containment, sample handling, waste disposal, and personnel safety.

This requires a comprehensive approach, integrating biosafety level (BSL) guidelines, risk assessments, and personal protective equipment (PPE) standards. Safety management in virology laboratories also includes the use of robust organizational structures to monitor compliance, ensure the proper use of equipment, and facilitate regular training. The all-encompassing goal is to minimize the risk of contamination, protect laboratory personnel, and prevent accidental release of viruses into the environment.

7.1 Organizational Management

7.1.1 Structural Management

7.1.1.1 Laboratory Hierarchical Setup

A clear organizational structure is necessary to define roles, responsibilities, and lines of authority within the laboratory. Typically, this includes a laboratory director, virologists, medical lab scientists or technologists (MLS or MLT), lab technicians, and support staff. Each role should have defined duties to ensure that all laboratory functions are covered.

7.1.1.2 Departmental Divisions

Virology laboratories can be divided into various sections based on the types of services provided, including diagnostic virology, molecular virology, viral serology, and research and development. A specialized team should manage each section to enhance efficiency and expertise. Laboratory networks should also be established and maintained in tropical countries to strengthen diagnostic capacity, disease surveillance, and outbreak response. For this purpose, the structure and activities of the World Health Organization (WHO) laboratory network can serve as a useful model for the development and coordination of national and regional laboratory systems.

Structure and activities of the WHO laboratory network[1, 2]

The World Health Organization (WHO), for example, through the Global Polio Laboratory Network (GPLN), plays a critical role in the global effort to eradicate polio. The network is structured to support polio surveillance, monitoring, and research, ensuring that polioviruses are quickly and accurately detected and characterized to guide immunization and containment strategies (see Figure 7.1).

Structure of the WHO Polio Laboratory Network

The WHO Polio Laboratory Network comprises over 145 laboratories globally, categorized into different tiers to ensure efficient coordination:

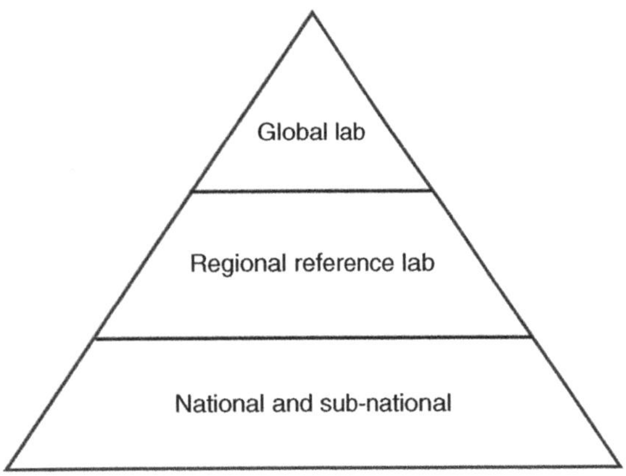

Figure 7.1 Structure of the WHO Laboratory Network for Polio.

1. **WHO National Laboratories (National Polio Laboratories [NPLs])**
   - These are the primary level of laboratories, located in countries with endemic or at-risk populations.
   - For example, NPLs are responsible for initial screening, detection, and identification of polioviruses from stool samples of suspected polio cases. Countries like Nigeria, Ethiopia, Uganda, and Ghana have active NPLs as part of their polio surveillance efforts.
2. **WHO regional reference laboratories (RRLs)**
   - RRLs provide support to NPLs in their respective regions.
   - They perform more detailed testing, including molecular sequencing and serotyping, and offer technical support. Some of the RRLs in tropical countries are located in Senegal, South Africa, Kenya, Nigeria, India, Thailand, Brazil, and Indonesia.
3. **Global specialized laboratories**
   - These top-tier laboratories perform advanced molecular and genetic analyses.
   - They support global surveillance efforts, track virus transmission patterns, and monitor genetic variations in circulating poliovirus strains.
   - Specialized laboratories play a critical role in detecting and characterizing viruses.

Key activities of the WHO Polio Laboratory Network

1. **Virus detection and isolation**
   - The primary function of the network is the timely detection of polioviruses from clinical samples, particularly stool specimens from suspected Acute Flaccid Paralysis (AFP) cases.
   - The laboratories use advanced techniques such as virus culture, real-time polymerase chain reaction (PCR), and molecular diagnostic tools to isolate and identify polioviruses.
2. **Environmental surveillance**
   - Many laboratories within the network are involved in environmental surveillance, testing sewage samples for the presence of poliovirus.
   - This approach helps detect silent transmission in communities and identifies outbreaks before clinical cases arise.
3. **Serotyping and genetic sequencing**
   - Once a poliovirus is detected, the laboratory determines the serotype (types 1, 2, or 3) and whether it is a wild or vaccine-derived strain.
   - Genetic sequencing is performed to trace virus transmission pathways, monitor mutations, and identify geographic origins.
4. **Quality assurance (QA) and training**
   - The WHO ensures high standards across the network through regular quality assessments, proficiency testing, and on-site evaluations.
   - Training programs are conducted to update laboratories on the latest techniques, ensuring the network remains equipped to handle the evolving demands of polio eradication.
5. **Vaccine efficacy monitoring**
   - Laboratories assess the effectiveness of the oral polio vaccine (OPV) by tracking changes in virus epidemiology and conducting tests on postimmunization samples.
6. **Support in outbreak response**
   - In the event of an outbreak, the network laboratories rapidly detect and characterize the virus, helping public health authorities implement appropriate vaccination campaigns.
   - They play a key role in containing virus spread by providing timely data on viral circulation.

7.1.2 Operational Management

Operational management in a medical virology laboratory is critical to ensuring efficient, accurate, and timely diagnostic testing for viral pathogens. This involves overseeing the day-to-day activities, resource management, quality control, and adherence to safety protocols to maintain high standards of laboratory operations.

7.1.2.1 Standard Operating Procedures (SOPs)

SOPs are critical in maintaining consistency and accuracy in all laboratory processes.

They should cover every aspect of the laboratory's operations, from sample collection and handling to testing, reporting, and waste disposal. More details about SOP can be found below. A good SOP should contain the following Ws: who, what, why, where, when.

Creating an SOP is crucial for ensuring consistent, safe, and effective processes in any laboratory or organization. An SOP outlines the step-by-step instructions to perform specific tasks or procedures and serves as a reference for training, compliance, and safety.

Best practices for SOP preparation:

- **Clarity:** Write in simple, direct language, avoiding jargon unless well-defined.
- **Consistency:** Use consistent formatting, terms, and instructions to make the SOP easy to follow.
- **Testing:** Before finalizing, have a colleague follow the SOP to ensure clarity and completeness.
- **Review and approval:** Get the SOP reviewed by relevant experts (e.g., lab supervisors and biosafety officers) before it is implemented.
- **Updates:** Regularly update the SOP to reflect changes in protocols, equipment, or safety regulations.

Step-by-step guide on how to prepare an SOP in a virology laboratory

1. **Title and purpose (What)**
   Provide a clear and concise title that reflects the procedure. And briefly state why the SOP is necessary. This could include objectives, scope, and the importance of the procedure to ensure uniformity and safety.
   Example
   - **Title:** RNA extraction from virus-infected samples.
   - **Purpose:** To describe the standardized procedure for extracting RNA from virus-infected cells to ensure accuracy, reproducibility, and biosafety compliance.
2. **Scope and responsibilities (Who and where)**
   Define the applicability of the SOP. Who should follow it, and in what settings? Specify the types of personnel (e.g., lab technicians, Med Lab scientists, and researchers) and areas where the procedure applies. Define the roles and responsibilities of the personnel involved.
   Example
   - **Scope:** This SOP applies to all laboratory personnel working with virus-infected cell lines in the RNA extraction process within the Biosafety Level 2 (BSL-2) virology lab.
   - Responsibilities
     Lab technicians: Execute the RNA extraction procedure as outlined.
     Lab scientists: Ensure proper training and adherence to the SOP.
     Biosafety officers: Verify that safety protocols are followed throughout the procedure.
3. **Required materials and equipment (What)**
   List all necessary reagents, consumables, equipment, and PPE required to perform the procedure. Be specific to avoid ambiguity.

Example

- **Materials:** TRIzol reagent, RNAse-free tubes, centrifuge, pipettes, and sterile tips.
- **PPE:** Gloves, lab coat, and safety goggles.

4. **Procedure (When)**
   Write the step-by-step instructions in chronological order. Ensure that each step is clear, concise, and easy to follow. You may include safety notes, critical checkpoints, and specific measurements. If the process is complex, consider breaking it down into sections or using subheadings (e.g., sample collection, RNA extraction, and sample storage).
5. **Safety considerations (Where)**
   Highlight any safety precautions or PPE requirements. Reference relevant biosafety guidelines, especially when working with infectious agents or hazardous chemicals.
   Example
   All procedures involving virus-infected cells must be performed in a BSL-2 containment area. Personnel must wear PPE, including gloves, lab coats, and face shields, and ensure proper waste disposal procedures for contaminated materials.
6. **Quality control and troubleshooting**
   Detail any checks or controls that must be performed to ensure the accuracy and quality of the procedure. If applicable, include troubleshooting tips for common issues that may arise during the process.
   Example
   Quality Control: Verify RNA concentration and purity using a spectrophotometer (260/280 ratio).
   Troubleshooting: If RNA yield is low, ensure complete lysis of the sample during the TRIzol incubation.
7. **Documentation and records**
   Specify how the process will be documented. This could include recording data in a lab notebook, completing a form, or entering information into a digital system.
   Example
   Document RNA concentration and quality in the RNA extraction logbook. Record any deviations from the SOP and report to the lab supervisor.
8. **References (Why)**
   Cite any external sources, manuals, or scientific papers referenced to create the SOP. Include references to regulations, such as biosafety guidelines (e.g., WHO, Centers for Disease Control [CDC], or institutional protocols).
9. **Definitions and abbreviations**
   Include any specific terms, technical jargon, or abbreviations used within the SOP. This ensures clarity for all users.
   Example
   RNA: Ribonucleic acid, BSL: Biosafety level
10. **Revision history**
    Include a section at the end to track revisions to the SOP. Indicate the revision number, date, and the nature of the changes made to the procedure.

Example
Revision 1.0, September 2024 – Initial version. Revision 1.1, November 2024 – Updated reagent concentrations.

7.1.2.2 Quality Assurance (QA)

QA is a systematic process that ensures all products, services, or activities consistently meet specified quality standards and regulatory requirements. In the context of laboratories, including virology labs, QA refers to all the planned and systematic activities implemented to ensure that testing processes and outcomes are accurate, reliable, and consistent. Implementing a robust QA program in a virology laboratory ensures the accuracy and reliability of test results. This includes regular calibration of equipment, participation in proficiency testing programs, and internal audits to identify and correct any discrepancies.

Accreditation and compliance

- **Follow international standards:** Ensure the lab complies with accreditation bodies such as International Organization for Standardization (ISO) 15 189 for medical laboratories or ISO 17025 for testing laboratories.
- **Regulatory compliance:** Adhere to national and international biosafety and biosecurity guidelines (e.g., WHO, CDC).
- **Regular audits:** Conduct internal and external audits to ensure adherence to standards and identify areas for improvement. Virology labs must follow guidelines from relevant health organizations like the WHO or national CDC for sample processing and data reporting. Regular audits and validation of techniques such as PCR, viral culture, or sequencing are vital for reliable results.

7.1.2.3 Quality Control (QC) Programs

Quality control (QC) programs are essential components of laboratory management that ensure the accuracy, reliability, and consistency of test results. In a laboratory setting, particularly in virology, QC programs involve systematic processes and checks that monitor the performance of diagnostic tests, equipment, and reagents to meet specified quality standards.

QC programs are designed to identify any errors or inconsistencies during testing by using control materials, standards, and routine monitoring of procedures. These programs involve comparing laboratory results against established benchmarks to ensure tests produce valid and reproducible outcomes. If deviations are detected, corrective actions should be taken to rectify the issues before they affect clinical or research outcomes.

Types of QC

1. **Internal QC**: Implement control samples and standards to check the accuracy and precision of tests, such as PCR assays, enzyme-linked immunosorbent assay (ELISA), and virus culture.
2. **External QC**: Participate in external quality assurance (EQA) programs,

such as proficiency testing offered by WHO, CDC, or other accrediting bodies. Compare results to ensure the lab meets international benchmarks.

3. **Use of controls**: Include both positive and negative controls in all diagnostic assays to verify the results.

7.1.2.4 Inventory Management

Proper management of laboratory supplies and reagents is crucial to prevent shortages and ensure that all necessary materials are available when needed. This involves keeping accurate records of inventory levels, expiry dates, and reorder points.

Key components of inventory management to remember

1. **Accurate record-keeping**
   - Maintain detailed records of all supplies, reagents, and consumables.
   - Use inventory management software or laboratory information management systems (LIMS) to track items in real time, including quantities, lot numbers, storage locations, and expiry dates.
   - Records should include the usage rate of each item, allowing the lab to predict when supplies need to be reordered.
2. **Tracking expiry dates**
   - Expiry date monitoring is essential to ensure that reagents and materials are used within their valid shelf life. Using expired or degraded reagents can compromise the accuracy of experiments and diagnostic tests.
   - Implement a system to flag items nearing their expiration date, ensuring timely use or disposal before they become unusable.
3. **Reorder points**
   Establish minimum stock levels or reorder points for critical supplies, ensuring there is enough lead time to place orders before supplies run out.
   For high-usage items, set higher reorder thresholds to prevent stock-outs that could delay crucial testing or research.
4. **First in, first out (FIFO) system**
   Implement the FIFO method to ensure that older supplies are used before newer ones, minimizing waste due to expiration.
   This is particularly important for perishable items, such as culture media, antibodies, and diagnostic kits.
5. **Storage conditions**
   Ensure proper storage conditions for all reagents, including temperature control (e.g., refrigeration, freezing) and protection from light or humidity as required by manufacturers.
   Improper storage can degrade reagent quality, affecting experimental results.
6. **Supplier management**
   Develop relationships with reliable suppliers to ensure timely deliveries and access to high-quality products.
   Keep a list of alternative suppliers in case of disruptions, shortages, or delays with the primary provider.

7. **Inventory audits**
Conduct regular inventory audits to verify the accuracy of records and check for any discrepancies between what is recorded and what is physically available.
Audits can identify overstocking, understocking, or unaccounted-for losses, allowing corrective measures to be implemented.

7.1.3 Human Resource Management

7.1.3.1 Staff Recruitment and Training

Recruiting skilled and knowledgeable personnel is critical for maintaining the high operational standards of virology laboratories. Ensuring that the team has the right mix of expertise in molecular biology, microbiology, virology, and laboratory management is essential for the lab's success. The recruitment process should prioritize candidates with not only the necessary academic qualifications but also hands-on experience in relevant virology techniques such as viral culture, PCR, serology, and bioinformatics.

After recruitment, continuous training and professional development programs must be in place to ensure staff are kept updated with the latest advancements in the field. This includes:

- Training on new technologies and diagnostic tools as they are introduced into the lab.
- Workshops and seminars on emerging viruses, antiviral therapies, and updated safety protocols.
- Competency assessments to ensure that all staff are proficient in handling biohazardous materials, understanding biosafety levels (BSL), and adhering to laboratory safety regulations (such as WHO or CDC standards).

Cross-train staff in various lab tasks ensures operational flexibility, especially during peak periods or emergencies.

7.1.3.2 Workload Management

Workload management is crucial to ensuring that the laboratory functions efficiently while maintaining a high quality of results. Effective workload distribution prevents staff burnout and errors caused by fatigue, which is particularly important in high-stakes environments like virology labs, where accuracy is critical.

Key strategies for managing workload include

- Task scheduling to balance routine diagnostic testing with research projects and urgent requests (e.g., outbreak response or high-priority patient samples).
- Automate where possible, using technologies such as liquid handlers or automated PCR systems to reduce manual workload, improve turnaround time, and minimize human error.
- Task prioritization based on the critical nature of the samples or projects. For instance, samples from suspected high-risk pathogens (e.g., Ebola virus, Lassa fever virus) should take priority

if the lab is equipped to work on them. Normally, trained personnel should be designated to work on urgent and high-risk pathogens.

- Resource allocation to ensure that the necessary equipment, reagents, and staff are available for the tasks at hand, preventing bottlenecks and delays in lab processes.
- Effective communication and collaboration among staff to adjust the workload dynamically in response to fluctuating sample volumes or changes in project deadlines.

Well-distributed tasks and effective workload management not only enhance productivity but also ensure the well-being of staff, leading to higher morale and reduced error rates.

7.1.3.3 Performance Evaluation

Regular performance evaluations are essential for maintaining the quality and efficiency of lab operations. These evaluations should focus on both technical proficiency and adherence to safety and quality protocols, ensuring that staff consistently deliver accurate results promptly. Key aspects of an effective performance evaluation process include:

- Objective assessment criteria such as accuracy of test results, adherence to SOPs, troubleshooting skills, and laboratory safety practices.
- Self-assessment opportunities, allowing staff to reflect on their performance, identify their strengths and areas for improvement, and set personal goals.
- Feedback mechanisms that are constructive and actionable. Performance feedback should be clear and linked to opportunities for professional development. Where necessary, additional training or mentoring should be provided to address skill gaps.
- Recognition of exceptional performance can motivate staff and create a positive work culture. Highlighting achievements such as breakthroughs in research, innovations in testing, or exemplary adherence to safety protocols encourages high performance.
- Evaluations should be followed by a discussion between staff and supervisors to establish goals for improvement and career growth, as well as strategies to address any challenges identified during the review process.

7.1.4 Data Management and Reporting

7.1.4.1 Laboratory Information Management System

An LIMS is essential for managing data efficiently. It allows for the tracking of samples, results, and reports, ensuring that information is stored securely and can be accessed easily when needed. Only authorized personnel should have access to patient data. Access controls, such as passwords, encryption, and secure login systems, should be implemented to prevent unauthorized access to sensitive information. This information system is essential in facilities above the level of a district laboratory.

7.1.4.2 Result Reporting

Timely and accurate result reporting is a critical function in virology laboratories, particularly for diagnostic purposes and clinical decision-making. Laboratories must ensure that reports are not only delivered promptly but also contain comprehensive information for proper interpretation by healthcare providers.

Key aspects of result reporting include

- **Clarity and precision:** Reports should be clear, free from technical jargon, and easily understandable by clinicians.
- **Inclusion of essential details:** Reports must contain key information such as patient identification, sample type, test performed, results, reference ranges, and interpretative comments when necessary.
- **Timeliness:** Delays in reporting can have significant implications, especially in the case of highly infectious diseases or critical patients. Ensuring that testing is completed within acceptable timeframes is crucial.
- **Digital reporting systems:** Implementation of LIMS for automated and real-time reporting can help streamline the reporting process, reduce human errors (post-analytical error), and ensure that reports are available as soon as testing is completed.

7.1.4.3 Confidentiality

Maintaining the confidentiality of patient data is of utmost importance; however, applicable government laws on confidentiality should be followed. The laboratory must have policies in place to protect sensitive information from unauthorized access.

Confidentiality in Public Health in Reporting Epidemiological Data

> *When reporting data for public health purposes, such as during outbreaks or surveillance activities, it is important to balance the need for public health information with the protection of individual privacy. Data should be aggregated and anonymized where possible to prevent the identification of individuals.*
>
> *While transparency is important in public health, it must be accompanied by safeguards and rules to protect sensitive information. Clear guidelines should be established for what information can be shared publicly and what must remain confidential.*

7.2 Personal and Laboratory Safety Precautions

All biological samples in the lab are potential biohazards; consequently, they should be handled safely and treated accordingly. It is the responsibility of the sender to package the sample properly, and this should be double-checked upon reception in the lab. Spillage, contamination, and unprotected exposure should be avoided among all laboratory users (see Figure 7.2). Basic personal

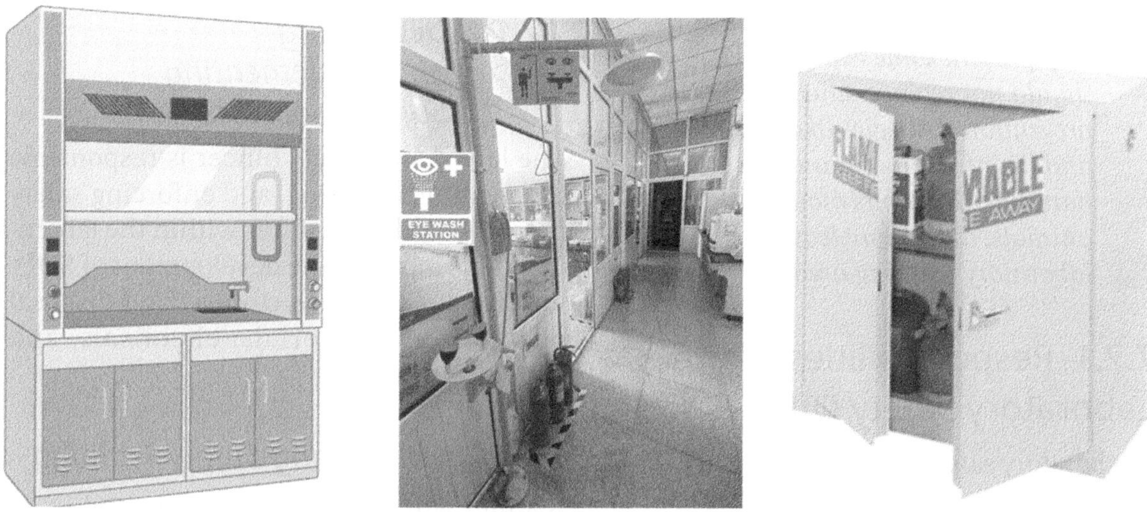

Figure 7.2 Examples of safety equipment: Fume cupboard, eyewash station, and cabinet for storing flammable reagents (left to right).

Source: Kewaunee Scientific Corporation India Pvt. Ltd.

safety precautions in a virology laboratory include, but are not limited to:

- Use and dispose of gloves (latex or nitrile) properly when holding, handling, and transporting all biological samples. If necessary, double gloves should be worn.
- Change gloves between patients and replace them with a fresh pair.
- Always wear PPE (lab coats, gloves, and safety eyewear) when collecting and processing all samples. For biosafety containment level 3 or 4, a respirator mask with a High-Efficiency Particulate Air (HEPA) filter is required for all samples that potentially contain Risk Group 3 and 4 pathogens, e.g., respiratory samples.
- When potentially infectious aerosol is released (outside a biological safety cabinet [BSC]) into the environment, all personnel should evacuate from the affected area immediately, and any exposed persons should be referred for medical advice.
- Properly discard used needles directly into the sharps box; do not recap used needles. All needles, broken bottles, or disposable materials exposed to infectious agents (used or not) should be discarded.
- All sample bottles must be opened under the biosafety cabinet.
- It is not advisable to sterilize and reuse any disposable material.
- Organize work area carefully to avoid a mess-up.
- Separate all contaminated materials, equipment, instruments, and wastes (dispose of them or place them where decontamination takes place).

All laboratory users must be taught proper handling, usage, and disposal of PPE. While personal protective equipment can help prevent some exposures, it should not replace other personal prevention strategies like washing and disinfection of hands after handling infectious materials and animals, and before leaving the laboratory working areas.

7.2.1 Responsibilities of the Laboratory Safety Officer

A laboratory safety officer plays a crucial role in ensuring the safety and well-being of all personnel within a laboratory setting. Their responsibilities encompass a wide range of tasks aimed at maintaining a safe working environment and preventing accidents. Some of the responsibilities of a laboratory safety officer are:

7.2.1.1 Developing and Implementing Safety Protocols

The laboratory safety officer is responsible for creating, updating, and enforcing safety protocols and procedures within the laboratory. This includes the development of SOPs for the safe handling, storage, and disposal of hazardous materials, chemicals, and biological agents.

7.2.1.2 Conducting Safety Training

Regular safety training sessions are to be conducted by the laboratory safety officer to educate laboratory personnel on proper safety practices. This includes training on the use of PPE, emergency response procedures, and the safe operation of laboratory equipment (see Figure 7.3).

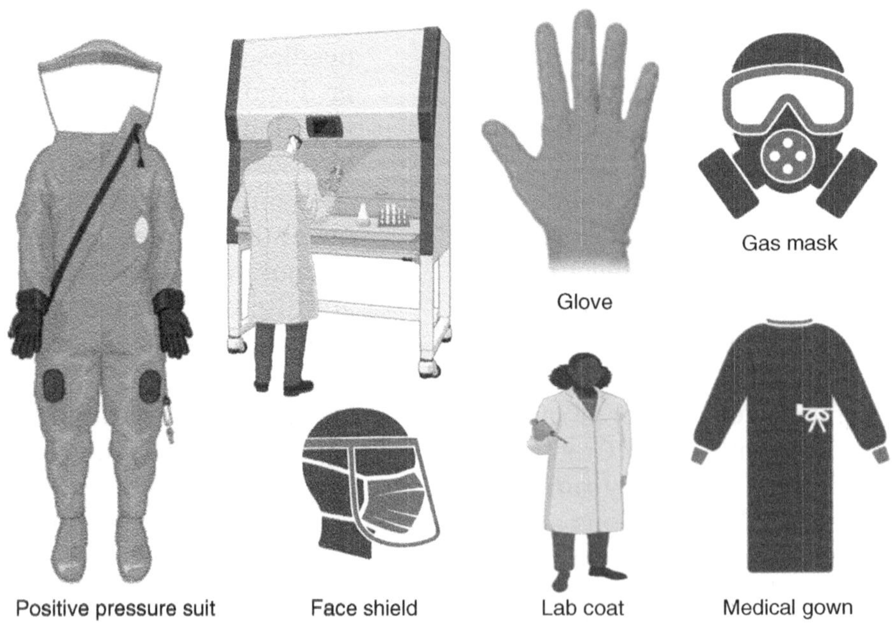

Figure 7.3 Illustrative picture of personal protective equipments.
Source: Created using Biorender.

7.2.1.3 Monitoring Compliance

The laboratory safety officer ensures that all laboratory activities comply with local, national, and international safety regulations. This involves conducting regular safety audits and inspections to identify potential hazards and ensure corrective actions are taken promptly.

7.2.1.4 Risk Assessment and Hazard Identification

It is the responsibility of every laboratory personnel to pay attention to the identification and assessment of potential risks within the laboratory. The laboratory safety officer then conducts thorough risk assessments to determine the likelihood and severity of accidents and implements measures to mitigate these risks by placing appropriate warnings (see Table 7.1).

Table 7.1 Classification of chemical agents commonly used in virology labs, categorized by their hazard types, along with examples.

Hazard type	Examples of agents	Description	Warning symbols
Biohazard	Viruses, bacteria, fungi, parasites, prions, bloodborne pathogen-containing substances	Biohazards refer to biological substances that pose a threat to human health or the environment. These include infectious agents that can cause diseases in humans, animals, or plants, as well as materials like blood or bodily fluids that can carry such pathogens. Handling biohazards requires strict safety protocols to prevent exposure and contamination.	 Source: Dee-sign/ Adobe Stock Photos
Flammable	Ethanol, isopropanol, chloroform	Flammable liquids can ignite easily when exposed to heat, sparks, or open flames.	 Source: Nur khayati/Adobe Stock Photos
Carcinogenic	Ethidium bromide	Chemicals known to cause or increase the risk of cancer after prolonged or repeated exposure.	 Source: Vandhira/ Adobe stocks Photos

(*Continued*)

Table 7.1 (Continued)

Hazard type	Examples of agents	Description	Warning symbols
Corrosive	Sodium hydroxide (NaOH), hydrochloric acid (HCl)	Substances that can cause severe damage to tissues and surfaces upon contact.	 Source: AH/Adobe Stock Photos
Toxic	Chloroform, phenol, acrylamide	Chemicals that can cause harmful effects if inhaled, ingested, or absorbed through the skin.	 Toxic Source: viktorija reut/Adobe Stock Photos
Irritant	Acetic acid, ammonium hydroxide	Chemicals that cause reversible inflammation or discomfort in contact with skin or mucous membranes.	 Source: merly69/ Adobe Stock Photos
Oxidizing	Hydrogen peroxide, sodium hypochlorite	Agents that can cause or enhance the combustion of other materials, potentially leading to fire.	 Source: Marty's Art / Adobe Stock Photo
Mutagenic	Ethidium bromide	Chemicals that can cause genetic mutations, potentially leading to cancer or reproductive harm.	 Source: Nanotril lion/Adobe Stock Photos

Table 7.1 (Continued)

Hazard type	Examples of agents	Description	Warning symbols
Allergy	Latex proteins (in gloves), formaldehyde	Substances that can cause allergic reactions after repeated exposure.	 Source: Alain Besançon/Adobe Stock Photos
Environmentally hazard	Mercury compounds	Substances that can cause long-term damage to aquatic life or the environment.	 Source: Fast_Cyclone/Adobe Stock Photos
Explosive	Picric acid (when dry)	Substances that can explode when exposed to heat, shock, or friction.	 Source: Fast_Cyclone/Adobe Stock Photos

7.2.1.5 *Incident Investigation and Reporting*

In the event of an accident or near-miss, the laboratory safety officer is responsible for investigating the incident to determine its cause. They must also document the incident, report it to relevant authorities, and recommend corrective actions to prevent future occurrences.

7.2.1.6 *Emergency Preparedness and Response*

The laboratory safety officer develops and implements emergency response plans for various scenarios, including chemical spills, fires, and biological exposures. They also conduct regular drills to ensure that all personnel are prepared to respond effectively in an emergency.

7.2.1.7 *Maintenance of Safety Equipment*

The laboratory safety officer oversees the maintenance and regular inspection of safety equipment such as fire extinguishers, eyewash stations, fume hoods, and first aid kits. Ensuring that this equipment is in good working order is essential for laboratory safety.

7.2.1.8 *Waste Management*

Proper disposal of laboratory waste is critical to preventing environmental contamination and exposure to hazardous materials. The laboratory safety officer is responsible for developing waste management protocols and ensuring that all laboratory personnel follow these procedures.

7.2.1.9 *Communication and Coordination*

The laboratory safety officer acts as a liaison between laboratory personnel, management, and external regulatory bodies. They communicate safety concerns, updates on regulations, and changes in safety protocols to all relevant parties.

7.2.1.10 *Continuous Improvement*

A laboratory safety officer continuously seeks ways to improve safety practices within the laboratory. This involves staying informed about the latest safety standards and technologies and integrating them into the laboratory's safety program.

7.2.2 Common Causes of Possible Accidents in Virology Laboratories

7.2.2.1 *Accidental Exposure to Sharps*

Accidental exposure to sharps presents a significant risk in laboratory environments, particularly when potentially infectious materials are involved. Injuries can occur from needles, scalpels, lancets, scissors, blades, broken glass, or any object contaminated with blood or blood products. In the event of accidental, unprotected contact with potentially infectious material or a sharps injury, even if there is no visible blood or the skin appears intact, the following recommended first aid steps should be taken immediately:

- Allow the wound to bleed gently while rinsing it under clear running water.
- Thoroughly wash the area with plenty of soap and running water.
- For splashes to the eye or mouth, rinse the area with water or sterile saline.
- Dry the wound and, if necessary, cover it with a dressing or waterproof plaster.
- Apply an antiseptic and cover the wound with a sterile bandage.
- Report the incident to a supervisor or biosafety officer right away.
- Document the incident, including details about the exposure, the virus or pathogen involved, and the circumstances leading to the injury.
- Seek immediate medical attention to assess the risk of infection.

- Follow up with blood tests and monitoring as advised by a healthcare provider.
- Continue with regular medical checkups as recommended to monitor for any signs of infection.
- Review and update lab safety protocols if necessary to prevent future incidents.

7.2.2.2 Accidental Exposure to an Infectious Agent

Accidental exposure to infectious agents in a medical virology lab poses significant risks to lab personnel, as these agents may include high-risk pathogens that can cause serious or life-threatening diseases. Such exposures can occur through a variety of means, including spills, splashes, inhalation of aerosols, or contact with contaminated surfaces and materials. The following are the primary types of accidental exposures and the essential steps to take in response:

i. **Common types of accidental exposures**
   - **Spills or splashes**: Spills or splashes of infectious material onto the skin, mucous membranes (eyes, nose, mouth), or open wounds.
   - **Inhalation:** Inhalation of aerosols generated during procedures involving viral cultures or samples.
   - **Sharp injury:** Needlestick injuries or cuts from contaminated sharps, such as needles, broken glass, or scalpels.
   - **Physical contact:** Contact with contaminated surfaces, such as lab benches, equipment, or PPE.

ii. **Immediate steps to follow accidental exposure**
   - **For skin exposure or needlestick injury:** Immediately wash the affected area thoroughly with soap and water and allow the wound to bleed gently if necessary to help flush out potential contaminants. Apply an antiseptic and cover the area with a sterile bandage or dressing.
   - **For splashes to the eyes or mucous membranes:** Rinse the affected area immediately with plenty of water or sterile saline. Flush the eyes for at least 15 minutes if exposed.
   - **For aerosol exposure:** Evacuate the area if possible and report the incident immediately. Seek medical attention to assess the level of exposure and determine required follow-up.
   - **For spills on surfaces:** Alert others and restrict access to the contaminated area. Then, follow established spill response protocols to clean and disinfect the area with appropriate chemical disinfectants.

iii. **Reporting and medical follow-up**
   - Notify a supervisor or biosafety officer immediately after the exposure.
   - Document the incident in detail, including the type of exposure, the infectious agent involved, and the circumstances.
   - Seek medical evaluation promptly, especially when high-risk pathogens are involved.

- Follow up with blood tests or other monitoring as recommended to detect potential infection or illness.
- Adhere to treatment protocols, such as post-exposure prophylaxis (PEP), if necessary, depending on the pathogen involved.
- Monitor for symptoms over time and follow the advice of healthcare providers for further care.

Following the report of injury to the supervisor and physician-in-charge, immediately collect a baseline blood sample from the exposed personnel and, if possible, collect a blood sample from the patient or individual who had used the needle. Blood tests can be used to test for HIV, hepatitis B, and C viruses, and others. In rare cases, there may be a risk of other infections. Blood samples can be tested for cytomegalovirus and Epstein–Barr virus.

7.2.2.3 Accidental Exposure to Chemical Agents

Accidental exposure to chemical agents in the lab can be hazardous, leading to immediate and long-term health effects. These exposures might occur during the handling, mixing, or disposal of chemicals used in various laboratory procedures through inhalation, ingestion, or contact. Once this occurs, take the following immediate steps:

i. **Evacuate the area:** If there is a spill, ensure that the area is ventilated and evacuate, if necessary, to prevent inhalation of fumes.

ii. **Remove contaminated clothing:** Carefully remove any clothing that has been stained with the chemical to prevent further exposure.

iii. **Flush the affected area:** For skin contact, rinse the affected skin with copious amounts of water for at least 15 minutes. Use soap if the chemical is oily.
   - **Eye contact:** Rinse eyes immediately with plenty of water for at least 15 minutes. Ensure that the eyelids are held open to flush out the chemical completely.
   - **For inhalation:** Move the exposed person to fresh air immediately. If they are having trouble breathing, seek medical attention immediately.
   - **For ingestion:** Do not induce vomiting unless instructed by medical personnel. Seek immediate medical attention and provide details of the chemical involved.

iv. **Seek medical attention:** Regardless of the exposure level, consult a healthcare provider immediately to assess the severity of the exposure and begin appropriate treatment.

7.2.2.4 Accidental Exposure to Equipment

Accidental exposure to equipment-related dangers, such as centrifuges, autoclaves, and cryogenic systems, and various equipment hazards can pose significant risks to laboratory personnel and lead to serious injuries if proper safety protocols are not followed.

Causes of accidents in the lab

1. **Centrifuges:** Centrifuges are used to separate components in liquid samples. Improper use, such as imbalance or overloading, not covering the lid, can cause the rotor to fail, leading to the release of hazardous aerosols and potential injury from flying debris.

   Prevention: Ensure the rotor is balanced and the lid is securely fastened before operation. Regular maintenance and inspections should be conducted to detect any signs of wear or damage.
2. **Autoclaves:** Autoclaves sterilize equipment and biological waste using high-pressure steam. If not operated correctly, they can cause burns, scalds, or even explosions.

   Prevention: Always follow the manufacturer's guidelines for loading and operating the autoclave. Allow sufficient time for the pressure to release and the temperature to drop before opening the door.
3. **Cryogenic systems:** Cryogenic systems, such as liquid nitrogen tanks, are used for preserving biological samples at extremely low temperatures. Direct contact with cryogenic liquids or uninsulated surfaces can cause severe frostbite or burns.

   Prevention: Wear appropriate PPE, such as insulated gloves and face shields, when handling cryogenic materials. Ensure proper ventilation to avoid asphyxiation risks.
4. **Electrical equipment:** Electrical hazards can arise from faulty wiring, improper grounding, or the use of damaged equipment. These can lead to electric shocks, fires, or explosions.

   Prevention: Regularly inspect and maintain all electrical equipment. Use equipment with proper grounding and avoid overloading circuits. Ensure that all personnel are trained to recognize and respond to electrical hazards.

Make sure that all laboratory personnel receive adequate training to operate equipment in the lab safely.

7.3 Laboratory Waste Management

Improper handling and disposal of laboratory waste can significantly endanger both laboratory personnel and the surrounding community. All laboratory users must be trained in proper waste management practices to minimize the risk of contamination. Adhering to standard waste disposal and decontamination protocols is essential for safeguarding oneself and others. Laboratory wastes targeted for handling, decontamination, and disposal precautions can be categorized as follows:

- **Liquid wastes** such as fluid samples like blood and blood products, urine, sputum, culture media
- **Solid wastes** such as anatomy wastes from cells and tissues, used gloves, aprons, gowns, plastics, and glass bottles.
- **Gaseous waste/aerosols,** airborne particles generated during laboratory processes.

These three categories of waste can be classified as general waste, hazardous waste, and offensive waste (see Figure 7.4).

7.3.1 General Waste

These are nonhazardous wastes that do not pose any significant risk of infection, chemical, or radiological hazards. Despite its nonhazardous nature, general waste must still be managed properly to maintain a clean and organized laboratory environment, prevent contamination, and adhere to waste management regulations.

Examples: Paper products, noncontaminated plastic and glassware, and general packing material.

Management of general waste

- **Segregation:** General waste should be separated from other types of waste (e.g., infectious, hazardous) at the source to prevent cross-contamination and ensure proper disposal.
- **Color code:** The color code for general waste bins varies depending on local regulations and guidelines. However, in many healthcare and laboratory settings, black or gray bins are typically used for general waste. These colors help distinguish general waste from other types of waste, such as infectious (often red or yellow), hazardous (generally yellow), and recyclable materials (generally blue or green).
- **Avoid contamination:** It is crucial to ensure that general waste does not contain infectious or hazardous materials. If any doubt arises regarding the contamination of general waste, it should be treated as hazardous or infectious waste.
- **Disposal:** General waste is usually disposed of in regular trash bins, which are then collected and managed by the facility's waste management system. It does not require special handling or disposal methods.
- **Recycling:** Where possible, general waste such as paper, plastic, and glass should be recycled. Recycling helps reduce the environmental impact and promotes sustainability within the lab.

7.3.2 Hazardous Waste

This type of waste poses a significant risk to human health and the environment. It includes materials that are toxic, infectious, radioactive, corrosive, flammable, or reactive. Examples in a laboratory setting include chemicals, solvents, contaminated sharps, human and animal blood, and body fluids.

Management of hazardous wastes

- **Segregation:** Hazardous waste should be separated from other types of waste at the point of generation. Exposure to chemical hazardous waste can be toxic, corrosive, flammable, or radiative and may cause burns, explosions, or physical injury if mishandled. Furthermore, exposure to biological hazardous waste can lead to infections or disease outbreaks. Radiation exposure can lead to radiation sickness, genetic damage, or cancer.

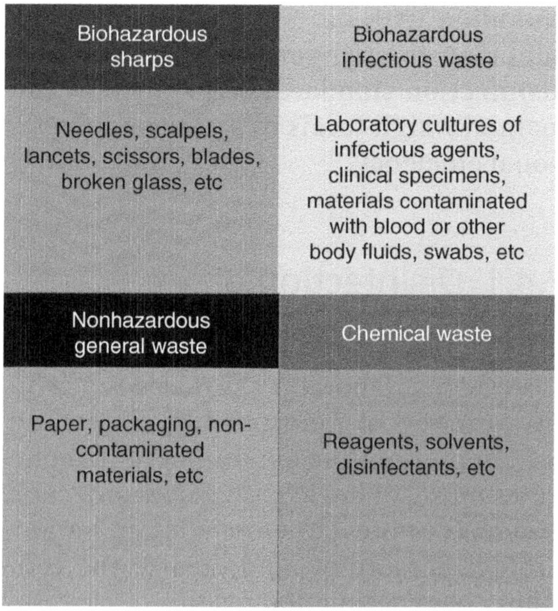

Figure 7.4 Common color code for different types of wastes.

- **Color code:** Use clearly labeled containers with appropriate color coding. The color code for hazardous waste can vary depending on regional regulations and the specific type of hazardous waste, but commonly, yellow is often used for chemical waste. Red is typically used for biohazardous waste, which includes infectious materials. Purple or orange may be used for cytotoxic waste (e.g., waste containing pharmaceutical agents) (Figure 7.4).
- **Containment:** Hazardous waste should be stored in leak-proof, puncture-resistant containers. Containers should be sealed and labeled with the contents, hazard symbols, and the date of disposal.
- **Handling:** Personnel handling hazardous waste must be trained in safe handling procedures, including the use of PPE like gloves, lab coats, and face shields. Spills should be cleaned up immediately using appropriate spill kits, and the area should be decontaminated.
- **Disposal:** Hazardous waste should be disposed of according to local regulations and guidelines.

Chemical waste may need to be neutralized or treated before disposal.

Biological waste should be autoclaved or incinerated to ensure the decontamination of infectious agents.

Radioactive waste must be handled by licensed professionals and disposed of according to radiation safety protocols.

- **Documentation:** Maintain records of hazardous waste generation, handling, and disposal. This includes logs of waste types, quantities, and dates of disposal. Ensure that the waste disposal contractors the lab uses are certified and comply with all regulatory requirements.

7.3.3 Offensive Waste

Offensive waste is noninfectious waste that may offend those who come into contact with it. While it does not pose a direct risk of infection, it can be unpleasant due to its appearance or smell.

Examples include sanitary waste, diapers, and waste from animal bedding. Any "general or offensive" waste contaminated by blood and blood products, body fluids, tissues, or other potentially infectious materials should be considered infectious, hazardous, or radioactive, respectively. Waste from laboratory animals or animal products (e.g., virus propagation in fertile chicken eggs) used for experiments or testing must be treated the same way as human hazardous wastes.

7.4 Sterilization, Decontamination, and Disinfection in a Virology Laboratory

Sterilization and disinfection are critical processes in virology laboratories, because handling of infectious agents poses a significant risk to laboratory personnel, the environment, and the wider community. These procedures are essential for ensuring biosafety by eliminating or inactivating viruses and other pathogens present on surfaces, equipment, and biological materials.

Sterilization refers to the complete eradication of all forms of microbial life, including viruses, bacteria, and spores, typically through methods such as autoclaving, chemical sterilization, or radiation. Disinfection, on the other hand, involves reducing the microbial load to safe levels, particularly on surfaces and equipment, using chemical agents or heat.

In virology laboratories, where work often involves high-risk pathogens, the use of appropriate sterilization and disinfection protocols is paramount to prevent cross-contamination, laboratory-acquired infections, and the accidental release of infectious agents. These practices ensure a safe working environment, contribute to the accuracy of experimental results, and align with biosafety regulations. By adhering to stringent sterilization and disinfection standards, virology labs can maintain high levels of safety and operational efficiency.

7.4.1 Disinfection of Laboratory Space and Surfaces

Disinfecting laboratory surfaces is a critical practice in maintaining a safe and contamination-free environment. Laboratories must adhere to stringent disinfection protocols to prevent cross-contamination, protect personnel, and ensure the reliability of experimental results.

Laboratory space includes laboratory floors, furniture, equipment, and bench surfaces. While surfaces and other materials can be decontaminated with liquid disinfectants, equipment and furniture, and the air space, require both liquid and gaseous disinfectants, respectively.

Steps for disinfecting laboratory surfaces:

- Physically remove any visible dirt, dust, or residues from surfaces using appropriate cleaning agents and methods. This step is crucial because organic matter can interfere with the effectiveness of disinfectants.
- Choose a disinfectant that is effective against the specific microorganisms or contaminants of concern. Common disinfectants include chlorine-based compounds, alcohols, phenolic compounds, and

quaternary ammonium compounds (see Table 7.2).

- Follow selection and read the manufacturer's instructions or laboratory guidelines for concentration, contact time, and application method, of any chosen disinfectants, to ensure effective disinfection.
- After cleaning and selecting the appropriate disinfectant, apply the disinfectant to all surfaces, including those that may have been touched or come into contact with potentially contaminated materials.
- For surfaces without visible spillage and/or with low possibility of contamination by infectious agents, wipe clean with an absorbent material soaked with chemical germicides used as disinfectants and or antiseptics.
- For surfaces with visible fluid, blood, or other body fluid spills (these have with a high possibility of contamination by infectious agents), apply disinfectant very liberally and cover with a towel or any absorbent material for 30 minutes (or as prescribed by the laboratory SOP).
- Clean surfaces with a high possibility of contamination using different absorbent materials, separate from others.
- All the absorbent materials used for cleaning must be disinfected before disposal.

Table 7.2 Categories of chemical agents that can be used as disinfectants.

Chemical agent	Type	Activity	Limitations
Chlorine, chlorine dioxide, sodium hypochlorite	Oxidizing agents	Effective against most viruses	Corrosive to aluminum and metal surfaces; can bleach equipment and fabrics; higher concentrations may be needed for tissues protecting viruses.
Formaldehyde	Gas	Effective against most viruses at temperatures above 20 °C	Slow acting; requires higher temperatures for effectiveness.
Phenolic compounds	Disinfectants	Effective against vegetative bacteria and some enveloped viruses	Most viruses are relatively resistant to phenolic compounds.
Alcohols (ethanol, isopropyl alcohol)	Solvents	Effective against enveloped viruses, vegetative bacteria, and fungi	Non-enveloped viruses are resistant to alcohol treatments.
Hydrogen peroxide and peracids	Oxidizing agents	Potent broad-spectrum germicidal activity	Can be corrosive to aluminum and metal surfaces; capable of bleaching fabrics; and safer than chlorine for humans and the environment.

- Allow the disinfectant to remain on the surface for the recommended contact time specified by the manufacturer (usually 15–60 minutes). This ensures that the disinfectant effectively kills or inactivates microorganisms.
- Let the surfaces air-dry after disinfection to avoid recontamination. In some cases, you may need to wipe the surfaces dry with a clean cloth, depending on the disinfectant used.
- Dispose of any used cleaning materials, gloves, or wipes according to laboratory waste disposal protocols. Checking for residues ensures that no residues from the disinfectant or waste remain on the surfaces, as this can be harmful if not fully removed.

Laboratories need to conduct in-house disinfectant efficiency tests to assess the performance of the chemical of choice. Contact time, concentration of the disinfectants, temperature, relative humidity, and presence or absence of organic materials are factors influencing chemical decontamination processes.

Note that not all instrument surfaces may be suitable for some products, as some, such as those containing sodium hypochlorite (bleach), can be corrosive at high concentrations. Disinfectants for emerging viral pathogens could include options like strong disinfectants. Always verify that the product is registered for use against the specific pathogen in question.

7.4.2 Decontamination of Liquid Biological Waste

7.4.2.1 *Sewer Decontamination of Liquid Biological Waste*

Sewage treatment aims to remove pathogens from waste liquid. However, this kind of treatment may not eliminate pathogens from the liquid waste. Mixing liquid waste in the sewer with chemical decontaminating agents may increase the efficiency of decontamination.

Some local or national regulations may prohibit the disposal of biological and chemical liquid waste into the public sewage system. In such cases, it is crucial to follow the guidelines for decontaminating liquid waste within the sewage system, as failure to comply could lead to legal consequences.

7.4.2.2 *Chemical Decontamination of Liquid Biological Waste*

Chemical treatment can be a standard means of decontaminating all biological liquids before disposal if done properly. Chlorine, chlorine dioxide, and sodium hypochlorite are the most common chemicals used for decontamination of liquid waste. However, laboratory waste technicians must take into consideration the chemical components of the liquid waste to avoid the adverse effects of mixing incompatible chemicals.

Some liquid waste may not be suitable for sanitary sewer disposal even after chemical decontamination; therefore, it must be disposed of through the National Chemical Waste Regulation Agency. In this case, package the sewage in strong containers with a lid, properly labeled with the name of the chemical constituent or content.

7.4.2.3 Autoclave Sterilization of Liquid Biological Waste

Autoclaving can be considered the most effective and reliable method of sterilizing laboratory materials, including liquid wastes; however, this must be done with great caution, especially when sterilizing liquid waste that has been previously treated with chemical decontaminants. Chemicals like ethanol, hypochlorite, and formaldehyde are volatile at high temperatures. Therefore, autoclaving liquid wastes, previously decontaminated with high volumes of these compounds, must be done with caution.

When considering using an autoclave with a specific liquid cycle for the decontamination of liquid waste, the following can be considered:

- Fill liquid waste in containers (preferably an autoclave bottle, e.g., Pyrex bottle) not more than ¾ of their capacity.
- Loosen bottle caps or use vented closures to allow penetration of steam into the container.
- Cover non-autoclave bottles that are not made of safety glass with aluminum foil to avoid breakage. Not all plastic containers are suitable for autoclaving.

7.4.3 Decontamination of Solid Biological Waste

Chemical decontamination of solid wastes is a rarely preferred option for solid biological wastes; it is routinely achieved in combination with autoclaving and, in some instances, supplemented by incineration. This is because it is difficult to ensure the contact of the chemicals with all the surfaces.

7.4.3.1 Autoclave Sterilization of Solid Biological Waste

Autoclaving (steam sterilization) is an effective method for decontaminating solid biological wastes. In this case, materials are sterilized in saturated steam under high pressure for a particular period.

However, the technician must ensure the use of suitable secondary containment or bags (like polypropylene and polycarbonate bags) for all solid items being autoclaved. If unsure of the suitability of the bag, demonstrate it by autoclaving it alone before using it.

To ensure that all parts of the waste are exposed to the steam, leave space between items to allow steam circulation. Additional water can also be added to plastic pans or solid dry waste before autoclaving; this may ensure proper decontamination under specific conditions.

Where possible, consider using an autoclave with a specific solid cycle ("slow exhaust") for decontamination of solid waste. For materials containing infectious materials, set the cycle time and temperature, appropriately. Unless decontamination operation conditions at a shorter time can be validated, ensure that the time and temperature settings for decontamination for infectious or biohazardous waste are set for not less than 30 minutes and 121 °C. If sterilization is not achieved, rerun the load.

Sterilization of potentially infectious animal carcasses can be challenging. To achieve sterilization of animal carcasses, consider prolonging the time of sterilization (compared with typical solid waste) and ensuring an effective temperature of 121 °C is achieved. The weight of the carcasses must be taken into consideration to ensure that moist heat penetrates throughout all the carcasses. Appropriate procedures for the sterilization of animal carcasses must be validated in all individual laboratories handling animal carcasses (see Figure 7.5).

7.4.3.2 Alkaline Hydrolysis-Based Solid Tissue Digestion

Alkaline hydrolysis-based tissue digestion decontaminates animal carcasses in an alkaline solution of sodium or potassium hydroxide under high temperature and pressure. This process can inactivate all viruses and prions to a noninfectious substance and digest animal carcasses into a sterile solution or dried material that is more environmentally friendly and could be disposed of directly into the sanitary sewer.

7.4.3.3 Incineration of Solid Biological Wastes

Incineration involves the combustion of solid waste under direct high temperatures. This is a highly efficient method that ensures that all solid laboratory wastes (with or without prior decontamination) are sterilized and converted into ash. This is different from the mere burning of waste in pits and furnaces. Incineration is performed in a technologically designed machine called an incinerator (see Figure 7.5).

NOTES: Personnel handling the incinerator should be properly trained.

- Waste transport process (especially the ones without prior decontamination)

Autoclave   Incinerator

Figure 7.5 Images of laboratory autoclave and incinerator.

Sources: Photo by Prof. Zhaohua Zhong and Waterman Engineers Australia, respectively.

to the incinerator must be done in an orderly manner, especially if the incinerator is not built within the same building, as is mostly the case.

- Transport all materials for incineration in leak-proof containers.
- Decontaminate all the containers used for the transportation of all waste.
- All ash must be disposed of in strict compliance with national or local regulations.
- Traditionally, pits and furnaces sealed with clay, cement, or concrete are commonly used in some rural settings. The use of these methods is discouraged for the decontamination of infectious wastes because of the health risks associated with exposure to physical harm, particles, hazardous smoke, or harmful combustion products by the operator and the community

7.5 Risk Assessment and Safety in the Virology Laboratory

In many countries, human infectious agents/viruses are classified into risk groups based on their degree of relative risk of infectivity, pathogenicity, the availability of prophylaxis, treatment, and/or vaccines, and potential damage to the environment. No database contains a comprehensive list of all biological agents and their corresponding risk groups; therefore, laboratories must refer to and comply with institutional guidelines regarding the handling of specific viruses.

In general, infectious agents are classified into four risk groups, and biological laboratories that can handle such are classified into four corresponding safety levels. When handling samples or analyzing biological samples that may contain these viruses, it is crucial to strictly follow safety protocols designed to ensure the protection of personnel (the user), the integrity of the product (the sample), and the safety of the environment.

7.5.1 Biosafety Level 1

Risk assessment: Risk Group 1. Required for work on samples that contain potential organisms that are unlikely to cause human disease. Organisms are enzootic and do not produce observable disease in healthy humans. Samples containing these organisms should still be regarded as hazardous (see Figure 7.6).

Examples: Adeno-associated virus (AAV), all serotypes, avian sarcoma virus, and other plant and animal viruses that are known not to cause illness in humans.

Typical work area/sample collection site: Open bench/open area

Laboratory type: Basic teaching and research labs

7.5.2 Biosafety Level 2

Risk assessment: Risk Group 2. Required for work on samples that contain potential organisms that can cause disease in humans, but the disease is treatable or preventable. They are hazardous to employees but unlikely to spread to the community. Usually, attenuated types or specific strains are used for

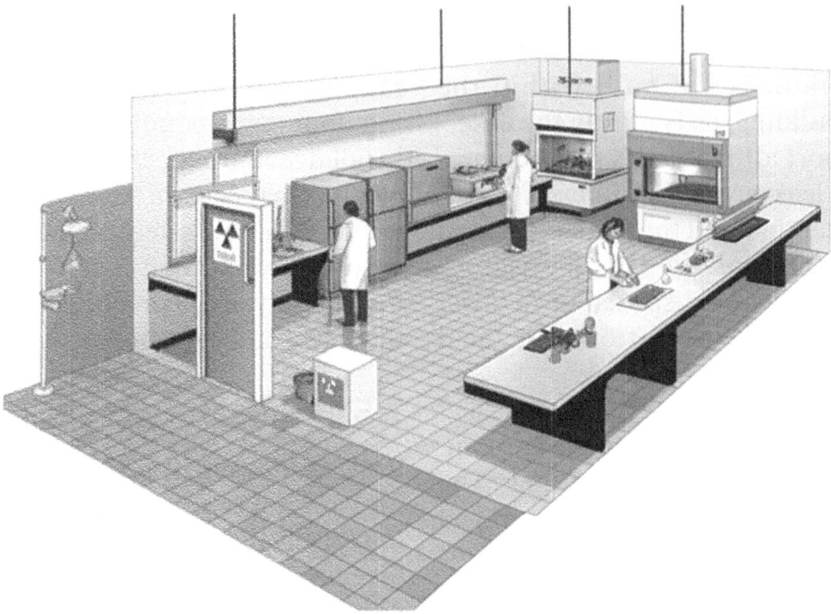

Figure 7.6 Putative picture of a typical biosafety level 1 laboratory.

research. Access to the laboratory is restricted when work is being done (see Figure 7.7).

Examples: All viral agents posing a moderate hazard, like herpesvirus, acute hemorrhagic conjunctivitis virus (AHC), and human adenovirus, all types.

Typical work area/sample collection site: Class 2 biosafety cabinet or laminar flow hood/confined well-ventilated room. An autoclave should be on-site.

Laboratory type: Research, secondary healthcare, and diagnostic labs.

7.5.3 Biosafety Level 3

Risk assessment: Risk Group 3. Required for work on samples that contain potential viral agents that may cause serious or potentially fatal disease, usually through the inhalation route of exposure. Effective treatments, prophylaxis, or vaccines for these diseases may exist. Laboratory users must be under medical surveillance and might need to receive immunizations for the virus they work with (where available). Access to the laboratory is restricted and always controlled (see Figure 7.8).

Examples: West Nile virus, Lassa fever virus, HIV (when culturing or dealing with high volume), Chikungunya virus, Hantaviruses, and Rift Valley fever virus.

Typical work area/sample collection site: Class 3 biosafety cabinet. The laboratory must be isolated from general traffic. An autoclave must be available

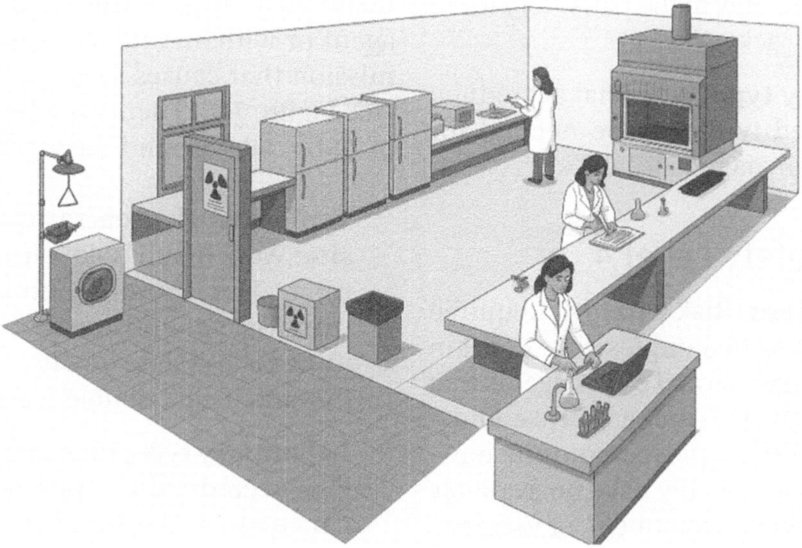

Figure 7.7 Putative picture of a typical biosafety level 2 laboratory.

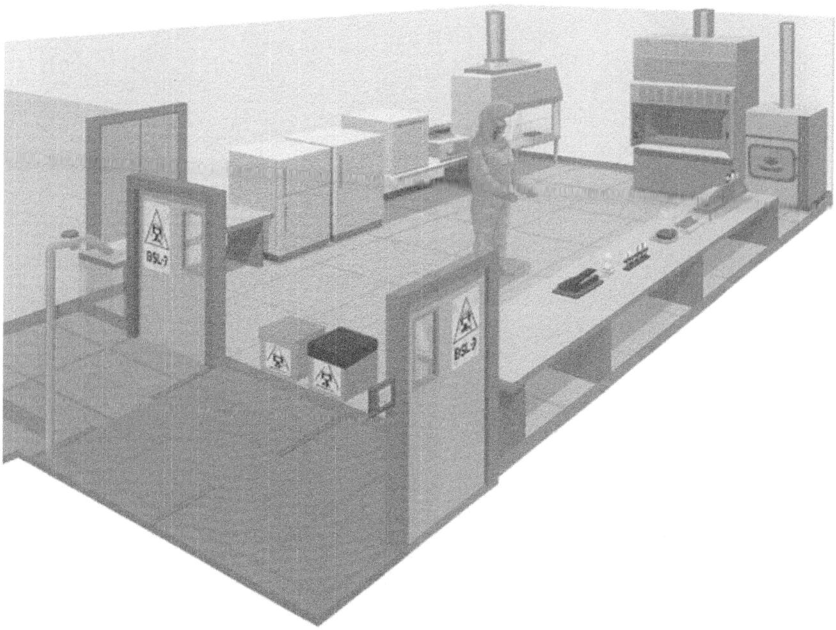

Figure 7.8 Putative picture of a typical biosafety level 3 laboratory.

on-site (preferably in the laboratory room and double-jacketed).

Laboratory type: A special accredited or approved type may be needed for research and diagnostic labs.

7.5.4 Biosafety Level 4

Risk assessment: Risk Group 4. Required for work related to samples that contain potentially dangerous and exotic agents that pose a high, severe human disease and a serious hazard to employees and the environment. Viruses usually have no available effective prophylaxis, treatment, or vaccines. Required for work related to an unidentified agent or with an unknown route of transmission that causes serious, fatal diseases (see Figure 7.9). There are a limited number of BSL-4 labs around the world.

Typical work area/sample collection site: Work in full isolation suits/isolation room. The laboratory must be isolated from the general traffic. An autoclave must be available on-site, in the laboratory room, and double-jacketed.

Laboratory type: Research or diagnostic labs accredited or approved to handle dangerous pathogens.

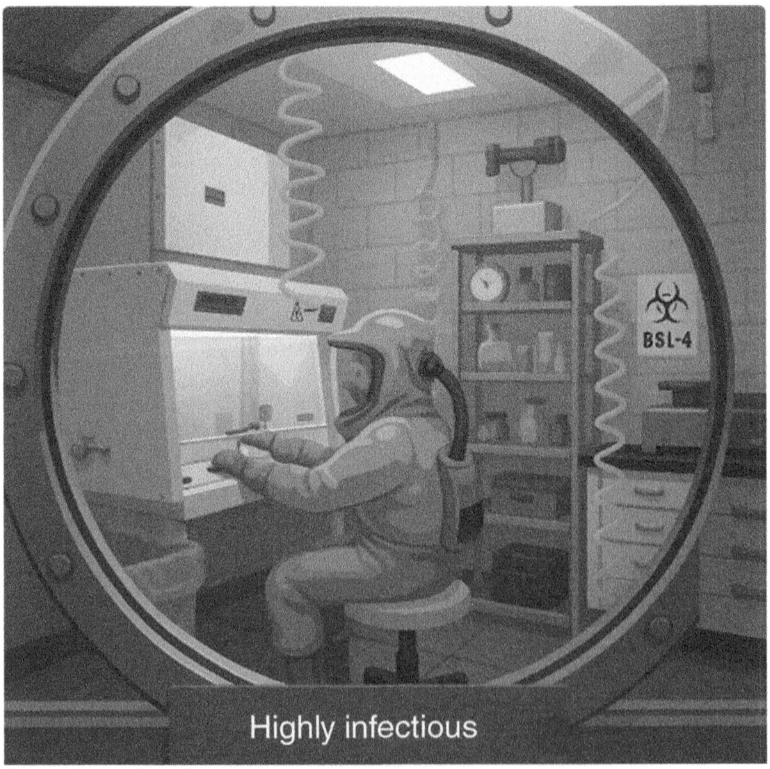

Figure 7.9 Outline of a typical biosafety level 4 laboratory.

7.6 Procedure for the Use of Biological Safety Cabinets (BSCs)

BSCs are designed to protect laboratory workers, the environment, and materials from contamination when handling biohazardous substances. To ensure maximum protection, proper operating procedures, including the optional use of ultraviolet light, must be followed (see Figures 7.10 and 7.11).

1. **Preparation before use**
   - **Wear appropriate PPE**: Before working, put on PPE, including gloves, lab coats, and eye protection.
   - **Check certification**: Ensure the BSC has been certified within the last 12 months and is functioning correctly.
   - **Turn on the BSC**: Turn on the cabinet and let it run for 10–15 minutes to establish proper airflow before starting work.
   - **Inspect the airflow**: Ensure the airflow indicator shows normal functioning airflow.
   - **Disinfect surfaces**: Wipe down all interior surfaces with an appropriate disinfectant, like 70% ethanol or 10% bleach, before placing materials inside.
   - **UV light**: If applicable, turn off the UV light before entering the BSC. UV light should only be used when the cabinet is not in operation to avoid exposure to harmful UV radiation.

Figure 7.10 Picture of a scattered work environment in a biosafety cabinet (intentionally scattered for illustration).

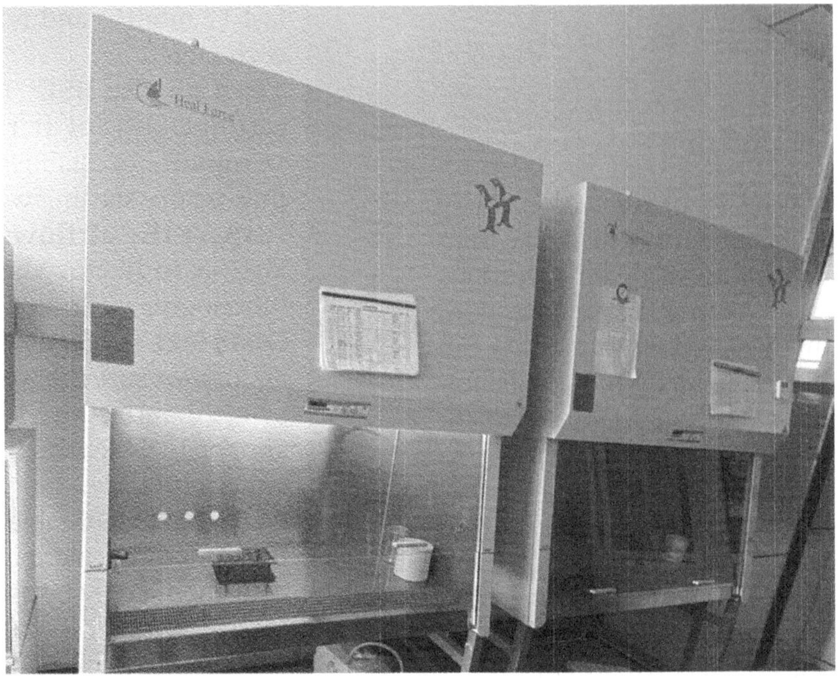

Figure 7.11 Picture of a biosafety cabinet level 2.
Source: Photo by Prof. Zhaohua Zhong.

2. **Loading materials**
   - **Place equipment carefully**: Only introduce necessary equipment into the BSC to avoid overcrowding, which can disrupt airflow.
   - **Position equipment correctly**: Place equipment in the center and toward the back of the BSC, ensuring air can flow freely around it.
   - **Avoid blocking grilles**: Ensure no materials obstruct the air intake or exhaust grilles.
3. **Working in the BSC**
   - **Minimize movements**: Work slowly and avoid rapid or sweeping movements that can disrupt airflow.
   - **Work cleanly**: Arrange work from clean to contaminated areas, keeping separate zones for clean materials and waste.
   - **Handle biohazardous materials carefully**: Open containers cautiously and work with slow, controlled techniques to minimize aerosol production.
   - **Avoid flames**: Open flames should not be used inside BSCs as they can disrupt airflow and pose fire hazards. Use electric alternatives when necessary.
   - **Pipetting**: Use slow, careful pipetting techniques to avoid splashing or aerosol generation.

4. **Waste handling**
   - **Proper containers**: Place all waste in appropriate biohazard containers inside the BSC to avoid cross-contamination outside the cabinet.
   - **Decontaminate waste**: Ensure all biological waste is properly sterilized or disinfected before removal.
5. **Finishing work**
   - **Seal and remove samples**: Once work is finished, securely close all containers and decontaminate the outside surfaces before removing them from the cabinet.
   - **Clean surfaces**: Disinfect all surfaces within the BSC using an appropriate disinfectant after work is completed.
   - **Run the BSC fan**: Allow the BSC to run for an additional 10–15 minutes after work to clear any residual particles or aerosols.
6. **UV Light use for decontamination**
   - **Post-use UV sterilization**: Once the cabinet has been fully cleaned and all materials have been removed, UV light can be used for additional decontamination. The UV light should only be turned on when the BSC is unoccupied.
   - **Duration**: UV light should typically run for 15–30 minutes to allow sufficient exposure for decontamination, depending on laboratory protocols.
   - **Safety precautions**: Never operate the UV light while personnel are working in or near the cabinet. UV radiation can damage the skin and eyes. Always ensure the sash is closed and leave the room if the light is on.
   - **Turn off UV light**: Ensure the UV light is turned off before starting the next work session or before opening the BSC.
7. **Shutting down**
   - **Turn off the BSC**: After proper decontamination and UV light use, switch off the cabinet's fan and record usage.
   - **Remove PPE**: Remove gloves, lab coats, and other PPE carefully and dispose of biohazardous materials according to safety protocols.
8. **Maintenance and cleaning**
   - **Regular cleaning**: Perform routine maintenance and cleaning of the BSC. Replace filters and clean the interior surfaces regularly to maintain cabinet performance.
   - **Annual certification**: Ensure that the BSC is certified annually to meet operational safety standards.

What if a sample with potential Risk Group 3 or 4 organisms is sent to your lab?

To determine the risk group of an organism, laboratory personnel must consult reliable sources, the public health agency of the country, state, or regional government. Not all facilities can handle samples with confirmed or potential Risk Groups 3 and 4 viruses. Most viruses in these groups cause serious and deadly diseases that constitutes a potential hazard to health personnel and the community. The majority of countries have published a list of all the laboratory facilities equipped for handling, analyzing, and storing a variety of microorganisms. The list of facilities can be checked on national or WHO databases and other reliable sources.

There should be no compelling reason for laboratory personnel to work on samples (with potential risk group level 2, 3, and 4 organisms) not accredited or equipped for. Do not do it. Find an alternative. Handling samples requires extremely stringent safety precautions, specialized lab facilities, and training.

Laboratory considerations before collecting samples with potential Risk Group 3 or 4 viruses

Aside from needing an accreditation, before a laboratory can handle any sample with potential Risk Group 3 or 4 viruses for testing, the following are important considerations that should guide decisions:

- **Personnel and community protection consideration:** The code of ethics of the country's Medical Laboratory Services, which sets forth the principles and standards to which Medical Laboratory Professionals operate, must be strictly adhered to. This protection should be extended to other laboratory support staff.
- **Personnel competence and experience:** Laboratory personnel must have the competence and experience in testing-related services for handling, testing, isolating, and identifying infectious diseases and organisms that affect human populations or pose environmental hazards.

Other technical considerations: How well can the potential hazard be managed? How well can the tests be performed and standardized in the laboratory? To answer this, it is important to take into consideration the availability of equipment, reagents, standards, controls, and consumables required to perform tests. In some tropical countries, stability, storage, cost of consumables, power and water requirements, complexity, safety, maintenance, and local repair facilities of equipment need to be further considered. Likewise, the availability of facilities for safe handling, transport, storage, and disposal of samples should be considered.

7.7 Storage of Biological Samples, Chemicals, Materials, and Apparatus

Effective storage practices for biological samples, chemicals, materials, and apparatus are vital in virology laboratories to maintain the integrity of experiments, ensure safety, and comply with regulatory standards. Each category requires specific considerations for optimal preservation, safety, and accessibility.

i. **Storage of biological samples**
Before storage, all samples should be clearly labeled with pertinent information, including the sample type, date of collection, patient

identifier, and processing notes. Utilizing a tracking system can help maintain organization and prevent mix-ups.

- **Storage at ambient temperature:** Glassware, some chemicals (that are inflammable, like alcohol, isopropanol, salts), plastics, and small equipment can be stored on clean shelves and drawers away from moisture and direct sunlight. Ensure to follow the instructions labels of the manufacturer.
- **Storage at 2–8 °C temperature:** For reagents like culture media, buffers, and physiological solutions, a standard refrigerator for storing laboratory reagents for storing reagents at 2–8 °C should be used. However, the domestic refrigerator is an adequate alternative.
- **Storage below – 20 °C temperature:** Most reagents that require to be frozen can be stored at –20 °C to –40 °C. Domestic freezers are cheaper alternatives to laboratory freezers. Some reagents, like cell, antiviral inhibitors, and virus isolates, can be stored in an ultra-deep freezer (i.e., –80 °C freezer) for up to 6 months. Do not store cells longer than 3 months in an ultra-deep freezer because their viability decreases when they are stored above this temperature.

ii. **Storage of Chemicals**

Chemicals used in virology labs, including reagents, solvents, and disinfectants, must be stored following strict safety protocols.

- **Chemical classification:** Chemicals should be stored according to their hazard classification, such as flammable, corrosive, or toxic substances, to minimize the risk of reactions or accidents.
- **Proper containers:** Use original or compatible containers that are clearly labeled with the chemical name, concentration, hazard warnings, and expiry dates.
- **Temperature sensitivity:** Certain chemicals may require specific temperature conditions for storage to maintain their efficacy and safety. For instance, flammable solvents should be stored in cool, ventilated areas to prevent vapors from accumulating.
- **Ventilated storage areas:** Chemicals should be stored in well-ventilated areas or chemical cabinets designed to handle volatile substances. Fume hoods may be necessary for highly hazardous chemicals.
- **Material safety data sheets (MSDS):** Access to MSDS for all chemicals should be readily available, providing essential information on handling, storage, and emergency procedures.

iii. **Storage of apparatus**

Laboratory equipment and apparatus must be stored in a way that preserves their integrity and functionality. The following should be taken into consideration:

- **Clean and dry storage:** Equipment should be cleaned and dried before storage to prevent contamination and damage. This includes pipettes, glassware, and any apparatus that may come into contact with biological samples.
- **Temperature sensitivity:** Some equipment, such as certain centrifuges or analytical instruments, may have specific temperature requirements for optimal functioning. Ensure that such equipment is stored or operated in environments that meet these requirements.
- **Calibration and maintenance:** Ensure that all equipment is regularly calibrated and maintained according to manufacturer specifications, with proper documentation of maintenance activities.
- **Safe handling practices:** Store equipment securely to prevent damage. Utilize protective covers or cabinets for sensitive instruments and avoid stacking items that could fall and cause injury.

Care of glassware and reusable plastic containers

Disposable plastics, although more expensive, are preferred to glassware for example, clean glass surfaces used for suspension cultures are not perfect materials for cells to adhere to.

- Sterile glassware
- Multi-well plates (6, 12, 24, 48, 96 wells)
- Racks
- Beakers (500–1000 mL)
- Glass bottles (different ranges1 L)
- Centrifuge tubes (1.5, 2, 5, 15, 50 mL)
- Disposable Pasteur pipettes
- Automatic pipettes and tips, and
- Sterile Petri dishes (10 cm or 15 cm diameter)
- Trypsin zing flasks – used for the separation of tissue fragments and homogenates into cell suspensions by trypsin dissociation.
- Tissue culture flasks – available in "vented" and "non-vented" caps. The former, unlike the latter, has a filter integrated into it that allows the exchange of gaseous contents without having to loosen the cap itself.

Sterile glassware and plastics should be stored at room temperature on a clean shelf, drawer, or cabinet.

Procedures for washing glassware

- Wash glassware as quickly as possible after use. Also, wash all new glassware (except otherwise stated by the manufacturer) as they may contain toxic products derived from manufacturing chemicals.
- Autoclave the glassware (where possible, aim to use borosilicate glassware).

- Wash thoroughly with tap water several times.
- Soak in a 4% solution of hydrochloric acid in water overnight.
- Rinse with tap water.
- Soak in detergent solution overnight.
- Scrub all parts of the glass with a soft brush or foam and avoid any scratching of the glassware.
- Rinse several times with tap water and soak overnight in clean water to remove residual detergent.
- Rinse several times with double-distilled water.
- Dry slowly to avoid trapped condensed liquid in the bottle. Glassware intended for tissue culture use should be dried after sterilization.

7.8 Study Questions

7.1 Identify and describe the essential PPE that should be worn in a BSL-3 virology laboratory. How does each piece of PPE contribute to personal safety?

7.2 A researcher observes a small tear in his/her gloves while handling viral samples. What immediate actions should be taken to ensure safety and prevent potential exposure?

7.3 Evaluate the potential risks associated with improper use or malfunction of a BSC in a BSL-3 laboratory. What steps should be taken to mitigate these risks?

7.4 A virology laboratory has recently upgraded to BSL-3 containment due to an increase in research involving a highly pathogenic virus. The laboratory is conducting studies on the virus transmission and pathogenicity. The laboratory manager is reviewing safety protocols and risk assessments to ensure compliance with biosafety standards. What are the key biosafety containment measures that should be implemented in a BSL-3 laboratory to minimize the risk of exposure to a highly pathogenic virus?

7.5 You are a biosafety officer in a laboratory that handles high-risk pathogens. An incident occurs where a large volume of liquid biological waste, potentially contaminated with a hazardous pathogen, needs to be decontaminated. Describe the immediate steps you would take to ensure safe decontamination of the liquid biological waste. Evaluate the effectiveness of different decontamination agents for liquid biological waste.

Objectives

7.1 **In comparison to chemical disinfectants, what is the primary advantage of heat treatment (e.g., autoclaving) for decontaminating liquid biological waste?**
(a) Faster action time
(b) Broader spectrum of microbial inactivation, including spores
(c) Lower operational costs
(d) Minimal environmental impact

7.2 **Which factor is most critical in determining the concentration of bleach required for effective decontamination of liquid biological waste?**
(a) pH of the waste
(b) Temperature of the waste

(c) Type and concentration of pathogens present
(d) Volume of the waste

7.3 **Which of the following agents is most effective against viral pathogens in liquid biological waste?**
(a) Ethanol
(b) Sodium hypochlorite
(c) Hydrogen peroxide
(d) Iodine-based compounds

7.4 **Which method of decontamination is often used in combination with other agents to ensure complete inactivation of pathogens in liquid biological waste?**
(a) Chemical disinfection
(b) Filtration
(c) Heat treatment
(d) Radiation

7.5 **For effective decontamination of biological waste containing bacterial spores, which of the following agents is preferred?**
(a) Sodium chloride
(b) Hydrogen peroxide
(c) Peracetic acid
(d) Ethanol

7.6 **In the context of virology labs, which specific procedure will not likely to lead to aerosolization of infectious agents?**
(a) Using a biosafety cabinet with a malfunctioning HEPA filter
(b) Performing centrifugation without sealing the rotor properly
(c) Handling cultures in an open workspace
(d) Regularly disinfecting the work surface with approved agents

7.7 **Which advanced biosafety measure cannot mitigate the risk of accidental exposure to high-risk pathogens during the transfer of infectious samples?**
(a) Using double-layered gloves
(b) Implementing automated sample transfer systems
(c) Ensuring HEPA filtration in all rooms
(d) Employing manual transfer techniques with PPE

7.8 **In virology laboratories, which specific type of equipment malfunction is most likely to result in a breach of containment protocols?**
(a) A malfunctioning autoclave that fails to achieve proper sterilization
(b) A refrigerator that maintains incorrect temperature settings
(c) A centrifuge with an balanced rotor
(d) A fume hood with a nonfunctional light

7.9 **Which of the following situations represents a potential critical failure in decontamination protocols in a virology lab?**
(a) Using a disinfectant that is not validated for the specific virus present
(b) Decontaminating surfaces after spills with a standard bleach solution
(c) Properly documenting decontamination procedures and outcomes
(d) Conducting routine cleaning with approved disinfectants

References

1. Organization, W.H, Polio laboratory manual. 2004.
2. Diop, O.M., et al., *The Global Polio Laboratory Network as a Platform for the Viral Vaccine-Preventable and Emerging Diseases Laboratory Networks*. J. Infect. Dis., 2017. **216**(suppl_1): p. S299–S307. doi: 10.1093/infdis/jix092 PMID: 28838192; PMCID: PMC5853949. https://pmc.ncbi.nlm.nih.gov/articles/PMC5853949/.

8 Basics of Cell Culture and Cell Line Maintenance: *For Resource-Limited Settings*

Cell culture refers to the cultivation of cells in an artificial medium that supports their growth, survival, and function as they would in their natural environment. This method allows for the manipulation and analysis of tissues or cells under controlled conditions, which can be referred to as "*in vitro* micropropagation of cells." The cultured cells can be grown as a single monolayer, in suspension, or as part of a three-dimensional tissue or organ.

Cells for culture can be obtained directly from living tissue through enzymatic dissociation using enzymes such as trypsin, collagenase, or proteinase, or through mechanical dissociation. Additionally, cells can be sourced from established cell lines.

8.1 Types of Cell Culture

8.1.1 Primary Cell Culture

Primary cell are derived directly from host tissue and cultivated under specific conditions until they reach confluence (large quantity). These are finite cell lines. They closely represent the genetic integrity of the original tissue because they are unmodified genetically. This makes them potentially sensitive to various viral infections. However, primary cell cultures have a finite lifespan and eventually undergo senescence (aging) after a limited number of divisions. Examples of primary cell types include epithelial cells, fibroblasts, endothelial cells, muscle cells, hematopoietic cells, and mesenchymal stem cells, typically sourced from neonatal or young animals.

Initially heterogeneous, primary cell cultures can be purified to predominantly contain a single cell type of interest using various methods based on shared characteristics such as size, shape, magnetic properties, or surface protein expression. Common techniques for cell purification include centrifugation and various cell sorting methods.

The following are the procedures for obtaining a primary cell line from a suckling mouse heart for the culture of coxsackievirus

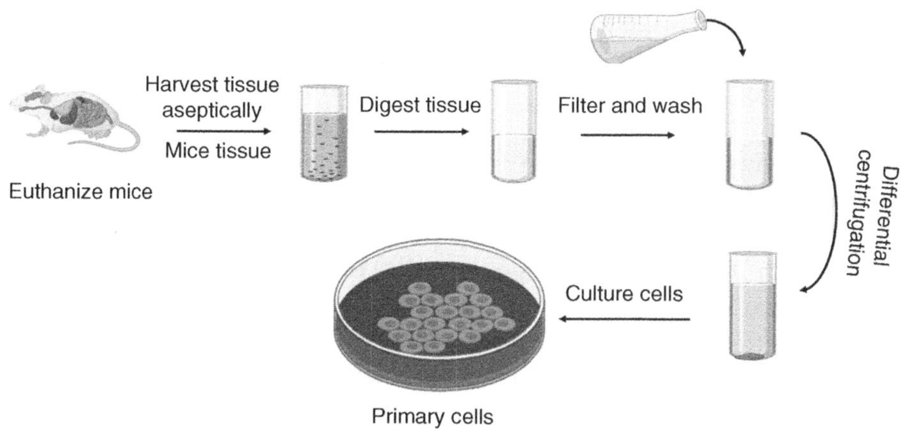

Figure 8.1 Systematic illustration of the isolation of primary cells from animal tissue.

B (this can be adapted for other cell types) (see Figure 8.1).

- **Materials and equipment**
  Suckling mouse: Typically, 1–2 days old
  Dissection tools: Scissors, forceps, and sterile dissection pads.
  Culture media: DMEM (Dulbecco's Modified Eagle Medium) or another suitable medium, supplemented with fetal bovine serum (FBS) and antibiotics
  Enzymes: Collagenase, trypsin, or other proteolytic enzymes.
  Cell culture dishes/flasks: Sterile 6 mm plate.
  Incubator: Set at 37 °C with 5% CO_2.
  Centrifuge: For cell pelleting
  Sterile phosphate-buffered saline (PBS): Prepare or purchase

  Ensure all tools, media, and equipment are sterile to prevent contamination.

- **Sacrificing and dissecting the mouse**
  Euthanize the suckling mouse humanely, following ethical guidelines and institutional regulations. Place the mouse on a sterile dissection pad and use sterile scissors and forceps to open the abdominal cavity. Gently remove the heart (ensure that no other tissue, artery, or vein is attached) and place it in a sterile Petri dish or a tube containing cold PBS to rinse off any blood.
- **Tissue dissection**
  Minimize harsh tissue handling to prevent degradation and distortion. Dissect the heart into small pieces (1–2 mm^3) using sterile scissors or a scalpel.
- **Tissue digestion**
  Prepare the enzyme solution (e.g., 0.1% collagenase in PBS or 0.1% trypsin) and add the enzyme solution to the tissue pieces in a sterile tube or dish. Incubate at 37 °C for 20–30 minutes with gentle shaking or agitation

to ensure thorough digestion. This step helps to dissociate the cells from the tissue matrix.

- **Cell isolation**
  a. Stop the enzymatic reaction by adding an equal volume of complete culture medium (e.g., DMEM with 10% FBS) to the digestion mixture.
  b. Gently pipette the mixture up and down to further dissociate the cells.
  c. Filter the cell suspension through a sterile mesh or cell strainer (70 µm) to remove undigested tissue fragments.
- **Cell centrifugation**
  a. Centrifuge the cell suspension at 300–400 × g for 5–10 minutes to pellet the cells.
  b. Discard the supernatant and resuspend the cell pellet in a fresh complete culture medium.
- **Cell culture and maintenance**
  a. Plate the cells in a sterile culture dish or flask.
  b. Incubate at 37 °C with 5% CO_2.
  c. Change the medium after 24–48 hours to remove non-adherent cells and debris.
  d. Monitor the cells under a microscope for morphology and growth.
  e. Passage the cells when they reach confluency using standard cell culture techniques (e.g., trypsinization).

8.1.2 Secondary Cell Culture

Secondary cell culture refers to the cells that are derived from a primary cell culture after it has been subcultured or passaged. This means that after the primary culture (initial culture from tissue or dispersed cells) has grown to confluence or a desired density, it is divided and transferred to new culture vessels. They have the same number of chromosomes as the primary cells (semi-continuous) of origin. Secondary cell cultures, like primary cultures, have a limited number of cell divisions before they undergo senescence, meaning they are finite cell lines. However, they can be subcultured multiple times before reaching this stage.

8.1.3 Cell Line

A cell line broadly refers to a population of cells that can be cultured in vitro (outside of a living organism) for an extended period. These cells are typically derived from primary cells that have been subcultured, allowing them to grow and divide over multiple generations.

While all continuous cell lines are immortalized, meaning they can proliferate indefinitely, not all of them are cancerous. Immortalization can occur through various mechanisms, and cancer is just one of them.

Cell lines can be either finite (having a limited lifespan) or infinite/continuous cell lines (able to divide indefinitely). Finite cell lines are cells that have a limited number of divisions before they enter senescence and stop proliferating. They retain many characteristics of the original tissue and are often used in research for a relatively short time. Continuous cell lines are cell that have undergone immotalization process, allowing them to divide indefinitely. This immortality is typically achieved through spontaneous mutation or deliberate genetic manipulation, such as the introduction of viral oncogenes. Cell lines, whether cancerous or not, may exhibit abnormal numbers of chromosomes (heteroploidy) or altered chromosomal morphology compared to the primary cells from which they originated. (see Tables 8.1–8.3).

Table 8.1 Types of cell lines and their features.

Features	Finite cell line	Continuous cell lines
Definition	Cell lines with a limited number of divisions before senescence	Cell lines that have been transformed to divide indefinitely
Lifespan	Limited (finite)	Indefinite (immortal)
Genetic stability	Generally, maintains genetic stability similar to the original tissue	May undergo genetic changes during transformation
Senescence	Undergoes senescence after a set number of divisions	Does not undergo senescence; can proliferate indefinitely
Origin	Derived from primary cells, like the tissue of origin	Derived from primary cells but transformed via spontaneous or induced methods
Use in research	Short-term studies; maintaining tissue characteristics	Long-term studies; large-scale production and consistent results
Examples	Primary fibroblasts, keratinocytes, melanocytes, and endothelial cells	HeLa (derived from human cervical cancer cells)
Ease of maintenance	Requires regular replenishment and careful maintenance	Easier to maintain due to indefinite growth
Applications	Studies requiring near-native cell properties	Drug testing, vaccine production, cancer research, and long-term experiments
Availability	Limited availability due to finite lifespan	Widely available and extensively used

Table 8.2 Examples of cell lines used for virus isolation.

Cell line	Morphology	Origin	Virus susceptibility	Medium
HeLa cells	Epithelial	Cervix carcinoma	Used for culturing a wide variety of viruses.	DMEM, 10% FBS
Human cervix carcinoma: A derivative of HeLa (HeLa B)	Epithelial	Human	Used for culturing a wide variety of viruses.	DMEM, 10% FBS
Vero E6 cells	Kidney	African green monkey	Highly susceptible to SARS-coronavirus-2	DMEM, 10% FBS
HEK-293T cells	Fibroblast	Embryonic kidney	Used for culturing a wide variety of endogenous viruses.	MEM, 10% FBS
Rhabdomyosarcoma (RD) cell lines	Epithelial/ Spindle	Human (muscle tissue)	Surveillance of polioviruses and other enteroviruses.	DMEM, 10% FBS
Medical Research Council cell strain 5 (MRC-5) cells	Fibroblast	Human (fetal tissue)	Enteroviruses and cytomegalovirus	MEM, 10% FBS
Madin-Darby canine kidney (MDCK)	Epithelial	Dog (canine)	Widely used for the isolation and propagation of human influenza viruses.	MEM, 10% FBS
Rhesus monkey kidney cells (RhMK)	Kidney	Rhesus monkey	Varicella-zoster virus and herpes simplex virus	DMEM, 10% FBS
Rhesus monkey kidney epithelial cells (LLC-MK2)	Epithelial	Rhesus monkey	For the isolation of influenza and parainfluenza viruses, human enteroviruses, and mumps virus	DMEM, 10% FBS
Human foreskin fibroblast cells (HuF)	Fibroblast	Human	Herpes simplex virus	DMEM, 10% FBS
Human alveolar epithelial cells (A549)	Epithelial	Lung carcinoma	Adenovirus, rhinovirus, cytomegalovirus, and herpes simplex virus	F-12 K, 10% FBS
Hep-2 monkey kidney	Epithelial	Monkey (African green)	For a variety of fecal viruses, respiratory syncytial viruses, and Herpes virus	DMEM, 10% FBS
Jurkat cells	Lymphoblast	Human (lymphoma)	HIV, human T lymphocyte virus studies	RPMI-1640, 10% FBS

Table 8.2 (Continued)

Cell line	Morphology	Origin	Virus susceptibility	Medium
Normal rabbit kidney epithelial cells (RK13 Line)	Epithelial	Rabbit	Susceptible to herpes simplex, vaccinia, and rubella viruses.	MEM, 10% FBS, 2 mM glutamine
Baby Hamster Kidney (BHK cells)	Fibroblast	Hamster	Used for studying a variety of viruses, including rabies and polio	DMEM, 10% FBS
Madin-Darby Canine Kidney (MDCK)-SIAT1 cells	Epithelial	Dog (canine)	Highly susceptible to human influenza viruses and other respiratory viruses.	MEM, 10% FBS
C6/36 cells	Epithelial	Aedes mosquito	Used for culturing flaviviruses such as dengue and West Nile virus.	MEM, 10% FBS
Retinal Pigment Epithelial-1 (RPE1) cells	Epithelial	Human retinal pigment epithelium	Used for studying cell cycle and cellular aging. And for studying Herpes simplex virus 1 (HSV-1)	DMEM, 10% FBS
substrate-adherent (S-type) human neuroblastoma (SHEP) cells	Epithelial	Human (neuroblastoma)	Used for neuroblastoma research and studies on neuroblastoma-associated viruses. Particularly HSV-1	DMEM, 10% FBS
SW13 cells	Epithelial	Human (adrenal carcinoma)	Used for studies on adrenal carcinoma and virus interactions.	MEM, 10% FBS
Chinese Hamster Ovary, subclone K1 (CHO-K1) cells	Epithelial	Chinese hamster ovary	Widely used for recombinant protein production and virus research.	DMEM, 10%

8.1.4 Cell Strain

A cell strain refers to a subset of a cell line that has been isolated and maintained under specific conditions to preserve certain characteristics. Unlike a cell line, which is a broad term that encompasses various subpopulations, a cell strain is typically a more homogeneous population derived from a cell line or a primary culture.

Table 8.3 Advantages and disadvantages of the use of the cell culture system.

Advantages	Disadvantages
Large-scale production: Enables large-scale production of cells that can be stored and reused at a convenient time.	**Skilled personnel required:** Requires highly skilled personnel and special working conditions.
Distinct cytopathic effects: Produces distinguishable cytopathic effects characteristic of different viruses.	**Strict asepsis needed:** Must be performed using strict aseptic techniques; antibiotics are often used, but some microorganisms may be resistant.
Manipulability: This can be easily manipulated, such as for knock-out experiments, radiolabeling, and studying virus pathogenesis and multiplication.	**Expensive environment control:** A highly controlled physicochemical environment is costly and cumbersome to maintain, especially in rural areas.
Defined medium: Grown in a chemically defined medium, minimizing interference from host cell components like antibodies and innate infection.	**Survival of animal cells:** Some animal cells may not survive in isolation from their natural host.
Simplicity: Simpler compared to studying the whole host organism.	**Morphological and functional changes:** Some cell lines may change morphology and function after long passages and lose gene expression, leading to spontaneous differentiation.
Consistency and reproducibility: Produces relatively consistent and reproducible results, which could mimic the natural host environment.	**Less sensitive:** Less sensitive compared to molecular methods of virus detection and can only identify viable viruses.

Characteristics of cell strains

Homogeneity: Cell strains are generally more uniform in their genetic and phenotypic characteristics compared to the broader cell line from which they are derived.

Stability: They are often selected for traits or behaviors and may be maintained under conditions that enhance these traits.

Examples of cell strains

HeLa S3: A subline of the HeLa cell line, HeLa S3 cells were selected for their growth characteristics and have been widely used in research [1].

293 T Cells: A strain of HEK 293 cells, 293 T cells are modified to express the large T antigen from SV40, which enhances their ability to produce recombinant proteins [2].

MCF-7/Luc: A derivative of the MCF-7 breast cancer cell line, MCF-7/Luc cells are engineered to express a luciferase reporter gene, allowing for bioluminescence imaging in research [3].

8.1.5 Stem Cells

Stem cells are undifferentiated cells with the remarkable ability to differentiate into various specialized cell types [4]. They are broadly classified into two types: embryonic stem cells (ESCs) and adult stem cells.

ESCs are derived from embryonic cells and are pluripotent; therefore, they can give rise to nearly all cell types in the body [5]. In contrast, adult stem cells are typically multipotent, that is, they can only differentiate into a limited range of cell types. An example of commonly used adult stem cells is human induced pluripotent stem cells (hiPSCs). HiPSCs are adult cells that have been genetically reprogrammed to an embryonic stem cell-like state. This reprogramming involves the upregulation or downregulation of some specific genes and factors that allow these cells to regain pluripotency, making them capable of differentiating into specific cell types in the human body through several stages (Figure 8.2). HiPSCs can be derived from different adult cell types, such as skin or blood cells.

Protocol for the derivation of hiPSCs [6–8]

- **Consumables**
  - Matrigel (Corning)
  - mTeSR™1 medium or E8 medium
  - ROCK inhibitor (Y-27632)
  - PBS (no calcium, no magnesium)
  - 6-well or 12-well tissue culture plates (TC-treated)
  - Pipettes and tips (sterile)
  - 15 mL and 50 mL sterile conical tubes
  - Cryovials
- **Methodology**

  For example, neural stem cells (NSCs) or cerebral organoids will be generated from the iPSC line (iPS15) using the dual SMAD inhibition protocol

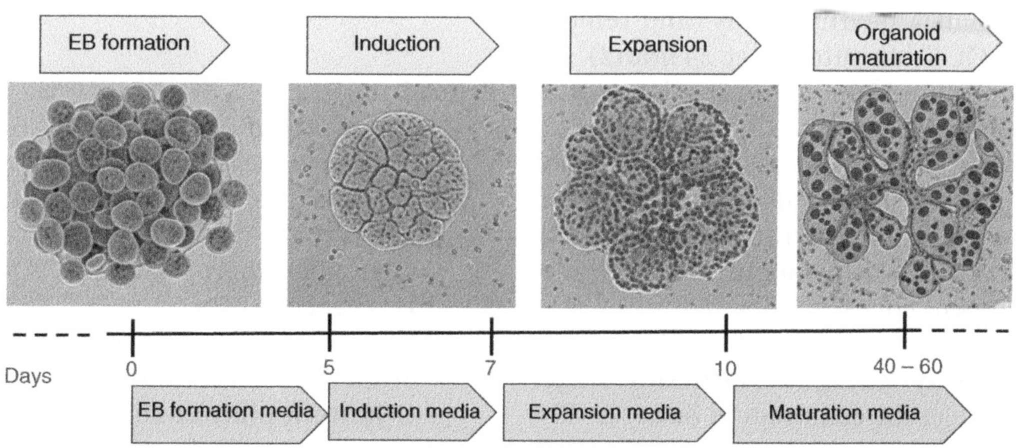

Figure 8.2 Putative diagram of the stages of development of organoids from embryoid bodies (EBs) formation stage to the organoid stage.

with SB431542 and LDN-193189. Stem cells can be validated by immunocytochemistry (Sox2, Nestin for NSCs) and flow cytometry.

Organoid development: There are four stages

Day 0 - Day 5: (Stage 1: Embryoid bodies formation [EBs] stage)

a. Pre-warm Matrigel-coated plates (coated 1 hr. prior or overnight at 4 °C).
b. Thaw vial of frozen iPSCs in 37 °C water bath (1–2 minutes).
c. Transfer contents to 10 mL mTeSR1 + 10 μM Y-27632.
d. Centrifuge at 300 × g for 5 minutes.
e. Discard supernatant, gently resuspend in fresh medium (EB seeding medium containing 10 μM Y-27632).
f. Seed approximately (100 μL per well) 9000 cells per well into an ultralow-attachment 96-well plate (Plate on Matrigel-coated plates) for EB formation and incubated at 37°C with 5% CO_2.
g. Change medium daily (and remove ROCK inhibitor after 24 hours). EB become 100–200 μm in diameter at this stage.
h. Day 2 and 4 days. Add 100 μL of freshly prepared medium to each well. EBs will range from 400 to 600 μm, with smooth, round, translucent edges.

Day 5 to Day 9: Stage 2. Induction stage. At this stage the cells should be at 70–80% confluent. This stage transitions EBs into neural induction conditions to promote the formation of neuroectodermal tissue.

Materials needed

I. Low-attachment 24-well plate (e.g., Corning ultra-low attachment)
II. Sterile forceps or wide-bore pipette tips
III. Induction medium (composition listed below)
IV. DMEM/F-12 for washing
V. Sterile PBS (Ca^{2+}/Mg^{2+} free)
VI. Optional: Gentle Cell Dissociation Reagent (if pre-processing is needed)
VII. 15 mL conical tubes
VIII. Centrifuge
IX. Biosafety cabinet and incubator (37 °C, 5% CO_2)

Induction medium composition

- DMEM/F-12 (1X)
- 1% N2 supplement
- 1X GlutaMAX
- 1X nonessential amino acids (NEAA)
- 1 μg/mL heparin
- Optional: 1% penicillin–streptomycin

Procedure

1. **Preparation**
   Pre-warm the induction medium and PBS to 37 °C.
   Label the 24-well ultra-low attachment plate appropriately.
2. **Harvesting EBs**
   Carefully collect EBs from each well of the previous culture (e.g., 96-well U-bottom plate) using a wide-bore pipette or sterile forceps.
   Transfer EBs to a sterile 15 mL conical tube.

3. **Washing**
   Gently add 5 mL of pre-warmed sterile PBS to the EBs.
   Allow them to settle by gravity or gently centrifuge at 100 × g for 3 minutes.
   Carefully aspirate the supernatant without disturbing the EBs.
   Repeat the PBS wash once more to remove residual media.
4. **(Optional) Dissociation or debris removal**
   If EBs are sticking together or have excess dead cells/debris
   - Incubate briefly with Gentle Cell Dissociation Reagent for 5–7 minutes at room temperature.
   - Gently pipette up and down once or twice (not to dissociate but to remove loosely bound cells).
   - Wash again with DMEM/F-12 and centrifuge gently.

   Note: It is not meant to fully dissociate the EBs into single cells, but rather to loosen loosely attached cells or debris (such as dead cells or protein clumps). Gently remove any early surface cell layers that might inhibit neuroectodermal differentiation.
5. **Transfer to the induction medium**
   - Resuspend the washed EBs in fresh induction medium.
   - Transfer 1–2 EBs per well into a low-attachment 24-well plate, ensuring they remain suspended and not sticking to the wall.
   - Add 500–700 µL of induction medium per well.
   - Place the plate in the incubator at 37 °C with 5% CO_2 for 2 days.
6. **Culture maintenance**
   - Do not disturb for the next 2–3 days to allow neural ectoderm formation.
   - No media change is usually needed on Day 6; start checking morphology from Day 7. It should be visible to the naked eye. With smooth, translucent edges. Indicative of neuroepithelium formation.

Day 10: Stage 3. Differentiation to Neural Lineage (Expansion Stage)

Embedding EBs in Matrigel to provide a supportive 3D matrix environment that facilitates neuroepithelial bud formation and structural integrity, and then transfer them into an expansion culture condition.

Matrigel preparation

Thaw Matrigel® (Corning, Cat. #354-277) slowly overnight at 4 °C (do not freeze-thaw repeatedly).
Keep Matrigel on ice at all times to prevent premature gelation.
Chill pipette tips, tubes, and embedding sheets before use.

Organoid embedding sheet prep

Place a chilled Parafilm® sheet or a specialized organoid embedding sheet (pre-cooled) into a clean Petri dish or 100 mm culture dish.
Label spots or zones for tracking EBs (optional).

Steps:

1. **Harvesting EBs**
   Using a wide-bore 200 µL pipette tip, gently transfer 4–6 intact EBs per well from the Day 5 induction plate.

Wash once in cold PBS or culture medium to remove residual induction media.

2. **Embedding in Matrigel**

On ice, gently suspend each EB in 15 µL of cold Matrigel.

Carefully place the Matrigel–EB droplet onto the pre-cooled embedding sheet.

Repeat for 4–6 EBs per group.

Allow the droplets to partially solidify on the sheet for 10–15 minutes in a 37 °C incubator (to initiate gelation) (Figure 8.3).

3. **Transfer to a culture plate**

Once the Matrigel has partially gelled, carefully transfer the embedded EBs into an ultra-low-attachment 6-well plate using a wide pipette tip or spatula. Place 4–6 embedded EBs per well.

4. **Add expansion medium**
   - Add 3 mL of neural expansion medium per well. This medium supports neuroepithelial expansion and tissue budding.

5. **Incubation**

Culture the embedded organoids at 37 °C in a 5% CO_2 humidified incubator. Avoid disturbing the gelled Matrigel drops for at least the first few hours.

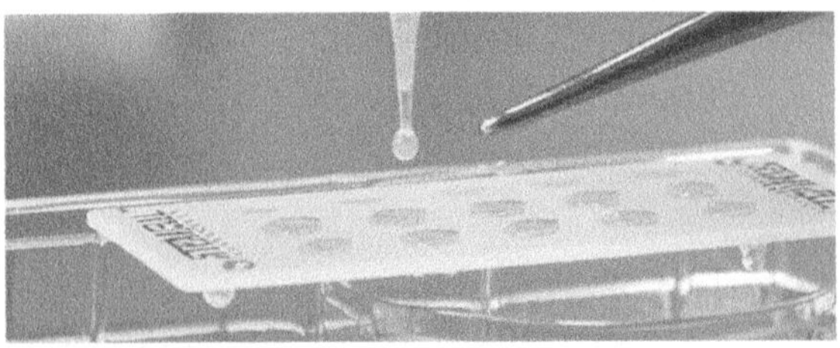

Figure 8.3 Putative diagram of embedding in Matrigel.

Change medium every 3–4 days, half medium.

After 3 days, there will be budding morphology, indicative of neuroepithelium expansion.

Day 10 to Day 60+: Stage 4– transition to maturation

1. **Prepare maturation medium**
   - Use neural maturation medium (can be STEMdiff™ Cerebral Organoid Maturation Medium or equivalent).
   - Warm to 37 °C.
2. **Transfer the embedded organoids to the shaker**
   - Use wide-bore pipette tips or a spoon/spatula to gently lift Matrigel-embedded organoids from the 6-well plate.
   - Place 4–6 embedded organoids per well into a 6-well ultra-low-attachment plate (or spinner flask, depending on the available system).
3. **Add maturation medium**
   - Add 3–4 mL of maturation medium per well.
   - Ensure organoids are fully submerged but not too crowded.
4. **Place on orbital shaker**
   - Use a rotating orbital shaker inside a 37 °C, 5% CO_2 incubator.
   - Set rotation speed to 65–85 rpm (optimal speed may vary by incubator and volume).

Maintenance from Day 10 to Day 60+

- **Medium change**

  Perform half-media changes every 3–4 days.

  - Carefully aspirate half of the old medium, avoiding suctioning organoids.
  - Gently add fresh warm maturation medium.
- **Monitoring growth**
  Check under the microscope for:
  - Tissue folding (neural tube–like structures)
  - Healthy appearance (no major dark centers = necrosis)
- **Optional: Staining or sectioning (Post-Day 30+)**
  - Immunostaining for markers (e.g., MAP2, Tuj1, Sox2, Pax6).

The rocking speed must ensure that the organoid does not rub against the wall of the well.

For virus infection of cerebral organoids

1. Choose organoids at Day 25–40 (depending on virus and study design).
2. Harvest and transfer organoids (3–6 per well) to a low-attachment 24-well plate (static).
3. Wash once with warm PBS or medium.
4. Add virus diluted in a small volume (e.g., 100–200 μL per well, multiplicity of infection [MOI] depending on virus).
5. Incubate 2–6 h at 37 °C, 5% CO_2.
6. Gently wash 2–3 times to remove unbound virus.
7. Transfer back to a 6-well ultra-low-attachment plate with maturation medium.

8.2 Facility, Material, and Methods for Cell Culture

8.2.1 Laboratory Workspace

A separate, well-ventilated, aseptic work environment designated for this purpose is a major requirement for setting up a cell or tissue culture facility in a virology laboratory. However, a segregated area within the large laboratory space can be used, depending on the biosafety level of the virus being handled. When using a separate room, the room or space designated for cell culture must be large enough to accommodate all or some basic necessary equipment for sterile handling, incubation, storage, reagents, and media of cell cultures.

8.2.2 Basic Equipment and Reagents

8.2.2.1 Biosafety Cabinet

This offers a ventilated, enclosed, protective aseptic environment to the user, cells, and reagents. The airflow characteristics of the safety cabinet must be taken into consideration, such that it can provide significant protection for the personnel, the culture, and the environment.

Important notes on biosafety cabinets for virus culture

Always clean up spilled culture medium immediately, since it contains high glucose and amino acids that can encourage bacterial contamination.

Do not work with different cell lines in the safety cabinet at the same time.

Preferably, designate a single safety cabinet for virus cultivation and cell passage, respectively.

Divide the workspace within the cabinet into clean and dirty areas (with clean containers to hold waste).

Disinfect the items or the container holding the items with 70% alcohol before placing them in the cabinet.

Arrange all the required reagents, pipettes, and medium into the cabinet appropriately before starting the work, and avoid awkward positioning that will encourage contamination.

Where available, disinfect the inside of the cabinet and all the materials placed in it with ultraviolet for 30 minutes beforehand.

8.2.2.2 Incubators

This provided optimum environments for the growth of the cells and a space large enough to accommodate the work requirements or laboratory needs. CO_2 and humidity essentially affect the buffering system of the culture medium. Cell culture incubators must provide stringent regulation of temperature (37 °C ± 0.2), CO_2, and humidity requirements of 5–10% for the culture of most mammalian cells (Figure 8.4)

Two types of incubators can be used – dry and humid CO_2 incubators. The former is more economical and may requiry that sterile distilled water be placed at the bottom of the incubator to maintain humidity without precise control. To prevent evaporation of the culture medium (when humidity is not required), cell culture placed in this type of incubator must be placed in a sealed flask. The humid CO_2 incubator has an in-built, well-controlled atmospheric humidity. It can be used to incubate cells cultured in wellplates and Petri dishes. Most standard incubators often use for cell culture requires to reach 90–95% humidity.

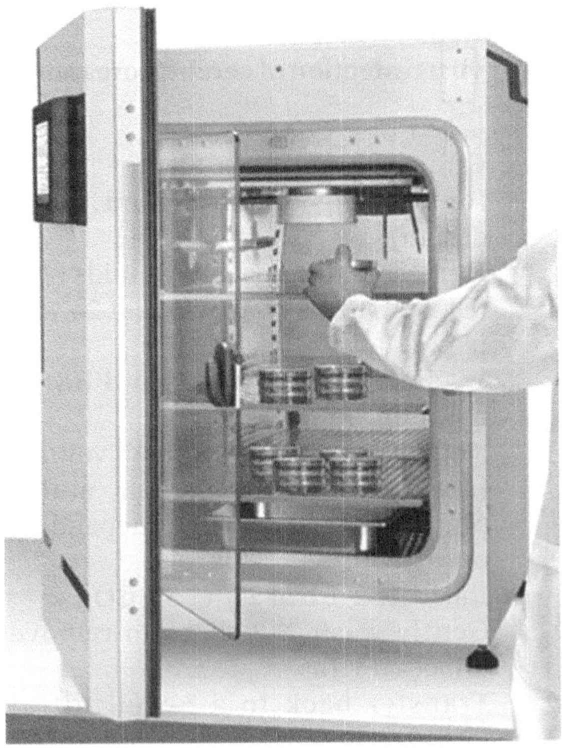

Figure 8.4 Cell culture incubator.
Source: Photo by RWS Life Science Inc.

The racks, incubator doors, and trays/plates (containing water used to maintain humidity) can grow mold or become contaminated by bacteria. Frequently clean these surfaces with appropriate germicides.

Most incubators require an external source of CO_2 supply; therefore, it is essential to consider potential linkages.

8.2.2.3 Cell Culture Maintenance Medium

Synthetic Maintenance Medium Synthetic media are composed of chemically defined components or ingredients (synthetic) designed to meet nutritional requirements to support cellular growth [9]. Nowadays, synthetic media are commonly used in most laboratories because of their convenience.

Natural Maintenance Medium Natural media consist exclusively of naturally occurring biological components derived from tissue or animals [10]. The components or ingredients are not chemically defined. Examples include biological fluids (like serum, plasma, lymph, and amniotic fluid) and tissue extracts (see Table 8.4).

8.2.2.4 Basic Components of Culture Media

Amino Acids Amino acids are the building blocks of protein and are vital biomolecules for the proliferation and growth of cells and viral proteins. All cultures must contain amino acids (essential amino acids); however, in different compositions to meet the nutritional requirements of the cells.

Conditionally essential amino acid like L-glutamine (although not an essential amino acid) is required in most culture media. It serves as a nutritional source of nucleotide nitrogen, and secondary energy sources like nicotinamide adenine dinucleotide (NAD) and nicotinamide adenine dinucleotide phosphate (NADPH) [11, 12]. Different culture media contain varying concentrations of L-glutamine; therefore, the amino acid nutritional requirements of different viruses should be considered when preparing or purchasing (chemically defined) culture media.

Nonessential amino acids may be added to some culture media to replace depleted ones used up during growth, enhance growth, or prolong cell viability.

For example, lysine, a nonessential amino acid, is not a nutritional requirement for propagation of the herpes simplex virus in human cells. It causes partial inhibition of virus replication. Herpes simplex virus-infected cells deficient in arginine do not support virus-induced cytopathic effects and replication.

Proteins and Peptides For protein requirements, FBS will take care of this requirement because it is a rich source of proteins like albumin, fibronectin, aprotinin, transferrin, and fetuin. However, it is important to supplement cells cultured in serum-free media with these proteins.

- Albumin binds, sequesters, and stabilizes various important molecules (fatty acids, hormones, and vitamins), ions (salts), and toxic substances [13, 14]. Its binding capacity helps to deliver essential biomolecules into the cells and also protects them from oxidizing ions and toxic substances.

Table 8.4 Types of culture media and their applications.

Media	Examples	Applications	Examples of suitable cell lines
Basal media	Minimum Essential Medium Eagle (MEM)	Used for maintaining primary and diploid cell cultures for virus infection and tumorigenicity studies.	Human cervix carcinoma: A derivative of HeLa (HeLa B)
	Dulbecco's Modified Eagle Medium (DMEM)	Commonly used medium; a modified MEM with increased levels of amino acids and vitamins. Supports a wide range of cell types, including for virus infection and transfection studies.	HeLa, Vero cells, 293T cells, COS-7 cells, BALB/3T3 clone A31
	Glasgow Minimum Essential Medium (GMEM)	Modified MEM defined for Baby hamster kidney-21 (BHK-21) cells. Used for virus studies susceptible to BHK-21 cells, e.g., human adenovirus D, reovirus 3, and vesicular stomatitis virus.	BHK-21 cells
Complex media	Roswell Park Memorial Institute Medium (RPMI) 1640	Originally designed to culture human leukemic cells. Suitable for various mammalian cells used for virus isolation and studies.	HeLa and Jurkat cells (for HIV, human T lymphocyte virus studies)
	Leibovitz L-15	Designed for cell culture without a CO_2 system. Used for the growth of cells for the isolation of many human viruses from clinical samples.	Hep-2 monkey kidney (for a variety of fecal viruses), Rhesus Monkey Kidney Epithelial Cells (LLC-MK2) (for the isolation of influenza and parainfluenza viruses, human enteroviruses, mumps virus)

Table 8.4 (Continued)

Media	Examples	Applications	Examples of suitable cell lines
	Iscove's Modified Dulbecco's Medium (IMDM)	Contains additional amino acids, vitamins, and other components to support high-density cell cultures.	Hybridoma, macrophage, and a variety of primary cells
	F-12 K Nutrient Mixture	Designed for the growth of human diploid cells and other mammalian cell lines.	A549 cells, CHO cells, and primary rat hepatocytes
Serum-free media	EX-CELL 293 Serum-Free Medium for Human Embryonic Kidney 293 (HEK 293) Cells (EX-CELL 293)	Chemically defined, serum-free medium optimized for high-density suspension cultures.	HEK-293 cells
	Chinese hamster ovary (CHO)	Chemically defined protein and serum-free medium for Chinese hamster ovary (CHO) cells.	CHO-K1, CHO-S
	QBSF-60	Serum-free medium; xeno-free expansion medium designed for the growth of $CD34^+$ human hematopoietic stem/progenitor cells.	Stem/progenitor cells.
Balanced salt solutions	Phosphate-buffered saline (PBS)	Basis of many complex media. Used to maintain osmotic pressure and pH. Routinely used as a wash buffer for washing cells.	N/A
	Hanks' Balanced Salt Solution (HBSS)	Composed of inorganic salts and supplemented with glucose. Used to maintain osmotic pressure and pH.	For short-term maintenance of cells outside of growth medium. These media are now commercially available in powdered and liquid forms.

(*Continued*)

Table 8.4 (Continued)

Media	Examples	Applications	Examples of suitable cell lines
	Earle's Balanced Salt Solution (EBSS)	Contains inorganic salts and glucose. Used to maintain osmotic balance and supply essential nutrients. For short-term maintenance of cells outside of growth medium.	For short-term maintenance of cells outside of growth medium; HeLa cells, fibroblasts, and primary cultures

- Aprotinin, an antifibrinolytic, is a broad-spectrum competitive inhibitor of proteases, especially trypsin [15, 16].
- Fibronectin is a glycoprotein of the extracellular matrix that plays a vital role in cell-to-cell attachment [17].

Carbohydrates Carbohydrates are a type of macronutrient that serves as an energy source for the cells. It is the main source of energy in most cells [18]. Most media contain glucose and galactose. Some media contain fructose or maltose [19, 20]. Glucose-depleted cells could reduce viral yield beyond the amounts formed in a medium containing glucose [21].

Fatty Acids and Lipids Cells can synthesize most of the lipids required for primary cellular functions. However, linoleic and linolenic acids, which are essential lipids, are generally present in serum.

Vitamins Vitamins act as cofactors for many enzymes and are essential for cellular growth and proliferation [22]. Vitamin-deficient cell culture medium may cause a decrease in cell viability. Likewise, the serum contains many vitamins, essential for cell growth, but vitamins must be substituted for in serum-free media. The presence of vitamins in sera varies widely among species and with storage and handling conditions.

Some vitamins (for example, B12) are chemically unstable, so there may be a need to supplement the levels in different culture mediums. Vitamin B12 is required for the growth and genetic stability of the cells. Most commercially available culture media contain varying formulations or no vitamin B12 in their constituents; therefore, this must be taken into consideration during purchase. In serum-free media, all essential vitamins must be added to mimic the nutritional environment provided by serum.

Inorganic Salts Inorganic salts help to regulate membrane potential and osmolality balance. Examples include sodium, potassium, and calcium ions. This requirement must be considered when choosing the type of media to purchase.

Basic and Trace Elements Trace element that most cells need in minute amounts for proper cell growth are iron, magnesium, and zinc, and to some extent copper and selenium. Serum contains most of the needed trace elements for cell growth; trace element is often supplemented in serum-free media if needed.

Buffering Systems This is an acid–base homeostatic mechanism that resists a change in the pH of the culture medium. The buffering system used in cell culture must be suitable to maintain the pH in the solution at 7.2–7.4; otherwise, a change in the pH will cause harmful effects on the cells.

The main cause of changes in the pH of the culture medium is the presence of CO_2 and H_2O in an airtight environment. This increases H^+ concentration. Most commercially available culture media use $NaHCO_3$ buffer systems to prevent this change in pH. $NaHCO_3$ acts as a weak buffer that maintains the desired pH range in an air atmosphere with 5–10% CO_2. Refer to the equation below.

$CO_2 + H_2O \leftrightarrow$ H2CO$_3$ (Carbonic acid) $\leftrightarrow$ $H^+ + HCO_3^-$ (weak conjugate base)

Example: 4-(2-hydroxyethyl)-1-piperazineethane sulfonic acid (HEPES) buffer

HEPES is a chemically stable zwitterionic organic buffer that has a superior capacity to maintain the pH of the culture medium in the range of 7.2–7.4 [23]. It is a good alternative buffering system where a CO_2 incubator is not available because it does not need a controlled gaseous atmosphere to maintain the desired pH range.

Although it is relatively expensive, its usage in culture medium varies from 10 mM to 25 mM, as concentrations above this may be toxic to some cell types.

How to prepare HEPES buffer (1 M, 7.5 pH)

Prepare 800 mL of sterile double-distilled H_2O in a suitable container.
Measure 238.3 g of HEPES powder and add to the water.
Adjust by adding NaOH or HCL dropwise to the desired pH (7.5).
Make up the volume to 1 L with sterile double-distilled H_2O.
Sterilize by filtration and store at 4 °C for later use.

pH Indicator Phenol red (phenolsulfonphthalein) is a commonly available pH indicator that allows constant monitoring of pH for cell culture. Its color changes gradually from bright red ("normal" color) to yellow (acidic condition) within the pH range of 7.2 – < 6.5. Above pH 8.2, the color turns to bright pink and then orange. At the optimum pH for cell culture (7.2–7.4), the medium containing phenol red is bright red. Waste products produced by dying cells, overgrown cells, or contaminants (like bacteria or yeast) will cause a decrease in the medium pH and a change of color of the medium to yellow (see Figure 8.5).

Culture media contain varying concentrations of phenol red formulations (11–15 mg/mL); therefore, changes in pH could be more pronounced in different media. A common problem with phenol red indicator is that it interferes with or reacts with:

- Flow cytometric analysis (spectral interference) [24]. Therefore, cells should be washed with saline solution before analysis
- Sodium–potassium homeostasis in serum-free medium.

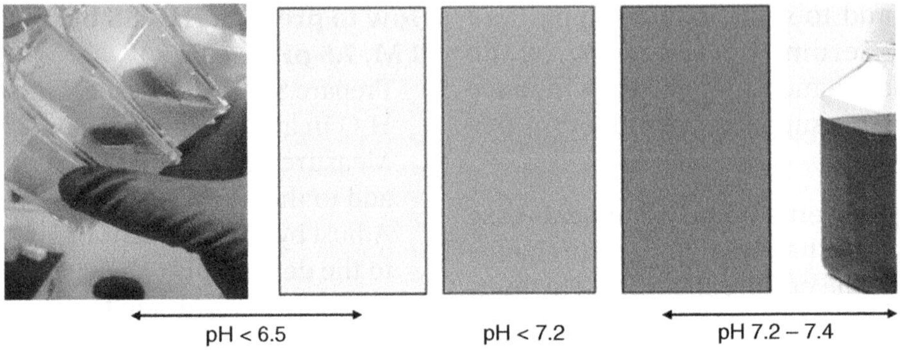

Figure 8.5 Image showing culture medium changing color from red (first two on the right) to yellow (last two on the left) due to a change in pH.

- Myeloperoxidase is expressed in the majority of promyelocytic leukemia cell lines, participating in redox reactions within the culture medium. This enzyme activity can influence cell viability at concentrations around 15 mg/mL.
- It is a weak estrogen, thereby enhancing the growth of estrogen-sensitive cells [25].

Preparation of phenol red (0.5%)

- Place 0.5 g phenol red in a pestle and mortar and grind with a small amount of 0.20 N NaOH till it becomes completely dissolved.
- Add sterile double-distilled water to make up 100 ml.
- Filter the solution with a filter.
- Sterilize in an autoclave.

Notes

- Store at + 2 °C to + 8 °C.
- Do not use the product after its expiry date.
- Protect the product from light.
- Manipulate under aseptic conditions.

8.2.2.5 Preparation of Chemically Defined Culture Media Locally

All chemically defined culture media can be made locally from their constituents. For convenience, avoidance of the risk of contamination, and to save time, most laboratories performing cell culture use commercially prepared media. These media are available in three varieties:

- **Powdered form:** This is the least expensive and comes in powdered form, requiring preparation and sterilization by the laboratory personnel.
- **Concentrated form:** This comes in a variety of concentrated forms (e.g., X2, X10) that must be diluted by the laboratory personnel to the required concentration.
- **Working solution:** It is the most convenient and easiest form of media commonly used in most laboratories. They are designed to be used without direct further manipulation by laboratory

personnel. Depending on the test to be performed, it may be necessary to add antibiotics, serum, and other constituents and filter when appropriate (Figure 8.6).

If purchasing the powdered and concentrated media, ensure that the water used to reconstitute the powder or dilute the concentrated media is distilled (preferably double-distilled), sterile (autoclaved), and tissue/cell culture-grade; pH of the medium must be adjusted appropriately; store media according to the manufacturer's instructions.

Use appropriate storage conditions that protect the media from light, and at +2 °C to +8 °C. Do not use the product after its expiry date.

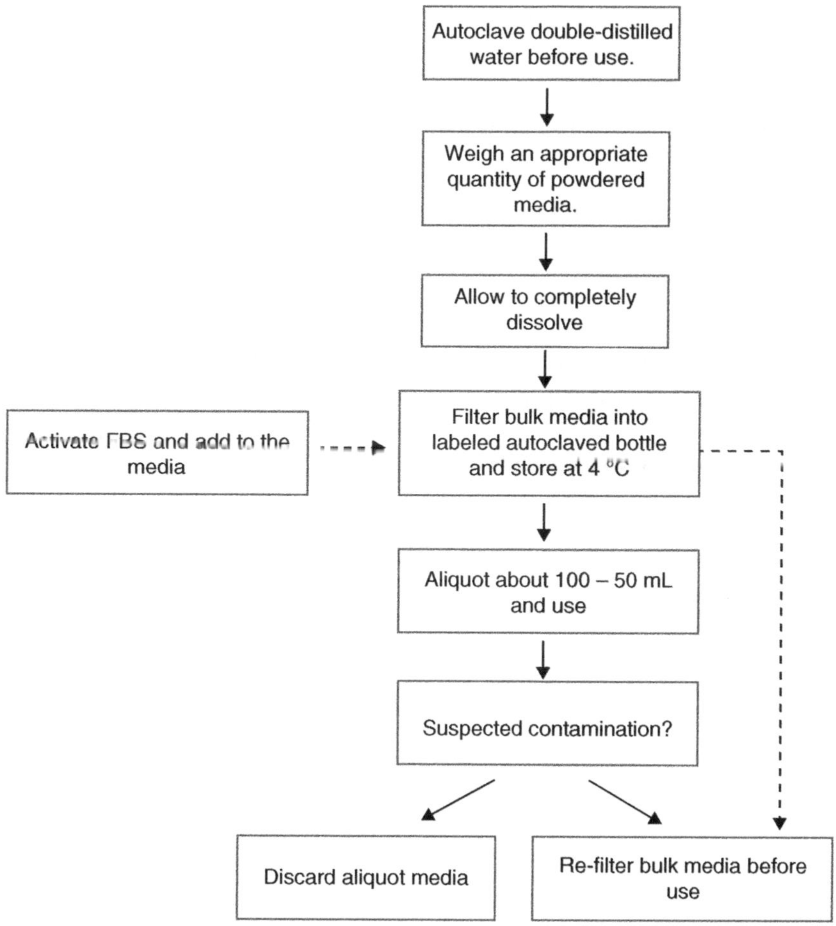

Figure 8.6 Flow chart of the use of culture media. FBS: fetal bovine serum.

How to Prepare Phosphate-Buffered Saline

Component (1 L, 10X)	Weight (g)
NaCl	80 g
KCl	2.0 g
Na_2HPO_4	14.4 g
KH_2PO_4	2.45 g

1. Measure out 900 mL of double-distilled water.
2. Measure the appropriate weight of the components and add them to the distilled water while stirring. Stir until dissolved.
3. Adjust the pH to 7.4, using 1 N HCl or 1 N NaOH, as appropriate.
4. Make up the final volume of double-distilled water to 1 L.
5. Sterilize by autoclaving.
6. Store at 4 °C for further use.

Note: After dilution to 1x PBS, the pH of the resultant volume should be adjusted to 7.4.

How to Prepare Hank's Balanced Salt Solution (HBSS) Without Phenol Red

Components (1 L, X1)	Weight (g)
Calcium chloride (anhydrous)	0.1396
Sodium phosphate dibasic (anhydrous)	0.04788
Magnesium sulfate (anhydrous)	0.09767
D-Glucose	1.0
Sodium chloride	8.0
Potassium phosphate monobasic (anhydrous)	0.06
Potassium chloride	0.4

1. Measure out 900 mL of double-distilled water (15–20 °C)
2. Measure the appropriate weight of the components and add them to the distilled water while stirring. Stir until dissolved.
3. Add 0.35 g sodium bicarbonate and stir until all the powder dissolves. Alternatively, add 4.7 mL of sodium bicarbonate solution (7.5%w/v)
4. Adjust the pH to 7.2–7.4, using 1 N HCl or 1 N NaOH, as appropriate.
5. Make up the final volume of double-distilled water to 1 L.
6. Filter the salt solution (using a membrane with a porosity of 0.22 μm).

Notes: Anhydrous salts are hygroscopic; ensure they are protected from moisture.

Use the entire contents immediately after preparation or opening.

Culture supplements can be added to the salt solution, and ensure the solution is filtered before use.

Precipitate or particulate formation change in pH, a color change, or a cloudy appearance of the salt solution are signs of deterioration.

How to Prepare 0.25% Trypsin

Component (1 L, 10X)	Weight (g)
Trypsin powder	0.25 g
EDTA-Na_2	0.02 g
PBS	100 mL

1. Measure the appropriate weight of the components and add to 90 mL of PBS X1. Stir until dissolved.

2. Make up the final volume of buffered saline to 100 mL.
3. Adjust the pH as appropriate.
4. Sterilize using a membrane filter in the safety cabinet.
5. Store at 4 °C for further use.

How to Prepare 1X DMEM from Powder Stock

1. Check the manufacturer's instructions for the exact amount of DMEM powder required to make 1 L of 1X DMEM. Weigh the appropriate amount of DMEM powder and dissolve it in approximately 800 mL of distilled sterile water.
2. Aseptically, stir the solution using a magnetic stirrer or a stirring rod until the powder is completely dissolved.
3. Add sodium bicarbonate (if not included in the powder). Typically, 3.7 g of sodium bicarbonate is added per liter of medium.
4. Once the DMEM powder and sodium bicarbonate are fully dissolved, add distilled water to bring the total volume up to 1 L.
5. Check and adjust the pH. The pH of DMEM should be around 7.2–7.4.
6. Filter the solution through a 0.22 µm sterile filter to remove any potential contaminants and collect the filtered medium into sterile bottles.
7. Label the bottles with the date of preparation and any other relevant information. Store the prepared 1X DMEM at 4 °C until use.

8.2.2.6 Media Supplements

Serum Supplement Serum is a crucial component of cell culture media, containing a complex mixture of amino acids, albumins, carbohydrates, fatty acids, vitamins (such as fat-soluble vitamins A, D, E, and K), minerals, trace elements, hormones, various growth factors, and growth inhibitors. FBS is the most commonly used serum for many cell types in virology laboratories. Additionally, horse serum and bovine calf serum can be utilized. However, it is essential to screen different types of serum to determine which best supports cell growth.

a. **Advantages of serum in culture media**
   - **Inhibition of proteases:** Serum contains protease inhibitors that protect cells from proteolysis. Epithelial and myeloid cells, which release protease enzymes during replication, are safeguarded by these inhibitors. Additionally, during cell passage, serum neutralizes the proteolytic effect of trypsin.
   - **Enhancement of cell growth and proliferation:** Hormones and growth factors in serum, including epidermal growth factor (EGF), platelet-derived growth factor (PDGF), and fibroblast growth factor (FGF), promote cell proliferation and differentiation.
   - **Promotes cell attachment and adherence:** Proteins such as fibronectin and laminin in serum facilitate cell-to-cell attachment, adhesion to the substrate, and the spread and adherence of cells *in vitro*.
   - **Buffering capacity:** Serum can act as a buffer, maintaining the stability of the culture environment.

b. **Disadvantages of serum in culture media**
- **Contamination:** Serum is highly prone to contamination; therefore, it must be filtered before use.
- **Undefined constituents:** The constituents of each batch of serum purchased may lack uniformity in their composition.
- **Mixed constituents:** Serum contains both growth factors and growth inhibitors. These inhibitors can be harmful to certain cell types.
- **Interference:** The presence of serum in media may interfere with virus attachment. Also, there is the possibility of contamination by the presence of antibodies against the virus, which is being cultivated. Therefore, it is important to standardize requirements for virus infection.

Preparation of FBS Locally FBS can be prepared from the blood of healthy bovine fetuses, which is obtained through a closed collection system after the fetuses are removed from freshly slaughtered cows.

- Collect blood from freshly slaughtered cows and allow it to coagulate by keeping it at room temperature (25 °C) for about 1–2 hours.
- Fractionate cell components by centrifugation (at 3000 rpm for 30 minutes) into serum (containing a large number of nutritional and macromolecular factors), and clot red cells with fibrin and clotting factors.
- Freeze serum in −20° of chain filtration using a Seitz depth filter of 0.1-μm membrane (cellulose fibers, asbestos-free).
- Store in aliquots (in 50 mL tubes, to avoid repeated freeze and thaw cycles) at −20 °C after testing for its sterility.

Heat inactivation: *For inactivation of complement proteins and some potential unknown growth inhibitors, FBS can be heat-inactivated at 56–58 °C for 30–60 minutes in a water bath. However, different cells have different responses toward heat-inactivated FBS because heat could cause the inactivation of some vital constituents in the serum. Ensure to determine the need for heat inactivation for different cell types.*

Quality control testing of serum: *FBS may be tested for microbes (screening for mycoplasma and virus, and bacteria), endotoxin, immunoglobulin G, etc.*

Commercially Prepared FBS Several suppliers provide FBS with different grades. It is a good idea to examine variations in suppliers, grades, and lots/batches to select the most suitable FBS for your laboratory by conducting cell growth performance testing.

Antibiotic Supplement Maintaining a standard cell culture protocol with aseptic techniques is crucial, making the use of antibiotics often unnecessary. While prophylactic use of antibiotics or antimycotics at appropriate concentrations can prevent or minimize bacterial or fungal contamination, laboratory personnel need to be aware that adding antibiotics to culture media may be toxic to some cell lines or result in a change in gene expression [26, 27]. Therefore, evaluating the effectiveness, toxicity, and cytotoxic concentration of antibiotics before their application in culture media is important. Commonly used antibiotics for cell culture are gentamicin and PenStrep – a combination of penicillin and streptomycin.

Potential side effects of antimicrobial agents include induced apoptosis, oxidative stress, mitochondrial damage, morphological and ultrastructural changes, inhibition of protein synthesis and degradation, and antiproliferative effects.

There is a wide variety of antibiotics and antimycotics to control or prevent bacteria, fungi, mycoplasma, or yeast contamination of culture media. The following points can guide the choice of antibiotics for cell culture:

- Does it eliminate or inhibit the microbial contaminants? Bactericidal antibiotics are preferred over bacteriostatic antibiotics.
- Does it interfere with cellular growth and metabolism?
- Is it relatively safe or toxic?
- Is it compatible with the experimental procedure (e.g., compatibility with transfection) and other components of the culture medium?
- Does it confer antimicrobial protection throughout culturing the cells?

8.3 Cell Passage, Subculture, or Splitting

Cell passage, subculture, or splitting is the technique by which cultured cells that have reached a certain confluence are divided and transferred into a fresh growth medium. This enables the continuous propagation of the individual cells for an extended period.

Cell lines or cultures must adhere to specific guidelines, including details on the number of passages, incubation conditions, recommended media, and splitting methods. They should be uncontaminated. Cultures developed in the lab should follow written protocols and be thoroughly documented with the following details: cell culture identification; source and date of acquisition; passage number; growth medium used; maintenance medium; seeding density; and storage conditions.

8.3.1 What to Consider Before Passaging Your Cells?

8.3.1.1 Confluency and Cell Density

Adherent cultures should be passaged while they are still growing and spreading (mid-logarithmic phase of growth) below 80–90% confluency. They stop growing (lag phase) when they reach contact inhibition (100% confluency). If adherent cells have available space, they replicate rapidly and freely. Also, suspension cultures should be passaged while they are still growing (log phase) before they reach 100% confluency. When suspended cells stop growing, they clump together.

__Confluency:__ This is the term that describes the percentage growth of cells in a cell culture dish, well plate, or flask. For example, a cell culture that has 50% confluency means half of the surface of the culture dish, well, plate, or flask is covered by the growth or spread of the cells.

__Cell density:__ This is the term used to describe the relative number of cells present in a cell culture dish, well plate, or flask.

In general, an estimation of confluency can be translated into the number of cells (cell density) for each culture vessel type.

8.3.1.2 *Depletion of Medium and Nutrients*

Overgrowth cultures that exceed the capacity of the medium can cause the accumulation of cellular, metabolic by-products or rapid depletion of culture medium and nutrients that support further growth. To avoid cell deterioration or death, cultures should be passaged, whether or not they reach 100% confluency, when there is a depletion of medium and nutrients, or if the medium appears turbid or yellow (due to a drop in the pH of the growth medium).

Always keep cell sub-culture log details. The details that record the initial seeding concentration, morphological characteristics of the cells, number of splits, and contaminations.

Practice a strict cell culture schedule to be able to monitor cell growth patterns.

8.3.2 Subculturing of Adherent Cells

8.3.2.1 *Materials for Subculturing of Adherent Cell Passage*

1. **Disposables:** Sterile tubes (15 mL), cryogenic storage vials, pipette tips (1000 μL or 5 mL), and tissue-culture-treated flasks, plates, or dishes (see Table 8.5 and Figure 8.7).
2. **Chemical:** Balanced salt solution like PBS 1X.
3. **Dissociation materials:** Trypsin with or without phenol red, trypsin with collagenase, dispase, or cell scraper.
4. Reagents and equipment to determine cell viability and cell counts, like an automated cell counter, trypan blue, a hemacytometer (counting chamber), and a coverslip.
5. Appropriate culture media with serum.

8.3.2.2 *Method of Subculturing Adherent Cells*

- **Step 1: Washing of adherent cells** This step aims to remove culture media, serum, and serum proteins, inorganic elements, and cellular debris to expose cell receptors and aid the activity of the dissociation reagent. To wash adherent cells

1. Discard the culture media from the culture vessel.
2. Cover the adherent cells with the appropriate volume of balanced salt solution (e.g., PBS) sufficient to cover the culture vessel.

Table 8.5 Recommended culture medium volume capacity for different tissue-culture-treated flasks, plates, or dishes.

Flask (surface area cm²)	Approximate volume of culture medium (mL)	Dishes (surface area cm²)	Approximate volume of culture medium (mL)	Plates (surface area cm²)	Approximate volume of culture medium (mL)
T-25 (25)	2.5	35 mm (8.8)	2	6 Wells (9.6)	1.5–2
T-75 (75)	10	50 mm (21.5)	5	12 Wells (3.5)	1
T-175 (175)	20	100 mm (56.7)	10	24 Wells (1.9)	0.5
T-225 (225)	30	150 mm (145)	15	48 Wells (1.1)	0.5
				96 Wells (0.32)	0.2

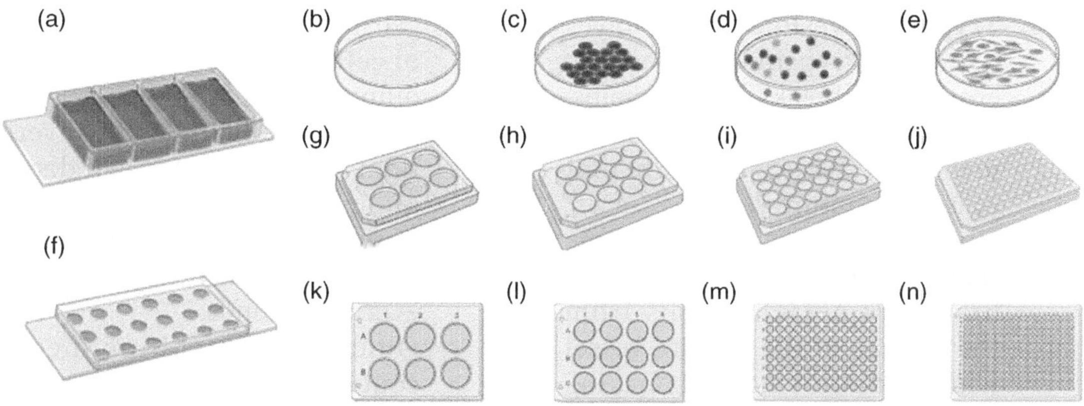

Figure 8.7 Illustration of different cultures of different tissue-culture-treated flasks, plates, or dishes. A. Chamber slide (4 wells). B–E. Petri dish, G and K. 6-well plate; F. Chamber slide (18-well); H and L. 12-well plates; I. 24-well plate; J and N. 96-well plate.

Source: Created using Biorender.

3. **Note:** To avoid disturbing the cell layer, place the wash solution at the side of the culture vessel.
4. Firmly adherent cells can also be washed with trypsin solution.
5. Rock the vessel back and forth gently and carefully.
6. Aspirate the wash solution and discard.
7. Repeat steps 1–3 if the cells are known to adhere firmly.

Ensure that all chemicals, solutions, and equipment used for cell culturing and passage are clean and sterile. If the chemicals or reagents are stored in the refrigerator, remove them and keep them at room temperature or 37 °C before usage.

- **Step 2: Dissociation of adherent cells**
  This aims at detaching the cells from the surface of the culture vessel. Cell dissociation can be done by using two methods.

1. **Enzymatic method of dissociation of adherent cells**
   - **Trypsin:** Trypsinization involves the use of trypsin, a proteolytic enzyme that breaks down proteins. It is the commonest enzymatic method of dissociating adherent cells from culture vessels.
   - **Dispase:** This is a gentle enzyme that does not damage cell membranes. It detaches cells from the surface of the culture vessel without destroying or damaging the cell membranes. It can be used for the detachment of cell colonies and the dissociation of tissue pieces into small clumps of cells. Generally, dispase can be added to suspension cells to avoid clumping.
2. **Mechanical method of dissociation of adherent cells**
   - **Shake-off:** This can be used for non-firmly adherent cells or loosely attached cell lines
   - **Scraping:** This can be used for cell lines that are sensitive to protease dissociation or loosely attached cell lines. This method can damage some firmly adherent cells.

To detach adherent cells:

I. Cover the surface of washed adherent cells with the solution trypsin or dispase.
II. Rock gently and discard the solution.
III. Incubate at 37 ° C for 2–10 minutes. Note that some cells can be placed at room temperature (on the bench). The time of incubation varies with cell lines. Avoid over-trypsinization of cells, as it may destroy the viability of the cells.
IV. Examine cells under an inverted light microscope for cell dissociation. If there are still up to 20% of the cells attached to the culture vessel, incubate further and check every 30 seconds.
V. Quickly add a culture medium containing serum (enough to cover the surface of the culture vessel) to prevent further trypsinization of cells.
VI. Flip the bottom of the culture vessels gently and mix vigorously by pipetting to ensure that the cells are not attached.
VII. Optional:
   a. Transfer the cells to a 15-mL conical tube and centrifuge at 4 °C, 800 × g for 5 minutes. Note that centrifugation separates

particles into fractions, and it is important to remove cell debris. The speed and time of centrifugation vary depending on the cell type.

b. Discard the supernatant and aspirate the remaining culture medium.
c. Add a small volume of fresh culture medium to resuspend the cell pellet.

VIII. Remove a small portion of the cell suspension and determine cell density and viability.
IX. Add appropriate volume (based on the recommended seeding density of the cell type) of fresh culture medium.
X. Split the culture into multiple culture vessels if needed and discard the initial vessels.
XI. Cover the culture vessel and incubate at appropriate incubation conditions.

Observe cellular morphology daily under the microscope. It is common to detect slight changes in the cell morphology; this should not be a reason for concern.

Seeding density differs per cell type.

Change the culture flask, plate, or dish after three passages or subculturing.

Where necessary, seed cell cultures at higher densities than normal, as some cells may not survive in the new culture environment. It may increase doubling times and the number of viable cells to further passage.

Employ a consistent split schedule to prevent the cells from overgrowing, especially during the weekend.

8.3.3 Subculturing Suspension Cells

A variety of cells can also be grown naturally in suspension cultures without attaching to the culture vessel surface (e.g., cell lines derived from peripheral blood components like lymphocytes). Other naturally adherent cells can be modified to be able to grow in suspension cultures.

Suspension cells can grow to a higher density than adherent conditions would allow. Cells may grow as single cells or in clumps. Unlike adherent cells, subculturing suspension cells is relatively easy and less complicated because the cells are already suspended in a growth medium, and do not require enzymatic or mechanical treatment to detach them from the culture surface. These cells are simply maintained by dilution of culture media with fresh growth media until they reach the desired confluency or density.

8.3.3.1 Materials for Subculturing of Suspension Cells

1. Disposables: Sterile roller bottles, shaker flasks without baffles, or spinner bottles. Conventional culture flasks and petri dishes can be used.
2. Appropriate culture medium. Remove from the refrigerator and place at room

temperature for at least 30 minutes before subculturing.

3. Reagents and equipment to determine cell viability and cell count, like an automated cell counter, trypan blue, a hemacytometer (counting chamber), and a cover slip.
4. Incubator with appropriate incubation conditions affixed with a magnetic stir plate (when a spinner flask is used), roller rack (when a roller bottle is being used), or shaker (when a conventional culture flask is being used) (see Figure 8.8).

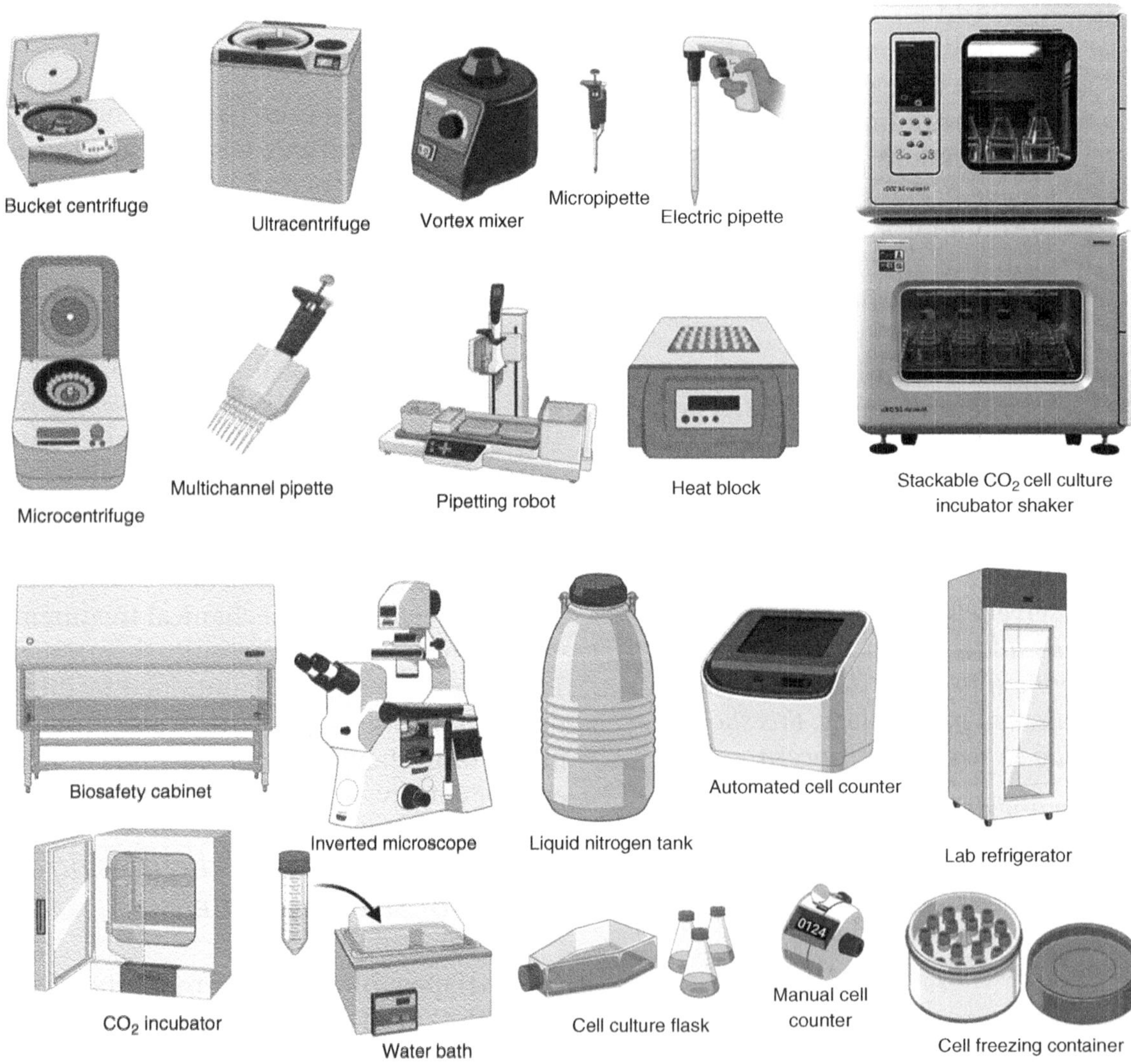

Figure 8.8 Illustration of various equipment and disposables used in a virology laboratory.
Source: Created using Biorender.

8.3.3.2 *Methods of Subculturing Suspension Cells*

1. Remove culture vessels from the incubator and take an aliquot sample (100–200 μL of the culture medium), using a sterile pipette. Ensure that the cells are evenly distributed before taking the aliquot sample.
2. Determine the cell density (cells/mL) and viability from the aliquot sample.
3. Dilute the culture with an appropriate volume of fresh culture medium (prewarmed) to the desired or recommended seeding density.
4. Return the culture vessel to the incubator. Ensure that the caps (if it is not gas-permeable) are loosened before being placed in the incubator for proper gas exchange.

Notes

- Refer to the data sheet supplied with the cell line for the recommended shaking speed and seeding density.
- Maintain seeding of cells every 2–3 days until the desired density is reached. And replace the whole culture medium once every 1–3 weeks.
- Resuspend the cell pellet in fresh growth medium to the recommended seeding density once every 3 weeks to reduce the accumulation of cell debris and metabolic waste.
- Some cell lines may grow into large clumps, which are difficult to count. Ensure that the cells are reconstituted into a single-cell suspension by filtration or centrifugation. Resuspend the cell pellet in a fresh culture medium.
- Passage cells at mid-log growth phase, before they reach confluency.
- Maintain aseptic conditions. Wipe the outside of the vial with 70% ethanol before placing it back into the safety cabinet.

8.4 Measuring Cell Growth and Growth Curves

Automatic electronic cell counters and hemacytometers may be used to count the number of viable cells in adherent and suspension culture medium, hence indicating cell growth. While the electronic cell counter, which may be obtained from various laboratory suppliers, is quicker and faster, the hemacytometer is an undeniably economical and cumbersome method for counting cells. It can be used to provide data as accurate as that obtained by an electronic counter.

8.4.1 How to Use a Hemacytometer to Count Cells

Materials

Trypsinized adherent cells or suspension cells
Trypan blue (0.4%, buffered PBS, pH 7.2–7.3).
Sterile tubes (1.5 mL)
Counting camber
Hemocytometer
Coverslip
Hand tally counter

Method

Preparing a cell suspension

1. Aseptically pipette 50 μL of suspended cells. Note: Gently mix the culture vessels to ensure that the cells are evenly distributed. Before the cells settle down (while they are evenly distributed), quickly take a sample into a sterile tube (1.5 mL tube).
2. Into another tube, pipette 10 μL of the well-mixed cell suspension and add 1 μL of 0.4% trypan blue. Mix gently and incubate at room temperature on the bench

Preparing adherent cells for measuring cell growth

1. Trypsinize adherent cells and resuspend in a culture medium with serum.
2. Aseptically suspend cells.
   Note: Pipette up and down the culture medium to ensure that the cells are well mixed. Before the cells settle down (while they are evenly distributed), quickly take a sample into a sterile tube (1.5 mL tube).
3. Into another tube, pipette 10 μL of the well-mixed cell suspension and add 1 μL of 0.4% trypan blue. Mix gently and incubate at room temperature.

Counting the cells

1. To count cells, clean the counting chamber with alcohol and fix the coverslip in position on the hemocytometer
2. Place 10 μL of the well-mixed sample.
3. Place the hemocytometer on a moist surface and allow the cells to settle for 2–3 minutes.
4. Observe under a microscope by focusing on the grid lines of the hemocytometer with a 10X objective. Count the number of live (unstained) cells in all four outer central gridded squares ($1\,mm^2$) and divide by four (average number of cells/square).
5. Record your counts and perform the necessary calculations.

Calculations

The grid lines divide the hemocytometer into nine main squares of $1 \times 1\,mm$ ($1\,mm^2$).

The four corner squares (big squares) are further subdivided into 16 grids. Each of which is $0.25 \times 0.25\,mm$ ($0.0625\,mm^2$). The raised edges of the hemocytometer chamber hold the coverslip to a height of 0.1 mm. This gives each square a defined volume of $0.1\,mm^3$ or 10^{-4} ml.

8.4.2 How to Calculate the Total Number of Cells Using a Hemocytometer

Example

For instance, if you add 1 μL of trypan blue stock solution to 10 μL of cell suspension (giving a dilution factor of 1:10) and load this mixture onto a hemocytometer, then count the total number of cells (including both live and dead cells, with the dead cells stained blue) in the four corner squares (16 grids), and the counts were 48, 45, 52, and 49. The average count would be $(48+45+52+49)/4 = 48.5$ cells.

To calculate the total number of cells per mL:

Total number of cells = Average number of cells counted $\times$ 10,000 (constant) $\times$ 1.1 (dilution factor) = cells/mL.

So, the number of cells = $48.5 \times 10{,}000 \times 1.1 = 533{,}500$ cells/mL

Dilution factor:

- µL of trypan blue to 10 µL of cell suspension.
- The total volume of the mixture is 11 µL.

The dilution factor should be calculated as follows:

Dilution Factor = Total Volume/Volume of Cell Suspension = 11 µL/10 µL = 1.1

8.4.3 How to Calculate Total Cell Viability

Take the average cell count (living cells) of the four corner squares (16 grids) and multiply by the dilution factor and the constant.

For example: Average of the living cells $(40 + 43 + 50 + 48) = 45.25$ cells

Total viable cells = Number of viable cells counted $\times$ 10,000 $\times$ 1.1 = cells/mL culture

$$= 45.25 \times 10{,}000 \times 1.1 = 497{,}750 \text{ cells/mL}$$

% of Viable cells = (Number of living cells/Number of total cells) $\times$ 100

$$= (497{,}750/533{,}500) \times 100$$
$$= 93\% \text{ viability.}$$

Live cells do not take up trypan blue stain.

Count cells that are set within a square and on the right-hand or lower boundary lines (but not both).

To get the percentage viability, count the number of dead cells (stained with trypan blue) to get the total number of cells per chamber.

8.5 Cryopreservation of Cells

Cryopreservation of cells is an important integral laboratory procedure widely used for the long-term storage of cells. It preserves cells at low temperatures (−80 °C to −196 °C), which dramatically reduces biological and chemical reactions or suspends cellular metabolism in living cells. As soon as cell lines are available in abundance (especially after two passages), it is advisable to preserve some cells for the future to avoid:

- **Microbial contamination:** Continuous growth and cell passaging make cell lines susceptible to microbial contamination.
- **Changes in genetic composition:** Continuous growth and passaging of continuous cell lines or finite cell lines make them prone to genetic drift or senescence, respectively.
- **Loss of valuable cell line:** Laboratories can experience equipment errors and failures. Cryopreservation

of cells preserves quality-controlled cell lines for future and long-term use.

8.5.1 Cryoprotective Agents

Cryoprotective agents can be divided into intracellular and extracellular cryoprotectants.

- Intracellular cryoprotective agents are small molecules capable of penetrating the cell membrane and preserving the intracellular components of the cell. Examples include commonly used dimethylsulfoxide (DMSO), glycerol, ethylene glycol, and propylene glycol.
- Extracellular cryoprotective agents are large molecules that possibly serve as a source of nutrients and a buffer to the preserved cells. They are added to the intracellular cryoprotective solution. Examples include FBS, sucrose, and dextrose.

The most common cryopreservation medium is a mixture of 90% FBS or DMEM + 10% DMSO. It can be used to preserve a diverse number of cell types. There are several commercially prepared specialized cryopreservation media designed specifically for certain cell types; therefore, each laboratory needs to optimize the cryopreservation process.

Whichever cryoprotective agents the laboratory decides to choose, the laboratory must take note that a good cryoprotective agent must be able to reduce the freezing point of the medium, allow the medium to cool/freeze slowly, and it must be able to reduce the risk of ice crystal formation within the cell, which can cause cell damage and death.

When handling DMSO or containers that may retain product residue, there is a potential risk of exposure. While DMSO is not classified as a carcinogen, mutagen, or teratogen, repeated exposure, ingestion, or inhalation at high concentrations may lead to low toxicity exposure. Therefore, it should be opened only in a laminar flow hood and in the dark.

8.5.2 How to Prepare Cells for Cryopreservation

Materials

Cultured cells (adherent or suspension cells)
Cryoprotective agents like DMSO, FBS
Disposable sterile 15-mL conical tubes, cryotubes or cryovials
Isopropanol chamber
Liquid nitrogen storage tank

Methods

1. Prepare cryopreservation medium suitable for your cell type. For example, to preserve 10×10^6 HeLa cells in a 2ml cryogenic vial (or cryovials), add 900 µL of FBS and, 100 µL of DMSO (90% FBS + 10% DMSO) to the tube.
2. Store the cryopreservation medium at 2–8 °C until use.
3. Harvest the cells from the culture vessels (as described above) and centrifuge at 800–1000 rpm for 5–10 minutes.

Note that the centrifugation speed and time depend on the cell type.

4. Remove the supernatant thoroughly and carefully to avoid disturbing the cell pellet
5. Resuspend the cell pellet in the prepared cryopreservative and mix gently to maintain a homogeneous cell suspension.
6. Freeze the cells.
   a. **Long-term storage of cells:** Cells can be stored in an isopropanol-containing cryo-freezing chamber or temperature-controlled rate freezing container at −80 °C overnight. Then transfer to a liquid nitrogen tank at −135 °C (see Figure 8.9).
   b. **Short-term storage of cells:** Cells can be stored in an isopropanol-containing cryo-freezing chamber or a temperature-controlled freezing container at −80 °C. *Maximum* of 1–2 months.
   c. Alternatively, the temperature of cells can be reduced gradually by first cooling it at 2–8 °C for 20 minutes, −20 °C for 20 minutes, and then transferring to −80 °C.

Freeze cells at optimum concentration and a low passage number. A very high concentration could lead to loss of cells or clumping.

Ensure that the cell viability is above 90% before cryo-freezing.

Ensure proper labeling of cryovials for traceability and record keeping. Use a marker resistant to both alcohol and liquid nitrogen.

Figure 8.9 Liquid nitrogen tank.

Source: Photo by Prof. Zhaohua Zhong, Harbin Medical University, China.

8.5.3 How to Recover Cryo-Persevered Cells

Maintain proper aseptic techniques to recover cryo-persevered cells after freezing.

Materials

Cryovial containing frozen cells
Cell culture media (pre-warmed)
Appropriate culture vessels

Water bath set to the appropriate temperature (37 °C)
Disposable sterile 15-mL conical tubes

Methods

1. Remove the cryovial containing the frozen cells from storage (liquid nitrogen or – 80 °C freezer) and place it into a 37 °C water bath. Do not rub cells together between your palms to avoid stings like frostbite.
2. Allow the cells to thaw by gently swirling the vial in the 37 °C water bath.
3. Disinfect the outside of the vial with 70% ethanol and transfer it into the safety cabinet before opening.
4. Transfer the thawed cells slowly into sterile 15-mL conical tubes and add enough appropriate culture medium.
5. Centrifuge cells at 800–1000 rpm for 5–10 minutes. Disinfect the outside of the vial with 70% ethanol and transfer it back to the safety cabinet before opening.
6. Decant the supernatant and avoid disturbing the cell pellet.
7. Resuspend the pellet with a small volume of appropriate culture medium, transfer it, and incubate in the recommended culture environment and avoiding disturbing the cells.

8.6 Cell Culture Contamination and Infections

Cell culture contamination and infection are common phenomena in most virology laboratories. It denotes the presence of microorganisms (mostly bacteria and yeasts or even unintended virus infection) among cells or in culture media. Once this happens, it affects cell culture health, and virtually all parameters of the study become affected, and this could result in the generation of non-reproducible results or eventual cell death and loss. Therefore, any data or result generated from a contaminated cell culture, even with "just a little contamination," could be misleading and not helpful.

8.6.1 Sources of Cell Contamination and Infections

8.6.1.1 Chemical Source of Cell Culture Contamination

This type of contamination can be best defined as the presence of undesirable nonliving substances or the presence of desirable nonliving substances in an inappropriate proportion (for example, essential nutrients can become toxic at higher concentrations than required), which may cause unwanted effects in cultured cells. Sources of chemical contamination include culture media, serum, water sources, storage vessels, photochemical lights, and air sources.

Ideally, chemical contamination from certified, commercially prepared, or laboratory-prepared cell culture media rarely occurs, provided the preparation, storage, and usage conditions are followed.

General precautions to prevent chemical contamination

- Ensure the purchase of appropriate media certified for diagnostic or research purposes, as appropriate.

- Read reagent bottle labels and instructions before preparation or usage.
- For laboratory-prepared media, weighing errors are a common source of contamination. Ensure that the various constituents of the media are properly weighed.
- Always use laboratory-grade water for preparing media, solutions, or buffers and for reconstitution of lyophilized reagents.

Serum as a Source of Chemical Contamination The major source of chemical contamination in serum is the variation in the constituents of different batches of serum, the presence of antibodies, growth factors, and inhibitors that can be harmful to some cell types.

To prevent this, ensure proper testing of sera for toxicity before purchase, and consider bulk purchase of serum to prevent uncontrollable lot-to-lot variation of constituents.

Water as a Source of Chemical Contamination Water can be a major source of chemical contamination, especially during in-house culture media preparation. However, it is unusual to find significant water contamination in most laboratories. This is because most laboratories use ultrapure water. Modern water purification systems typically employ multiple steps, including UV radiation to eliminate microorganisms, as well as reverse osmosis, ion exchange, and ultrafiltration to remove organic and inorganic contaminants. As a result, the presence of trace metals, heavy metals, and dissolved organic compounds is rare in laboratories that use such systems. However, poorly maintained water treatment facilities can harbor significant levels of viable bacteria or their endotoxins.

Water used for making media or physiological saline may become contaminated by steam (buildup in pipes) generated in autoclaves; this may contain potentially toxic metals.

Prevention

- Ensure special and frequent care of the water source and water purification system.
- Ensure that containers containing water or liquid for steam sterilization are properly covered or wrapped in aluminum foil to prevent contamination by steam generated in autoclaves.
- Always use water that is appropriate for media or solutions preparation for washing, rinsing, and soaking laboratory glassware and containers for media or solutions.

Storage Vessels as a Source of Chemical Contamination Toxic metals and ions can accumulate on disposable plastic vessels, glassware, or metal pipes, potentially contaminating culture media, buffers, or saline solutions used for cell culture. Similarly, chemical residues from disinfectants, detergents, or washing agents used on glassware can remain on the surfaces of bottles or caps and become toxic contaminants.

Prevention

- Ensure that used or new vessels are properly washed and rinsed in several changes of water, sterilized, and dried slowly before usage.

- Ensure proper handling and storage of sterile vessels.
- Purchase high-quality plastic vessels certified for specific tests or experiments.

Photochemical Lights as a Source of Chemical Contamination Photochemical light exposure, like fluorescence or UV irradiation of media, can result in changes to the composition and concentration of substrate components. Some culture media can be affected by exposure to direct sunlight. Extensive exposure of culture media containing photoproducts components like riboflavin, HEPES, or tryptophan to UV irradiation or fluorescent light can produce a cytotoxic effect on the cells [28].

Prevention

- Avoid long-term exposure of culture media or reagents containing photosensitive components to light. Keep media away from direct sunlight.
- Reagents containing photosensitive components can be stored and wrapped with aluminum foil to avoid direct exposure to sunlight.

Gases as a Source of Chemical Contamination Gassing of cell culture incubators involves the use of gas-tight containers to flush in prepared gas mixtures to maintain a proper atmosphere for growing cells. This gas mixture may contain impurities or other gases previously used in the same storage tank.

Similarly, chemical disinfectants used to clean or disinfect the incubator can leave behind volatile chemical residue that can be potentially toxic to the culture cells.

Another main source of air contamination is the humidity tray and incubator fan in humidified incubators. Likewise, plastic guards in fan-assisted air circulator incubators can accumulate dust, which may contaminate cultured cells.

Prevention

- Whenever possible, duplicate important cultures in separate incubators, each equipped with a different CO_2 source.
- To reduce airborne contamination, the standard recommended practice is to use non-fan-assisted incubators.
- Constantly replace evaporated water in the humidity tray with sterile, laboratory-grade water.

8.6.1.2 Biological Source of Cell Culture Contamination

Cell culture contamination from biological sources is a major concern. This is because, in the cell culture system, the living organisms metabolize and replicate together with the growing cells. This could affect their biological function. Biological infection of one culture vessel can increase the incidence of infection in others; therefore, contaminated cells should be removed immediately.

Sources of biological culture contamination can be difficult to identify and eliminate. For example, bacteria, molds, and yeasts are usually easy to detect either by microscopic observation or cause direct effects in culture media, often visible to the unaided eye (e.g., pH shift – as observed in color change, turbidity, or cloudiness of the culture media) [26]. The ease of detection of other contaminants is often more difficult, which makes detection and control cumbersome.

Infection of cell lines by non-cytolytic viruses, mycoplasmas, and protozoa could result in potentially serious problems usually characterized by observations like unexplained cell detachment, nonadherence of adherent cell types, slow or poor growth, or cell death. The major causes of biological contamination are bacteria and fungi, viruses, protozoa, and cross-contamination of other cell cultures.

Bacteria and Fungi as Sources of Cell Culture Contamination Bacteria and fungi (yeast and molds) are ubiquitous organisms. Due to the nutrient-rich environment of the cell culture media, they can grow, replicate, and even colonize the relatively unprotected cell culture system. Some fast-growing bacteria and yeast can flourish and become readily detected in culture within a few hours or days of contamination. They can be detected either by direct microscopic observation or by causing visible effects in culture media. This ease of detection may result in potentially less serious cultural problems, as amendments can be readily made. However, slow-growing bacteria and molds may be difficult to detect and become persistent indefinitely in culture media. By the time they are detected, they could have caused significant defects in the cell growth, morphology, function, and metabolism, capable of compromising the reliability and reproducibility of experimental or test results.

Antibiotics and antifungals are routinely used in culture to reduce the incidence of bacterial or fungal infection, respectively. However, like in humans, organisms may develop resistance to these agents.

Mycoplasmas as a Source of Cell Culture Contamination Robinson and colleagues were the first to report the contamination of cells by mycoplasmas [29, 30]. Except for fluorescence labeling, mycoplasma, unlike other bacteria and yeast, cannot be detected or visualized by a light microscope; they require an electron microscope. Therefore, it is very difficult to completely remove and sterilize mycoplasma-contaminated media and serum by membrane filtration.

Mycoplasma contaminants are extremely detrimental to any cell culture. They grow slowly, causing a wide range of structural, functional, metabolic (synthesis of nucleic acid), and genetic (chromosomal aberration and damage) defects in infected cells [31]. Like some viruses, they often grow and replicate in cell culture systems without causing direct visible effects (like turbidity, pH change) or changes like cytopathic effects in cultured cells.

Most laboratories do not routinely test for mycoplasma infection, which makes the detection, containment, and elimination difficult. Testing for mycoplasma should be incorporated into the routine activities of diagnostic laboratories and should be done more frequently if contamination has once been detected. Also, ensure that all new media brought into the laboratory and cell stocks are tested for mycoplasma before being frozen or stored. DNA staining or polymerase chain reaction (PCR) is a common method for the detection of Mycoplasma.

Viruses as a Source of Cell Culture Contamination Virally contaminated cells may behave similarly to mycoplasma in that they may not show visible evidence of their existence in cell cultures and are not visible under a light microscope. Special precautions should be taken in a virology laboratory to prevent unintended virus infection of cells, as this could be regarded as contamination. Some virus contaminations of cells tend to be self-limiting and may cause no apparent changes or adverse effects in infected cells; as a result, they will go undetected for weeks to cause a more serious problem.

More importantly, laboratory personnel handling viruses capable of causing serious potential health hazards to personnel must ensure safety precautions when handling tissues or cells, even though they have not been intentionally infected.

8.6.2 Routes of Microbial Contamination and Prevention

8.6.2.1 Nonsterile Disposable Supplies, Media, Serum, and Solutions

Most supplies of pipette tips, centrifuge tubes, and culture vessels are sterilized by the manufacturers. However, improper storage, opening, and resealing can cause serious contamination outbreaks.

Also, improper maintenance, operation of sterilization methods (for example, short sterilization cycle), storage of glassware, and storage bottles used during routine cell culture procedures can constitute a major source of biological contaminants.

8.6.2.2 Airborne Particles and Aerosols

Laboratory personnel and machine activities can generate several particles and microbial loads. For example:

a. Talking, coughing, and sneezing can generate relatively high amounts of aerosols that contain a variety of microorganisms, including Mycoplasma.
b. Dirty lab coats or street clothes. Placing a dirty sleeve into a safety cabinet during cell culture procedures can introduce dust particles.
c. Machine activities like air conditioners, fans, pipetting devices, aspirators, freezers, refrigerators, centrifuges, and incubators can trap and generate significant amounts of particulates and aerosols containing microbial contaminants.
d. Small insects and other invertebrates can temporarily visit or reside in cell culture rooms, enter unsealed culture plates and dishes, and loosen caps of culture flasks.

8.6.2.3 Prevention of Microbial Contamination of Culture Media

a. Ensure that sterile laboratory disposables are properly stored in an insect- or dust-free area to prevent recontamination.
b. Ensure that commercially purchased media and serum supplies are always being tested for sterility in-house before use.

c. It is advisable to filter (through 0.2-μm membranes) liquid media or serum products halfway to their use to ensure that they have not been contaminated.
d. Clearly label solutions, media, or other supplies if they are sterile or not.
e. Ensure proper handling and opening of culture vessels and maintain aseptic procedures.
f. Ensure that the laboratory floor, cabinets, incubators, and refrigerators are clean, dust-free, rodent- and insect-free.
g. Clean work area (with 70–75% ethanol spray) and environments before and after each experiment.
h. Ensure that animals used as a source of tissue or cell culture are infection-free.

8.6.3 Cell Line Cross-Contamination in the Culture System

Cross-contamination of cells occurs when there is co-culture of a cell line with another undesired cell in a culture medium [32]. This could also result from the incorrect designation of culture cells. Although this phenomenon is not as common as microbial contamination, when this occurs, it could lead to a serious problem that poses difficulty for the quality, reliability, and reproducibility of research or test results, required to diagnose infection, thus leading to wrong research or diagnostic decisions.

While a particular cell line may be fully free from all sorts of microbial contaminations, the presence of another cell line or its product due to accidental co-culture may result in inherent instability in the population, genotypic, or phenotypic properties of the desired cell line.

Unacceptable degrees of genetic and phenotypic instability may affect the resulting outcome. Unless a specifically designed medium that encourages balanced growth of specific cell lines is used, it is usually difficult to readily identify different types of cells that grow together under the microscope.

An important, fast, practical (but not reliable) approach to identifying cross-contamination could rely on cell morphological characteristics like shape. For example, contamination of normal human fibroblasts (orbital-shaped) with HeLa cells (rounded cell colonies) can be easily identified.

8.6.3.1 Sources of Cross-Contamination of Cell Culture

a. Cell culture-related accidents. This involves the inoculation or transfer of different cell lines together. For example, dripping of liquid from the pipette tip can drop into another cell suspension or culture vessel.
b. Cell line misidentification. This can occur as a result of mislabeling or misreading labels on culture flasks or stock
c. Culture medium contamination. This occurs when one reservoir of medium is used for more than one cell line, or when the same pipetting tool (especially pipette tips) is used while working on different cell lines.
d. Retrieving and thawing the wrong bottle of cell stock.

8.6.3.2 *Prevention of Cell Culture Cross-Contamination*

a. Ensure that labels on stock bottles and culture vessels are properly written to avoid thawing the wrong frozen cell stock. Avoid using acronyms, where possible.
b. Avoid working with different cell lines at a time. Keep one cell line in the safety cabinet at a time.
c. Use one reservoir of medium for one cell line.
d. Test cell line samples from other laboratories or colleagues thoroughly before using them.
e. To minimize the thawing of wrong bottles, use standardized record-keeping forms for all cells retrieved from storage (liquid nitrogen tank or refrigerators).

8.7 Sterility Testing for Cell Culture Contamination and Infections

8.7.1 Testing Methods for Detecting Bacterial and Fungal Contamination of Cell Culture

Gross bacterial and fungal cell culture contamination can be identified with the naked eye or with a microscope. However, detection of low-level infections and slow-growing bacteria or fungi (that may take weeks to appear) may require special traditional microbial culture procedures. Microscopic observation of bacterial or fungal contamination of cells, culture medium, or solution provides morphological characteristics of contaminants.

Also, the rate, burden, and characteristics of the contaminant can be identified and characterized in traditional methods that involve culturing microbes using a range of nonselective and selective enrichment culture media (agar and or broths).

Materials

1. Culture media: Aerobic and anaerobic nutrient broth and agar
2. Controls: Positive control organisms, e.g., *Bacillus subtilis*, *Candida albicans*
3. Microbiological safety cabinet
4. Incubator set at 32 °C

Methods

1. Grow test cells in an antibiotic-free medium from at least three passages.
2. Bring adherent test cells into suspension using a cell scraper (not trypsin). Suspension cell lines may be tested directly.
3. Inoculate culture media (broths and agar) with enough suspension test samples, cell culture medium, or test solution.

Examples of media (agar or broth) that can be used for the isolation of bacterial contaminants include thioglycolate broth, brain heart infusion agar, and blood and chocolate agar.

4. Incubate the culture medium appropriately and examine it every day for growth.

5. Indirect detection, such as nucleic acid amplification tests (NAAT) (e.g., PCR), DNA fluorochrome test, and serological testing, like enzyme-linked immunosorbent assays (ELISA), can provide a rapid detection method of both intra- and intercellular microbes.

Evaluation of cells or culture media for the presence of microbial contamination should be carried out in a microbiology/bacteriology laboratory away from the cell culture laboratory.

8.7.2 Testing Methods for Detecting Virus Contamination of Cell Culture

The presence of endogenous and contaminant viruses in cell culture can be difficult to identify, especially for viruses that do not produce noticeable cytopathic effects. These viruses can remain undetectable (except by serological or biochemical methods or using electron microscopy) in cell cultures without necessarily causing an obvious morphological change in the cell culture system.

8.7.3 Testing Methods for Detecting Mycoplasma Contamination

Mycoplasmas are bacteria that are fastidious, small microbes (0.3–0.8 μm in diameter), which lack cell walls. These characteristics make detection under careful light microscopic observation impossible. Detection of cells contaminated by these microbes can be done by two methods.

8.7.3.1 *Detection of Mycoplasma Contamination*

This method involves growing mycoplasma by incubating the suspected contaminated sample in a special nutrient medium (agar and broth) that promotes the growth. This process is energy- and time-consuming. Some of the methods for detection of mycoplasma in cell culture include:

- Standardized methods of NAATs for the detection of mycoplasma DNA or RNA are commercially available. This targets a variety of mycoplasma genes [33]. Conventional NNAT-based techniques like nested, real-time PCR, and isothermal amplification technologies can be used. These tests are faster and can detect all species of Mycoplasma, but lack the specificity as compared to direct culture.
- Serological methods, in particular ELISA, can be used to detect mycoplasma contamination in cell culture. However, they have some limitations, as they requires proper interpretation of the serological test results and have low sensitivity.
- Fluorescent staining methods can also be used to detect mycoplasma [34, 35]. This method relies on the ability of Hoechst 33258 bisbenzimide dye to bind specifically to DNA. It is commonly used for routine screening of cell culture and cell culture reagents.

Rapid Fluorescent Staining Methods for the Detection of Mycoplasma Contamination

Materials

1. Hoechst 33258, 1 mg/mL stock solution
2. Fixative: acetic acid and methanol (freshly prepared 1:3 ratio)
3. Mounting solution (50% glycerol in water)
4. Glass slide and coverslip
5. PBS
6. Fluorescence microscope with UV epifluorescence and oil immersion lens
7. Chamber slides

Methods

1. Grow the reporter cells in an antibiotic-free medium for at least three passages.
2. On the day before the sample test, seed test cells in chamber slides in an antibiotic-free medium and incubate at appropriate conditions. Place the chamber slide in a Petri dish for easy handling. Seed the cells so that more than 50% confluence (exponential phase of growth) can be reached in 24 hours.
3. The next day, take the chamber from the incubator and check if the cells are attached. Change the culture medium.
4. Add 0.5–1 mL of the medium from the test cells, test culture medium, or solution to the medium of the reporter cells and incubate for 2–5 weeks.
   Note: Bring adherent test cells into suspension using a cell scraper before adding them to the reporter cells. Also, suspension cell lines may be used directly to infect the reporter cells.
5. On the day of examination, aspirate the culture medium from the reporter cells. Avoid cross-contamination while aspirating.
6. Wash the cells with PBS twice.
7. Add the fixative to the cultured cells, enough to cover the cells, and incubate at room temperature for 10 minutes. This step can be repeated.
8. Aspirate the fixatives and rinse with double-distilled water.
9. Cover the slide with Hoechst 33258 working solution (0.5–1 μg/mL stock) for 25–30 minutes.
10. Wash twice with double-distilled water.
11. Place one drop of the mounting solution at the edge of the slide. Using sterile forceps, place a cover slip on the mounting solution and allow it to spread through.
12. Examine the slide under a fluorescence microscope (excite at 360 nm and emission at 490–500) under 100X objectives.

Positive and negative control slides that are commercially available should be used when interpreting the test results. These slides are safe to use since they have already been fixed.

Mycoplasma detection might be difficult. It is possible to have a false negative result. Successful detection of Mycoplasma may requiry several carefully combined methods- both the traditional direct (culture-based) and indirect test methods.

8.7.4 Testing Methods for Detecting Cell Line Cross-Contamination

Testing for cell line cross-contamination may be the most difficult type of contamination to detect in a routine virology laboratory. There may be widespread contamination of this type of contamination for years, with a lack of vigilance. Most cell line cross-contaminations are detected by cytogenetic analysis (e.g., karyotyping) and genetic profiling (by isoenzyme analysis and short tandem repeat profiling).

- **Isoenzyme analysis:** It is an electrophoresis technique that is used to study the band patterns of migration of isoenzymes present in cell lysates. Different cell lines (even the same species) may express different isoenzyme patterns. Therefore, the method depends on the isoelectric separation of a specific set of intracellular enzymes. It can be used for the determination of inter-species cross-contamination. It is easy to perform, but can have low reproducibility [36, 37].
  Commonly studied enzymes for this purpose include glucose-6-phosphate, lactate, and malate dehydrogenase [38].
- **Karyotyping:** Karyotyping analysis is a test technique that involves the use of stained chromosomes to determine the number of chromosomes (pairing and ordering) in a sample of cells. It is important as it helps in determining the stability of cell lines [39].
- **DNA fingerprinting:** It is a DNA profiling technique that is based on the identification of a variable number of tandem repeat (VNTR) loci or short tandem repeat loci amplified in cells using PCR. It can be used to detect both intra- and interspecies contamination [40].

8.8 Dealing with a Contaminated Cell Culture Room

Once contamination has been detected in the cell culture, to prevent the contamination from spreading to other cultures, it is important to deal with contaminated cultures immediately by

1. Removing all contaminated culture vessels to an appropriate room where cell or tissue culture experiment does not occur. Check other cells cultured at the same time as the contaminated cells, even though there is no evidence of contamination, to avoid low-level contaminants.
2. Where possible, discard all culture medium and solutions used for the cultivation of the contaminated cells.
3. Fill the contaminated disposable culture vessel with appropriate disinfectant (e.g., 10% sodium hypochlorite). Disinfect and autoclave nonreusable vessels immediately.
4. Wipe the outside and inside of the incubator thoroughly with 70% isopropanol.
5. Wash the humidity tray properly and allow it to air-dry and disinfect with 70% isopropanol before placing it in the incubator.

It may be important to test the cell stock, especially when a cell has recently been revived from storage.

If there is persistent contamination, discard all the cells, aliquots of antibiotics and FBS, and any open bottles of culture media and solution. Fumigate the entire cell culture room. Put all gowns in the laundry and ensure that they are freshly laundered before using them again.

It is important to keep samples of previous supplies and standardized procedures to investigate contaminations that are suspected to have occurred due to operational errors (because of changes in SOPs) or changes in suppliers.

Accurate and comprehensive record keeping of product information (like supplier's name, product or company's name, date supplied, product code, batch number, expiry date, and date opened) is vital to any lab; this can help trace contamination that is likely to have originated from a change in the stock of reagents or consumables.

8.9 Curing Irreplaceable Contaminated Cell Cultures

Occasionally, there may be cases where valuable cells that are irreplaceable become contaminated. In such instances, attempts can be made to save such cells. This might be possible depending on the degree and nature of contamination, as the cells could have lost important characteristics; therefore, further attempts to "clean up" the contaminant may fail.

8.9.1 Curing Irreplaceable Contaminated Cell Cultures due to Microbial Infections

- If the contamination is a result of bacterial or fungal infection, a clean-up attempt might be difficult, especially if the infection is from a human or animal source. In this case, the organism is likely to have developed resistance to commonly used antibiotics.
- With extensive washing, replating, and treating with antibiotics and antifungal agents, an attempt to recover contaminated cell cultures might appear successful. However, commonly used antifungal agents like nystatin and amphotericin B are unlikely to kill the fungi but could prevent their further growth.
- If the contamination is caused by Mycoplasma, commonly used antibiotics for bacteria, like penicillin and streptomycin, will not affect it. Most cleanup attempts against mycoplasma involve treating cells with antibiotics like tetracyclines, macrolides, and quinolones. This may be beneficial in the early stages of infection for a limited period. Subsequently, there is a risk of further spread later as mycoplasma may show resistance to these

antibiotics. While it is important to use antibiotic formulations recommended and optimized specifically for mycoplasma decontamination in cultured cells, routine use and long-term over-reliance on such could allow cryptic contaminants, including the same Mycoplasma, to grow undetected and may affect cell proliferation and metabolism.

- Cells may also suffer from cytotoxicity because of the continuous use of antibiotics. If antibiotics are used for a longer period, antibiotic-free cell cultures should be maintained in parallel as a control for the detection of cryptic contaminants.

During the Recovery of Irreplaceable Contaminated Cell cultures

It may be beneficial to add fresh antibiotics directly to the culture. Mix the culture medium thoroughly to ensure optimal distribution.

To ensure that trapped bacteria have physical access to the antibiotics, mix the culture medium properly and break cell clumps and clusters.

To compensate for possible toxicity or growth-inhibitory effects of the antibiotic, culture cells at a higher density and grow with higher concentrations of FBS.

To encourage the growth of residual bacteria to a detectable level, after the treatment periods, grow the cells in antibiotic-free media for at least five passages and recheck.

8.9.2 Curing Irreplaceable Contaminated Cell Cultures due to Cell Line Cross-Contamination

If the contamination is because of cell line cross-contamination, an attempt can be made to recover a pure culture of the initial cells by diluting the cells and plating it at low density or single-cell density in a 96-well plate. After the recovery period, cells should be characterized to ensure that the altered cell culture population has been restored. Karyotype analysis, DNA fingerprinting, and isotype analysis can confirm if the pure culture has been recovered.

While it might be impossible to avoid cell culture contamination, every laboratory must ensure that it has fast, detailed, and systematic standard procedures for dealing with contamination. This is instrumental in reducing the burden and consequences of contamination.

8.10 Study Questions

8.1 In your virology laboratory, you are tasked with conducting experiments to study the replication of a newly identified virus. The laboratory uses various cell lines for viral propagation and assays. Recently, a batch of experiments has had inconsistent viral yields and unusual cell morphology changes.

(a) Analyze the potential causes of inconsistent viral yields observed in the cell culture experiments.
(b) Propose a troubleshooting strategy for addressing unusual cell morphology changes observed during viral infection studies.
(c) How does the choice of cell line influence viral replication efficiency and the interpretation of experimental results?
(d) What specific characteristics of cell lines (e.g., receptor expression, growth rate) are important for studying a particular virus?
(e) How should cell line characteristics be considered when designing experiments to study viral behavior?

8.2 A virology laboratory is conducting a critical experiment or test involving a newly isolated viral strain for vaccine development. During routine maintenance, the laboratory encounters sporadic issues with cell cultures showing unexpected growth patterns and signs of contamination. Sterility testing is necessary to ensure that the observed issues are not affecting experimental results.

(a) Design a comprehensive sterility testing strategy for the affected cell cultures.
(b) How will you interpret the results of the sterility tests to differentiate between contamination, infection, or normal cell growth variations?

Objectives

8.1 Which of the following statements correctly describes the difference between primary cell cultures and established cell lines?

(a) Primary cell cultures are immortalized and can be propagated indefinitely, while established cell lines are only used for a limited number of passages.
(b) Primary cell cultures are derived directly from tissues and have a limited lifespan, while established cell lines are derived from primary cultures and can be propagated indefinitely.
(c) Primary cell cultures are typically more homogeneous than established cell lines.
(d) Established cell lines require more specialized media compared to primary cell cultures.

8.2 In the context of cell culture, what is the primary purpose of using a feeder layer?

(a) To provide a substrate for cell adhesion and growth
(b) To supply essential nutrients and growth factors that are depleted in the medium
(c) To enhance the genetic stability of the cultured cells
(d) To provide a source of secreted proteins that aid in cell differentiation

8.3 Which method is most used for the long-term preservation of cell lines?

(a) Cryopreservation in liquid nitrogen with DMSO
(b) Storage at −80 °C without any cryoprotectant

(c) Desiccation in a vacuum chamber
(d) Storage in a standard refrigerator

8.4 When passaging a cell line, why is it important to use trypsinization?
(a) To remove the cells from the culture vessel by digesting extracellular matrix components and cell surface proteins
(b) To replenish the nutrients in the culture medium
(c) To induce cell differentiation
(d) To increase the expression of surface receptors on the cells

8.5 What is the primary difference between adherent and suspension cell lines?
(a) Adherent cell lines grow attached to a surface, while suspension cell lines grow freely in the culture medium.
(b) Suspension cell lines require a solid surface for growth, while adherent cell lines do not.
(c) Adherent cell lines are generally less sensitive to changes in temperature compared to suspension cell lines.
(d) Suspension cell lines are used to produce monoclonal antibodies, while adherent cell lines are not.

8.6 Which of the following is a critical factor in the selection of culture media for a specific cell line?
(a) The color of the media
(b) The pH level and nutrient composition are tailored to the cell line's specific needs
(c) The source of the media (commercial vs. homemade)
(d) The temperature at which the media is stored

8.7 In a cell culture experiment, why is it important to maintain a sterile environment?
(a) To prevent contamination from microorganisms that could alter cell growth and experimental results
(b) To ensure cells adhere more effectively to the culture vessel
(c) To enhance the metabolic activity of the cells
(d) To increase the rate of cell division

8.8 Which of the following characteristics is NOT typically used to distinguish between different types of cell lines?
(a) Growth pattern and morphology
(b) Genetic markers and karyotype analysis
(c) Protein expression profiles
(d) The size of the culture vessel used for maintenance

8.9 In which scenario is the use of conditioned media particularly beneficial?
(a) When trying to achieve a specific differentiation of primary cells
(b) When culturing cells that do not adhere to plastic surfaces
(c) When studying the effect of secreted factors on cell behavior or growth
(d) When cryopreserving cell lines

8.10 What is a potential consequence of overpassaging a cell line?
(a) Increased cellular heterogeneity and potential loss of specific cell functions
(b) Enhanced cell growth rate and proliferation

(c) Improved genetic stability and decreased mutation rates
(d) Decreased demand for growth factors in the culture medium

References

1. Adey, A., et al., *The haplotype-resolved genome and epigenome of the aneuploid HeLa cancer cell line*. Nature, 2013. **500**(7461): p. 207–211.
2. Shaw, G., et al., *Preferential transformation of human neuronal cells by human adenoviruses and the origin of HEK 293 cells*. FASEB J., 2002. **16**(8): p. 869–871.
3. Wang, D., et al., *Differential protein expression in MCF7 breast cancer cells transfected with ErbB2, neomycin resistance and luciferase plus yellow fluorescent protein*. Proteomics, 2004. **4**(7): p. 2175–2183.
4. Page, N.S.C.I.H *Stem cell basics*. 08.06.2025; Available from: https://stemcells.nih.gov/info/basics/stc-basics.
5. Liang, G. and Zhang, Y., *Embryonic stem cell and induced pluripotent stem cell: an epigenetic perspective*. Cell Res., 2013. **23**(1): p. 49–69.
6. Gunhanlar, N., et al., *A simplified protocol for differentiation of electrophysiologically mature neuronal networks from human induced pluripotent stem cells*. Mol. Psychiatry, 2018. **23**(5): p. 1336–1344.
7. Anastasaki, C., et al., *Generation of human induced pluripotent stem cell-derived cerebral organoids for cellular and molecular characterization*. STAR Protoc., 2022. **3**(1): 101173.
8. Technologies, S *How to Grow Cerebral Organoids from Human Pluripotent Stem Cells*. 2025; Available from: https://www.youtube.com/watch?v=SvnX9NJ1Zuo.
9. Basu, S., et al., *Evolution of bacterial and fungal growth media*. Bioinformation, 2015. **11**(4): p. 182–184.
10. Verma, A., Verma, M., and Singh, A., *Animal tissue culture principles and applications*, in *Animal Biotechnology*. 2020. p. 269–293. 10.1016/B978-0-12-811710-1.00012-4 Epub 2020 Jun 26. PMCID: PMC7325846. https://pmc.ncbi.nlm.nih.gov/articles/PMC7325846/.
11. Jagušić, M., et al., *Stability of minimum essential medium functionality despite L-glutamine decomposition*. Cytotechnology, 2016. **68**(4): p. 1171–1183.
12. Rubin, H., *Deprivation of glutamine in cell culture reveals its potential for treating cancer*. PNAS, 2019. **116**(14): p. 6964–6968.
13. Mishra, V. and Heath, R.J., *Structural and biochemical features of human serum albumin essential for eukaryotic cell culture*. Int. J. Mol. Sci., 2021. **22**(16): p. 8411.
14. Sigma, M. *Bovine Serum Albumin (BSA Protein)*. 2025; Available from: https://www.sigmaaldrich.com/US/en/products/cell-culture-and-analysis/cell-culture-supplements-and-reagents/albumin?s-and-transport-proteins/bovine-serum-albumin?srsltid=AfmBOop4aC3C6WB7MNeObtTejuYZAYgrOwdzVhfoilu_Cc1Zp66Srpt6.
15. Lisman, T., et al., *Aprotinin inhibits thrombin generation by inhibition of the intrinsic pathway, but is not a direct thrombin inhibitor*. TH Open, 2021. **5**(3): p. e363–e375.
16. Brannon, J.R., et al., *Inhibition of outer membrane proteases of the omptin family by aprotinin*. Infect. Immun., 2015. **83**(6): p. 2300–2311.
17. Dalton, C.J. and Lemmon, C.A., *Fibronectin: molecular structure, fibrillar structure and Mechanochemical signaling*. Cells, 2021. **10**(9): p. 2443.
18. Jéquier, E., *Carbohydrates as a source of energy*. Am. J. Clin. Nutr., 1994. **59**(3 Suppl): p. 682s–685s.

19. Kitahata, S., et al., *Effect of the sugar present in the culture medium on the preservation of human RPE cell suspensions*. Transl. Vision Sci. Technol., 2025. **14**(2): p. 1.
20. Science, T.N.C *Which cell culture media is right for you?* [cited 2025]; Available from: https://www.technologynetworks.com/cell-science/articles/which-cell-culture-media-is-right-for-you-331552.
21. Eagle, H. and Habel, K., *The nutritional requirements for the propagation of poliomyelitis virus by the HeLa cell*. J. Exp. Med., 1956. **104**(2): p. 271–287.
22. Godoy-Parejo, C., et al., *Roles of vitamins in stem cells*. Cell. Mol. Life Sci., 2020. **77**(9): p. 1771–1791.
23. Liu, P., et al., *Zwitterionic betaines over HEPES as the new generation biocompatible pH buffers for cell culture*. Bioact. Mater., 2023. **24**: p. 376–386.
24. PromoCell. *Impact of phenol red in cell culture and solutions*. 28.01.2025 [cited 2025]; Available from: https://promocell.com/us_en/blog/impact-of-phenol-red-in-cell-culture-and-solutions.
25. Berthois, Y., Katzenellenbogen, J.A., and Katzenellenbogen, B.S., *Phenol red in tissue culture media is a weak estrogen: implications concerning the study of estrogen-responsive cells in culture*. PNAS, 1986. **83**(8): p. 2496–2500.
26. Weiskirchen, S., et al., *A beginner's guide to cell culture: practical advice for preventing needless problems*. Cells, 2023. **12**(5): p. 682.
27. Ryu, A.H., et al., *Use antibiotics in cell culture with caution: genome-wide identification of antibiotic-induced changes in gene expression and regulation*. Sci. Rep., 2017. **7**(1): p. 7533.
28. Wang, R.J., *Effect of room fluorescent light on the deterioration of tissue culture medium*. In Vitro, 1976. **12**(1): p. 19–22.
29. Roth, J.S., et al., *Keeping it clean: the cell culture quality control experience at the national center for advancing translational sciences*. SLAS Discov., 2020. **25**(5): p. 491–497.
30. Barile, M.F., Hopps, H.E., and Grabowski, M.W., *Incidence and sources of Mycoplasma contamination: A brief review*, in *Mycoplasma Infection of Cell Cultures*, G.J. McGarrity, D.G. Murphy, and W.W. Nichols, Editors. 1978, Springer US: Boston, MA. p. 35–45.
31. Nikfarjam, L. and Farzaneh, P., *Prevention and detection of Mycoplasma contamination in cell culture*. Cell J., 2012. **13**(4): p. 203–212.
32. Capes-Davis, A., et al., *Check your cultures! A list of cross-contaminated or misidentified cell lines*. Int. J. Cancer, 2010. **127**(1): p. 1–8.
33. Loens, K. and Ieven, M., *Mycoplasma pneumoniae: current knowledge on nucleic acid amplification techniques and serological diagnostics*. Front. Microbiol., 2016. **7**: p. 448.
34. Ligasová, A., et al., *A new sensitive method for the detection of mycoplasmas using fluorescence microscopy*. Cells, 2019. **8**(12): p. 1510.
35. Masover, G.K. and Becker, F.A., *Detection of mycoplasmas in cell culture by fluorescence methods*, in *Mycoplasma Protocols*, R. Miles and R. Nicholas, Editors. 1998, Humana Press: Totowa, NJ. p. 217–226.
36. Nims, R.W., et al., *Sensitivity of isoenzyme analysis for the detection of interspecies cell line cross-contamination*. In Vitro Cell Dev. Biol. Anim., 1998. **34**(1): p. 35–39.
37. Ramya, R., et al., *Identification of cross-contaminated animal cells by PCR and isoenzyme analysis*. Cytotechnology, 2009. **61**(3): p. 81–92.
38. Raymond W. Nims, P.a.C.J.H *Cell line authentication using isoenzyme analysis*. 2005 [cited 2025]; Available from: https://www.biopharminternational.com/view/cell-line-authentication-using-isoenzyme-analysis.

39. Dittmar, K.E., et al., *Quality of cell products: authenticity, identity, genomic stability and status of differentiation*. Transfus Med. Hemother., 2010. **37**(2): p. 57–64.

40. Matsuo, Y., Nishizaki, C., and Drexler, H.G., *Efficient DNA fingerprinting method for the identification of cross-culture contamination of cell lines*. Hum. Cell, 1999. **12**(3): p. 149–154.

9 Methods of Laboratory Diagnosis of Viral Infection from Clinical Samples: Virus Cultivation, and Characterization

The diagnosis of viral infections from clinical samples is the cornerstone of effective patient care, outbreak response, and public health surveillance. Common diagnostic methods include antigen and / or antibody detection (serology), molecular techniques such as polymerase chain reaction (PCR), and real-time reverse transcriptase quantitative PCR (RT-qPCR). These techniques are widely used in clinical settings due to their turnaround time, sensitivity, and ability to confirm active or past infection.

In addition to routine diagnostics, advanced techniques such as next-generation sequencing (NGS), immunofluorescence assays, are frequently used in research laboratories and the biotechnology industry. These tools are essential for studying viral pathogenesis, monitoring viral evolution, and supporting the development of vaccines and antiviral therapies. They also contribute to the identification of novel or emerging viral strains, which is crucial for pandemic preparedness and response.

Unlike the above methods that do not necessarily require the isolation or cultivation of viable viral particles, the ability to recover, isolate, and characterize viable viruses or viral products from biological samples is not only important for diagnostics but also for epidemiologic investigations and biopharmaceutical production. This includes the manufacturing of biological products such as vaccines, antiviral drugs, and recombinant proteins.

All newly identified strains or genotypes of viruses isolated in the laboratory must be submitted to the National Virus Bank for extensive characterization and documentation.

As discussed in Chapter 1, viruses are obligate intracellular pathogens and cannot replicate in non-living culture media. Thus, virus cultivation and propagation require living systems, such as cell cultures, embryonated eggs, or laboratory animals, depending on the virus and the intended use of the isolate.

To successfully isolate viable viruses, several factors must be carefully considered:

the type and timing of sample collection, proper transport and storage conditions, and the selection of appropriate host cells or culture systems that support viral replication. These systems must provide an enabling environment with the necessary nutrients, ions, pH, temperature, moisture, and gas composition to support infection and propagation.

Once isolated, viruses can be further characterized through morphological observation (e.g., electron microscopy), cytopathic effect assessment, molecular genotyping, and functional assays to study infectivity, antigenicity, and response to therapeutic agents.

9.1 Selection, Collection, and Transportation of Clinical Samples

Overview

Typically, samples for diagnosis of viral infection, identification, and isolation should be collected at the early onset of illness, preferably after 24 hours and not more than 4–7 days after the onset of symptoms [1]. This is because the viral load and antibody titer are typically highest during this period, which increases the likelihood of successful virus detection and isolation. Collecting samples early in the course of illness also helps ensure accurate diagnosis and timely intervention.

Most viruses are stable in a proper transport medium at 4 °C up to ambient temperature for a maximum of 3 days. Except for virus quantification assays, where proper transportation (temperature) conditions are not available, virus samples not sent to a refrigerator or on wet ice are worthwhile for analysis. Some samples may withstand longer transport times; however, samples should not be exposed to extreme environmental conditions.

9.1.1 Selection of Appropriate Virological Samples and Investigations

To reach the best diagnostic efficiency for each patient, it is important to select an appropriate sample and investigation at the appropriate time. This aspect requires collaboration between laboratory personnel, clinicians, public health officers, and every other person involved in the management of the patient.

9.1.1.1 Laboratory Test Request Form

Information that will accompany a request form should be clear and detailed. The format should be agreed on by all the laboratory users. As much as possible, little information, including the name and type of samples, should be written on the sample's container. All forms should be filled out electronically or with adequate ink. Where possible, the officer collecting the sample should avoid the cancellation of any information. In rural areas where a standard request form may not be available, a simple request form that contains the basic information about the patients and the type of sample should be made locally.

If possible, computerized request forms in the form of a QR or bar code can be attached to sample containers. However, due to the limited internet supply in some areas, it may be necessary to accompany it with a handwritten form (see Tables 9.1 and 9.2).

Table 9.1 Example of basic required information on a request form and sample container.

Laboratory test request form for virus diagnosis	Include this information in the form	Include this information on the sample bottle
Patient information		
Name	Yes	Yes
Date of birth	Yes	Yes
Gender	Yes	No
Patient ID/Medical record number	Yes	Yes
Address (include state, province, county, etc.)	Yes	No
Phone number of the facility (not of the patient)	Yes	No
Email of the facility (not of the patient)	Yes	No
Requesting clinician information		
Name	Yes	No
Contact information (phone/email)	Yes	No
Department/Unit	Yes	No
Address (for outside the facility requests)	Yes	No
Phone number	Yes	No
Requesting clinician signature	Yes	No
Date	Yes	No
Additional notes	Yes	No
Clinical information		
Date of sample collection:	Yes	Yes

(*Continued*)

Table 9.1 (Continued)

Laboratory test request form for virus diagnosis	Include this information in the form	Include this information on the sample bottle
Time of sample collection	Yes	Yes
Type of samples (e.g., blood, swab, urine, etc.)	Yes	Yes
Site of sample collection (e.g., nasopharyngeal, blood, etc.)	Yes	No
Clinical history (brief):	Yes	No
Suspected diagnosis or symptoms	Yes	No
Date of onset of symptoms	Yes	No
Recent travel history	Yes	No
Contact with known cases.	Yes	No
Vaccination status (if relevant)	Yes	No
Current medications:	Yes	No
Allergies:	Yes	No
Test requested		
Specify tests (e.g., PCR, ELISA, viral culture, etc.)	Yes	Yes
Priority status (routine/urgent/routine/STAT)	Yes	Yes
Laboratory information		
Personnel receiving the sample	Yes	No
Address of the lab	Yes	No
Special instructions	Yes	No
Billing information	Yes	No
Insurance provider	Yes	No
Policy number	Yes	No
Authorization	Yes	No

Table 9.2 Recommended procedure for collecting samples and the conditions for their transport.

Sample	Procedures
Blood	**Materials required:** EDTA bottle for anti-coagulated blood samples (plasma). Plain bottle for serum. **Patient preparation and sample collection:** Ask the patient to make a fist. Clean the venipuncture site with 70% isopropyl alcohol or tincture of iodine in swab. Do not palpate the vein at this point. Once a major vein is visible, gently insert the needle through the skin and collect 5–10 mL. After venipuncture, place gauze immediately on the puncture site and apply adequate pressure. **Handling and transport:** No transport medium is required. Send the sample to the lab at 4 °C or up to ambient temperature for a maximum of 3 days. Serum or plasma samples can be separated and transported at −20 °C. **Remarks:** For safety purposes, use vacutainer bottles, and the bottle should not be opened. If a non-vacutainer bottle is used, it should be tightly capped after sample collection.
Faces	**Materials required:** Sterile, leakproof, wide-mouth container. **Patient preparation and sample collection:** The sample should be passed directly into the bottle (freshly passed stool). 1–3 mL liquid or 1–3 g solid (pea-size). **Handling and transport:** No transport medium is required. Send the sample to the lab at 4 °C, a maximum of 3 days. **Remarks:** For infants and some children, use a sterile spatula (made of synthetic fiber) to scoop the fecal sample into the sample container.
Rectal swab	**Materials required:** Swab stick, virus transport medium (VTM). **Patient preparation and sample collection:** Ask the patient to squat or lift one leg onto a raised surface. Partially open the swab package. Do not touch the soft tip of the swab with your hands or other surfaces. Moist in sterile saline. Insert the soft tip about 3–4 cm into the anus and gently rotate for about 10 seconds. Take the swab out and put it in the tube. Break the swab at the score line while placing it inside the VTM. **Remarks:** Send the sample (in a transport medium) to the lab at 4 °C, a maximum of 3 days.
Nasopharynx swab	**Materials required:** Swab stick, VTM. **Patient preparation and sample collection:** Insert the soft tip of the swab into the nostril (not more than 1.5 cm), closer to the floor of the nasal canal. Slowly rotate the swab and allow secretions to absorb for 15 seconds; then carefully remove the swab. Break the swab at the score line while placing it inside the VTM. Repeat for another nostril using a fresh swab. Place both swabs in the same transport tube. **Remarks:** Send the sample (in a transport medium) to the lab at 4 °C, a maximum of 3 days.

(*Continued*)

Table 9.2 (Continued)

Sample	Procedures
Oral and throat swab	**Materials required:** Swab stick, VTM. **Patient preparation and sample collection:** For an oral visible wound, firmly sample the base of the lesion(s) with the soft tip of the swab and carefully remove it. For a throat swab, depress the tongue (using a tongue depressor) to prevent contamination with saliva. Rub the soft tip of the swab on the surface of one or both tonsils, the tonsillar pillars, and the posterior pharyngeal wall, and place it in the VTM. **Handling and transport:** Send the sample (in a transport medium) to the lab at 4 °C, a maximum of 3 days.
Urethral swab	**Materials required:** Swab stick **Patient preparation and sample collection:** The patient should not have urinated at least 2 hours before collection. Gently insert the soft tip of the cotton swab about 4 cm into the urethra. Turn and rotate the swab 2–3 times and remove the swab. Break the swab at the score line while placing it inside the VTM. **Handling and transport:** Send the sample (in a transport medium) to the lab at 4 °C, a maximum of 3 days.
Urine	**Materials required:** Sterile, leakproof, wide-mouth container. **Patient preparation and sample collection:** 5 mL of midstream clean, voided urine in a sterile container. No transport medium is required. The sample should be sent to the lab at 4 °C, a maximum of 3 days.
Cervical, vaginal, or penile swab	**Materials required:** Swab stick **Patient preparation and sample collection:** Uses sterile gauze to clean the site of the sample collection if lesions are present. If lesions are not present, remove mucus from the cervix with a swab and discard. Gently insert the soft tip of the swab into the opening of the penis, endocervix, or vagina and turn and rotate the swab 2–3 times, and remove the swab. Break the swab at the score line while placing it inside the VTM. **Handling and transport:** Send the sample (in a transport medium) to the lab at 4 °C, a maximum of 3 days.
Cerebrospinal fluid (CSF)	**Material: Sterile screw-cap tube.** **Patient preparation and sample collection:** Follow the facility's standard procedure for the collection of cerebrospinal fluid. **Handling and transport:** Send the sample to the lab at 4 °C, preferably on the same day.

Table 9.2 (Continued)

Sample	Procedures
Vesicular fluid or rash	**Material:** Swab stick and VTM **Patient preparation and sample collection:** Clean the site of sample collection with sterile saline. Gently open the fresh vesicle/plaque with a needle or scalpel blade. Collect fluid and cellular material by sampling the base of the lesion with the soft tip of the swab. Break the swab at the score line while placing it inside the VTM. **Handling and transport:** Send the sample (in a transport medium) to the lab at 4 °C, a maximum of 3 days. **Remark:** Older crusted or ruptured vesicles/blisters/plaques may not contain viable virus.
Dried blood spot	**Material:** Dried blood spot cards **Patient preparation and sample collection:** Clean the site of the sample collection with 70% isopropyl alcohol or tincture of iodine in swab. Use the lancet to apply a finger prick. For newborn babies or infants, the recommended location for blood collection is the heel-stick. Allow two drops of blood to fall freely on the filter paper of the blood spot card. Let the blood spot card dry. Afterward, keep the card in a dry plastic container at ambient temperature. **Remarks:** A high-quality blood spot is round and fills at least the whole circle. Do not allow the two drops to touch each other. Dried blood spot testing for HIV is not considered sensitive enough.
Conjunctiva swab/drainage	**Material:** Swab stick and VTM **Patient preparation and sample collection:** Swab (tip moistened with sterile saline) firmly along the lower, rotating the swab to collect any discharge. Break the swab at the score line while placing it inside the VTM. **Handling and transport:** Send the sample (in a transport medium) to the lab at 4 °C, a maximum of 3 days.
Tissue biopsy	Tissue biopsy should strictly follow clinical procedures. Place the tissue inside the VTM. If rabies is suspected and brain samples are collected, freeze unfixed samples immediately after collection. Formalin-fixed samples are also useful and may be transported at ambient temperature. **Handling and transport:** Send the sample (in a transport medium) to the lab at 4 °C, a maximum of 3 days.

Where possible, it is necessary to contact the laboratory staff before collecting and sending the samples for special studies.

Label the sample bottles or containers before collecting the samples

To avoid further spread of the infection and to maintain the patient's confidentiality, ONLY trained personnel should handle both the samples and the request form.

While designing the request form, include the phone numbers of the laboratory. If necessary, a catalog containing the price of each test should be made available to all laboratory users.

Samples that are not labeled or properly labeled, dry-out samples (where not applicable), not collected in the appropriate containers, hemolyzed samples, and poor sample volume should not be accepted for virological investigations.

Samples required for immediate patient treatment and management or of public health importance should be labeled "urgent". Where possible, a telephone call should be placed to the laboratory by the requesting officer.

All samples from confirmed or suspected cases of viral hemorrhagic fever or any other infection of public health importance must be discussed with the laboratory personnel for risk assessment before collection. Healthcare providers can contact their local health department and the Centers for Disease Control to organize a risk assessment and determine if testing is necessary.

9.1.1.2 General Principle of Sample Selection

Once a viral infection or outbreak is suspected, reported, or detected, the laboratory users and personnel should be contacted before the team or the physician collects the sample (see Table 9.4). General principles for the selection and collection of samples include:

- Take samples that are representative of the disease process. For example, when respiratory disease is suspected, respiratory samples are more appropriate than blood samples, although blood samples can be used for serological studies or diagnosis of the disease.
- Identifying the optimal timing of collection for the different types of samples and the type of sample required, as this is essential for the successful isolation of viable viruses. For example, samples that are not suitable for molecular detection of a virus may not be appropriate for virus isolation.

- Samples from the actual site of the suspected infection or the disease process should be collected where possible.
- One or more sample types may be required from a patient to identify the cause of a disease or outbreak.
- An adequate quantity of samples should be collected, especially for samples like vesicle fluid and cerebrospinal fluid (CSF).
- The sample collection procedure must be the best possible service to patients. Procedures with a high risk of complications should be reconsidered if there is an alternative to this.
- Samples obtained in the acute phase of the disease before administration of antimicrobial drugs are preferred.
- Consider the logistical requirements for sampling, storage, and transport equipment.
- Personnel collecting the sample must be trained for the collection, handling, and transport of the required sample.
- When in doubt, key contact laboratory personnel must be consulted for advice on the selection of sample type and collection materials or container(s).
- During outbreaks, many samples might be collected for screening; hence, the laboratory disposal and reagents may be overwhelmed. It is therefore essential to determine whether a supply of specialized media, equipment, and reagents is available and whether decontamination procedures for such samples are established in advance before sample collection.

9.1.1.3 Sample Collection

A standard operating procedure, in the form of a written instruction, and a request form should accompany all the sample containers/materials that will be given out. Ensure that the procedure is properly explained to the patient.

> *If VTM is absent, swab can be placed in sterile saline. Swabs with wooden shafts, calcium alginate swabs, or swabs with wooden shafts are not advisable. These swabs can contain substances that inactivate some viruses or inhibit molecular tests.*

9.1.1.4 Sample Transportation and Cold Chain Logistics

The World Health Organization (WHO) recommends that commercially produced, United Nations (UN)-approved packaging systems be used for the transportation of all biological materials [2]. Consult relevant information. When this is not available, the following can be considered (see Table 9.3):

Transportation of Samples to be Maintained at 4–8 °C Fit an insulated sample transport box with four or more ice packs (the number may vary according to the season) around another (secondary) container that contains the sample. This method is capable of maintaining the sample at refrigeration temperature for 2–3 days.

Transportation of Samples to be Maintained at −20 °C Pack about 2 kg of dry ice into a large, insulated sample transport

Table 9.3 Summary of the transportation and classification of infectious substances.

Feature	UN 3373	UN 2814	UN 2900	UN 3291
Full name	Biological substance, category B	Infectious substance, affecting humans	Infectious substance, affecting animals	Clinical/biomedical waste
Category	Category B	Category A	Category A	Not assigned (but may contain A/B)
Risk level	Moderate	High (to humans)	High (to animals)	Variable
Purpose	Diagnostic or clinical specimens	Serious human pathogens (culture and isolates)	Serious animal pathogens (culture and isolate)	Waste from treatment, diagnosis, or research
Examples	Nasal swabs, blood for testing	Ebola, SARS-CoV-2	Foot-and-mouth disease virus	Used PPE, discarded cultures, and sharps
Label	UN 3373 diamond	Class 6.2 + UN 2814	Class 6.2 + UN 2900	Class 6.2 + UN 3291
Packaging	Triple, absorbent, tested	Triple, high standard	Triple, high standard	Leakproof, rigid containers (e.g., UN-approved waste drums or boxes)
Shipper's declaration	Required	Required	Required	Required if by air
Training required	Basic awareness	Dangerous Goods (Class 6.2)	Dangerous Goods (Class 6.2)	Waste transport/ Dangerous Goods
Maximum quantity	≤4 L or 4 kg	Limited by type	Limited by type	Varies by mode of transport
Shipping purpose	Diagnosis or research	Culture/shipping of high-risk pathogens	Veterinary pathogens	Waste disposal (e.g., to an incinerator)

Source: Adapted from Ref. [5].

box (that can permit the release of carbon dioxide gas) around the sample. Dry ice is available as pellets or blocks. Samples should be frozen before packaging. This method can maintain the sample at −20 °C for 1–2 days.

Transportation of Samples to be Maintained Below at −70 °C This involves the use of dry ice or liquid nitrogen. Availability and/or requirements governing the use (at facilities of the laboratory sending and receiving the sample) and transportation (by logistic companies) of liquid nitrogen must be taken into consideration. When liquid nitrogen is used, cryotubes are required for the storage and transport of samples. This method is capable of maintaining the sample for a longer time, depending on the volume and the number of samples.

In tropical areas or where the temperature is extreme, it is important to pack the sample transport box with more cooling materials. Re-ice the box, if needed, and important.

9.1.1.5 Packaging of Sample

Primary Container for Transporting Virus Samples The primary container for transporting virus samples must be leak-proof (for liquid samples) or a shift-proof (for solid samples), waterproof, and impermeable container that is in direct contact with the sample. Usually, it should be a hollow container, preferably capable of entering a secondary container, and must be able to withstand a wide range of temperatures and pressures (see Figures 9.1 and 9.2).

Some samples may require being wrapped with absorbent material capable of absorbing liquid in case of leakage or damage to the primary container.

Secondary Container The secondary container, as well, should be leak-proof, shock-proof, and waterproof. It must be large enough to enclose and protect the primary container. Several primary containers can be placed individually wrapped in a secondary container if they have similar transportation conditions.

Tertiary or Outer Container This is the outer container. It must have appropriate strength and space to bear the first and second containers while the package is in transit.

Packing sample with coolant: *When packing a sample that requires refrigeration, coolant must be placed between the secondary and outer packaging.*

Reuse of packing containers: *Outer packaging containers can be reused. Whenever a packaging container must be reused, it should be disinfected or sterilized, whether there is a visible stain or not. Ensure that the previous label is cleaned or removed. Packaging material must not be damaged or weak in strength.*

Marking: *Mark all the containers legibly with the shipper's name, receiver's name, and the UN number of infectious substances. For example,* ***CATEGORY A*** *is for an infectious substance capable of affecting humans and animals. It should be marked as UN2814 or UN2900 for only animals.* ***CATEGORY B:*** *Biological substances should be marked UN 3373 (Table 9.4).*

Labeling: *The nature of the sample must be appropriately labeled.*

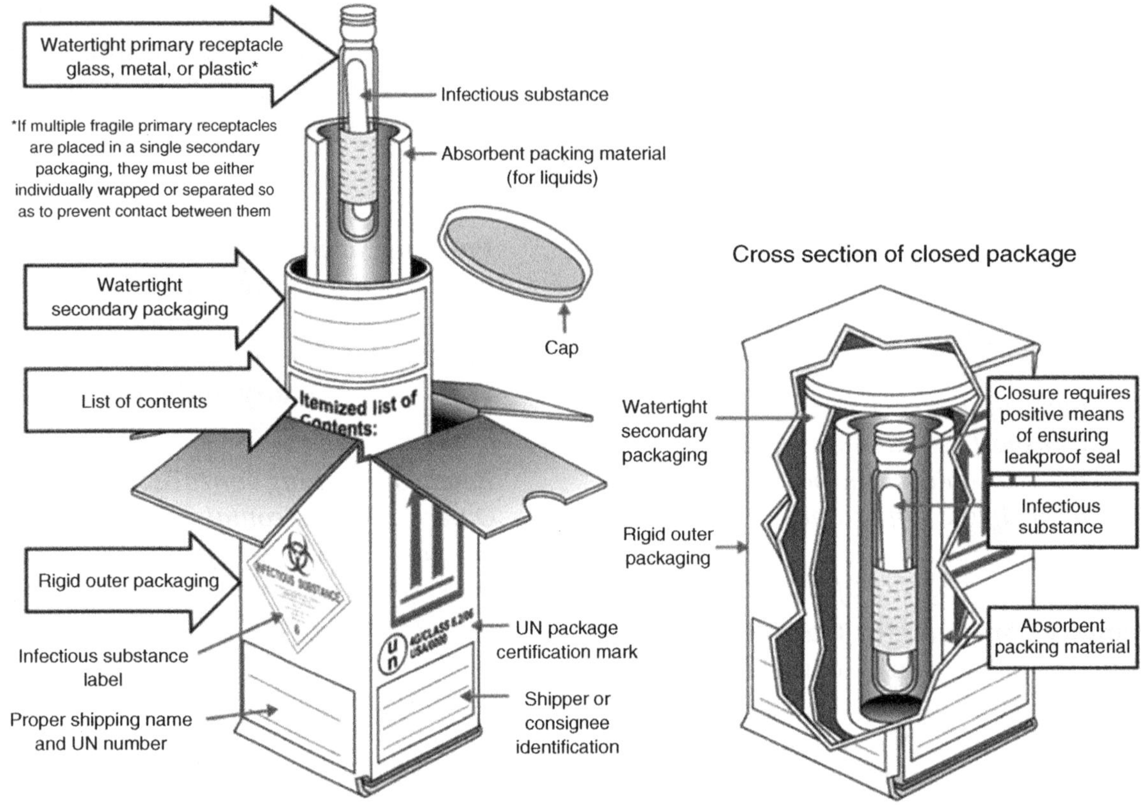

Figure 9.1 Packing and shipping category A clinical specimens.
Source: Ref. [3]/ United States Centers for Disease Control and Prevention/Public Domain.

9.1.1.6 Cold Chain Logistics

Locally trained individuals or biological courier service providers who specialize in the delivery of urgent shipments of medical samples with specific temperature requirements should be used. For international shipment, the International Air Transport Association (IATA) regulations and instructions should be reviewed for any restrictions [4, 5].

Recommendations when considering cold logistics

- **Cost:** Where necessary, review the cost of transportation with the patient. Shipping biological samples for domestic and international delivery can be significantly more expensive than a standard courier.
- **Transiting time:** International time zone differences, weekends, and national and local holidays should be planned to minimize transit time.
- **Validation of protocol:** If shipping samples like cell lines, whole blood, or serum, the laboratory may consider checking the viability of cells or other parameters (at different transport conditions) locally to determine the proper shipping condition of the sample.

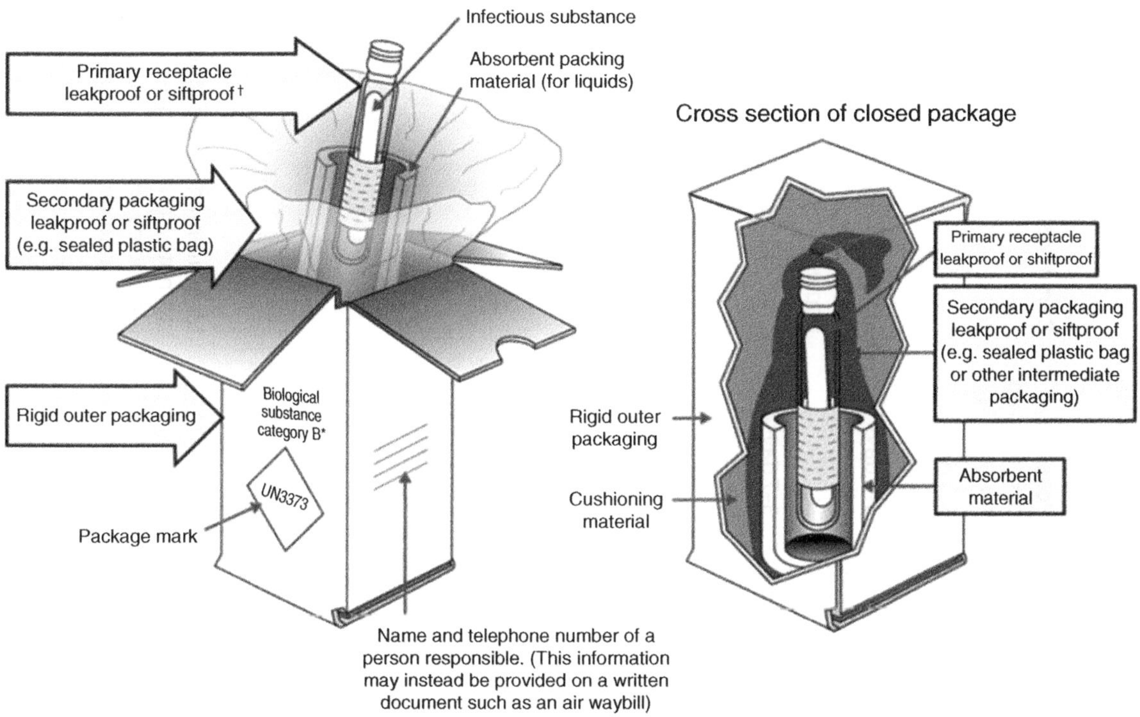

Figure 9.2 Packing and shipping category B clinical specimens.
Source: Ref. [3]/United States Centers for Disease Control and Prevention/Public Domain.

- **Accompanying materials:** Request forms or other paperwork that contain sample type, size, color, and weight of the sample transport box, type of sampling bottle (e.g., cryotube), and expected day and time of delivery (where applicable) should be recorded to determine consistency.

9.1.1.7 *Recording Receipt of Specimens*

Upon receipt of a specimen, the laboratory should meticulously document the following additional information to ensure proper tracking, handling, and analysis:

- **Specimen identification:**
  - Unique identifier or accession number.
  - Patient or subject identification details (e.g., name, date of birth, medical record number).
  - Date and time of receipt. Record the exact date and time the specimen was received in the laboratory.
  - Type of specimen: Specify or confirm the nature of the specimen (e.g., blood, urine, tissue, swab).
  - Source of specimen: Indicate the anatomical site or body fluid source from which the specimen was collected.

Table 9.4 Types of virological investigations, samples needed, and possible turnaround time.

Serology Virus	Test	Sample(s)	Possible turnaround time (TAT)	Remark(s)
Hepatitis A virus	Immunoglobulin M (IgM) and IgG antibodies	Clotted whole blood (10 mL)	3–7 days	
Hepatitis B virus	Hepatitis B total antigen (envelope, core)	Clotted whole blood (10 mL) or dried blood spot	3–7 days	Small blood samples are blotted on an absorbent paper and allowed to dry for analysis. Suitable for infants
	Hepatitis B antibody (envelope, core, and surface)	Clotted whole blood (10 mL) or dried blood spot	3–7 days	
	Hepatitis B antigen confirmation by neutralization (surface and envelope)	Clotted whole blood (10 mL)	3–7 days	Confirmatory assay uses the principle of specific antibody neutralization to confirm the presence of HBsAg
Hepatitis C virus	Hepatitis C virus antibody/antigen	Clotted whole blood (10 mL) or dried blood spot	3–7 days	
Parvovirus B19 Rubella virus Mumps virus Measles virus Human enterovirus	IgM and IgG antibodies	Clotted whole blood (10 mL)	3–7 days	
Cytomegalovirus	IgG avidity assays	Clotted whole blood (10 mL)		Measurement of the binding strength between IgG antibodies and the virus. Important during pregnancy

Varicella zoster virus	IgG antibody (quantitative assay)	Clotted whole blood (10 mL)	3–7 days	A positive IgG result coupled with a negative IgM result indicates previous vaccination
	IgM antibody (qualitative)	Clotted whole blood (10 mL)		
HIV-1 and 2	HIV-1 and 2 antibody/antigen	Clotted whole blood (10 mL), dried blood spots, or saliva	1–3 days	
Flaviviruses	Serology (IgM/IgG ELISA)	Blood, serum	2–7 days	IgM indicates recent infection; IgG indicates past exposure. Cross-reactivity with other flaviviruses may occur
Gastrointestinal tract viruses – norovirus, enterovirus, rotavirus	Rotavirus: Enzyme immunoassay (EIA) or latex agglutination tests. Norovirus: Some rapid immunochromatographic tests exist, but they are less sensitive than PCR. Enterovirus: Less commonly diagnosed using antigen tests: PCR is preferred.	Stool samples, whole blood, or CSF (for enteroviruses in systemic infections)	1–3 days	
Other serological tests that can be requested				
Premarital screening – STI test: HIV, hepatitis B, hepatitis C virus	Hepatitis surface antigen, Hepatitis C antigen, and HIV 1 and 2 antibodies	Non-coagulated whole blood (10 mL)	1–3 days.	HIV antibody confirmation may be done

(*Continued*)

Table 9.4 (Continued)

Virus	Test	Sample(s)	Possible turnaround time (TAT)	Remark(s)
Serology				
TORCH screening – STI test: toxoplasmosis, rubella, cytomegalovirus, herpes simplex, and HIV		Non-coagulated whole blood (10 mL)	1–3 days	Tests for CMV, rubella virus, and parvovirus B19 are typically carried out in endemic areas or suspected cases
Viral screen before transplantation	HCV antibody, CMV IgG, EBV IgG, treponemal IgG, Toxoplasma IgG, HTLV 1 and 2 IgG, VZV IgG, HB surface antigen, HB core total antibody, HIV 1 and 2 antibody / antigen	Non-coagulated whole blood (10 mL)	3–7 days	
Nucleic acid amplification tests (NAAT)				
HIV-1 and 2	RNA qualitative assay	10 mL EDTA whole blood. Other samples like genital swabs, endocervical swabs, or urine in universal bottles; urethral swabs; eye/conjunctival/corneal swabs; and CSF. Dried blood spot cards for infants	1–3 days	
	RNA quantitative assay	10 mL EDTA whole blood	2–3 days	Viral load measurement is primarily used for the management/monitoring of HIV-infected individuals

	HIV-1 proviral DNA –qualitative		3–7 days	It is used as a screening test to identify HIV-infected individuals.
	HIV-1 C-C chemokine receptor type 5/C-X-C chemokine receptor type 4 (CCR5/CXCR4) tropism assay		Rarely carried out in most labs. TAT varies depending on the availability of reagents	Used to determine the patient's HIV-1 strain for selection of CCR5 co-receptor antagonist therapy
HIV 1 and 2	HIV-1 and 2 antiretroviral resistance test			Used to monitor the efficacy of an antiretroviral treatment regimen
HVB	HBV DNA – quantitative and qualitative		3–7 days	
HBV	HBV DNA antiviral drug resistance		Rarely carried out in most labs. TAT varies depending on the availability of reagents	Used to monitor the efficacy of an antiretroviral treatment regimen
HCV	HCV RNA – quantitative	10 mL EDTA or Clotted whole blood (10 mL)	3–7 days	
	RNA viral – qualitative	10 mL EDTA, clotted whole blood (10 mL), dried blood spot		
Respiratory viruses – influenza viruses, parainfluenza viruses, rhinovirus, respiratory syncytial virus (RSV), adenovirus, and coronaviruses.	RNA qualitative and quantitative	10 mL EDTA blood Broncho-alveolar lavage, bronchial lavage, nasopharyngeal aspirate Nose and throat swab and Endotracheal aspirate	3–7 days	Swabs should be kept moist in a viral transport medium

(*Continued*)

Table 9.4 (Continued)

Serology Virus	Test	Sample(s)	Possible turnaround time (TAT)	Remark(s)
Adenovirus	Adenovirus DNA (qualitative assay)	Eye/conjunctival/corneal swab, CSF, urine, blood, feces (stool), rectal swab, bronchoalveolar lavage, bronchial lavage, nose, and throat swab, nasopharyngeal aspirate, endotracheal aspirate	3–7 days	The appropriate sample type depends on the suspected infection
Gastrointestinal tract viruses – norovirus, enteroviruses, rotavirus.	RT-PCR is the gold standard	Feces (stool) and rectal swab	3–7 days	
Flavivirus and alphaviruses	RT-PCR (reverse transcription polymerase chain reaction)	Blood, serum, CSF	24–72 hours	Highly sensitive and specific; early detection is crucial. Useful for detecting viral RNA during the acute phase of infection

Note: Serum specimen not analyzed within 24 hours should be frozen to maximize virus RNA recovery and serological test.

- **Clinical information:**
  - History: Relevant clinical history or diagnosis associated with the specimen.
  - Reason for testing: Routine screening, diagnostic investigation, research.
  - Specimen condition: Note the physical condition of the specimen upon receipt (e.g., intact, hemolyzed, contaminated).
  - Handling instructions: Record any special handling instructions or requirements for the specimen (e.g., temperature requirements, light sensitivity).
  - Collection date and time: Document the date and time when the specimen was collected to assess stability and appropriate processing time.
  - Collector Information: Identify the person or entity that collected the specimen (e.g., healthcare provider, facility).
  - Transportation conditions: Describe the conditions under which the specimen was transported or received (e.g., ambient, refrigerated, frozen).
- **Testing requested:** Specify the tests requested for the specimen analysis.
- **Quality control information:** Note any quality control measures or checks performed on the specimen before receipt.
- **Chain of custody:** If applicable, maintain a record of the chain of custody for the specimen, especially for forensic or legal cases.

Storing samples until they are processed
In general, for reliable detection of viruses (especially virus isolations), samples sent to the virology lab should be processed immediately. If it is not possible, samples should be properly stored following general recommendations. All samples collected in the virology laboratory sent through long-term transportation above 24 hours should be analyzed immediately. As explained previously, this is why a prior agreement must be made before receiving such samples.

9.2 Methods for the Recovery of Virus Particles from Samples

Once the appropriate samples arrive at the laboratory, virus extraction and recovery from samples are crucial steps in the medical laboratory diagnostic process. These techniques enable proper identification, quantification, and characterization of viral pathogens, which is essential for diagnosing viral infections. However, recovery of viable viruses may be difficult in some situations.

Always use instruments with closed systems that do not require any decontamination between patient specimens when analyzing biological samples. Optimize the procedure for each sample.

The following may be considered during virus recovery.

9.2.1 Virus Recovery from Swab Samples (e.g., Respiratory Secretions, Anal, Oral)

a. Record details of the sample in the sample database or book, including date, time, and patient ID if applicable.
b. Follow equipment-specific instructions and ensure all materials are sterile.
c. Transfer the swab into a sterile centrifuge tube or bottle with buffer or protein/amino acid-based solution, like Dulbecco's Modified Eagle Medium (DMEM) without serum, containing appropriate antibiotics (penicillin, gentamicin, or amphotericin B), and leave it on the bench at room temperature for one hour. The choice of antibiotic will be determined by the nature of the sample and the test required.
d. Vortex the sample at low speeds for one minute and allow the sample to stand further at room temperature for an additional 2–5 minutes for the materials to detach from the swab. If the sample bottle contained pooled samples (two or more samples from the same patient or more), remove the swab and treat each sample separately.
e. Clarify the homogenate by centrifugation at 3000 × g for 10 minutes at 4 °C.
f. Collect the supernatant. Filter the supernatant using a 0.2 μm filter if needed. Note that filtration may reduce the number of virus particles.
g. Store the supernatant at 4 °C for short-term use or at −70 °C for long-term storage until ready for inoculation onto cell cultures.

9.2.2 Virus Recovery from Unfixed Frozen Tissue Samples (Biopsy)

a. Record details of the sample in the sample database software or book.
b. Follow equipment-specific instructions.
c. Transfer the tissue into a centrifuge tube/bottle with buffer or protein-based solution (like DMEM without serum) containing antibiotics to make an approximate 10–20% (w/v) tissue (lung, lymph nodes, spleen). Keep adding liquid nitrogen when required.
   I. Place 0.5 g of tissue in 5 mL of phosphate-buffered Saline (PBS) or DMEM-containing antibiotics and homogenize using an automatic homogenizer if available. Followed by repeated freezing and thawing three times.
   II. Brain homogenates can be prepared in PBS containing antibiotics by repeatedly passing the brain tissue through gauge needles, followed by repeated freeze–thawing three times.
d. Clarify the homogenate by centrifugation at 3000 × g for 10 minutes at 4 °C.
e. Collect the supernatant. The sample can be further filtered (using a 0.2 μm filter). Filtering samples may decrease the number of virus particles.
f. Store supernatant at 4 °C for a shorter period, ready for inoculation onto cell cultures. Alternatively, store supernatant at −70° for a longer period.

9.2.3 Virus Recovery from Blood Samples

a. Record details of the sample in the sample database software or book, including date, time, and sample ID.
b. Follow equipment-specific instructions and ensure all materials are sterile.
c. Collect blood using appropriate anticoagulants (e.g., ethylenediaminetetraacetic acid [EDTA], heparin or plasma separator bottle) or without anticoagulants (plain bottle or serum separator bottle). Centrifuge to separate plasma or serum.
d. For plasma/serum, no additional homogenization is required. If using whole blood, ensure thorough mixing if necessary.
e. Clarify plasma/serum by centrifugation at 3000 g for 10 minutes at 4 °C if needed.
f. Collect the supernatant. Filter if necessary to remove any particulate matter.
g. Store plasma/serum at 4 °C for short-term use or at −70 °C for long-term storage.

9.2.4 Virus Recovery of Urine Samples

a. Record details of the sample in database software or a book, including date, time, and sample ID.
b. Follow equipment-specific instructions and ensure all materials are sterile.
c. Collect urine in sterile containers. Centrifuge to remove debris and collect the supernatant.
d. Homogenization is generally not required; filter if there is particulate matter.
e. Clarify the supernatant by centrifugation if needed.
f. Collect the supernatant. Filter if necessary to remove any particulate matter.
g. Store urine at 4 °C for short-term use or at −70 °C for long-term storage.

Challenges in virus recovery from urine

Urine is not a common specimen for diagnosis of viral infections. Several viruses (e.g. BK virus, JC virus) can be isolated from urine, especially those that are shed from the kidney or infect the urinary tract. Of note, urine specimens can be collected for PCR confirmatory test for congenital cytomegalovirus (CMV) infection in infants less than 21 days. However, several challenges can affect the efficient recovery of viruses from urine samples. In some infections, the viral load present in urine may be variable, making detection difficult. Additionally, urine contains RNases and DNases that can degrade viral nucleic acids, thereby compromising molecular detection methods of some viruses. The presence of PCR inhibitors in urine further reduces assay sensitivity and may lead to false-negative results. Moreover, variations in urine pH and salt concentration can impact virus stability, posing further obstacles during sample handling and processing.

9.2.5 Virus Recovery from Environmental Samples (e.g., Water, Air Filters)

a. Record details of the sample in the sample database software or book, including date, time, and sample ID.
b. Follow equipment-specific instructions and ensure all materials are sterile.
c. For water or sewage, concentrate viruses using methods like ultrafiltration or centrifugation. For air filters, elute viruses in a suitable buffer.
d. Homogenize if the sample contains particulate matter.
e. Clarify the homogenate by centrifugation at 3000 g for 10 minutes at 4 °C if needed.
f. Collect the supernatant. Filter if necessary to remove any particulate matter.
g. Store at 4 °C for short-term use or at −70 °C for long-term storage.

9.2.6 Virus Recovery from Stool Samples

a. Record details of the sample in the sample database software or book, including date, time, and sample ID.
b. Follow equipment-specific instructions and ensure all materials are sterile.
c. Collect approximately 1–2 g of stool or 1–2 mL of watery stool. Store at 4 °C.
d. Dilute the sample by adding 5–10 mL of sterile PBS (or saline) to the stool sample in a 15 mL or 50 mL conical tube. The stool should be suspended at a ratio of approximately 1:5 to 1:10 (w/v) stool to buffer.
e. Homogenize the stool by vortexing the tube vigorously in the buffer. If the stool is too solid, allow it to sit for 10–15 minutes to soften in the buffer.
f. Centrifuge the suspension at 3000 × g for 10–15 minutes at 4 °C to pellet large debris and bacteria. Carefully transfer the clear supernatant to a new sterile tube, avoiding disturbing the pellet.

Optional step for increasing viral yield: Concentrate the virus in the clarified supernatant by centrifuging the supernatant at 10,000–12,000 × g for 30–60 minutes at 4 °C to pellet viral particles.

Discard the supernatant and resuspend the pellet in a smaller volume of sterile PBS (1–2 mL).

a. Pass the supernatant through a 0.22 μm or 0.45 μm filter to remove bacteria and larger particles. Use a sterile syringe filter for small volumes.
b. For immediate use, store the virus concentrate at 4 °C for short-term use (within 24–48 hours). For long-term storage, aliquot the virus concentrates and store at −80 °C.

9.3 Virus Isolation or Propagation

Isolation and propagation of viral particles are crucial steps in virology for studying virus characterization, vaccine development, and drug screening. The process generally involves culturing viruses in suitable host systems, which include inoculation into embryonated eggs, cell culture, and animal inoculation.

9.3.1 Virus Propagation in an Embryonated Egg

Virus propagation in embryonated eggs is a traditional and effective method for isolating and amplifying viruses, particularly those that infect avian species (Table 9.5). It is the earliest documented means of isolation and identification of animal and human viruses. To date, it is still a primary convenient method for the isolation of some viruses and the development of some vaccines. This method is widely used for growing influenza viruses, some

Table 9.5 Advantages and disadvantages of the use of embryonated eggs for virus culture.

Advantages of virus propagation in embryonated eggs	Disadvantages of virus propagation in embryonated eggs
It is readily available, easily manipulated, and maintained.	Some birds are considered the natural reservoir for some pathogens, e.g., *Salmonella* and *Mycoplasma*. When these pathogens cross from infected birds to the eggs, they expose the growth environment for the virus to infect.
It is convenient and offers an economical source for propagating and testing viruses in living tissue.	Antibodies from vaccinated birds may cross into the yolk, which may interfere with the growth of some viruses.
It is relatively free from bacterial and many viral infections often latent in birds and mammals.	The age of the embryo and the site of inoculation vary with different viruses. Therefore, there is a need to validate growth and replication using different sites for a new strain of virus.
It is free from natural immune response, both specific and non-specific, capable of eliminating inoculated viruses.	Like other methods of virus cultivation, not all viruses grow in embryonated eggs, and not all eggs may be used for virus isolation.
It provides a naturally sterile environment for the propagation of viruses.	
It can grow many virus types because the cells are highly undifferentiated.	
It has different inoculating sites, which allow for the possibility of examining different characteristic lesions that may resemble those in the natural host. For example, the vaccinia group can form typical skin eruptions, and pocks or inclusion bodies appear when inoculated into the embryo and chorioallantoic membrane, respectively.	

arboviruses, and other viruses that have a preference for avian hosts.

List of viruses that can be propagated in a chicken embryonated egg

1. Influenza viruses – Influenza A virus (includes various subtypes such as H1N1, H3N2, H5N1, and H7N9) and influenza B Virus

   Site of inoculation: Allantoic cavity

   Method: Inoculate the virus directly into the allantoic cavity, where it replicates in the fluid surrounding the embryo.
2. Poxviruses – Smallpox virus (variola virus), though now eradicated. Historically used for the propagation of the vaccinia virus, a virus used as a smallpox vaccine.

   Site of inoculation: Chorioallantoic membrane

   Method: Inoculate the virus onto the chorioallantoic membrane, which provides a rich vascular bed for viral replication.
3. Arboviruses (arthropod-borne viruses) – Yellow fever virus (used in the production of the yellow fever vaccine).

   Site of inoculation: Allantoic cavity (e.g., yellow fever virus)

   Method: Like influenza, the virus is injected into the allantoic cavity.
4. Rabies virus (adapted strain): Certain strains can be propagated in embryonated eggs, although cell culture methods are more commonly used today.

   Site of inoculation (fertile 1-day-old hen's egg): Brain tissue, yolk sac or allantoic cavity

   Method: While traditionally propagated in brain tissue, some adapted strains of rabies virus can be grown in the allantoic cavity.
5. Other viruses

   Mumps virus: Historically propagated in eggs for vaccine production. Live attenuated vaccines against mumps virus have been traditionally produced by passing the virus through embryonated chicken eggs.

   Measles virus: Also, historically been propagated in eggs for vaccine development.

   Herpes simplex virus: Certain strains of herpes simplex virus (HSV)-1 and 2 can be propagated in embryonated eggs, though cell culture is more common. Chorioallantoic membrane infection with HSV-1 and HSV-2, and also chick embryo fibroblast (CEF) [6, 7], have been demonstrated.

9.3.1.1 Facility Required for Embryonated Egg Virus Culture

As previously noted, all the procedures involved in the handling of viruses should be done under high safety standards conditions. Areas designated for this process should be constantly kept clean and accessible to strictly laboratory personnel involved in the procedure.

Ensure that all laboratory staff involved in this procedure are familiar with and have adequate skills. They should have experience in handling and propagation of the virus using embryonated eggs. Properly train newly employed laboratory technologists or scientists before the commencement of the analysis.

Laboratory Space Required for Embryonated Egg Virus Culture A separate, well-ventilated darkened room capable of accommodating laboratory personnel during candling, and essential equipment must be provided. If this workspace is not available, install a black photographic curtain to divide in part the routine virus culture room. Ideally, sufficient space should be provided.

Two or more different working spaces or rooms should be built. To avoid the possibility of cross-contamination, in no instance should virus inoculation, incubation, and harvesting be done for different samples in the same safety cabinet, even if they are suspected to have the same virus.

Inoculating space must be provided with a suitable power source for the equipment needed to carry out the inoculation and incubation of eggs.

Basic Equipment Required for Embryonated Egg Virus Culture

- Biosafety cabinet, depending on the biosafety level class of the virus.
- Egg driller for making shell pores. If not available, sterile needles, forceps, or a rubber cork can serve as alternatives to gently dissect through the shell membrane.
- Incubator that can maintain temperature at 37 °C ± 2.
- Needles, sterile and Luer Lock syringes for inoculation.
- Candle lamp.

9.3.1.2 Methods of Embryonated Egg Virus Culture

Selection of Embryonated Eggs An embryonated egg is commonly referred to as an egg whose embryo is viable and developing. This is not a factor of the time the egg was laid. Although fertilization of the eggs occurs before they are laid, during this period, some early primitive developmental stages occur. Once the egg is laid, this development discontinues until the eggs are incubated by the bird; this is called "settling down." This can occur naturally or artificially by incubating the eggs in a temperature-controlled system. Subsequently, incubation of a fertilized egg (naturally or artificially) provides an appropriate environment that promotes further development of the egg contents into an embryo. Here, an embryonated egg is therefore referred to as a fertile egg that comprises the developing embryo and its membranes (chorioallantoic membrane, amniotic, and yolk) enclosed in a cavity. Depending on the choice of egg, the time for development of the embryo varies widely from one species of hen to another (see Figure 9.3).

Obtain specific pathogen-free (SPF) freshly fertilized avian eggs. Purchase all eggs that will be used for analysis, preferably from a suitable vendor who can supply all the necessary information about the eggs.

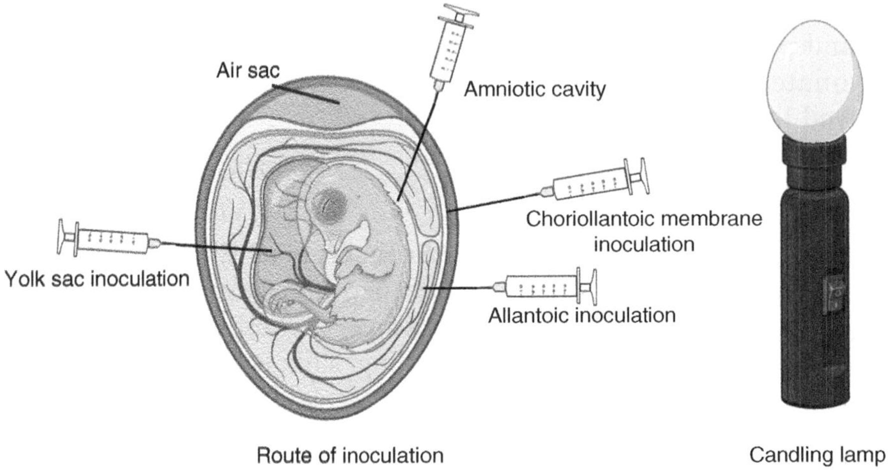

Figure 9.3 Different routes of inoculation for virus propagation using embryonic eggs and a candling lamp.

Selection of a Healthy and Viable Egg-Candling Candling is the process of determining and monitoring the viability of an embryo (growth and development) inside an egg. Candling can be done by holding a strong light above or below the egg to observe the embryo. It is done by placing the egg against a candling lamp aperture illuminated by electric light. Where a candling lamp is not available, try using an LED torchlight. All eggs must be candled before inoculation to check their viability. Dead and unfertilized eggs must be discarded. The eggs should also be visibly inspected for the appearance of cracks before virus inoculation.

Preferably, candling should be done in a dark room or area shielded by a black photographic curtain so that the shadow of the embryo and other structures, especially the blood vessels, will be visible. A viable embryo has a well-defined blood vessel (orange to yellow color) and moves in response to light.

If the laboratory personnel handling this procedure are unsure whether the embryo is viable, place the egg back in the incubator and recheck at a later day. It is preferable to candle an egg daily and on the proposed day of use.

Characteristics of dying or dead embryos

- Blood ring just under the shell and a small dark area: This indicates that the egg was fertile, and development had begun, but the embryo died (early death) during incubation, releasing the blood present in the embryonic vessels.
- "Dead germs": This indicates that the embryo died after growing large

enough to the extent that is visible when candled (late death). It might be difficult to identify. Look for disrupted blood vessels and a dark ring around the egg.
- Low, sluggish, or no movement: This indicates that the embryo is dying or is dead. Usually presenting with reduced or absent vessels.
- Clear egg: This indicates infertile eggs.

Incubation Condition

- **Temperature:** Maintain the temperature within the incubator at 38–39 °C. A commercially available incubator can be purchased.
- **Relative humidity:** Maintain the relative humidity within the incubator at 55–65%. The condition for humidity could be routinely validated. All eggs must be adequately spaced to allow proper humidity. A tray containing sterile water should be at the bottom of the incubator to maintain this level of humidity.
- **Positioning:** Place eggs in the incubator (humidified egg incubator) with the air sac on top and turn the eggs 2–3 times a day. Some incubators have an inbuilt automatic egg turner to rotate eggs regularly. Preferably, candle all the eggs daily and observe the development of the embryo or cytopathic effect.

Do not incubate dirty eggs; clean all stained eggs with a 70% alcohol solution wipe.

Purchase eggs with a white shell, as this will enhance candling.

Preferably, use newly laid fertilized eggs. Eggs that are not ready for use should be refrigerated at 4 °C ± 2 °C. However, prolonged storage (beyond 7 days) may influence embryonic development and cause early death. Some embryos may not develop after storage.

Change gloves often or disinfect with 70% ethanol to minimize the risk of contamination by pathogens. Turn off the lights when leaving the room.

Documentation: The following information about the eggs should be recorded:

- Date of order and delivery of eggs.
- Number of eggs ordered and delivered.
- Condition of eggs delivered – color, physical appearance, and damaged eggs.
- Date of storage or incubation.
- Number of eggs with viable embryos after incubation or storage.

Piecing Eggshell The site and the size of the opening required for virus inoculation vary depending on the route of inoculation of the virus. Piecing can be done using a suitable device like a dental drill, or vibrating engraver, or by punching a small hole through the shell using a needle (e.g., 18G needle). Before use, ensure that the tip of the engraver or drill is sterile or disinfected.

Egg Inoculation and Incubation Embryonated eggs can be inoculated by placing the virus onto various parts of the embryo's membrane or cavity. This process is referred to as the route of inoculation, which targets specific regions of the embryo. The primary routes of virus inoculation include chorioallantoic membrane, allantoic sac, yolk sac, and amniotic sac inoculations, along with three additional methods: intraembryonic, intracerebral, and intravenous (IV) inoculation. Different viruses may have a preference for a particular inoculation route, which corresponds to their preferred site of infection. Selecting the correct inoculation route for each virus is essential to ensure it reaches the appropriate cell types for optimal replication.

Inoculation in the Allantoic Sac This is the main inoculation route for the production of vaccines against the influenza virus [8, 9]. Embryonated 9–12-day incubated eggs (for eggs with a 21-day embryonation period) from avian species are commonly used for inoculation by this route. Eggs from other avian species, like duck and turkey eggs (with a 28-day embryonation period), can be incubated for up to 11–14 days.

Procedure:

1. Candle eggs to determine embryo viability and mark the air space of the eggs.
2. Place the selected eggs in a separate holder, placing the air sac up.
3. For virus isolation, withdraw 0.2 mL of the virus suspension into a sterile 1 mL syringe and attach an 18–22G needle.
4. Drill or punch a small hole over the air sac through the shell to penetrate the air sac. Be careful not to go too deeply to avoid stabbing the embryo or yolk.
5. Carefully place a needle into the hole, holding the syringe vertically and advancing toward the allantoic cavity. Be careful not to torch the portion of the egg below the air sac.
6. Inject 0.2 mL of the sample suspension into the allantoic cavity. Repeat the procedure for other eggs.
7. Seal the holes with molten paraffin wax to prevent desiccation and maintain sterility. Return the eggs to the incubator with the air sac up. Record inoculation.
8. Incubate the eggs for 3–7 days.
9. Discard waste safely and appropriately.
10. Daily examine the embryos and the allantoic fluid for the presence of virus (for embryo abnormalities and deaths) using a candling lamp and record observations.
11. Death of the egg within 24 hours of inoculation is usually due to bacterial contamination or trauma of inoculation. Discard.

It is important to candling all eggs before inoculation, to check if they are viable, and to mark the site of inoculation properly. Discard dead eggs, label viable eggs properly with a sample identification number, and mark the air space of the eggs.

Disinfect the egg with 70% ethanol, 3.5% iodine, or 1.5% sodium iodide, and allow disinfectant to dry up before piercing the shell. A negative control egg should be incubated with the test to ensure that the eggs are free of bacterial contamination.

Carry out all virus inoculation in a biological safety cabinet.

Inoculation of the Amniotic Sac This is the main inoculation route for the isolation of influenza and mumps virus [10, 11]. Embryonated 12–13-day-old egg-incubated eggs (21-day embryonation period) from avian species (typically from chicken) are commonly used for inoculation by this route. Duck and turkey eggs, which have a 28-day embryonation period, can take up to 18–22 days to reach the appropriate stage for inoculation.

1. Candle eggs to determine embryo viability and mark the air space of the eggs. Alternatively, mark the general location of the embryo (below the base of the air sac).
2. Place the selected eggs in a separate holder, placing the air sac up.
3. For virus isolation, withdraw 0.2 mL of the virus suspension, ready for inoculation, into a sterile 1 mL syringe and attach a 22 G needle.
4. Drill or punch a small hole over the air sac.
5. Carefully place the needle into the hole, holding the syringe vertically and advancing toward the embryo.
6. Alternatively, inoculate the egg in a dark room, holding it against the candler to visualize the inoculation. Insert the needle into the amniotic cavity to penetrate it, advancing toward the embryo. Be careful not to stab or puncture the embryo.
7. Inject not more than 0.2 mL of the sample suspension into the amniotic cavity. Repeat the procedure for the other 3–4 eggs.
8. Seal the holes with molten paraffin wax to prevent desiccation and maintain sterility. Return the eggs to the incubator with the air sac up. Record inoculation.
9. Incubate the eggs for 3–7 days.
10. Discard waste safely and appropriately.
11. Examine the embryos and the allantoic fluid for the presence of the virus daily (for embryo abnormalities and deaths) using a candling lamp and record observations. Death of an egg within 24 hours of inoculation is usually due to bacterial contamination or the trauma of inoculation. Discard.

Yolk Sac Inoculation This is the simplest route of inoculation, commonly used to grow the yellow fever virus and some other arboviruses, like Zika virus [12, 13]. Embryonated 4–7-day-incubated eggs (21-day embryonation period) from avian species commonly used for inoculation by this route. Eggs from other avian species, like duck and turkey eggs (28-day embryonation period), can take up to 18–22 days.

1. Candle eggs to determine embryo viability and mark the position of the embryo and the yolk sac.
2. Place the selected eggs in a separate holder, placing the air sac up.
3. For virus isolation, withdraw into a sterile 1 mL syringe 0.2 mL of the virus suspension and attach a 20–22G needle.

4. Drill or punch a small hole over the air sac (at the top of the air sac) through the shell.
5. Carefully place the needle into the hole, holding the syringe vertically and advancing perpendicular through the hole to a depth of at least 1 inch (almost to the hilt of the needle) and deposit the inoculum. Repeat the procedure for a minimum of 5–6 eggs.
6. Seal the holes with molten paraffin wax to prevent desiccation and maintain sterility, and return the eggs to the incubator with the air sac up. Record inoculation.
7. Incubate the eggs for 9–10 days.
8. Discard waste safely and appropriately.
9. Examine the embryos and the allantoic fluid for the presence of virus daily (for embryo abnormalities and deaths) using a candling lamp and record observations. The death of any egg within 24 hours of inoculation is usually due to bacterial contamination or trauma of inoculation. Discard.

Chorioallantoic Membrane Inoculation

This is the route of inoculation commonly used to grow poxvirus and herpes viruses [14–16]. The chorioallantoic membrane is a vascular membrane in embryonated eggs that provides a rich environment for viral replication and growth. Embryonated 11–12-day eggs (with a 21-day embryonation period) from avian species are commonly used for inoculation by this route.

1. Candle eggs to determine embryo viability, and the position of the air sac and mark. Also, mark the area of the chorioallantoic membrane that is well developed.
2. Place the selected eggs in a separate holder, with the air sac facing up.
3. For virus isolation, withdraw into a sterile 1 mL syringe 0.1–0.05 mL of the virus suspension and attach a 26–27G (1/4 inch) needle. Set aside and ready for inoculation.
4. Drill or punch a small hole through the marked area of the air sac. The hole should be as small as possible. Avoid puncturing the chorioallantoic membrane and the areas with large blood vessels.
5. Create artificial air by sucking the air through the hole in the air sac using a rubber bulb. Following suture, the chorioallantoic membrane will drop from the shell. This can be visualized by holding the egg against a candle to control the size of the artificial air space formed (usually 20–25 mm in diameter).
6. Gently rotate the egg (to the right or left) as soon as the artificial air space is formed, such that the newly formed air space moves to a well-developed area of the chorioallantoic membrane, and mark such area.
7. Drill or punch a very small hole through the area of the artificial air sac – this forms the site of inoculation.
8. Carefully place a needle into the hole, holding the syringe vertically. Evenly distribute 0.1–0.05 mL of the sample suspension over the area of the chorioallantoic membrane. Repeat the procedure for a minimum of 7–8 eggs.

9. Seal the holes with molten paraffin wax and return the eggs to the incubator with the artificial air space up. Record inoculation. Leaving the artificial air space unsealed may in no way affect egg viability.
10. Incubate the eggs for 3–4 days undisturbed.
11. Discard waste safely and appropriately.

9.3.1.3 Conditions for Incubation of Inoculated Eggs

Temperature Maintain the temperature within the incubator at 37 °C after inoculation of the virus. Although the optimum temperature for the development of the embryo is 38–39 °C, most viruses will not grow at temperature above 37 °C. Lowering the temperature for the period of virus inoculation will not have a significant effect on the development of the embryo.

Relative humidity Maintain the relative humidity within the incubator between 55 and 65%. The condition for humidity could be routinely optimized and validated.

All eggs must be adequately spaced to allow proper humidity for all incubated eggs. Where applicable, a tray containing sterile water should be at the bottom of the incubator to maintain this level of humidity.

Positioning This depends on the location of the site of inoculation of the virus or the site with the largest aperture.

Duration The duration of incubation depends on the type or nature of the virus, the concentration of inoculum, the site of inoculation, and the tissue to be harvested.

9.3.1.4 Evaluating Inoculated Eggs for the Presence of the Virus

Determine the presence/propagation of the virus in the inoculated eggs, during or after the incubation period, by observing for the following changes, which may indicate viral infection or propagation.

Atrophy of the Embryo This is characterized by the shrinkage or underdevelopment of the embryo.

Observation: The embryo may appear smaller than expected or exhibit signs of degeneration. This can be indicative of a viral infection impacting the embryo's growth and development.

Growth of the Embryo The embryo appears underdeveloped or stunted.

Observation: Compare the embryo's size and development to the normal (control) expected growth at that stage of incubation. Stunted growth or lack of normal development may signal a viral infection affecting embryonic development.

Volume of the Amniotic Fluid Changes in the volume of the amniotic fluid can be indicative of viral infection. The amniotic fluid volume may increase or decrease within the amniotic cavity.

Observation: An abnormal increase in amniotic fluid may suggest viral replication, causing inflammation or other pathological changes. A decrease in amniotic fluid volume could indicate embryo distress or abnormal fluid absorption.

Pork, Lesions, or Inclusion Bodies An infected chorioallantoic membrane may

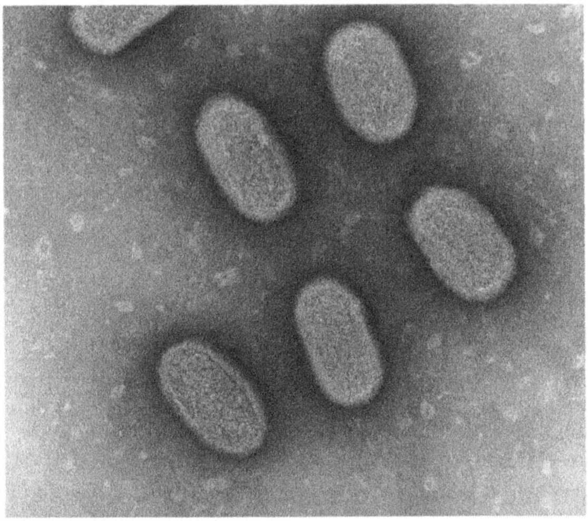

Figure 9.4 Putative electron microscopy of a virus isolated on the chorioallantoic membrane.

show distinctive lesions or inclusion bodies due to viral infection (see Figure 9.4).

Observation:

Pocks: Small, raised lesions that may appear on the chorioallantoic membrane, particularly for viruses like poxviruses.

Inclusion bodies: The presence of intracytoplasmic or intranuclear inclusion bodies within the chorioallantoic membrane cells, often observed using special staining techniques or microscopy.

Hemorrhages Hemorrhaging indicates bleeding within the embryo, subcutaneous tissues, or feather follicles.

Observation: Look for signs of bleeding in the following

- **Subcutaneous tissues:** Blood accumulation beneath the skin.
- **Feather follicles:** Presence of blood in or around feather follicles.
- **Embryo dark spot:** Bleeding inside the embryo, which may appear as dark spots or discoloration.

Embryo Death The embryo stops moving and developing.

Observation: Check for movement of the embryo through candling or other noninvasive techniques. An absence of movement or cessation of normal developmental processes may be a sign of embryo death, which could be due to severe viral infection.

9.3.1.5 Harvest of Viruses from Embryonated Eggs

Harvesting of Virus Inoculated in the Allantoic Cavity

1. Candle all eggs and discard the ones that died <24 hours of inoculation with viruses.
2. Disinfect the surface of the egg with 70% ethanol and transfer it into the biosafety cabinet.
3. Gently crack the shell above the air sac with forceps, scissors, or a spoon. Be careful not to disrupt the chorioallantoic membrane.
4. Open the chorioallantoic membrane with blunt forceps to expose the allantoic fluid. If necessary, use a spoon or spatula to depress the embryo and the yolk. Be careful not to break the yolk, as this will spill into the allantoic fluid.
5. Gently pipette or aspirate the allantoic fluid (5–10 mL) using an automatic pipette or a syringe with a needle. Be careful not to withdraw blood-stained fluid. Dispense the fluid into a sterile tube and place it on ice while you

aspirate the fluid from other eggs. Fluid can be pooled together, where necessary.

6. Centrifuge the fluid at 1000 g for 10 minutes at 4 °C. Collect the supernatant into another sterile bottle and discard the pellet debris.
7. Aliquot supernatant, store at −70° or in liquid nitrogen, and perform sterility test.
8. Discard waste safely and appropriately.

Harvesting of the Virus Inoculated in the Amniotic Sac

1. Candle all inoculated eggs to determine the eggs with the embryos that survived or died before the speculated period of incubation and record. Discard the ones that died <24 h of inoculation with viruses.
2. Disinfect the surface of the inoculated and control eggs with 70% ethanol and transfer them into the biosafety cabinet.
3. Gently crack the shell above the air sac with forceps, scissors, or a spoon to expose the egg. Further, expose the embryo within the amniotic cavity using blunt forceps.
   a. Collect the fluid by inserting a sterile syringe with a needle into the amniotic sac and slowly aspirating the amniotic fluid. The fluid should be clear to slightly cloudy, indicating the presence of the virus.
   b. Avoid contamination by ensuring that the needle does not touch the outer shell or other potentially contaminated surfaces during aspiration.
4. Aliquot the supernatant and perform a sterility test.
5. Discard waste safely and appropriately.

Harvesting of the Virus Inoculated in the Yolk Sac

1. Candle all eggs to determine whether the eggs' embryos survived or died before the speculated period of incubation and record. Discard the ones that died <24 h of inoculation with viruses.
2. Disinfect the surface of the inoculated and control eggs with 70% ethanol and transfer them into the biosafety cabinet.
3. Gently crack the shell above the air sac with forceps, scissors, or a spoon to expose the egg. Further, expose the embryo within the amniotic cavity using blunt forceps.
4. Carefully place a needle (20G) into the cracked hole, holding the syringe vertically and advancing perpendicular through the hole to a depth of at least 1 in. or more to aspirate the content of the yolk sac.
5. Centrifuge the fluid at 1000 g for 10 minutes at 4 °C. Collect the supernatant into another sterile bottle and discard the pellet debris.
6. Dispose of the remaining part of the egg into the appropriate biosafety waste bin.
7. Aliquot the supernatant and perform a sterility test.
8. Discard waste safely and appropriately.

Harvesting of Virus Inoculated on Chorioallantoic Membrane

1. Candle all eggs to determine if the embryos survive or die before the speculated period of incubation and

record. Discard the ones that died <24 h of inoculation with viruses.
2. Disinfect the surface of the inoculated and negative control eggs with 70% ethanol and transfer them into the biosafety cabinet.
3. Remove the shell above the artificial air sac (at the site of inoculation of the virus) with sterile forceps to further expose the embryo within the amniotic cavity using blunt forceps.
4. Rupture the chorioallantoic membrane and discard the contents of the egg into a Petri dish. Alternatively, create another crack at the other side of the shell to discard the content.
5. Gently strip off the chorioallantoic membrane from the shell and place it on a sterile dish containing saline.
6. Rinse the membrane and examine.
7. Discard waste safely and appropriately.

Important notes:

Always compare your observation with a control.

In some instances, the inoculation route of the virus does not determine the part of the egg that will be harvested. For example, the harvesting of embryonic hearts is related to viruses that are inoculated through the yolk sac.

Some viruses are non-embryolethal strains; virus replication may occur without apparent effects on the embryo.

Some strains of viruses require multiple passages in embryonated cells. It is important to determine and validate the test procedure for each virus.

9.3.1.6 Sterility Test for Embryonated Egg Virus Culture

This is important to determine bacterial contamination that occurs during inoculation or at the time of harvest. A sterility test should be done for all eggs with dead or living embryos at the time of harvest. Importantly, it should be done for eggs with dead embryos <24 h of inoculation with viruses. This will help to determine if the cause of death was due to bacterial contamination, injury during inoculation, or a high concentration of virus inoculum.

Method

- Swap the outside part of the shell using a sterile, moist swab stick or
- Take samples from the tip of the pipette (for liquid content within the egg) and the tip of a forceps (for tissues) used for harvesting the content of the cell.
- Inoculate the content into suitable anaerobic and aerobic bacterial culture media and incubate appropriately. Follow the standard bacteriological procedures.
- If bacterial growth is observed, discard the harvested content appropriately.
- If no bacterial growth, refrigerate the aliquot at −70 °C. Fresh tissues are required if immunohistochemistry is needed.

9.3.2 Virus Propagation on Cells

The discovery of cell culture has become a vital tool for propagating large quantities of most viruses, enabling a deeper understanding of how viruses cause disease, known as virus–pathogen interactions. Culturing viruses in cell lines (in vitro) not only confirms the presence of a virus potentially responsible for a disease but also allows for testing the virus's susceptibility to antiviral agents, where applicable.

Since viruses exhibit tissue tropism, meaning they only grow in cells that can support their replication, it is crucial to select the appropriate cell types. The choice of an appropriate cell culture system for virus propagation depends on the specific virus being studied. This has been discussed in the previous chapters. Other factors to consider include the cost and availability of cell culture materials, media, laboratory consumables, laboratory facilities, and trained personnel.

9.3.2.1 General Techniques of the Propagating Virus on Adherent Cells

In this section, the following common cell lines will be used as representative examples: HeLa, Rhabdomyosarcoma (RD), 293T, Vero, S3 cells, and J cells.

The Day Before the Virus Infection Grow the cells in an appropriate culture vessel for about 18–24 hours and examine the quality of the cells under the (inverted) microscope (see Figure 9.5). A general rule to follow when growing cells for virus propagation is to determine the appropriate seeding density of each cell. For example, the entire 6-well plate is approximately about ¾ the surface area of a T-75, about ¼ the surface area of a T-150, and about ⅙ the surface area of a 10-cm plate (see Table 9.6).

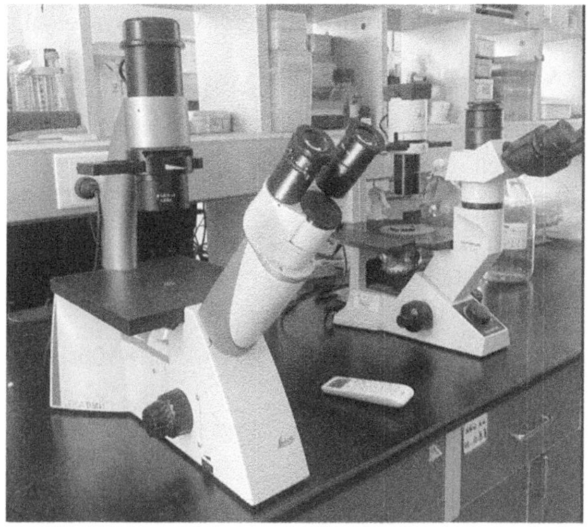

Figure 9.5 Picture of inverted microscopes.
Source: Photo by Department of Microbiology, HMU.

Also, specific required cell numbers at seeding should be optimized for each test and experiment. As a general guideline, seed cells at approximately one-third of the plate's capacity to allow for optimal replication of viruses as the cells divide (cell confluence should not be more than 25–35%) at first seeding.

Day of Virus Infection

1. Examine the cells to determine the confluence and viability. Usually, 40–50% confluence is advisable, and 90–95% viability (at the time of virus infection). If the cells are not ready, incubate further, and if necessary, change the culture medium.
2. Use high-quality cells and discard dead or contaminated cells.

Table 9.6 Useful information for various sizes of cell culture dishes and flasks.

Culture plates	Cells at confluency	Volume of cultural media
6-well	1.2×106	1–2
12-well	0.5×106	1–2
24-well	0.24×106	0.5–1.0
48-well	0.12×106	0.2–0.4
96-well	0.04×106	0.1–0.2
35mm	1.2×106	2
60mm	3.2×106	5
100mm	8.8×106	12
150mm	20.0×106	30

Source: Adapted from Ref. [18].

3. Retrieve the virus sample from storage and thaw it on ice. Disinfect the vial surface with 70% alcohol before placing it in the safety cabinet.
4. Prepare a dilution of the sample by adding an appropriate volume of culture medium (if the multiplicity of infection [MOI] is not known) without serum and place it on ice.
   - You may want to prepare different dilutions of the sample. For example, serial dilution of 10^{-1} to 10^{-6} or double dose dilution.
5. Optional: Aspirate previous growth media from the cell and wash the cell surface gently with pre-warmed PBS 2–3 times.
   - Alternatively, cells can be washed with serum-free culture media (however, this is not economical).
   - Leave the final washing liquid on the cells until just before the inoculation.
   - This step may not be necessary for some virus infection analysis.
6. Add the virus dilution (or concentration) into each well and label the cover of the well with the appropriate dilution.
   - Remember to change pipette tips for each virus dilution. Alternatively, start with the lowest concentration of dilution to avoid the need to change pipette tips between dilutions of the same sample.
7. Rock the plate gently and incubate the virus-infected cells at 37 °C in 5% CO_2 for virus attachment. Rock the plate every 15–30 minutes for 1 hour.
8. Optional: Remove the culture vessels and aspirate the infection media (place medium in a flask containing bleach).
9. Replace the serum-free culture media with culture media that contain serum (or other growth media).

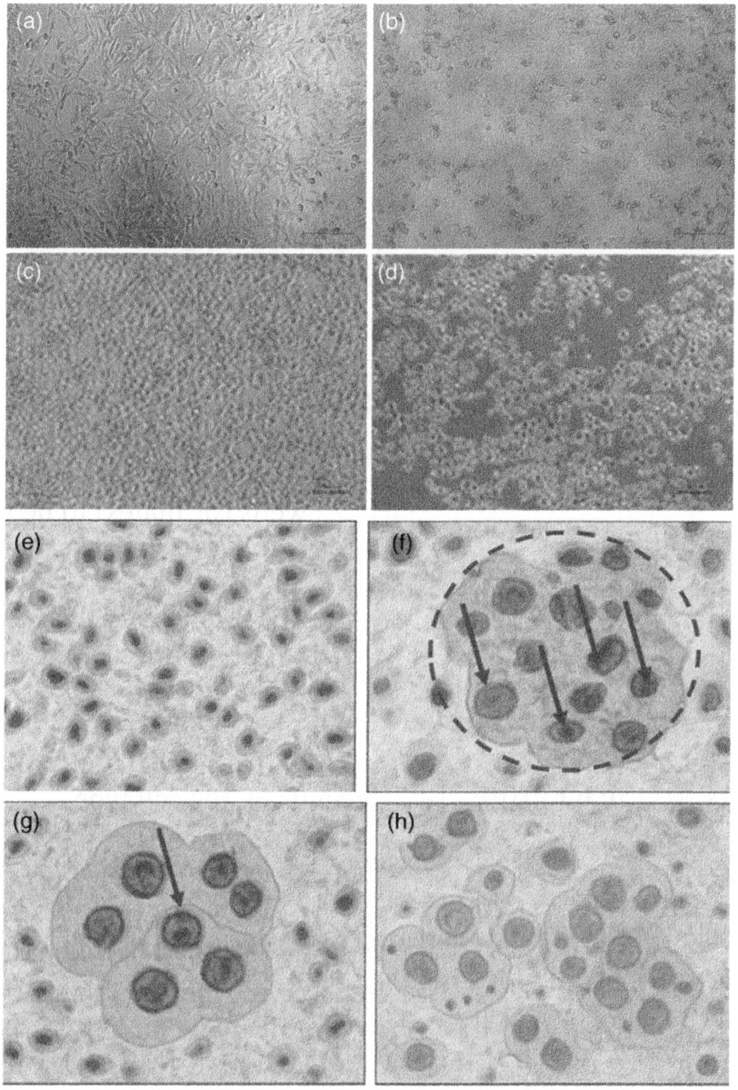

Figure 9.6 Rhabdomyosarcoma (RD) cells viewed under light microscope: (a) no virus; (b) infected with Coxsackievirus (Source: Photo by Wenran Zhao's lab); (c) no virus; (d) infected with coxsackievirus) (*Source: Ref,* [17]/Frontiers Media S.A./CC BY 4.0). Putative structure of stained (e) cell death; (f) giant cell formation; (g) syncytial formation; (h) vacuolation.

Incubate the virus-infected cells at 37 °C in 5% CO_2 for 24 hours.

Note: The duration of incubation depends on the nature of the clinical sample, type of virus, and cell line. Examples of the cytotoxic effect on infected cells can be seen in Figure 9.6.

10. Examine the cells daily in an inverted microscope for cytopathic effect (where applicable) (see Figure 9.4).
11. Replace the culture media if no cytopathic effect is observed (when expected) and incubate further. If a cytopathic effect is not observed, perform confirmatory tests such as an immunofluorescent assay or a flow cytometry assay.

9.3.2.2 *How to Calculate Multiplicity of Infection*

MOI is a convenient term used to describe the relative concentration of a virus that would lead to a controlled number of infections in a susceptible host. This is with the assumption that:

a. All the virions are adsorbed, attached, and infectious.
b. All the cells are susceptible to the virus.

To calculate the desired MOI, two key parameters must be known: the estimated concentration of the virus, plaque-forming unit (PFU)/mL, and the number of cells.

Equation 1: MOI = PFU/μL of virus (stock) used for infection ÷ number of cells.

Equation 2: Total PFU needed = Number of cells × desired MOI.

Equation 3: Total volume of virus needed (μL) = Total PFU needed ÷ PFU/μL of virus used for infection.

For example: Given a virus stock of 1.3×10^8 PFU/μL, a 6-well plate containing 0.3×10^6 cells and a desired MOI = 5, calculate the volume of virus needed:

Total PFU needed = $(0.3\times10^6)\times5 = 1.5\times10^6$ PFU

Total volume of virus needed = $1.5\times10^6/1.3\times10^8 = 0.01$ μL.

Notes

- You may need to multiply the volume of virus needed by the number of wells (or plates) to be infected. Preferably, prepare the same virus diluents for all the wells.
- The number of cells may be calculated based on the seeding density or confluence density.

The number of cells at confluency depends on the cell type.

9.3.2.3 *Identification of Viral Growth in Cell Culture*

Viral growth in cell cultures may be detected by observing the development of microscopic morphological changes in the cells. These changes result from various biochemical interactions between the virus and the cells, indicating an increase in virion particles and macromolecules within and between cells. The observed changes may include cytopathic effects, such as the deterioration or death of the monolayer cells, hemadsorption (where monolayer cells gain the ability to attach to mammalian erythrocytes), or hemagglutination (cells acquire the ability to clump mammalian erythrocytes). However, some viruses can reach high titers without producing visible morphological changes, and in such cases, alternative detection methods may be required.

> *When using cytopathic effect as the diagnostic tool, this must be handled by only an experienced and trained virologist.*
>
> *While recognizing cytopathic in monolayer cells may be possible without necessarily being stained, however, for diagnostic purposes, both stained and unstained cultures of virus-infected cells must be examined.*
>
> *Infected cells should be analyzed in parallel as a uninfected control to differentiate between cytopathic effects and age-related changes in the monolayer cells.*

Cytopathic Effect of Viruses This term is used to describe virus-induced cellular changes that result in microscopically visible changes in a cell's morphology. These changes can be easily observed under a light microscope using low magnification, even in unfixed, unstained monolayer cell cultures. This is based on the principle that when viruses infect a susceptible host, as the newly produced virions are released and spread, they cause deterioration in the structural pattern of the monolayer cells in which they replicate. Hence, the cells gradually develop visible evidence of damage or death. These changes are termed cytopathic effect. Examples of cytopathic effects include:

Plaques Formation A virus plaque is characterized by total or subtotal lysis or detachment of adherent cell area on a monolayer of host cells. It is a key method used to quantify and study viruses in cell cultures. Some picornaviruses, togaviruses, and paramyxoviruses may produce this cytopathic effect.

This process is fundamental in virology for understanding viral replication, determining viral titers, and isolating viral strains.

Mechanism of Plaque Formation When a virus is introduced to a monolayer of susceptible host cells, it infects individual cells. After a virus enters a host cell, it hijacks the cell's machinery to replicate its genome and produce new virion particles.

As the virus replicates, it eventually causes the host cell to lyse (break open), releasing new virions into the surrounding environment. These virions then infect neighboring cells, continuing the cycle of infection, repli cation, and cell lysis. Over time, the repeated cycle of infection and cell lysis results in a visible clear zone, or plaque, within the cell monolayer. Each plaque corresponds to an area where the cells have been destroyed by the spreading virus.

By counting the number of plaques formed, virologists can determine the approximate concentration of infectious virus particles in the original sample. This is often expressed as PFU per mL, indicating the number of infectious particles capable of forming plaques in the host cells.

Quantification of Viral Titers Using Plaque Assays Plaque assays are commonly used to quantify the number of infectious virus particles in a sample. The results provide a direct measure of the virus's ability to infect and destroy host cells. However, not all viruses form cytopathic effects; hence, plaque assay cannot be used to analyze their concentration (see Figure 9.7).

a. **Procedure and methods of plaque assays**

Materials

Cell culture plates (6-well or 12-well), suitable cells for the virus (e.g., Vero, HeLa), viral stock (unknown titer), cell culture medium e.g., DMEM with 10% fetal bovine serum (FBS), PBS, agarose or agar powder, overlay medium (e.g., 2X DMEM with 10% FBS), incubator (37 °C, 5% CO_2).

Step 1: Cell seeding and preparation

Seed the appropriate number of cells in each well of a 6-well or 12-well plate. The seeding density should be such that the monolayer is about 80–90% confluent by the next day (or day of virus infection).

Step 2: Virus dilutions

On the day of the virus infection, prepare 10-fold serial dilutions of the viral stock in sterile tubes. Label each tube with the dilution factor (e.g., 10^1–10^6, etc.).

For example, take 100 µL of the viral stock and add it to 900 µL of culture medium to make the first 10^1 dilution. Mix well, then take 100 µL of this dilution and add it to another 900 µL of culture medium to make the 10^2 dilutions. Repeat this process for several dilutions.

Step 3: Infect the cells

1. Carefully aspirate the growth medium from the cells in each well and infect it with the virus dilution, as appropriately labeled. Inoculate cells with the virus by adding 100–200 µL of each viral dilution directly onto the cell monolayer in separate

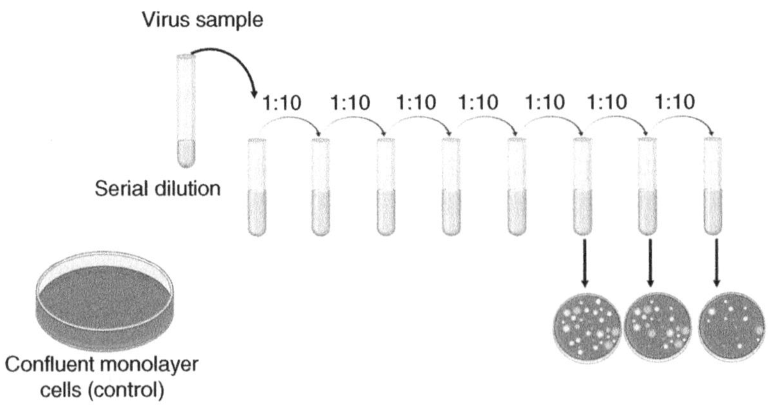

Figure 9.7 Systematic illustration of the plague assay.

wells. Rock the plate gently to ensure even distribution of the virus over the cell monolayer.

2. Incubate the plates at 37 °C for 1 hour to allow the virus to adsorb to the cells. Gently rock the plate every 15 minutes to keep the cells evenly covered with the viral inoculum.

Step 4: Overlay the cells

3. During the incubation, prepare the overlay medium by melting agarose or agar in water (1–2% w/v) and cooling it to 40–42 °C. Mix the melted agarose with an equal volume of 2X culture medium (e.g., DMEM with 10% FBS) to create the overlay medium.
4. After virus adsorption, carefully aspirate the viral inoculum from the wells without disturbing the cell monolayer.
5. Add 1–2 mL of liquid overlay medium to each well using a pipette, ensuring even coverage of the cells.
6. Allow the agarose to solidify and invert it upside down.

Step 5: Incubate for plaque development

Place the plates in the incubator at 37 °C with 5% CO_2. The incubation time will vary depending on the virus, but it typically lasts 48–72 hours.

During this period, plaques will form as the virus replicates and lyses the host cells.

Step 6: Stain and count plaques

Remove the overlay medium and agar gently, and stain the cells to visualize the plaques more clearly.

a. Stain solution such as neutral red (0.1% in PBS) or crystal violet (0.5% in 20% ethanol) can be used. Incubate for 30–60 minutes at room temperature.
b. Wash the wells with PBS to remove excess stains. Visualize stained plaques by placing them against a clear background or a light source.

Step 7: Count the plaques

Examine the plates under a light microscope or by eye. Usually, plaques are seen without using a microscope. Count the number of plaques (clear zone) in wells where the plaques are distinct and countable.

Only wells with a count between 10 and 100 plaques are considered countable. Count all plaques in those wells and record the numbers.

Step 8: Calculate the viral titer

Calculate PFU/mL:

Use the following formula to calculate the viral titer (PFU/mL) for each dilution:

$$\text{PFU / mL} = \left(\text{Number of plaques} \times \text{Dilution factor}\right) / \text{Volume of inoculum}\left(\text{mL}\right)$$

For example, if you counted 50 plaques in the 10^4 dilutions in wells where 100 μL (0.1 mL) of the virus was inoculated:

$$\text{PFU / mL} = \left(50 \times 10^4\right) / 0.1 = 5 \times 10^6$$

Different viruses and host cell types can influence plaque morphology and the efficiency of plaque formation. Some viruses produce clear, well-defined plaques, while others may produce diffuse or small plaques.

An agar or agarose overlay is often used in plaque assays to limit the spread of virus particles, ensuring that plaques remain distinct and countable.

The time of incubation can affect the size and visibility of plaques. Longer incubation times may result in larger plaques, while shorter times may only allow for the formation of smaller plaques.

Environmental conditions, such as temperature and pH, can impact viral replication and plaque formation. Optimal conditions must be maintained to ensure accurate results.

Swelling and Clumping This is characterized by the appearance of cells that are greatly enlarged, swollen, or clumped together to form grape-like clusters as a result of virus replication. These effects are observable under a microscope and are crucial indicators of viral activity within a cell culture. Swelling and clumping of cells are commonly seen and are often associated with specific viral infections, especially adenoviruses (see Figure 9.6f).

Fusion (Syncytia) Formation This is otherwise known as polykaryon or syncytia formation. It is characterized by the fusion of the cell membrane of virus-infected uninuclear cells with neighboring cells to form multinucleated, enlarged cells called syncytia. Syncytia formation can be better observed after staining with Hoechst 33342 dye or other commercially available fluorescence-based stains [19]. To differentiate between clumping cells and syncytia formation, careful examination of the cytoplasm of clumping or swelling cells will reveal distinct plasma membranes. Notably, viruses that cause fusion formation are HSV-1, HIV, measles virus, severe acute respiratory syndrome coronavirus 2 (SARS-CoV-2), and respiratory syncytial virus (RSV) [20, 21]. Unlike syncytial cells, giant cells are large, multinucleated cells and may arise from fusion or nuclear division without cytokinesis (see Figure 9.6g).

Vacuolization This is otherwise known as foamy degeneration. It is characterized by the formation of large and/or several vacuoles in the cytoplasm of virus-infected cells. Vacuolization may be difficult to identify without staining. Some paramyxoviruses, flaviviruses, and retroviruses can cause vacuolization of infected cells (see Figure 9.6h).

Foci Degeneration Some viruses, instead of causing a generalized destruction of the monolayer cells, produce localized areas (foci) of infection. Degenerative changes in monolayer cells provide evidence of viral replication and spread through cell-to-cell transfer of the virus rather than diffusion through the extracellular medium. To observe foci degeneration, it is important to infect the monolayer at low MOI rather than high MOI, which could produce generalized degeneration of cells.

Inclusion Body Formation Virus infection can induce the formation of inclusion bodies [22, 23]. Depending on the type of virus, these inclusion bodies can be intranuclear or intracytoplasmic, round or irregularly shaped, and show a variety of sizes – large or small, eosinophilic or basophilic stained (see Table 9.7 and Figure 9.6 (e-h) and 9.8).

Simple Method of Demonstrating Inclusion Bodies Using Giemsa Stain Giemsa stain is a Romanowsky stain that produces a strong contrast between cell nuclei (deep purple) and the cytoplasm (blue). It can be used to observe cell morphological appearance to give a contrast between experimental cells (i.e., virus-infected cells with cytopathic effect) and uninfected (reference cells) or virus-negative cells.

A. Demonstrating inclusion bodies using Giemsa stain in culture plates

Materials

PBS (without Ca^{2+} and Mg^{2+})
Double-deionized water (ddH_2O)
PBS/methanol (1:1)
Undiluted Giemsa
4% v/v diluted Giemsa in ddH_2O

Preparation of Bouin's fixative (1X)

Reagents	Volume (mL)
Picric acid (saturated aqueous)	75
Formalin (40% aqueous formaldehyde)	25
Glacial acetic acid	5

1. Remove the cultured cells from the incubator and examine the control and test wells under an inverted microscope.
2. Gently aspirate and discard the spent culture media into disinfectant liquid. Be careful not to distort the adherent cells.
3. Wash the monolayer cells with 2 mL of PBS (for 6-well plates) and discard PBS.
4. Fill the wells with methanol/PBS (1:1) for 10 minutes.
5. Aspirate the methanol and replace it with anhydrous methanol for 2 minutes and rinse with PBS.
6. Alternatively, the cells can be fixed with Bouin's fixative for 15 minutes.
7. Fill the wells with freshly prepared, undiluted, filtered Giemsa for 2 minutes. Ensure that the volume of Giemsa covers the entire monolayer of cells. Dilute the stain with ddH_2O and stain further for 2 minutes.
8. Alternatively, wells can be stained with 4% v/v diluted Giemsa in ddH_2O for 2–3 hours.
9. Discard the stain and wash gently under running water until the pink background stain is removed from the well. Ensure that the cells are not detached and washed off from the plate. This step is usually about 10–20 seconds.
10. Rinse the cells with ddH_2O and examine them under the light microscope.

B. Demonstrating inclusion bodies using Giemsa stain on coverslips

Materials

- Glass slides and coverslips with monolayer cells, previously infected
- Bouin's fixative

Table 9.7 Inclusion bodies in human cells and animal models.

Intranuclear inclusion bodies	Intracytoplasmic inclusion bodies
Area indicative of synthesis or aggregate of virion particles: Examples include adenoviruses.	**Area indicative of assembly or aggregate of virion particles:** Examples include **Negri bodies**, typical of rabies virus (diagnostic for rabies encephalitis) [24, 25].
Area indicative of virion nucleic acid synthesis: Presence of clumped chromatin in the nucleus. Examples include herpesviruses.	**Area indicative of virion protein synthesis:** Clumps or aggregates of specific proteins at the ribosomes. Examples include arenaviruses [26].
Cowdry type A inclusion bodies: Seen in herpes simplex virus (HSV) and varicella-zoster virus (VZV) infections [27]. **Cowdry type B:** Seen in infection with poliovirus.	**Area indicative of cytoplasmic viral factories:** Where viral replication and assembly occur. Examples include **Guarnieri bodies** (B-type inclusions), typical of smallpox (variola) and vaccinia-infected cells [28].
Owl's eye inclusions: Typical of cytomegalovirus (CMV) infections [29].	**Councilman bodies:** Found in hepatitis and yellow fever virus infection, indicative of hepatocyte apoptosis [30].
Joest–Degen bodies: Intranuclear and eosinophilic inclusions found in nerve cells caused by Borna disease virus (BDV) is a virus that can infect humans and other animals [31].	**Bollinger bodies:** Found in fowlpox virus infections [32].
Basophilic intranuclear inclusions: Seen in adenovirus infections [33].	**Henderson-Patterson bodies:** Eosinophilic intracytoplasmic inclusions: Found in poxvirus-infected cells, such as molluscum contagiosum and vaccinia [34, 35].
Lipshultz bodies: intranuclear acidophilic inclusion bodies found in cells affected by HSV ulcers. These are typically seen during primary HSV infections. HSV-1 is mostly linked to oral and eye infections, while HSV-2 is commonly associated with genital infections [36].	**Paschen bodies:** Typically found in liver hepatocytes, in individuals infected with variola or vaccinia viruses [37]. It is unclear whether these bodies act as the infectious agents themselves or simply serve as mechanical carriers of the virus.
Torres bodies: Eosinophilic intranuclear inclusion bodies found in liver cells infected with the yellow fever virus (YFV) [38].	

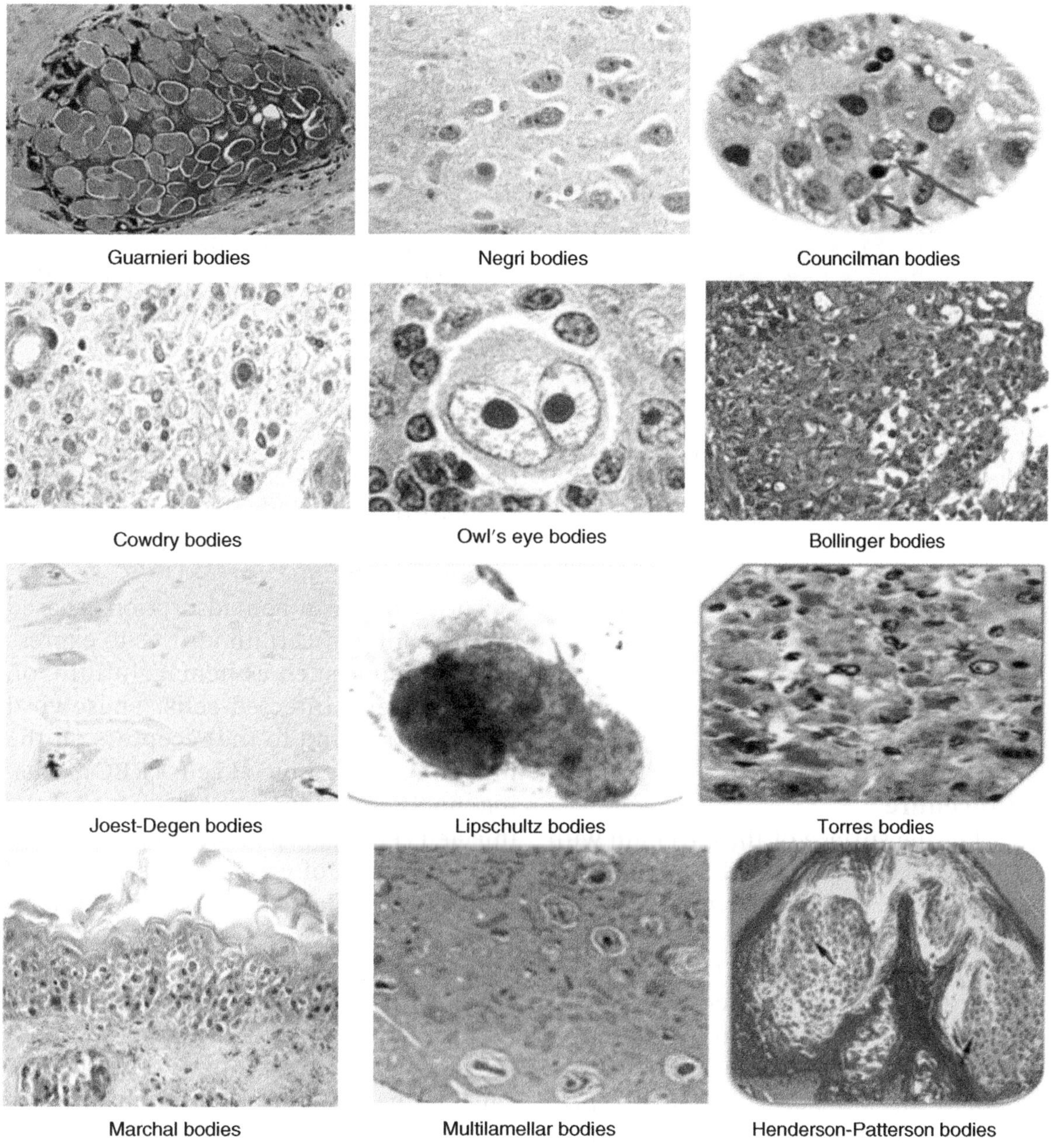

Figure 9.8 Images of inclusion bodies in human and animal cells.
Source: Photo by Dr. Philip Bryant.

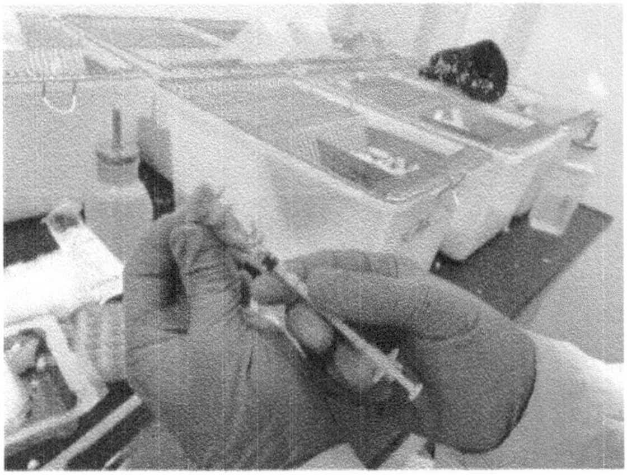

Figure 9.9 Image demonstrating virus inoculation at the intraperitoneal route of BALB/c suckling mice.

- Freshly prepared and filtered May–Grünwald stain working solution
- Giemsa buffer
- Acetone
- Xylene
- Permount mounting medium

Procedure

1. Fill the surface of the coverslip with monolayer cells previously infected with the virus, with methanol for 10–15 minutes.
2. Fill the surface of the coverslip with freshly prepared May–Grün wald working solution for 5 minutes and discard.
3. Fill the surface with freshly prepared Giemsa stain for 10 minutes and discard.
4. Rinse with ddH_2O and dry the slides in an upright position at room temperature.
5. Examine under a microscope.

Mounting for storage

1. Rinse with Giemsa buffer pH 6.8.
2. Rinse again with buffer and acetone (1:1).
3. Soak coverslips in two changes of acetone for 15 seconds each.
4. Soak coverslips in two changes of xylene for 15 seconds.
5. Mount and examine the coverslips with X100 objectives

Hemadsorption and Hemagglutination

Agglutination and adsorption of erythrocytes in virus-infected cells is a phenomenon in which some viruses cause erythrocytes to agglutinate or adsorb on the surface of infected monolayer cell cultures in a specific pattern.

The principle of hemadsorption is based on the ability of certain viruses to express viral proteins, such as hemagglutinin, on the surface of infected cells. These viral proteins can bind to the receptors on the membrane of red blood cells (RBC), causing the RBCs to adhere to the surface of the infected cells. This phenomenon results in the adsorption of erythrocytes to the cell monolayer, which can be observed microscopically.

Hemagglutination, on the other hand, is a related phenomenon where free, newly released virions in the supernatant of infected cultures agglutinate erythrocytes, causing them to clump together. This can be observed macroscopically. The presence of virions or viral proteins, such as hemagglutinin, in the supernatant can therefore be detected through hemagglutination.

Examples of viruses that can cause hemadsorption include the rubella, arboviruses, influenza A, and mumps virus (see Table 9.7) [39–41].

Hemadsorption and hemagglutination are non-specific tests. Under certain conditions, the presence of some hemagglutinating agents, like bacteria and chemicals, can cause RBC to agglutinate. Therefore, confirmation by identifying the virus may be demonstrated by observing characteristics specific agglutinating pattern, using neutralization (neutralizing antibody) or an inhibition test that prevents the hemagglutination or hemadsorption activity of specific viruses.

Hemadsorption and hemagglutination tests do not differentiate between infectious and non-infectious viral particles.

Hemadsorption and hemagglutination cannot be used to accurately determine the concentration of virus particles. However, it can be useful in comparing the relative viral titer of samples collected from different individuals or the same individual but at different time intervals.

Hemadsorption Test on Adherent Cells

Materials

1. Buffer solution, e.g., PBS at an appropriate pH
2. Test sample. That is a virus-containing sample
3. Negative and positive control virus-infected monolayer cell culture.

Methods

1. Remove the culture medium from the virus-infected monolayer cells and replace it with 0.4–0.5% of red cell suspension solution.
2. Incubate at room temperature or 4 °C for 20 minutes.
3. Wash the monolayer with appropriate buffer (e.g., PBS) to remove unattached red cells.
4. Examine the monolayer cells to determine red cell adherence. The RBC will adhere in a specific pattern if viral hemagglutinin is present in the cells. If the viral neuraminidase (which destroys the glycoprotein receptors on the erythrocyte surface) is active, it will cleave the hemagglutinin, and the red cells will detach if the culture is placed at 37 °C.

How to prepare Alsever's solution.

Reagent	Composition (g/L)
NaCl	4.2
D-Glucose	20.5
Citric acid (anhydrous)	0.55
Sodium citrate	8.0

Hemadsorption may not be a sensitive test to demonstrate the presence of small numbers of virions. Therefore, a negative test can be repeated using heavily infected culture cells.

How to prepare 1.0% RBCs

1. Collect whole blood of the animal into Alsever's solution or a heparin tube and allow the blood to sediment freely.
2. Use a sterile pipette to carefully collect packed RBC from the bottom of the tube and dispense them into another sterile clean tube.
3. Filter the packed cells through a sterile cotton gauze pad into a 50 mL conical centrifuge tube.
4. Resuspend the packed cells with PBS plus 0.5% bovine serum albumin (BSA) and mix gently by inversion. Avoid hemolysis of the red cells.
5. Centrifuge the cells at 2000 rpm for 5 minutes at 4 °C.
6. Gently decant or aspirate the supernatant and avoid disturbing the cell pellet.
7. Repeat steps 4–6 twice (for a maximum of three washes).
8. After the final wash, aspirate the PBS plus 0.5% BSA and store packed red cells at 4 °C for a maximum of 24 hours. Discard after use or after 24 hours of storage.
9. To prepare 1% RBC, add packed RBC and PBS and mix gently. For example, add 5 μL of packed RBC to 495 μL of PBS. RBC concentrations can be standardized by counting the number of RBCs and resuspending them with an appropriate volume of PBS.

Hemagglutination Test

Materials

1. Washed suspension of red cells, obtained from whole blood, collected from an appropriate animal species, collected in Alsever's solution or heparin.
2. Buffer solution or diluent, e.g., bovine albumin veronal buffer or PBS at appropriate pH.
3. Test sample, that is, a virus-containing sample, e.g., culture medium suspension from virus-infected cells.
4. Culture medium suspension from negative and positive control.
5. Micropipette and tips, glass Pasteur pipette.
6. Sterile V-bottom well 96-well plate.

Methods

Step 1

It is essential to examine the settling pattern of the suspended RBC by mixing them with the diluents or buffer solution before carrying out the hemagglutination test.

1. Wash the RBCs three times in PBS by centrifugation (500 × g for 10 minutes). Resuspend the cells to make a 0.5–1% cell suspension.
2. Mix gently by tapping the bottom of the tube.
3. Place the suspension into the 96-well plate and allow the RBC to settle.
4. Observe the settling pattern of the cell and check for agglutination or clumping.

Note

There should be no signs of hemolysis or autoagglutination in the red blood cell suspension. If this is observed, prepare a fresh red cell suspension and test again.

Step 2

i. Direct hemagglutination test (Qualitative)

1. Pipette cell culture supernatant into a 1.5 mL tube and keep it on ice. Add 25–50 μL of virus-infected cell supernatant (or sample) into a 96-well plate. Use a separate tip for each sample.
2. Alternatively, cell lysate samples obtained from repeated freeze-thawing cycles of virus-infected monolayer cells can be used.
3. Set up control by adding 25–50 μL of uninfected culture medium sample (negative control) and culture medium with known virus sample into another well (positive control), respectively.
4. Add 25–50 μL of buffer solution or diluents into two wells. This well will be the RBC control for autoagglutination.
5. Add an equal volume of red blood cell suspension to each well.
6. Mix gently by tapping the side of the plate. Incubate at room temperature or 4 °C for 30–45 minutes.
7. Observe the results macroscopically and microscopically to determine the settling patterns of single and agglutinated or clumped RBC

Expected results

Negative test and autoagglutination control wells contain cells that appear as single, non-agglutinated cells. Non-agglutinated cells roll down the sides of the well and settle at the bottom to form a teardrop shape due to gravity.

Positive test and control wells contain cells that appear as clumped or agglutinated cells. Agglutinated cells do not roll down the sides of the well and hence settle as a diffuse film at the wall of the wells.

Notes on the interpretation of results

1. A positive test result does not necessarily indicate whether the present virion particles are infectious or non-infectious.
2. A negative test result does not necessarily indicate the absence of virion particles. Virion particles present may be too low to reach the level needed to cause agglutination sufficient to be detected by the test.

ii. Direct hemagglutination test (Quantitative)

Materials

1. Washed suspension of red cells obtained from whole blood collected from an appropriate animal species, collected in Alsever's solution or heparin.
2. PBS at appropriate pH.
3. Test sample, micropipette, and tips, glass Pasteur pipette.
4. Sterile V-bottom well 96-well plate.

Method

1. Label the 96-well plates with the sample identification number. Each test can be carried out in duplicate.
2. Add 50 μL of PBS to all the labeled wells.
3. Add 50 μL of the test virus samples (test serum or culture supernatant)

into the first two columns (for duplicate samples). If there is more than 1 sample, use the rest of the rows as needed (in duplicate).

4. Perform serial 2-fold dilutions of the sample in each well by transferring 50 μL (of well-mixed PBS and sample) from columns 1 to 11. Discard 50 μL from column 11.
5. Add 50 μL of appropriate 0.5–1.0% RBC suspension to all columns in each roll, including column 12. The wells in column 12 will be the RBC control for autoagglutination. The wells in column 12 will serve as RBC control wells that contain only PBS and RBC.
6. Tap the plate gently and incubate at 4 °C and room temperature for 40–60 minutes.
7. Read by gently tilting the plate to observe RBC movement and agglutination. Record the results for each well.

Hemagglutination titer is recorded as the reciprocal of the dilution of the virus, e.g., ½ (2^{-1}), ¼ (2^{-2}), etc.

Negative test and control wells contain cells that appear as single, non-agglutinated cells. Non-agglutinated cells roll down the sides of the well and settle at the bottom to form a teardrop shape due to gravity.

Positive test and control wells contain cells that appear as clumped or agglutinated cells. Agglutinated cells do not roll down the sides of the well and hence settle as a diffuse film at the wall of the wells.

Identification of the endpoint of titration – HA unit

Hemagglutination end point of the titration is the well with the minimum dilution of virus that causes complete hemagglutination of RBC, and it is considered to contain 1 HA unit (see Table 9.8).

9.3.3 Virus Propagation and Isolation in Animal Models

Animals have been used for centuries for the isolation of viruses. However, since the advent of cell culture techniques, they are rarely the primary choice for cultivating viruses, especially for diagnostic purposes nowadays, but they still play an important role in the study of the pathogenesis of viral diseases and the evaluation of the safety and efficacy of therapeutic agents (see Table 9.9).

9.3.3.1 Ethical Considerations in the Use of Animal Models

The use of animal models in the study of viral diseases has made a valuable contribution to the identification and treatment of human and animal diseases. Animal research causes pain and involves the infliction of suffering. Therefore, when considering the use of animals for diagnostic and research purposes, it is important that the procedure is ethically acceptable and the benefits that come from it are significant and justifiable [42]. The existence of national or institutional policies and guidelines on the use of animals for research and diagnosis differs in many countries. Some countries have robustly enacted laws for animal welfare, while others have developed internal institutional guidelines on the prevention of cruelty against animal usage. Where there are no national or institutional

Table 9.8 Hemagglutination test: virus species with suitable RBC types.

Family	Virus (General/Species)	Appropriate RBC / Remarks
Rhabdoviridae	Rabies virus (some strains)	Goose RBC; HA is weak/strain-dependent; optimal at 0–4 °C and pH < 6.8; depends on viral glycoprotein integrity
Orthomyxoviridae	Influenza A, B, (C)	Chicken, guinea pig, or human group O RBC; HA and hemadsorption at 4 °C; elution at 37 °C due to neuraminidase
Bunyavirales	Some species	Goose or other avian RBC; HA is weak and strain-dependent; pH-sensitive
Reoviridae	Some orthoreoviruses	Human or animal RBC (strain-dependent); HA present in some members; rotavirus generally does not show HA
Parvoviridae	Some species	Guinea pig, hamster, pig, human (group O), monkey RBC; strain-dependent HA; typically at 4 °C
Polyomaviridae	Some species	Mammalian and avian RBC; guinea pig RBC often gives higher titers; performed at 4 °C; HA is strain-dependent
Matonaviridae	Rubella virus	Chicken, pigeon, or goose RBC; optimal pH 6.0–6.2; HA at 4 °C; mammalian RBC nonreactive; elution at 37 °C
Flaviviridae	Some species	Goose, pigeon, or chicken RBC; HA is variable and pH/temperature dependent; not a universal feature
Paramyxoviridae	Parainfluenza, mumps, Newcastle disease virus	Chicken, human, or guinea pig RBC; HA and hemadsorption at 4 °C; elution at 37 °C (especially with guinea pig RBC); measles virus HA/HAD may require primate RBC at 37 °C

Table 9.9 Advantages and disadvantages of using an animal model.

Advantages of the use of animals	Disadvantages of the use of animals
Animals provide a reliable model for studying and demonstrating typical viral pathogenesis and antiviral therapy.	Some human viruses cannot grow in lower animals. While some grow, they do not cause typical viral pathogenesis and syndromes or even cause disease in the animal.
It is used for the isolation of certain viruses.	Expensive, difficult to maintain, and require expertise.
It provides an avenue to produce antibodies against viral diseases.	Age, species, gender, and immunity of the animal may interfere with virus growth.

policies or guidelines on the use of animals, the following criteria could be considered:

1. **Principle of respect**: There must be respect for animals and their value. Scientific procedures in animal research cause pain and suffering; therefore, the purpose of the experiment that involves the use of animals must be justifiable. It is inhumane to cause avoidable death or suffering to animals than is necessary.
2. **Principle of replacement:** Researchers, technicians, and scientists must consider if there are alternatives and other options. The lab can also consider postponing the test or experiment until alternative replacements have been developed or found, where possible.
3. **Principle of reduction**: Where the use and suffering of animals cannot be avoided, personnel involved should minimize, where possible, the number of animals that will be used to achieve the scientific objective of the test/study. Besides, this will also reduce further unnecessary expenses to the cost of maintaining the animals, time, and purchase of consumables.
4. **Principle of responsibility:** Laboratory technicians, scientists, or researchers must provide good animal welfare that prevents extreme temperature or humidity, hunger, malnutrition, thirst, stress, and injury to the animal both outside and within the context of the experiment or testing. Likewise, when using animals for research purposes, the researcher must be willing to share experimental results publicly to avoid repetition of the same procedure.
5. **Principles of proportionality:** The use of animals can be considered acceptable if the objective for their usage is significant enough. This is such that the benefits that come from it are justifiable and provide commensurate benefits for humans, animals, and the environment.

If there is no constituted animal care ethical committee, individuals are encouraged to seek advice from other cooperating institutions or scientists.

9.3.3.2 Animal Husbandry

- Animal housing and caging system. Ideal housing of laboratory animals should allow the animals to freely move so that they are not aware of the procedure and be able to acclimate to the environment [43].
- It is appropriate to move the animals to another space or room different from where they are being kept when the procedure is to be carried out on the animal. This gives them more time to adjust to human presence. The optimal location to conduct the procedures must be appropriately determined.
- Animals housed in a restricted environment may display changes in locomotive activity, and this may have behavioral effects on the animals. Therefore, animal technicians or scientists must determine how often the

animals will be disturbed and what the magnitude and nature of the disturbance will be.

- Each laboratory should have appropriate protocols regarding feeding, cage cleaning of defecation, changing of animal beddings, and sanitation processes to prevent constant human interference and stress to the animals. Frequent routine changing of beddings and cleaning of cages can be detrimental to the animals because of the disruption of the environmental air and contamination among different groups.
- Animals, especially mice, are sensitive to odors; therefore, controlling the spread of such odors within the room where animals are kept is important. However, this might be impossible when several groups of animals are kept in the same room.
- Animals are capable of sensing seasonal variation and different photoperiods even when kept in a well-regulated windowless environment. These factors might affect the activities of the animal and cause variation in the results of some parameters. Therefore, it is important to include control groups during the study
- All personnel, including the animal technicians and laboratory scientists involved in the handling and manipulation of animals for experimental purposes, must be properly trained. Also, all persons visiting animal rooms must wear protective clothing dedicated to such purpose and should wash their hands with odorless soap before handling the animals.

9.3.3.3 Animal Housing and Caging

The housing system used for the cultivation of virus-infected mice must be designed primarily with disease control in mind. The protection of the personnel and the environment against viruses is crucial. Therefore, laboratory facilities involved in the cultivation of viruses in animal models must have stringent rules and standard acceptable working practices, which must be followed and obeyed.

Monitoring of the health and well-being of laboratory personnel should be done often to detect infection and related symptoms; once detected, it should be reported as quickly as possible to the relevant authorities.

The design and selection of small animal housing and caging systems for virus cultivation purposes are determined by several factors. Generally, housing and caging systems must be selectively designed and chosen to prevent transmission of disease to the community, interfere with the study parameters, and be comfortable for animal breeding. Generally, caging for laboratory animals should be designed such that:

- It provides the animal a conducive environment for engaging in species-specific behavioral activities, while it also allows for scientific study.
- Configuration of the housing and caging system can be done in such a way that it protects the animals against unintended pathogens and the spread of pathogens among animals and to humans. It must also provide significant protection against injury and factors that affect the well-being of the animals.

- The cages can be systematically placed without biases to external factors. Although the actual space needed for each animal and species differs, sufficient floor space and appropriate height that provide bio protection and are close to natural habitation must be provided. The housing system and space must be large enough to accommodate the entire test group and the control group cages at once. This is because cages placed at different shelf levels and locations could experience different environmental conditions, ventilation, and lighting gradients that could affect the experimental outcomes.
- Nothing should disrupt the diurnal rhythm of the animal, as this could affect some aspects of animal life.
- The animals should be able to access potable, uncontaminated water and food ad libitum when the study is not diet-related. It might be advisable to keep a batch of food for a set of experiments/test when analyzing critical components (e.g., immunoglobulin, glucose, fatty acid) that can be affected by variation in diet. If the duration of the study is prolonged, it may be helpful to know that food possibly could deteriorate at some point in long-term storage.
- The housing environment should have specialized infrastructure and integration like:
  - Temperature and humidity control system that could maintain the temperature of the facility within 1–2 °C of the ideal set point.
  - Air-conditioning and ventilation systems that maintain high levels of air changes to control odors and allergens in the animal rooms.
  - High-efficiency particulate air (HEPA) filters for controlling infectious airborne diseases.
  - Biosafety cabinets are applicable for use in an animal facility.
- The bedding in the cage used by animals for support and resting must be comfortable, clean, and absorbent to liquid. As much as possible, materials used for bedding must be inert or should not contain chemicals (like endotoxin and α- and β-pinene) that could interfere with the biochemical or pathological properties of the animal that is being analyzed [44]. It is also necessary to consider the potential effects of different bedding materials that interfere with the welfare of the animal strain in use. Highly efficient bedding is recyclable and has extremely absorbent capacity (unlikely to grow mold when in contact with liquid). Beddings should be inedible by the animals, less or non-dusty, and has good odor control. For example, wood shavings or sawdust can be dusty and may contain some chemical compounds like aromatic oils, which can cause gastrointestinal, respiratory, and skin diseases or allergies.
- The cage material should be intact and should not expose the animals to chemical compounds that may threaten their survival and the reproducibility of results. Conventionally, the use of cages made of polymer materials can be considered as being inert. However, constant exposure of

these materials to high heat, cleaning and sterilization procedures, alkaline treatments, and even longer duration of usage can cause increased leaching of chemicals from the materials to food, water, and the environment. This could cause harmful effects to the animals or interfere with any chemical-sensitive parameters.Use translucent thermoplastics cages, such as those made of polysulfone or polycarbonate, which afford easier observation of the animals without removal from the cage.

9.3.3.4 Choice of Animal and Route of Inoculation

The selection of the appropriate animal model is fundamental to accurately mimicking human disease or to studying specific aspects of viral infection. Several factors should be considered when choosing the appropriate animal (see Table 9.10), including:

i. **Susceptibility to the virus:** The animal must be susceptible to the virus, meaning it should allow for viral replication and pathogenesis. For instance, mice are often used for studying neurotropic viruses like poliovirus due to their susceptibility to neurological infection.
ii. **Immunological relevance:** The animal's immune system should be sufficiently similar to humans if the study is aimed at understanding immune responses. Nonhuman primates are frequently chosen for this reason in the study of viruses like HIV and the Zika virus.
iii. **Genetic background:** Inbred strains of animals, such as BALB/c mice, provide consistent and reproducible results due to their uniform genetic makeup, which is crucial for studying specific genetic responses to infection.
iv. **Ethical considerations:** The use of certain animals, especially higher primates, requires stringent ethical justification and compliance with animal welfare regulations.
v. **Availability and cost:** The practicality of using a particular animal model includes considerations of cost, ease of handling, and availability.

9.3.3.5 Route of Inoculation

The route of virus inoculation determines the initial site of infection and influences the spread and severity of the disease. The choice of route depends on the natural route of infection, the types of animal models being used, the test/study's goals, and the virus's characteristics. Examples of various routes of inoculation are explained below (also, see Figure 9.10).

i. **Intranasal route:** This route is often used for respiratory viruses like influenza and coronaviruses, as it mimics the natural infection route through the respiratory tract.
ii. **Intracerebral route:** Used for neurotropic viruses, such as certain strains of poliovirus, to directly introduce the virus into the central nervous system, facilitating the study of neurological effects.

Table 9.10 Guidelines for choosing the right animal for virus propagation.

Animals	Examples of viruses	Inoculation technique	Inbred/strain of animals	Organ/pathological observation
Mice	Hantaan virus	Intracranial route	Neonatal BALB/c mice (about 72 h old)	Brain (encephalitis, neuronal necrosis)
	Ebola virus	Intraperitoneal (IP)	Suckling BALB/c L mice	Liver (hepatitis), spleen (splenomegaly), hemorrhagic manifestations
	Dengue-2 viruses	Intracerebral	Suckling BALB/c mice	Brain (encephalitis)
	Coxsackieviruses	Intraperitoneal	Suckling BALB/c mice	Heart (myocarditis), pancreas (pancreatitis)
	Poliovirus	Intraspinal inoculation	C57BL/6, BALB/c, DBA/2, C3H/He	Spinal cord (paralysis), brain (encephalitis)
	Enterovirus 71	Intraperitoneal (IP)	BALB/c (1–7 days old) or ICR (1–14 days old)	Rear limb paralysis, spinal cord (neurodegeneration), brain (encephalitis), muscle (myositis)
			For BALB/c: The Virus can be isolated from hearts, livers, and spleens	
Rats	Lassa virus	Intraperitoneal	Outbred Wistar rats	Liver (hepatitis), spleen (splenomegaly), hemorrhagic manifestations
	West Nile virus	Intracerebral	Sprague–Dawley rats	Brain (encephalitis), spinal cord (paralysis)
	Respiratory syncytial virus	Intranasal	Brown Norway rats	Lungs (bronchiolitis, pneumonia)
Hamsters	Sin Nombre virus	Intranasal	Syrian hamsters	Lungs (pneumonia), liver (hepatitis)
	Hantavirus	Intraperitoneal	Syrian hamsters	Lungs (pneumonia), spleen (splenomegaly)

Guinea Pigs	Lymphocytic choriomeningitis virus (LCMV)	Intracerebral	Outbred Hartley guinea pigs	Brain (encephalitis), meninges (meningitis)
	Junín virus	Intraperitoneal	Hartley guinea pigs	Liver (hepatitis), spleen (splenomegaly), hemorrhagic manifestations
Rabbits	Herpes simplex virus	Corneal inoculation	New Zealand White rabbits	Cornea (keratitis), brain (encephalitis)
	Vaccinia virus	Intradermal inoculation	New Zealand White rabbits	Skin (lesions), lymph nodes (lymphadenitis)
Ferrets	Influenza A virus	Intranasal	Outbred ferrets	Lungs (pneumonia), trachea (tracheitis)
	Measles virus	Intranasal	Outbred ferrets	Lungs (bronchitis, pneumonia), brain (encephalitis)
Nonhuman primates	Simian immunodeficiency virus (SIV)	Intravenous	Rhesus macaques	Immune system (AIDS-like symptoms), lymph nodes, spleen
	Zika virus	Subcutaneous	Cynomolgus macaques	Brain (microcephaly in fetuses), liver, spleen
	Ebola virus	Intramuscular	Rhesus macaques	Liver (hepatitis), spleen (splenomegaly), hemorrhagic manifestations

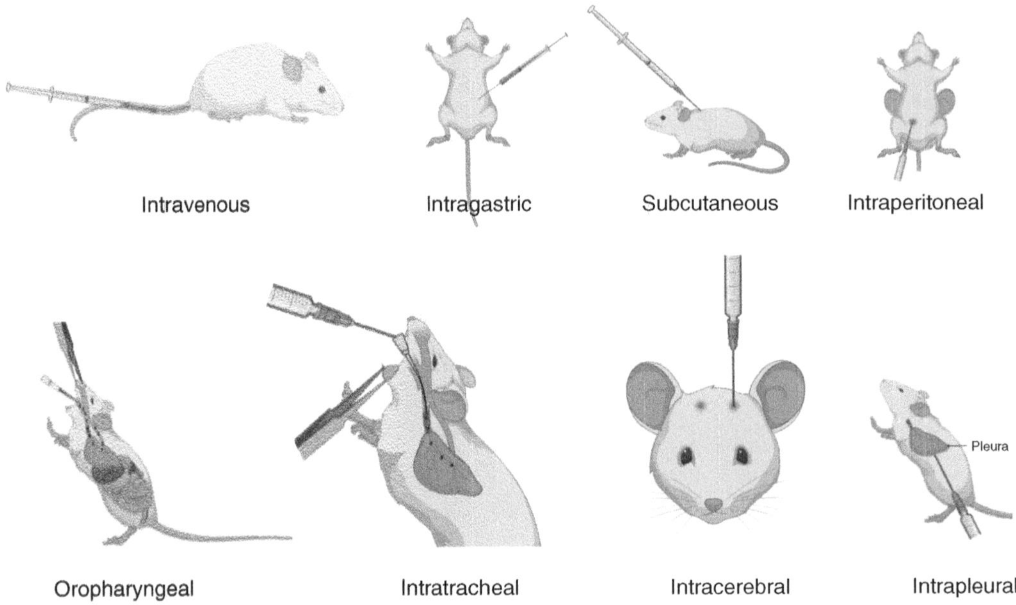

Figure 9.10 Illustration of different routes of virus inoculation of laboratory animals.

iii. **Intraperitoneal (IP) route:** Commonly used for systemic infections or when a large volume of inoculum is needed. It allows for rapid distribution of the virus throughout the body and is frequently used in studies with mice or hamsters (see Figure 9.9 and 9.10).
iv. **IV routes:** This route introduces the virus directly into the bloodstream, facilitating systemic infections and studies of viremia. It is often used in nonhuman primates for studying viruses like HIV or SIV.
v. **Subcutaneous:** This route is used when a slower absorption of the virus is required or to mimic the natural transmission route (to mimic natural mosquito-borne transmission), such as with arboviruses like Zika or chikungunya virus.
vi. **Oral or gastrointestinal:** Usually used for enteric viruses, such as rotavirus or enterovirus, to mimic natural infection via the digestive tract.
vii. **Intramuscular (IM):** Often used for vaccine studies or to simulate bites from rabid animals.

9.3.3.6 Determination of the Growth of Viruses in Inoculated Animals

When viruses are inoculated into animals, the growth of the virus can often be detected through various observable outcomes. These outcomes may include visible lesions, disease symptoms, or even death, depending on the type of virus, the animal model used, and the route of inoculation.

1. **Visible lesions:** This refers to any abnormal tissue change or damage in the animal that can be observed visually

or under a microscope. These are often a direct result of viral replication and the body's immune response to the infection. For example, in poxvirus infections, characteristic pustules or vesicles may appear on the skin. In some cases, lesions may develop on internal organs, such as the liver, spleen, or lungs, and can be detected during necropsy (postmortem examination).

The presence of lesions is often used as a (diagnostic) marker to confirm viral infection. The size, location, and type of lesion can provide insights into the pathogenesis of the virus and the extent of its spread within the host.

2. **Disease symptoms:** Symptoms like clinical signs exhibited by the animal may indicate an ongoing viral infection. These symptoms vary widely depending on the virus and the animal species. Common symptoms include:
   - **Respiratory symptoms:** Coughing, nasal discharge, and labored breathing are typical in respiratory viral infections. For example, in ferrets and Guinea pigs, influenza can cause coughing, sneezing, nasal discharge, and labored breathing. Infected Syrian hamsters with SARS-CoV-2 exhibit labored breathing, weight loss, and nasal discharge as symptoms.
   - **Neurological symptoms:** Viruses like rabies or certain flaviviruses can cause neurological symptoms such as tremors, paralysis, or seizures. For example, in dogs, rabies causes neurological symptoms like aggression, paralysis, and seizures. In horses, West Nile virus can cause ataxia, tremors, and muscle weakness. In mice, Japanese encephalitis virus (JEV) can cause encephalitis, leading to tremors, paralysis, or seizures.
   - **Gastrointestinal symptoms:** Diarrhea, vomiting, and dehydration may be observed in enteric viral infections. For example, rotavirus causes severe diarrhea and dehydration in piglets and mouse models, mimicking the infection in human infants. Norovirus induces diarrhea in mice, used to model human gastrointestinal infections.
3. **Death:** This is the most severe outcome of viral infection in animal models. It may occur rapidly in the case of highly virulent viruses or because of complications arising from the infection. Causes of death include:
   - **Direct viral cytotoxicity:** Some viruses cause direct damage to essential tissues or organs, leading to organ failure and death.
   - **Immune-mediated lesion:** In some cases, the body's immune response to the virus may be so intense that it causes widespread tissue damage, leading to death.
   - **Complications:** Secondary infections, dehydration, or severe metabolic imbalances resulting from the viral infection can also contribute to mortality.

Mortality is a critical endpoint in studies of viral pathogenesis and in evaluating the protective efficacy of vaccines or antiviral drugs. The time to death (survival time) after infection can be used to measure the aggressiveness of the virus and the effectiveness of interventions.

9.4 Molecular Diagnosis and Characterization of Viruses in Clinical Samples

Molecular characterization of viruses in clinical samples is a crucial aspect of virology that involves the identification and analysis of viral genomes to understand the nature, origin, and behavior of viruses and viral diseases. This process is essential for diagnosing viral infections, tracking disease outbreaks, studying virus evolution, and developing targeted therapies and vaccines. Molecular identification and characterization of viruses in clinical samples can be divided into nucleic acid testing (NAT) and protein-based assays (proteomics).

9.4.1 Nucleic Acid Testing

NAT includes methods that detect and quantify viral DNA or RNA directly.

9.4.1.1 Nucleic Acid Amplification Tests (NAATs)

NAATs are techniques that amplify nucleic acids (DNA or RNA) to detect low amounts of viral genetic material. Examples of NAATs include PCR, RT-PCR, and Loop-Mediated Isothermal Amplification (LAMP). To perform a NAAT, viral DNA or RNA must be isolated from the samples. There are several automated RNA/DNA isolation machines. Below is the manual method that can be performed in rural facilities. And where possible or necessary, isolated nucleic acids can be sent to a bigger facility for amplification or sequencing.

1. **Nucleic acid extraction**
   Viral RNA or DNA can be extracted from the clinical samples using various techniques, such as phenol-chloroform extraction, silica-based columns, or magnetic bead-based methods. The goal is to obtain high-quality nucleic acids suitable for downstream molecular analysis. There are numerous commercially available kits for DNA and RNA extraction.
   a. **RNA extraction using the TRIzol-chloroform method**
      Materials Needed:
      Biological sample (e.g., tissue, cells, blood), lysis buffer (e.g., TRIzol or another similar reagent commercially available), RNase-free water, chloroform, ethanol (70% and 100%), isopropanol, RNase-free tubes, centrifuge (preferably refrigerated), and pipettes (RNase- and DNase-free tips).
      Procedure:
      Step 1: Sample preparation
      Tissue samples: Homogenize the tissue using a mechanical homogenizer or mortar and pestle. Add an appropriate volume of lysis buffer (TRIzol or equivalent). Use approximately 1 mL of buffer per 50–100 mg of tissue.
      Cell culture: Lyse cells directly in the culture dish by adding lysis buffer. Use 1 mL of buffer per 10 cm^2

dish surface area. Scrape cells off and transfer to a tube.

Blood: Add an equal volume of lysis buffer to the blood sample and mix thoroughly.

Step 2: Incubation

Incubate the homogenized or lysed sample at room temperature or at 37 °C for 5–15 minutes to allow complete dissociation of nucleoprotein complexes.

Step 3: Phase separation

Add 0.2 mL of chloroform per 1 mL of lysis buffer used.

Cap the tube securely and shake it vigorously for 15 seconds, alternatively, vortex briefly.

Allow the tube to stand at room temperature for 10 minutes to allow the phases to separate.

Step 4: Centrifugation 1

Centrifuge the tube at 12,000 rpm for 15 minutes at 4 °C.

After centrifugation, the mixture will separate into three phases: The upper aqueous phase, which contains the RNA; the interphase, which contains DNA; and the lower organic phase, which contains proteins and other cellular debris.

Step 5: RNA precipitation

Carefully transfer the upper aqueous phase containing the RNA to a new RNase-free tube without disturbing the interphase.

Add 0.5 mL of isopropanol per 1 mL of initial lysis buffer used.

Mix the sample by inverting the tube several times.

Incubate at room temperature for 10 minutes.

Step 6: Centrifugation 2

Centrifuge the sample at 12,000 rpm for 10 minutes at 4 °C.

RNA will form a white gel-like pellet at the bottom of the tube.

Step 7: Washing the RNA pellet

Carefully remove the supernatant without disturbing the RNA pellet.

Add 1 mL of 70% ethanol (made with RNase-free water) to wash the pellet.

Vortex briefly to dislodge the pellet.

Centrifuge at 7500 × g for 5 minutes at 4 °C.

Step 8: Drying the RNA pellet

Remove the ethanol carefully without disturbing the pellet.

Air-dry the RNA pellet for 5–10 minutes. Avoid over-drying, as it can make the pellet difficult to dissolve.

Step 9: Dissolving the RNA

Resuspend the RNA pellet in RNase-free water or diethyl pyrocarbonate (DEPC)-treated water.

Incubate for 10–15 minutes at 55–60 °C to help dissolve the RNA pellet completely.

Step 10: Qualitative and quantitative assessment

Measure the concentration and purity of RNA using a spectrophotometer (e.g., NanoDrop) by taking readings at 260 nm and 280 nm.

Check the integrity of the RNA by running an aliquot on an agarose gel or using a bioanalyzer.

b. **DNA extraction using the phenol-chloroform method**

Materials needed: Biological sample (e.g., blood, tissue, cultured cells), lysis buffer (contains detergents to break down cell membranes), Proteinase K (enzyme to digest proteins), RNase A (enzyme to remove RNA, optional), Phenol-chloroform (for organic extraction, optional), isopropanol (for DNA precipitation), ethanol (70%) (for washing DNA), TE buffer or nuclease-free water (for dissolving DNA), centrifuge, microcentrifuge tubes, pipettes and sterile tips, vortex mixer, water bath or incubator (37 °C and 56 °C).

Procedure:

Step 1: Sample preparation

Blood or cultured cells: If using blood, add the appropriate volume (typically 200–500 μL) to a microcentrifuge tube. For cultured cells, pellet the cells by centrifugation, remove the supernatant, and resuspend the cell pellet in lysis buffer.

Tissue: Homogenize the tissue using a mechanical homogenizer or mortar and pestle. Add lysis buffer to the homogenized tissue.

Step 2: Cell lysis

Add lysis buffer (typically containing SDS or another detergent, for example, sodium dodecyl sulfate [SDS]-based lysis buffer of 1–2%; Tris–HCl: 10–50 mM; pH 7.5–8.0; EDTA: 1–10 mM; NaCl: 50–150 mM [optional]) to the sample. The buffer breaks down cell membranes and releases the cellular contents, including DNA.

Add Proteinase K to the sample to digest proteins, including nucleases that might degrade the DNA. Incubate the mixture at 56 °C for 30 minutes to 2 hours, or until the sample is completely lysed. For blood or soft tissues, this step might be shorter.

Step 3: RNA removal

Add RNase A to the lysate if RNA contamination is a concern and incubate at 37 °C for 15–30 minutes to degrade RNA.

Step 4: Protein removal

Organic extraction (optional): Add an equal volume of phenol-chloroform to the lysate. Mix well by vortexing.

Centrifuge at 10,000–12,000 × g for 5–10 minutes to separate the phases.

Carefully transfer the upper aqueous phase (which contains the DNA) to a new tube, avoiding the interphase (which contains proteins).

Step 5: DNA precipitation

Add isopropanol (0.6–1 volume) to the aqueous phase containing the DNA.

Mix gently by inverting the tube several times until the DNA precipitates out of the solution (you may see a white, stringy precipitate).

Centrifuge at 10,000–12,000 × g for 10–15 minutes to pellet the DNA.

Step 6: DNA washing

Carefully pour off the supernatant, and be careful not to dislodge the DNA pellet.

Add 70% ethanol to the pellet to wash away any remaining impurities.
Centrifuge briefly (5 minutes) and carefully remove the ethanol.
Air-dry the DNA pellet for 5–10 minutes, avoiding over-drying, as this can make the DNA difficult to dissolve.

Step 7: DNA resuspension
Resuspend the dried DNA pellet in TE buffer or nuclease-free water. The volume depends on the size of the pellet, typically 20–100 μL.
Incubate at 37 °C for 15–30 minutes to fully dissolve the DNA.

Step 8: Qualitative and quantitative assessment
Measure the DNA concentration and purity using a spectrophotometer (e.g., NanoDrop) by checking absorbance at 260 nm and 280 nm (see Figure 9.11).
Run an aliquot of the extracted DNA on an agarose gel to assess its integrity.

Polymerase Chain Reaction

PCR is a powerful and widely used molecular biology technique that allows for the amplification of specific DNA sequences.

Principle and Steps of PCR The principle of PCR is based on the ability of DNA polymerase to synthesize new strands of DNA complementary to the target sequence. The process involves repeated cycles of heating and cooling, which facilitate the separation of DNA strands, the binding of primers to target sequences, and the synthesis of new DNA strands (see Figure 9.12).

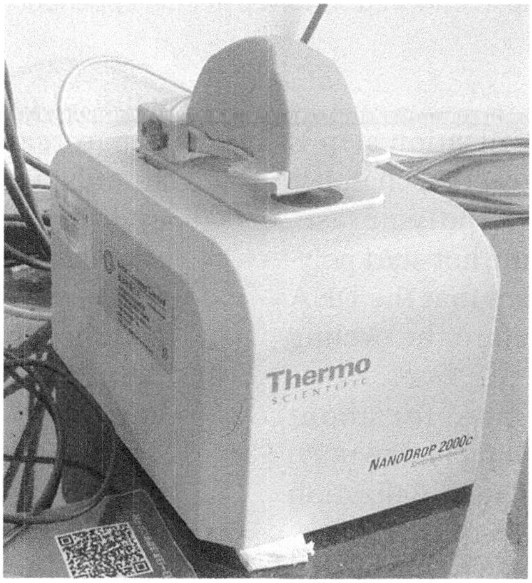

Figure 9.11 NanoDrop 2000c (Thermo Scientific) device used for assessing the quantity and quality of RNA or DNA extracts.

Source: Photo by Department of Medical Microbiology, HMU, China.

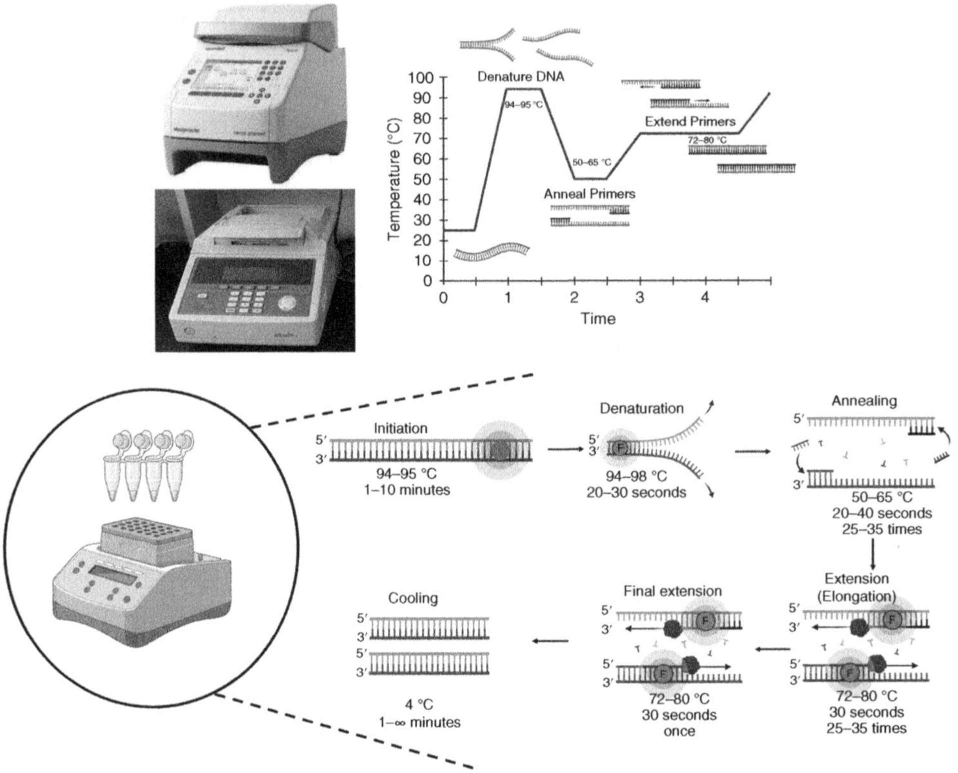

Figure 9.12 Overview of principles and steps involved in PCR. Eppendorf® Mastercycler® Nexus Thermal Cyclers.

Steps of PCR

i. **Initiation:** The initialization step is designed to heat the reaction chamber to activate the DNA polymerase, particularly when using hot-start polymerases, and to ensure that the DNA is fully denatured before the cycling process begins.
Temperature: 94–95 °C for about 1–10 minutes, depending on the polymerase being used. The initialization step in PCR does not require cycling. It is a single, initial step performed at the beginning of the PCR process.

ii. **Denaturation:** The reaction mixture is heated to around 94–98 °C for 20–30 seconds. This high temperature causes the double-stranded DNA to denature or separate into two single strands. This step is repeated in every cycle of PCR. It occurs once per cycle, usually repeated 25–35 times, depending on the desired amount of product and the efficiency of the reaction.

iii. **Annealing:** The temperature is lowered to 50–65 °C for 20–40 seconds. During this step, two short DNA sequences known as primers (forward and reverse) anneal to their complementary sequences on the single-stranded DNA. These primers are designed to flank the region of DNA that needs to be amplified. Typically, this step is set at 3–5 °C below the Tm of the primers (usually between 50 and 65 °C). Like denaturation, this step is repeated 25–35 times per PCR run. The exact number of cycles depends on the amount of starting material and the desired final product yield. Selecting an appropriate temperature for the annealing step is essential, as the efficiency and specificity of the PCR are significantly influenced by the annealing temperature.

iv. **Extension (Elongation):** The temperature is raised to 72–80 °C (which is the optimal temperature for the DNA polymerase enzyme, usually Taq polymerase) for 30 seconds to 1 minute per kilobase (kb) of target DNA, The DNA polymerase extends the primers by adding nucleotides to the growing DNA strand, synthesizing a new strand complementary to the template strand. This step is also repeated in every cycle of PCR, 25–35 times. The duration depends on the length of the target DNA sequence.

v. **Final extension:** This step occurs once at the end of the PCR cycling process to ensure that any remaining single-stranded DNA is fully extended. It is not cycled; it is performed once after the completion of all the PCR cycles.

vi. **Cooling:** The cooling step typically refers to the final stage where the reaction is brought down to a lower temperature after the completion of all cycles. This step is crucial for stabilizing the DNA products and halting the reaction. While it does not require cycling, the temperature is generally set around 4 °C for an indefinite period to preserve the integrity of the amplified DNA until further analysis or storage.

Why Does the Number of Cycles Matter?

- **Too few cycles:** This may result in insufficient amplification, especially if the starting material is low.
- **Too many cycles:** Can lead to non-specific amplification, depletion of reagents, and accumulation of PCR artifacts.

Typical PCR cycling number

Most PCR protocols use between 25 and 35 cycles. Fewer cycles (e.g., 25) might be used when the starting DNA concentration is high, while more cycles (e.g., 35) might be necessary for low-concentration samples to produce sufficient amplified DNA.

Basic PCR Set-up Reagents

1. **Template DNA:** This is the DNA sample containing the target sequence that is to be amplified.
2. **Primers:** These are short, single-stranded DNA sequences (usually 18–25 nucleotides long) that are complementary to the target DNA sequence. Primers define the beginning and end of the DNA sequence to be amplified. Two primers are required: a forward primer and a reverse primer. It is important to take note of the Tm of each primer.
3. **DNA polymerase:** This is an enzyme that synthesizes new DNA strands by adding nucleotides to the primers. The commonly used DNA polymerase in PCR is Taq polymerase, which is heat-stable. DNA polymerase is crucial for the replication of the target DNA sequence during the extension step.
4. **Deoxynucleotide triphosphates (dNTPs):** These are the building blocks of DNA, consisting of adenine (A), thymine (T), cytosine (C), and guanine (G). dNTPs are incorporated

into the new DNA strand by DNA polymerase during the extension step. Most commercially available PCR master mix reagents have dNTPs; in such case, there is no need to separately add the dNTPs.

5. **PCR buffer:** A solution that provides the optimal chemical environment for DNA polymerase activity, including the correct pH and the presence of necessary ions like magnesium (Mg^{2+}). The buffer ensures that the DNA polymerase functions efficiently throughout the PCR cycles.
6. **Magnesium ions (Mg^{2+}):** Magnesium ions are a cofactor for DNA polymerase and are essential for its activity. The concentration of Mg^{2+} can affect the specificity and yield of the PCR product. Often included in the PCR buffer or added separately as magnesium chloride ($MgCl_2$).
7. **Water (nuclease-free):** Water is used to adjust the volume of the PCR reaction mix.

For RNA viruses, an initial reverse transcription step is performed to convert RNA into complementary DNA (cDNA), after which conventional PCR can proceed. This process is called reverse transcription PCR (RT-PCR). It is a two-step protocol where the reverse transcription and PCR amplification are carried out in a single tube using a combined master mix containing both reverse transcriptase and DNA polymerase enzymes. It involves performing reverse transcription separately using a reverse transcriptase enzyme, gene-specific primers, or non-specific primers (like random hexamers, oligo-dT) to generate cDNA, which is then used as a template in a subsequent PCR reaction. RT typically takes 5–15 minutes at 42–55 °C, depending on the enzyme and reaction conditions used. Once complementary DNA (cDNA) has been synthesized, it is relatively stable and can be stored or transported to a reference laboratory for further analysis, such as PCR amplification or sequencing.

Running Quality Control and Tests of Purity

Running quality control (QC) of PCR reactions using agarose gel electrophoresis involves verifying the integrity and size of the amplified PCR products.

1. **Prepare agarose gel**
   Materials: Agarose powder, Tris-acetate-EDTA (TAE) or Tris-borate-EDTA (TBE) buffer, ethidium bromide (or alternative dye), gel tray, and comb.
   a. Weigh and dissolve the required amount of agarose (e.g., 1% gel requires 1 g agarose per 100 mL buffer).
   b. Heat (microwave) the agarose solution until completely melted.
   c. Allow it to cool slightly (around 50–60 °C) and add ethidium bromide or an alternative DNA stain if necessary.
   d. Pour the gel into the gel tray and insert the comb to create wells.
   e. Allow it to solidify (~20 minutes).
2. **Prepare samples for loading**
   a. PCR products: Mix RT-PCR reaction products with a loading dye (e.g., 6X loading dye). Use 1 part dye for every 5 parts sample. Adjust the volume according to the concentration of the loading buffer.
   b. DNA ladder/Marker: Prepare a DNA ladder of known molecular

weights to compare against your PCR products.

3. **Load the samples onto the gel**
   a. Carefully remove the comb from the solidified gel.
   b. Place the gel into the electrophoresis tank filled with TAE or TBE running buffer.
   c. Load samples (usually 5–10 µL of PCR product mixed with dye) into the wells.
   d. Load 5–10 µL of the DNA ladder into one well for size reference.
4. **Run the gel**
   a. Connect the gel box to a power supply.
   b. Set the voltage (e.g., 90–120 V) and run the gel for 10–30 minutes, depending on the size of the PCR products.
   c. Monitor the progress by tracking the migration of the dye front across the gel.
5. **Visualize the gel**
   a. Once the gel run is complete, carefully remove the gel from the tank.
   b. Place the gel on a UV transilluminator or blue-light transilluminator if using a safe DNA dye (e.g., SYBR Safe) to visualize the DNA bands.
   c. Capture an image of the gel for documentation (see Figure 9.13).
6. **Evaluate the results**
   a. **Positive control:** Check if the positive control band matches the expected size. A strong band indicates successful amplification.
   b. **Negative control:** Verify the absence of bands in the negative control to ensure there was no contamination.
   c. **Sample lanes:** Evaluate the size and intensity of the bands in the sample lanes.
   d. Bands should correspond to the expected size based on the DNA ladder.
   e. Strong, clear bands without smearing indicate good-quality amplification.

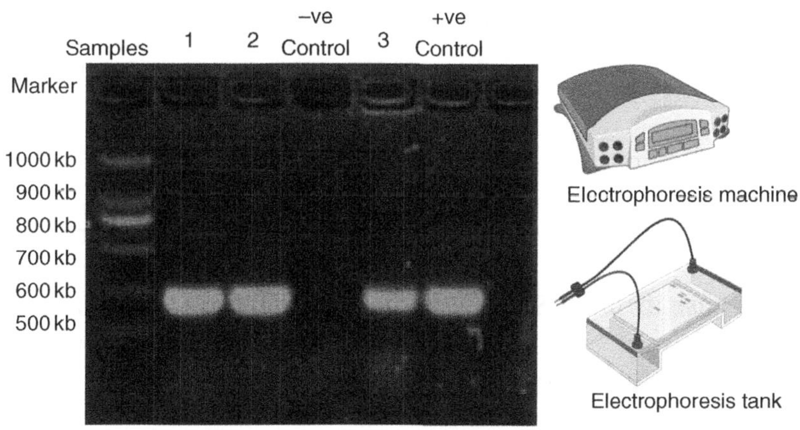

Figure 9.13 Ethidium bromide-stained agarose gel electrophoresis of PCR products (complementary DNA) from extracted total RNA.

7. **Analyze data**
   a. Compare the intensity and size of the sample bands with the DNA ladder.
   b. Verify that there are no primer-dimer formations or non-specific products by checking for unexpected band sizes (see Figure 9.13).
8. **Document the results**
   a. Save the image and note the gel results in your laboratory notebook.
   b. Record any abnormalities, such as smeared bands or unexpected sizes, for further troubleshooting.
9. **Conclusion and troubleshooting**
   a. If the PCR product is of the correct size and no contaminants or artifacts are present, the PCR is considered successful.
   b. If issues arise, consider optimizing reaction conditions or checking RNA integrity.

Preparing TAE running buffer (50X stock solution)

Materials needed: Tris base, acetic acid, EDTA, distilled water, pH meter or pH paper.

1. **Recipe (1 L)**
   242 g of Tris base
   57.1 mL of acetic acid
   20 mL of 0.5 M EDTA (or 7.45 g of EDTA disodium salt)
2. **Procedure**
   Dissolve Tris base in about 800 mL of distilled water.
   Add acetic acid carefully (it is an acid).
   Add EDTA.
   Adjust the volume to 1 L with distilled water.
   Check the pH (should be around 8.0) and adjust if necessary.
3. **Storage:** Label the buffer with the preparation date and concentration for reference. Store the 50X stock solution at room temperature. Dilute to 1X before usage.

Preparing TBE running buffer

Materials needed: Tris-base, boric acid, EDTA, distilled water, pH meter, or pH paper.

For 1 L of 10X TBE buffer:

1. **Recipe (1 L)**
   108 g of Tris base
   55 g of boric acid
   7.4 g of EDTA disodium salt
2. **Procedure**
   Dissolve Tris base and boric acid in about 800 mL of distilled water.
   Add EDTA.
   Adjust the volume to 1 L with distilled water.
   Check the pH (should be around 8.0) and adjust if necessary.
3. **Storage**
   Label buffer with the preparation date and concentration for reference. Store the 10X stock solution at room temperature. Dilute to 1X before usage.

Types of PCR used for the detection of viruses

Conventional PCR: Conventional PCR is used to amplify specific regions of the viral genome, allowing for the detection and identification of the virus in the sample. Primers are designed to target conserved regions of the viral genome.

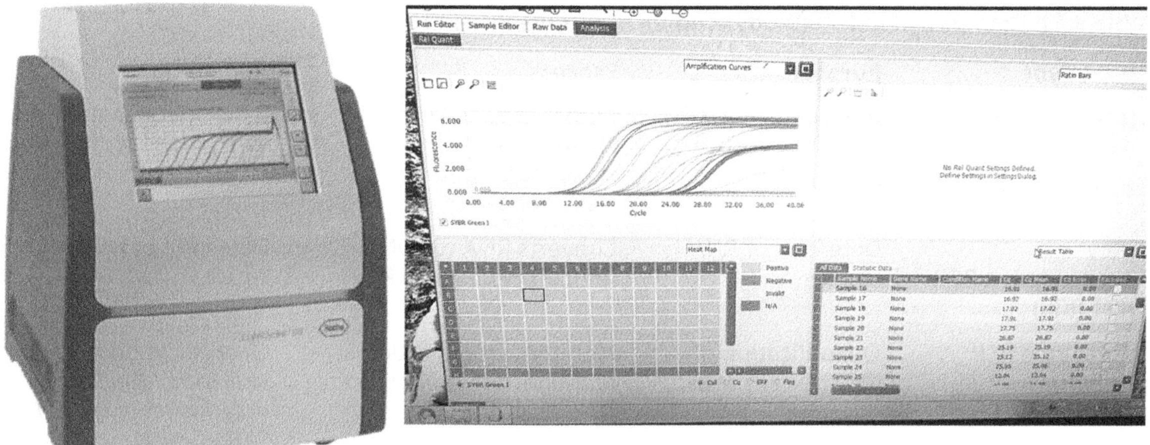

Figure 9.14 LightCycler® 96 Instrument for real-time quantitative PCR and quantitative graphs.
Source: F. Hoffmann-La Roche Ltd.

Real-time quantitative PCR: Real-time PCR allows for the quantification of viral load by measuring the amount of viral RNA or DNA in the sample. This method is particularly useful in monitoring the progression of viral infections and the response to treatment (see Figure 9.14).

Reverse transcription PCR (RT-PCR): For RNA viruses, reverse transcription is performed to convert RNA into cDNA, which can then be amplified by PCR. RT-PCR is the gold standard for detecting RNA viruses such as influenza, SARS-CoV-2, and HIV. More are listed in Table 9.11.

DNA and RNA Sequencing

DNA and RNA sequencing are essential techniques in molecular biology that allow scientists to determine the exact sequence of nucleotides in DNA or RNA. Both methods provide crucial insights into genetic information, gene expression, and genomic variations. This information is useful for the characterization of new virus types or strains, determining drug resistance, and molecular epidemiology. Here is an overview:

Overview of DNA/RNA sequencing in a virology laboratory

DNA sequencing is a powerful technique used to determine the precise order of nucleotides in a DNA molecule, allowing scientists to identify viruses, detect mutations, and study genetic variation across viral strains. The process typically begins with the receipt of clinical or environmental samples suspected to contain viral pathogens. Once received, the first critical step is the isolation of viral nucleic acid, either RNA or DNA, depending on the type of virus. For RNA viruses, reverse transcription may be performed to convert the RNA into cDNA before proceeding. The extracted genetic material is then purified and prepared for sequencing.

Table 9.11 Examples of PCR methods used for virus detection.

PCR Type	Purpose	Process	Application
Conventional PCR (Endpoint PCR)	Basic detection of viral DNA or RNA	Amplifies specific sequences; results analyzed by gel electrophoresis	Initial identification of viral presence
Reverse transcription PCR (RT-PCR)	Detection of RNA viruses	Converts RNA to cDNA, then amplifies with PCR	Detection of RNA viruses
Real-time PCR (qPCR)	Quantification of viral nucleic acids	Monitors amplification in real-time using fluorescent dyes or probes	Quantification of viral load; clinical diagnostics (e.g., HIV, SARS-CoV-2)
Multiplex PCR	Simultaneous amplification of multiple viral targets	Uses multiple primer sets for different targets	Detection of multiple viruses in a single sample (e.g., respiratory virus panels)
Digital PCR (dPCR)	Absolute quantification of viral nucleic acids	Sample partitioned into droplets; counts positive droplets	High-precision quantification, low-abundance targets
Hot Start PCR	Reduces non-specific amplification	Polymerase activation at higher temperatures	Enhanced specificity in complex samples
Touchdown PCR	Enhances specificity	Gradually decreasing annealing temperature in initial cycles	Improves specificity in cases of suboptimal primer binding

In the case of whole genome or targeted sequencing, library preparation involves fragmenting the DNA (if necessary), attaching adaptors, and amplifying the DNA fragments. These prepared libraries are then subjected to high-throughput sequencing using platforms such as Illumina, Oxford Nanopore (for direct RNA sequencing), or PacBio [45, 46]. The resulting sequence reads are compiled into digital formats such as FASTA or FASTQ, which can be further analyzed using bioinformatics tools.

A common first step in bioinformatics analysis is to perform a Basic Local Alignment Search Tool (BLAST) search [47], which compares the obtained sequence against known viral genomes in databases like GenBank. This helps in identifying the specific virus or strain present in the sample. To understand the evolutionary relationships of the identified virus, scientists often carry out phylogenetic analysis. This involves aligning the sequence with related viral genomes and constructing a phylogenetic tree to determine the virus's genetic similarity and lineage, revealing whether it clusters with known strains or represents a novel variant.

DNA sequencing plays a pivotal role not only in clinical diagnostics but also in epidemiological surveillance, mutation tracking, and vaccine development, especially when dealing with emerging or re-emerging viral pathogens.

Types of DNA sequencing technologies

- **Sanger sequencing:** This traditional method of sequencing is used to determine the exact nucleotide sequence of specific regions of the viral genome. It is useful for identifying mutations or genetic variations in the virus. It uses chain-terminating dideoxynucleotide triphosphate (ddNTPs) to terminate DNA synthesis at each nucleotide. The resulting fragments are separated by capillary electrophoresis, and the sequence is determined based on fragment lengths.

  It is often used for sequencing small DNA fragments, single-gene sequencing, and validation of NGS results.
- **NGS:** NGS technologies allow for the comprehensive sequencing of entire viral genomes, providing detailed information on viral genetic diversity, mutations, and evolutionary patterns. NGS is particularly valuable in outbreak investigations and studying emerging viruses.

  It is a high-throughput sequencing technology that allows rapid sequencing of large amounts of DNA or cDNA. It includes techniques such as Illumina sequencing, Ion Torrent, and sequencing-by-synthesis. NGS is used for whole-genome sequencing, exome sequencing, targeted sequencing, and metagenomics.

Types of RNA sequencing:

- **RNA-Seq:** Here, "bulk" RNA is converted into cDNA using reverse transcription, and then the cDNA is sequenced using NGS technologies. It provides information on transcript abundance, splicing, and gene expression.

  It is applicable for transcriptome analysis, gene expression profiling, differential expression analysis, and the discovery of novel transcripts in virus-infected cells.
- **Single-cell RNA sequencing (scRNA-Seq):** Like RNA-Seq but performed on individual cells. Therefore, it allows the study of gene expression at the single-cell level.

  Metagenomic sequencing: This technique involves sequencing all genetic material in a sample (RNA or DNA), allowing for the detection of known and unknown viruses without prior knowledge of the viral genome. It is useful in identifying novel or unexpected viral pathogens in clinical samples.

Phylogenetic Analysis Phylogenetic analysis is used to study the evolutionary relationships between viral strains by comparing their genetic sequences [48]. This helps in understanding the origin and spread of viruses, as well as the emergence of new variants. Also, by analyzing the genetic similarities and differences between viral strains, scientists can track the transmission routes of viruses and identify sources of outbreaks.

Phylogenetic analysis involves the use of genetic sequence data, where the genetic sequences of interest are aligned with reference sequences. This alignment allows for the comparison of genetic similarities and differences, which is essential for constructing a phylogenetic tree that illustrates the evolutionary relationships among the sequences. The process helps in identifying how closely related different organisms or strains of a virus are, based on their genetic makeup.

Basic steps for performing phylogenetic analysis

1. **Collection of sequences:** Obtain the nucleotide or protein sequences for the viruses isolated. Sequence data or reference sequence (control) can also be downloaded from public databases like GenBank.
2. **Sequence alignment:** Perform Multiple Sequence Alignment (MSA) using software tools like ClustalW, MUSCLE, or MAFFT to align sequences. Alignment arranges the sequences in a matrix so that homologous positions (e.g., nucleotides or amino acids) are compared across the sequences.
3. **Choose a substitution model:** By selecting an appropriate model of sequence evolution (e.g., Jukes-Cantor, Kimura 2-parameter), evolutionary changes that occur at different rates in different parts of the sequence can be accounted for.
4. **Phylogenetic tree construction:** Use software like MEGA, PhyML, RAxML, or BEAST to construct a phylogenetic tree based on aligned sequences. Common methods include:
   Neighbor-Joining (NJ): A distance-based method that groups sequences by minimizing the overall branch length.
   Maximum Likelihood (ML): A method that finds the tree most likely to produce the observed data, given a particular model of evolution.
   Bayesian Inference: Estimates the probability of a tree based on prior probabilities and the observed data.
5. **Tree visualization and analysis:** Use tools like FigTree, iTOL (Interactive Tree of Life), or Dendroscope to visualize the phylogenetic tree. These tools allow to label branches, adjust layouts, and interpret relationships.

   Perform a bootstrap analysis by resampling the data multiple times and reconstructing the tree each time. The bootstrap values on the branches indicate the confidence level of the inferred relationships.

   While interpreting the tree, look for clusters or clades that represent closely related organisms or strains. Compare these with known reference sequences to determine the evolutionary history or genetic relatedness.

How to conduct a molecular epidemiological surveillance during an outbreak

Case study

Amid ongoing global efforts to eradicate wild poliovirus (WPV), an outbreak of Acute Flaccid Paralysis (AFP) was reported in northern Nigeria. Although an active epidemiological surveillance and routine immunization campaign led by the WHO and the Nigerian government was already in place, a few AFP cases continued to be reported, raising concern about the possible silent circulation of WPV.

To address this concern, scientists initiated a focused molecular epidemiological investigation to better understand the etiological agents involved in the remaining AFP cases. Stool samples were collected from affected individuals as part of the AFP surveillance protocol. These samples were properly transported under cold chain conditions to the designated regional reference laboratory. Upon arrival, the samples underwent viral RNA extraction, and cDNA synthesis was performed using pan-enterovirus primers, which are designed to broadly amplify the enterovirus genus, including polioviruses and non-polio enteroviruses (NPEVs).

The resulting cDNA was then shipped to a reference sequencing facility in South Africa, where high-throughput sequencing was conducted. The generated nucleotide sequence data, provided in FASTQ format, was sent back to the regional laboratory for further analysis. Bioinformatic processing began with quality control and sequence alignment, enabling the assembly of the viral genome fragments. To identify the virus, a BLAST search was conducted against the National Center for Biotechnology Information (NCBI) nucleotide database. Sequences with 100% query coverage and 100% identity to a reference genome were considered definitive matches. In this case, the virus identified was not wild poliovirus, but a NPEV, effectively ruling out WPV circulation in the region.

To further understand the evolutionary relationship of the detected virus, a phylogenetic analysis was conducted. This step allowed the scientist to determine the virus's genetic lineage and its similarity to previously characterized strains, helping confirm whether it was a new introduction, a local variant, or part of an endemic strain (Figure 9.15).

This molecular approach provided clarity, reinforcing the conclusion that the observed AFP cases were not due to WPV but rather due to other circulating enteroviruses.

The methodology is a critical component of virological surveillance and can be extended to investigate other epidemiologically important viral pathogens, especially in regions undergoing eradication or control campaigns.

9.4.1.2 *Nucleic Acid Hybridization Tests*

These techniques rely on the hybridization of a labeled probe to a specific nucleic acid sequence but do not involve the amplification of nucleic acids. Major examples include:

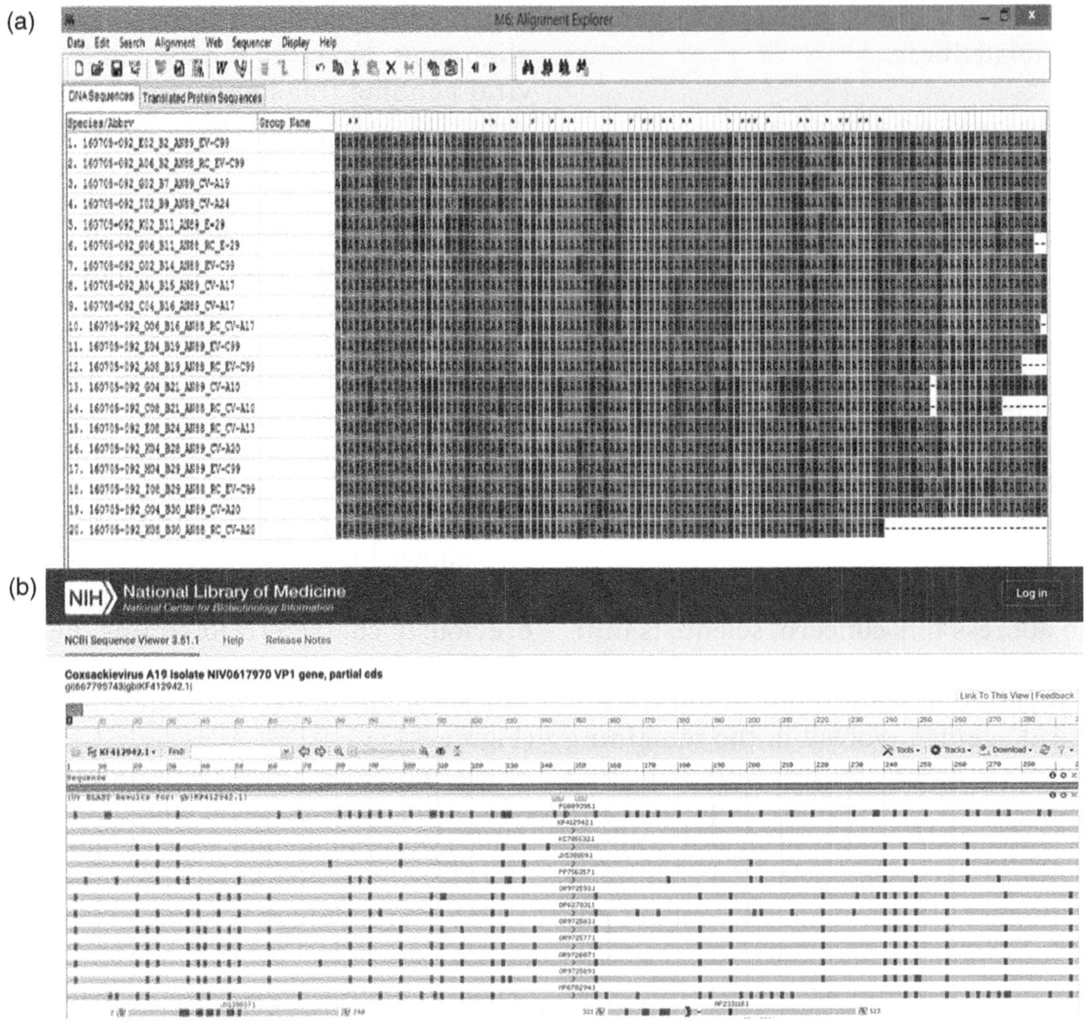

Figure 9.15 Alignment of one of the sequences (for CV-A20) using (a) MEGA 6 and (b) cluster W. Phylogenetic tree of (c) all isolates, (d) suspected PV1 isolates with other enteroviruses. The newly sequenced and reference strains are highlighted with black circle and triangle, respectively, and (e) phylogenetic analysis of suspected recovered CV-17 isolates. The newly sequenced and reference strains are highlighted with black circle and square, respectively.

Southern Blotting This technique is used to detect specific DNA sequences within a DNA sample. It involves the transfer of DNA fragments, separated by gel electrophoresis, onto a membrane, followed by hybridization with a labeled probe that binds to the target DNA sequence.

Northern Blotting This technique is like Southern blotting but is used for the

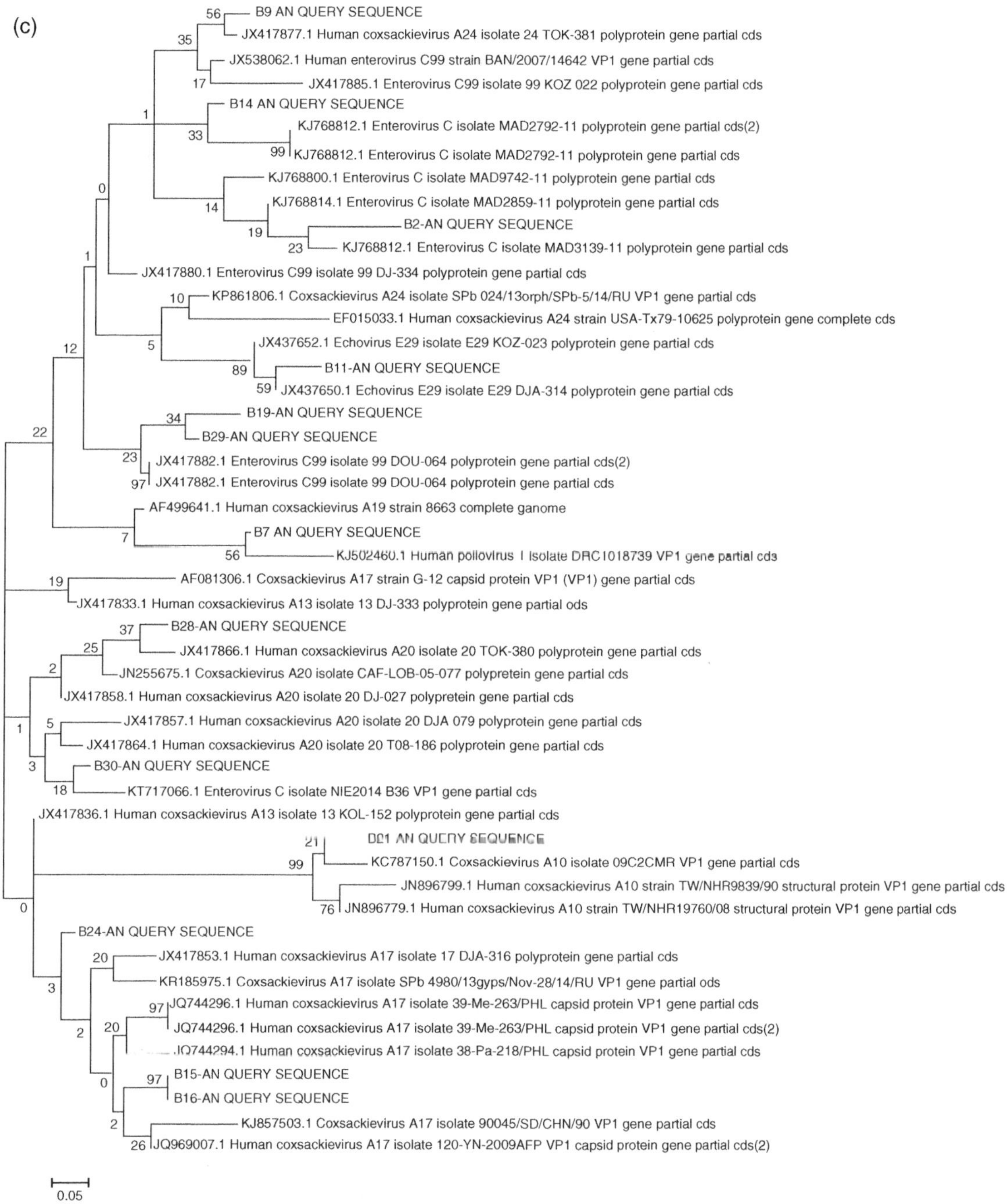

Figure 9.15 (Continued)

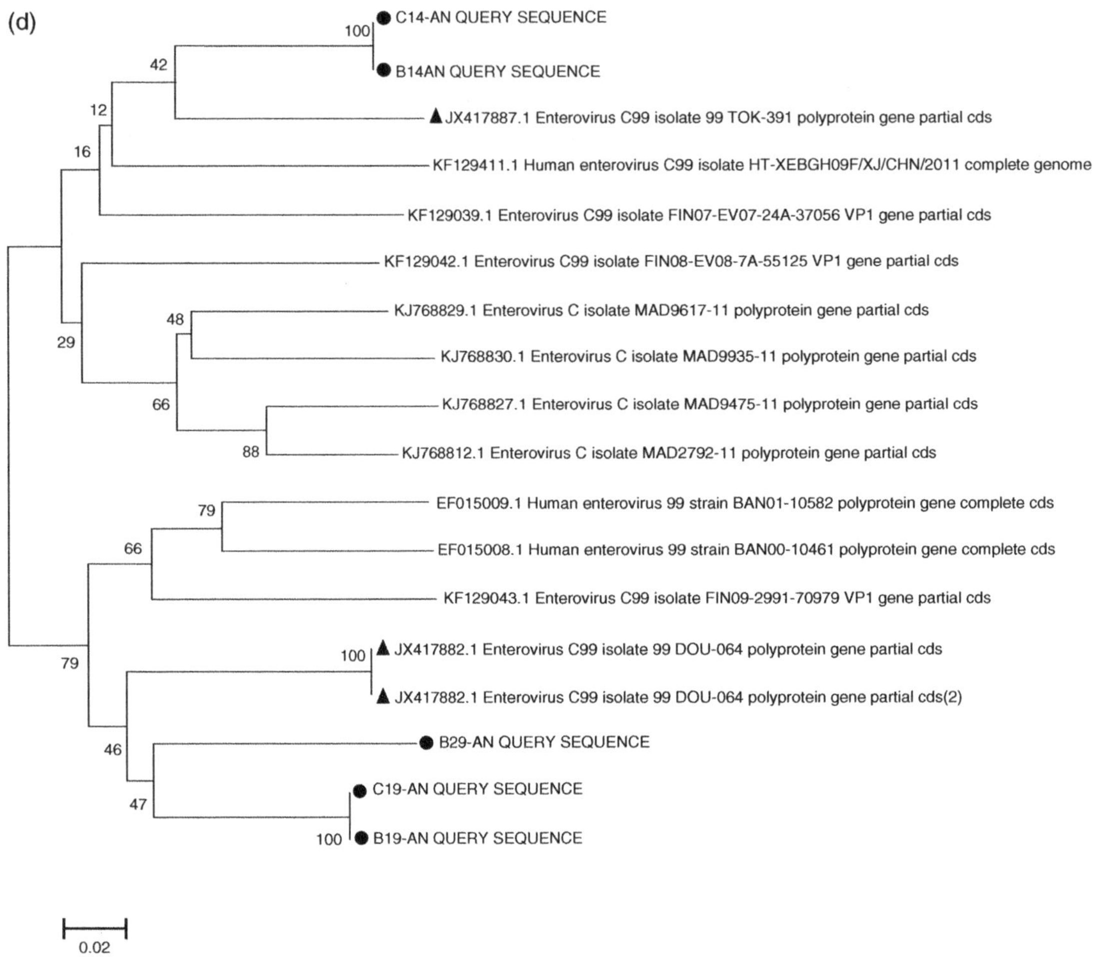

Figure 9.15 (Continued)

detection of specific RNA sequences. It involves the transfer of RNA from a gel onto a membrane, followed by hybridization with a labeled probe that binds to the target RNA sequence.

9.4.2 Western Blot (Proteomics)

Western blot, also known as immunoblotting, is a widely used analytical technique in molecular biology and biochemistry for detecting specific proteins in a sample. The process involves the separation of proteins based on their size using gel electrophoresis, typically sodium dodecyl sulfate-polyacrylamide gel electrophoresis (SDS-PAGE). Once separated, the proteins are transferred from the gel onto a membrane (usually made of nitrocellulose or polyvinylidene fluoride), where they are immobilized for further analysis.

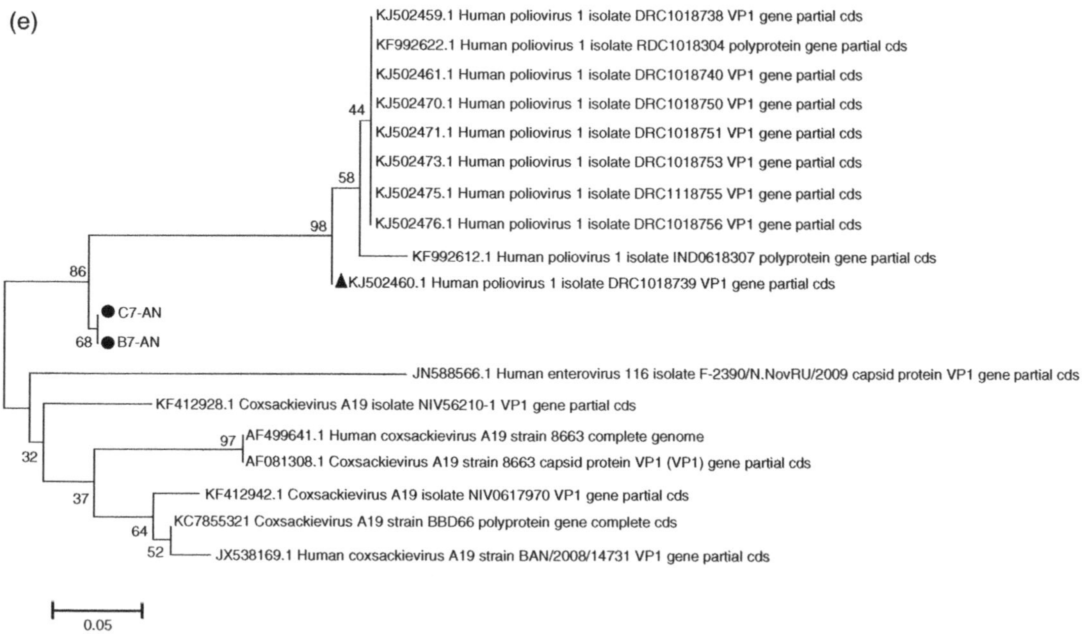

Figure 9.15 (Continued)

After transfer, the membrane is incubated with antibodies that specifically bind to the target protein. A primary antibody recognizes and binds to the protein of interest, and a secondary antibody, conjugated with a detection enzyme (such as horseradish peroxidase), binds to the primary antibody. The enzyme catalyzes a reaction with a substrate, resulting in a detectable signal, typically chemiluminescent or colorimetric, that reveals the presence and amount of the target protein.

In a virology laboratory, Western blotting is a critical technique used to detect and confirm the presence of specific viral proteins in infected samples. The method is often applied after virus propagation in cultured cells or after extraction from clinical samples such as blood, serum, or tissue. It allows virologists to identify viral proteins based on their molecular weight and confirm viral infections by targeting specific antigens.

One of the most common applications of Western blotting in virology is in HIV diagnosis, where it serves as a confirmatory test (or with combination of screening by NAT). Following a positive or indeterminate result from an ELISA, a Western blot is used to confirm the presence of HIV-specific proteins, such as gp120, gp41, and p24. The detection of these proteins is crucial for diagnosing and confirming HIV infection, as their presence indicates active viral replication and infection.

Running a Western blot involves several key steps to detect specific proteins from a sample (see Figure 9.16). Here is a detailed step-by-step guide:

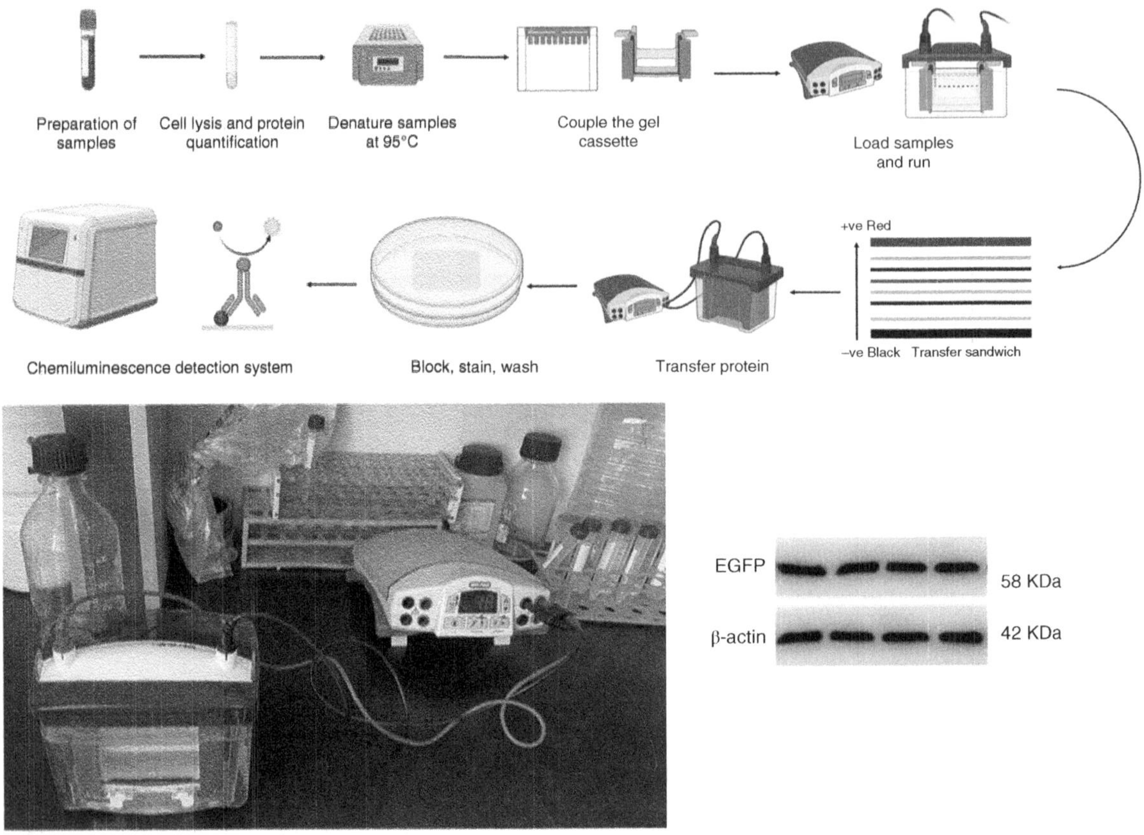

Figure 9.16 Systematic illustration of the steps in performing a Western blot.

1. **Sample preparation**
   - **Cell lysis:** Start by lysing the cells or tissues to release proteins. Use a lysis buffer containing protease inhibitors to prevent protein degradation. Serum can be used directly.
   - **Protein quantification:** Measure the protein concentration of the lysate using methods like the Bradford or Bicinchoninic Acid (BCA) assay.
   - **Denaturation:** Mix the protein samples with 2-4X loading buffer containing SDS, like Laemmli buffer. Then boil samples at 95 °C for 5–10 minutes to denature proteins, ensuring uniform migration in the gel.
2. **Gel electrophoresis**
   - **Prepare the gel:** Use a polyacrylamide gel with SDS-PAGE. Choose an appropriate gel percentage based on the molecular weight of the target protein (e.g., 10% for mid-sized proteins). Note that the gel contains the stacking layer and the running layer.

- **Load the gel:** Load equal amounts of the protein samples (e.g., 20–40 μg) into the wells of the gel. Also, load a molecular weight marker (protein ladder) in one well.
- **Run the gel:** Fix the gel cassette into the buffer tank and add the running buffer. Apply an electric current to the gel. Allow the protein to run at 80–90 V (20 minutes) at the stacking layer of the gel, and 100–120 V (50–70 minutes) at the running layer of the gel. This ensures that proteins move toward the positive electrode based on size.

3. **Protein transfer**
   - **Prepare transfer setup:** After electrophoresis, transfer the separated proteins from the gel onto a nitrocellulose or polyvinylidene difluoride (PVDF) membrane. Pre-wet or activate the membrane in methanol for 1 minute if using PVDF.
   - **Assemble transfer sandwich:** Place the gel and membrane between filter papers and sponges, forming a "sandwich."
   - **Transfer:** Place the sandwich into a transfer apparatus containing transfer buffer and run at 100 V (250A) for 1–2 hours or overnight at 20 V (wet transfer method). Semidry or dry transfer systems may have different voltage and time requirements.
4. **Blocking the membrane**
   - **Block the membrane:** Incubate the membrane in a blocking buffer (e.g., 5% nonfat dry milk or BSA in Tris-buffered saline with Tween 20 [TBST]) for 1 hour at room temperature or overnight at 4 °C. This step prevents non-specific binding of antibodies to the membrane.
5. **Antibody Incubation**
   - **Primary antibody incubation:** Incubate the membrane with an appropriate primary antibody specific to the target protein. Dilute the antibody in blocking buffer and incubate for 1–2 hours at room temperature or overnight at 4 °C.
   - **Washing:** Wash the membrane with PBST or TBST buffer (TBS + 0.1% Tween 20) for 5–10 minutes, three times, to remove excess antibodies.
   - **Secondary antibody incubation:** Incubate the membrane with a secondary antibody that is conjugated to an enzyme (such as HRP) or a fluorescent dye. This antibody binds to the primary antibody. Incubate for 1 hour at room temperature.
   - **Washing:** Wash the membrane again with TBST buffer three times to remove unbound secondary antibodies.
6. **Detection**
   - **Prepare substrate:** If using horseradish peroxidase (HRP)-conjugated antibodies, apply an enhanced chemiluminescence (ECL) substrate to the membrane. The substrate reacts with HRP to produce light.
   - **Capture signal:** Use a chemiluminescence detection system

(e.g., X-ray film or a digital imager) to capture the light emitted by the reaction. Alternatively, for fluorescent antibodies, use a fluorescence imager.

- **Analyze results:** The bands on the membrane represent the target protein(s). Compare their position to the molecular weight marker to confirm the protein size.

7. **Stripping and re-probing (optional)**: This is used to detect multiple proteins on the same membrane. Strip off the primary and secondary antibodies using a stripping buffer, then re-probe the membrane with another set of antibodies.
8. **Data analysis**
   - **Quantify bands:** Use imaging software to quantify the intensity of protein bands, comparing them to control samples or housekeeping proteins (e.g., β-actin or GAPDH) to normalize the results.

Recipe for 1X Tris-glycine transfer buffer (1 L)

a. Tris base (25 mM): 3.03 g
b. Glycine (192 mM): 14.4 g
c. Methanol (20%): 200 mL
d. Distilled water: Up to 1 L

Recipe for 1X Tris-glycine-SDS running buffer (1 L)

a. Tris Base (25 mM): 3.03 g
b. Glycine (192 mM): 14.4 g
c. SDS (0.1%): 1 g
d. Distilled water: Up to 1 L

9.4.3 Immunoassays

9.4.3.1 Principles of Immunoassays

Immunoassays are biochemical tests that measure the presence or concentration of a substance, typically an antigen (virus, virus protein, or nucleic acid) or antibody, using the principle of antigen–antibody interactions. The core principle of immunoassays is the specific binding between an antigen (a molecule capable of inducing an immune response) and its corresponding antibody (a protein produced by the immune system to neutralize the antigen). This binding is highly specific, like a lock-and-key mechanism, allowing the immunoassay to detect or quantify specific substances in a sample.

9.4.3.2 Direct Vs. Indirect Immunoassays

Direct Immunoassay In a direct immunoassay, the patient sample is tested for the presence of a viral antigen. The surface (e.g., a microplate well or membrane) is pre-coated with a capture molecule or sometimes directly with the patient sample. A labeled primary antibody is then added, which binds directly to the antigen (e.g., a viral protein) present in the sample. The label, commonly an enzyme, fluorescent tag, or chemiluminescent tag, produces a measurable signal.

Indirect Immunoassay In an indirect immunoassay, an unlabeled primary antibody binds to the target antigen (an antigen-coated surface is used to screen for an antigen or antibody). Then, a labeled secondary antibody, which specifically binds to the primary antibody, is added. The signal is generated by the label on the secondary antibody. Indirect immunoassays can be more sensitive because the signal is amplified by the binding of multiple secondary antibodies to each primary antibody. However, they involve more steps and can take longer to perform.

Noncompetitive Immunoassay In an indirect immunoassay, the surface is coated with a known viral antigen (not the patient's sample). The patient's serum or plasma sample is then added, and if the patient has antibodies against that viral antigen (indicating prior or ongoing infection), they will bind to it.

A labeled secondary antibody (e.g., anti-human immunoglobulin G [IgG] or IgM) is then applied, which binds to the patient's antibodies. The label on the secondary antibody generates a detectable signal.

Both direct and indirect immunoassays can be noncompetitive. For example, in a noncompetitive indirect ELISA, the antigen is first captured by an antibody on the plate, and then a secondary labeled antibody binds to the antigen to generate the signal.

9.4.3.3 Solid Phase Vs. Homogeneous Assays

Solid Phase Immunoassays These assays use solid support, such as a microtiter plate, bead, or membrane, to immobilize either the antigen or antibody. This allows for easy separation of bound from free components, improving assay sensitivity and specificity (see Table 9.12 and 9.13).

9.4.3.4 Types of Detection Methods of Immunoassay

i. **Colorimetric:** Involves the use of a color change system (e.g., enzyme) that can be measured by spectrophotometry (e.g., ELISA).
ii. **Fluorescent:** Uses fluorescent labels, with the signal detected by a fluorometer or fluorescent microscopy. See Figures 9.17 and 9.18.
iii. **Chemiluminescent:** Involves the emission of light during a chemical reaction, detected by a luminometer, e.g., Western blot, Northern blot.
iv. **Radioactive:** Use radioactive isotopes as labels, detected by scintillation counting (e.g., radioimmunoassay).

9.4.3.5 Limitations of Serological Tests in Viral Diagnosis

Serological tests detect antibodies (such as IgM or IgG) or viral antigens in a patient's blood or other body fluids. While these tests are valuable for understanding exposure history and immune response, they have several limitations, especially in the diagnosis of acute viral infections:

Window period (delay in antibody production): After infection, it takes time for the immune system to produce detectable levels of antibodies. This window period means serological tests may yield false-negative results in the early stages of infection when there is virus infection but antibodies are not yet present.

Table 9.12 Overview of the types of immunoassays.

Category	Type of immunoassay	Description	Applications
Non-microscopy-based immunoassay	Enzyme-linked immunosorbent assay (ELISA)	A technique that uses enzymes linked to antibodies to detect antigens or antibodies. Signal generation is typically through a color change.	Used for detecting and quantifying proteins, antigens, and antibodies in clinical diagnostics and research.
	Western blot (immunoblotting)	A method that separates proteins by gel electrophoresis, followed by transfer to a membrane and detection using specific antibodies.	Used to confirm the presence of specific proteins, often in conjunction with ELISA results.
	Rapid immunochromatographic tests	A quick and easy test using a membrane-based strip with an antigen or antibody, providing rapid results.	Used in point-of-care testing for infectious diseases, pregnancy tests, and drug detection.
	Neutralization assay	A test to measure the ability of antibodies to neutralize the infectivity of a virus or toxin.	Use in virology to determine the efficacy of vaccines and the immune response to viruses.
	Radioimmunoassay (RIA)	A sensitive technique that uses radioactively labeled substances to quantify antigen–antibody interactions.	Used to measure hormones, drugs, and other substances in clinical and research settings.
Microscopy-based immunoassay	Immunohistochemistry (IHC) and immunocytochemistry	Techniques that use antibodies to detect specific antigens in cells or tissue sections are visualized under a microscope.	Used in pathology to diagnose diseases by identifying specific markers in tissue samples.
	In situ hybridization (ISH) with immunoassay	A method that combines DNA/RNA hybridization with immunohistochemistry to detect specific genetic sequences in cells or tissues.	Use for detecting viral genomes, gene expression, and chromosomal abnormalities in clinical diagnostics.

Table 9.13 Examples of immunoassays.

Test Name	Target	Purpose	Interpretation
ELISA (IgM)	IgM antibodies	Detect recent or acute viral infection	Positive: Recent infection
			Negative: No recent infection or window period
ELISA (IgG)	IgG antibodies	Determine past exposure or immunity	Positive: Past infection or vaccination
			Negative: No past exposure
IgM/IgG combo ELISA	IgM and IgG	Stage of infection	IgM+/IgG–: Early infection
			IgM+/IgG+: Recent infection
			IgM–/IgG+: Past infection
Neutralization test	Virus-specific antibodies	Confirm protective antibodies	Positive: Neutralizing antibodies present
			Negative: No protection
Rapid immunochromatographic test	Viral antigens/ antibodies	Point-of-care diagnosis	Two lines: Positive
			One control line: Negative
			No control line: Invalid
Immunofluorescence assay (IFA)	Viral antigen	Detect viral antigen in tissues/cells	Fluorescence present: Antigen detected
			Absent: Antigen not detected
Western blot	Viral proteins	Confirmatory test for antibody detection	Bands match control: Positive
			No bands or mismatch: Negative or indeterminate
Latex agglutination test	Antigens/antibodies	Rapid screening	Agglutination: Positive
			No agglutination: Negative
Complement fixation test	Antigen–antibody complex	Historically, limited use	Complement used up: Positive
			Complement remains: Negative

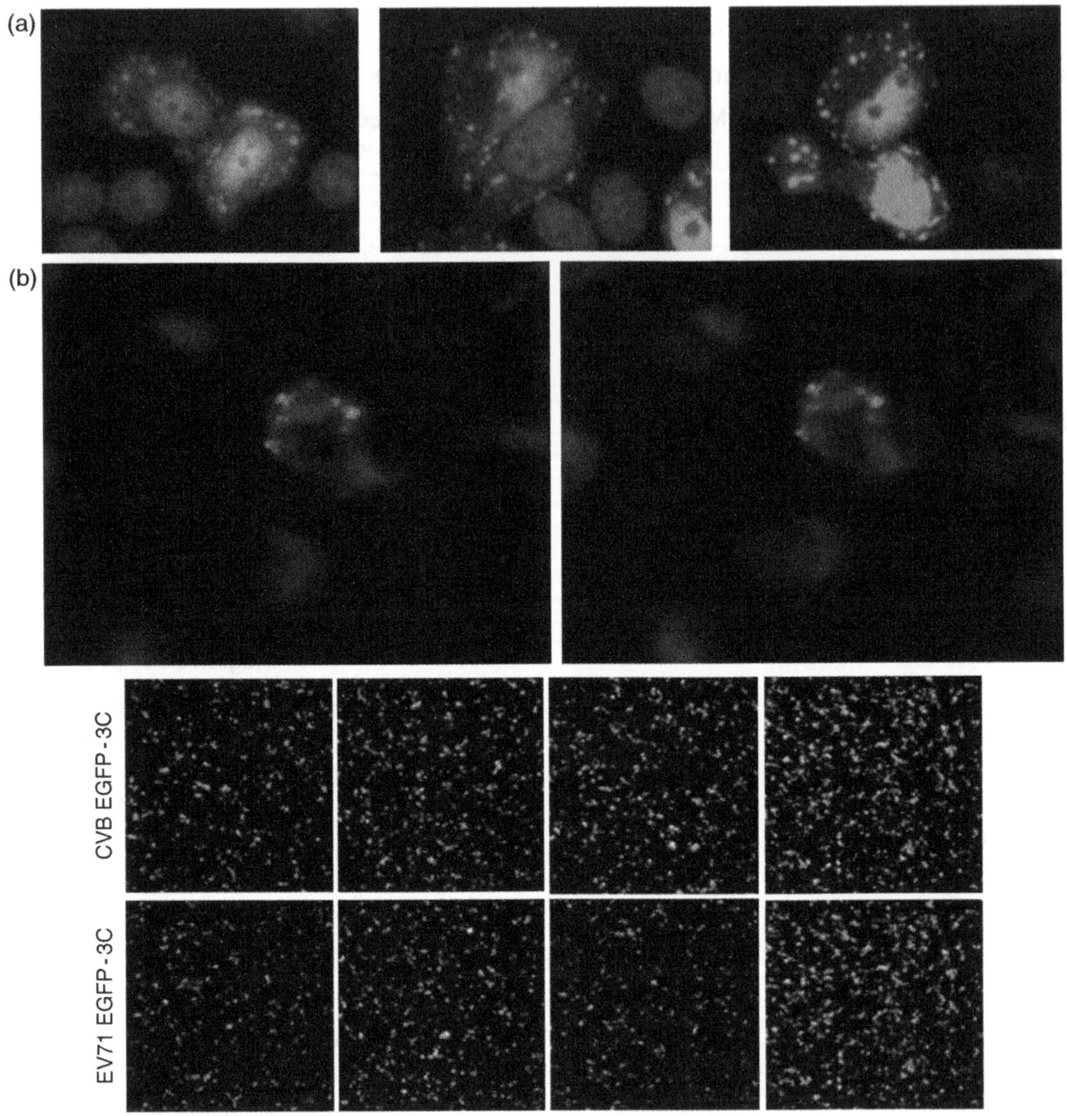

Figure 9.17 (a) Images of 293T cells transfected with a plasmid encoding the coxsackievirus B3 (CVB3) fluorescence protein. The nucleus was stained with Hoechst 33342 stain. Photo credit: Dr. Zhong's lab, HMU, China. (b) Images of 293T cells transfected with a plasmid encoding the coxsackievirus B3 (CVB3) fluorescence protein.

Source: Photo by Prof. Zhaohua Zhong, Harbin Medical University, China.

Cross-reactivity and false positives: Antibodies generated against one virus may cross-react with antigens of related viruses, leading to false-positive results. This is particularly problematic in areas where closely related viruses (e.g., flaviviruses like dengue, Zika, or yellow fever) co-circulate.

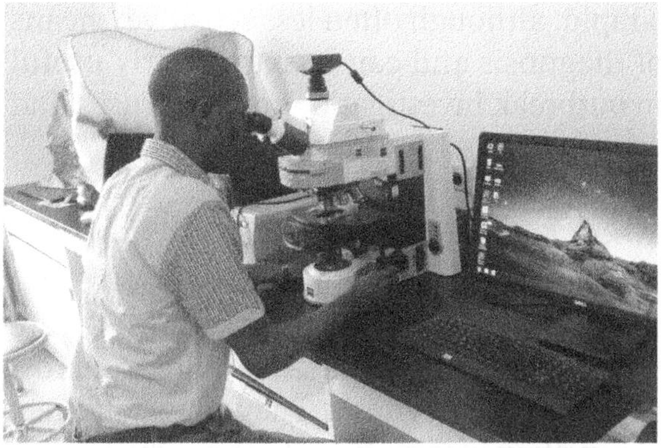

Figure 9.18 A scientist views a fluorescence microscope.
Source: Photo by Dr Aroge Temitope.

Inability to distinguish between active and past infection: The presence of IgG antibodies may indicate past exposure or vaccination, not necessarily an active infection. Thus, serological results must be interpreted cautiously and often need confirmation by molecular methods like RT-PCR.

Variable antibody response: Some patients, especially those who are immunocompromised or infected with certain viruses, may have a weak or delayed antibody response, leading to low sensitivity of the test.

Quality of test kits: The sensitivity and specificity of serological assays vary across manufacturers. Low-quality kits may yield inconsistent results, especially in low-resource settings.

Timing and interpretation complexity: The interpretation of IgM versus IgG levels can be complex. For example, IgM may persist for months in some viral infections, making it difficult to differentiate recent from older infections.

Limited use in early outbreak detection: Since serological tests are dependent on host response, they are less suitable for early outbreak detection or rapid containment strategies where immediate virus identification is crucial.

9.4.3.6 *Rapid Diagnostic Tests (RDTs)/Point-of-Care Testing in Virology*

Rapid diagnostic tests are immunochromatographic assays designed to detect viral antigens or antibodies within minutes, often at the patient's bedside or in low-resource settings. They are a form of point-of-care testing, aimed at providing immediate clinical decisions without the need for centralized laboratory infrastructure.

Advantages

Speed: Results are available within 10–30 minutes.

Simplicity: Requires minimal technical training; easy to perform.

Portability: Suitable for field settings and remote locations.

Early containment: Useful during outbreaks for screening and isolation.

Limitations

Lower sensitivity and specificity compared to molecular methods like RT-PCR.

Risk of false negatives, especially in early infection (before antigen/antibody levels are detectable).

Cannot provide viral load or genotype information.

Confirmation is often required by PCR or ELISA.

Examples in virology: HIV antibody tests (e.g., Determine HIV-1/2 test), Hepatitis B surface antigen (HBsAg) tests, SARS-CoV-2 antigen/antibody rapid tests (Figure 9.19).

9.4.4 Direct Microscopy

Direct microscopy of viruses is a powerful diagnostic tool that allows for the immediate visualization of viral particles in clinical specimens. This method can be particularly useful for the rapid identification of viral infections. There are a few common microscopy techniques used for viral detection. Although less commonly used compared to molecular techniques like PCR, it remains a valuable tool in specific scenarios, especially for viruses with distinct morphological or cytopathic characteristics. It provides a rapid, although often less sensitive, means of diagnosis and can be particularly useful in outbreak investigations or for viruses that cannot be easily cultured.

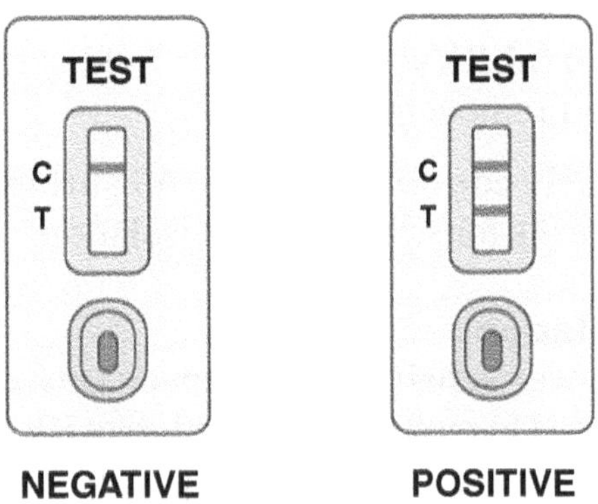

Figure 9.19 Putative example of a positive and negative result rapid diagnostic test.

9.4.4.1 Electron Microscopy (EM)

Transmission electron microscopy (TEM) and scanning electron microscopy (SEM) are the primary methods used for direct microscopy of virus diagnosis. They allow for direct observation of virus morphology, which helps in identifying viral families or genera. They can be used for diagnosing viruses that are difficult to culture (see Figures 9.20–9.22).

Limitations

Requires specialized equipment and expertise.

Low sensitivity compared to molecular techniques; it often requires a high viral load in the specimen for detection.

Typically used for identifying viruses with distinct shapes, such as herpesviruses, poxviruses, or rotaviruses.

9.4.4.2 Light Microscopy (Cytopathic Effects and Inclusion Bodies)

Light microscopy can also be used to observe cytopathic effects caused by viruses on infected cells or tissue samples. Certain viruses create distinct changes in host cells (e.g., syncytia formation, inclusion bodies), which can provide clues to their identification. Helpful in identifying viruses like CMV through observation of characteristic "owl's eye" inclusion bodies or rabies virus by the presence of Negri bodies.

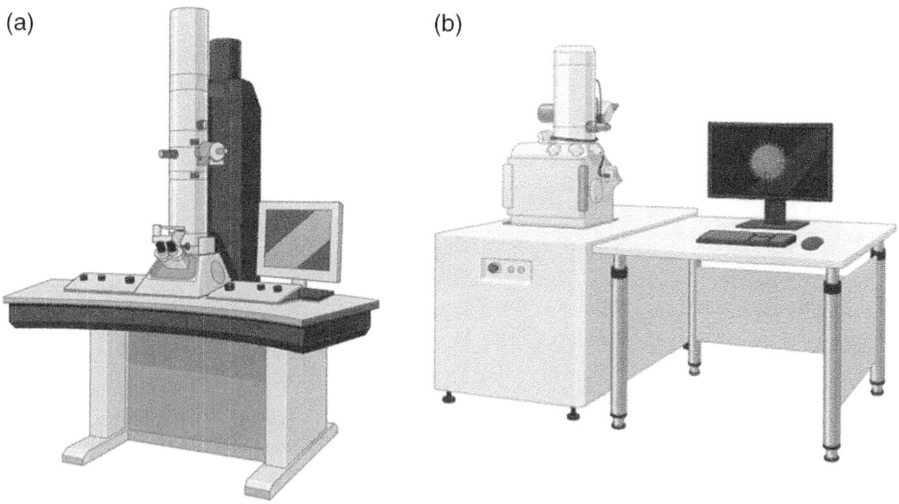

Figure 9.20 Illustrative image of (a) scanning electron microscope and (b) transmission electron microscope.

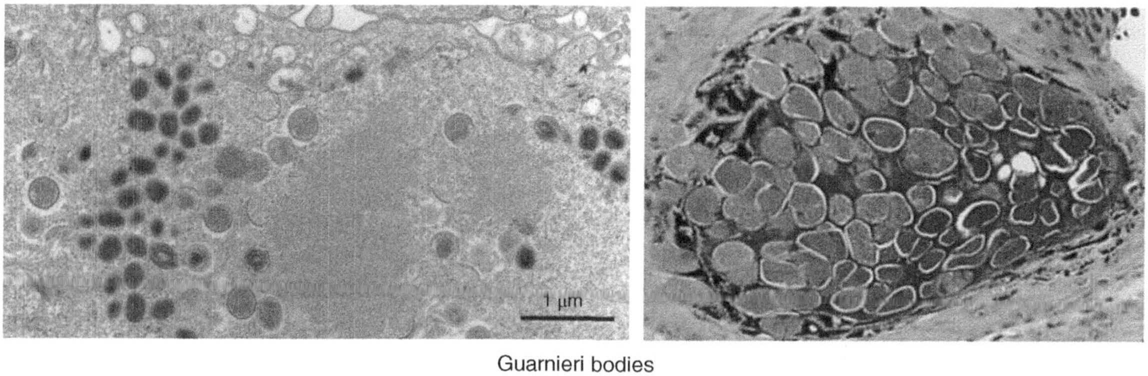

Figure 9.21 Electron microscopy and basophilic stain of Guarnieri bodies typical of smallpox (variola) and vaccinia-infected cells.

Source: Photo by Hans R. Gelderblom.

9.4.4.3 Immunofluorescence Microscopy (IFM)

Involves tagging viral antigens with fluorescent-labeled antibodies to directly visualize viruses in samples. It is more sensitive and specific than traditional light microscopy because it allows for the detection of specific viruses based on antigen–antibody interaction. Commonly used in the diagnosis of respiratory viruses, herpesviruses, and measles virus.

However, it requires specific antibodies and equipment (fluorescent microscope) and may not work for all viruses, depending on antibody availability.

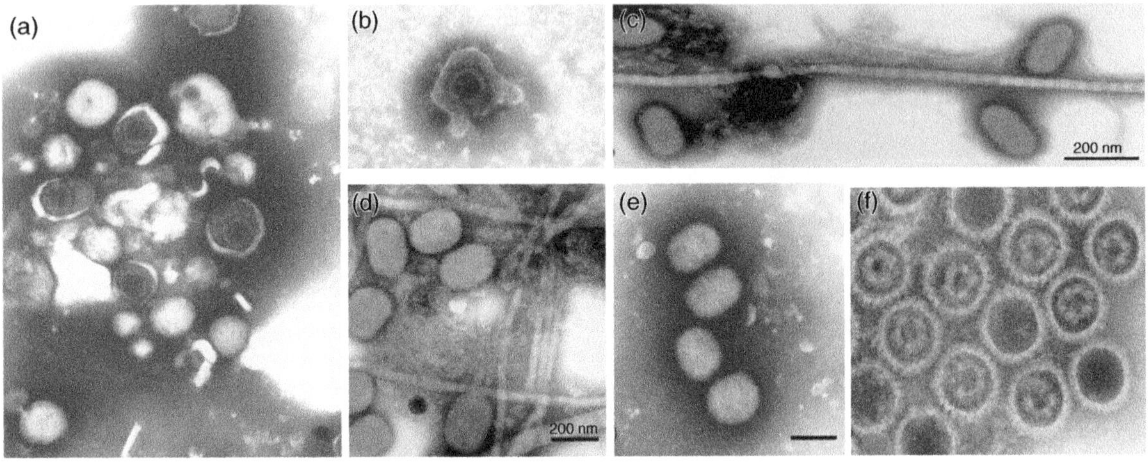

Figure 9.22 (a) Fluid collected from a vesicle in the later stages of the illness reveals clusters of herpesvirus particles, stained by PTA, along with some cellular debris. VZV infection. (b) TEM of biopsy sample, typical herpesvirus. (c) NS-TEM of biopsy showing Parapoxvirus (sample courtesy of Prof. Friedrich A. Bahmer, Bremen). (d and e) NS-DEM of the contents of MC papules, structure typical of orthopoxvirus. (f) Negative stain of Herpes capsule.

Source: Photo by Hans R. Gelderblom.

9.5 Recording, Reporting, and Interpretation of Results in a Virology Laboratory

Accurate recording and systematic reporting of laboratory results are fundamental components in the workflow of any virology laboratory. These practices not only ensure the integrity and reliability of data collected but also form the backbone of effective scientific communication. Precise documentation allows for unambiguous interpretation of results by researchers, clinicians, and other stakeholders, thereby reducing the risk of errors or misinterpretation that could compromise patient diagnosis or the validity of research findings.

The process of recording results involves carefully logging raw data, observations, and analytical outcomes in a timely and organized manner, using standardized formats and validated tools such as laboratory information management systems (LIMS). This meticulous approach facilitates traceability, allowing each result to be linked back to its original sample and testing conditions. Additionally, it supports reproducibility by providing a comprehensive record that other researchers can review or replicate under similar conditions. Effective documentation also promotes collaboration across multidisciplinary teams and institutions by providing a transparent and accessible record of experimental procedures and outcomes. This fosters peer review, data validation, and integration of findings into broader scientific knowledge, ultimately accelerating advancements in virology research and patient care.

Reporting laboratory findings extends beyond mere data presentation; it requires contextualizing results within the clinical or experimental framework, highlighting significant observations, and indicating limitations or uncertainties where applicable. Clear, concise, and standardized reporting formats are critical to ensure that results are easily understandable and actionable by healthcare providers or researchers.

Furthermore, adherence to established regulatory and accreditation standards, such as those set forth by national organizations or the International Organization for Standardization (ISO), is paramount. Proper recording and reporting protocols contribute to compliance, enabling laboratories to maintain certifications and operate within legal and ethical guidelines.

By committing to rigorous standards in recording and reporting, virology laboratories enhance their credibility, support evidence-based decision-making, and contribute meaningfully to public health initiatives and therapeutic development (see Table 9.14).

Interpretation of results in a virological laboratory

Interpreting laboratory results is a critical step in virology that transforms raw data into meaningful clinical or research information (see Table 9.15 for examples of interpretation of Hepatitis B virus serological test). Proper interpretation requires understanding the test purpose, the biology of the virus, the host immune response, and the characteristics of the diagnostic assay used. Inaccurate interpretation can lead to misdiagnosis, inappropriate treatment, or flawed research

Table 9.14 Summary of examples of laboratory results and reporting parameters in virology.

Serology tests		Nucleic acid base tests	
Qualitative assay	**Quantitative assay**	**Qualitative assay**	**Quantitative assay**
Positive	Appropriate values (greater than the reference value) in the SI unit	Positive	Appropriate values (greater than the reference value) in the SI unit
Negative	Appropriate values (lesser than the reference value) in the SI unit	Negative	Appropriate values (lesser than the reference value) in the SI unit
Indeterminate		Indeterminate	Indeterminate
Insufficient sample	Insufficient sample	Insufficient sample	Insufficient sample
Test to be repeated	Test to be repeated	Test to be repeated	Test to be repeated
Pending	Pending	Pending	Pending

Table 9.15 Interpretation of hepatitis B serological markers.

HBsAg	Anti-HBc (IgM)	Anti-HBc (total)	Anti-HBs	HBeAg	Anti-HBe	Interpretation
+	+	+	−	+	−	**Acute HBV infection:** Active infection with high infectivity; recent exposure; patient may be symptomatic.
+	−	+	−	+	−	**Chronic active HBV infection:** Ongoing viral replication, high infectivity, but not recent infection.
+	−	+	−	−	+	**Chronic inactive carrier:** Virus present, but low replication and infectivity; seroconversion to anti-HBe positive.
−	−	+	+	−	−	**Resolved HBV infection**: Past infection cleared; immunity developed.
−	−	−	+	−	−	**Immunity due to vaccination:** No infection, but protective antibodies present from vaccination.
−	−	−	−	−	−	**Susceptible individual:** No evidence of infection or immunity; vulnerable to HBV infection.
−	+	+	−	+	−	**Window period of acute infection:** HBsAg may be disappearing, anti-HBs not yet developed, and anti-HBc IgM positive.

conclusions. Virological tests often detect viral antigens, antibodies, or viral nucleic acids. The presence or absence, and the levels of these markers must be considered in the context of the patient's clinical history, vaccination status, and the timing of sample collection relative to infection or exposure.

Results should not be presented as Reactive/Non-reactive/Detected/ Not detected due to possible misinterpretation of results by clinicians.

"Reactive" can be misheard as "nonreactive" and vice versa. Where possible results should be reported as NEGATIVE OR POSITIVE.

9.6 Study Questions

9.1 A 45-year-old male patient presents to the outpatient department with a high fever, severe headache, muscle pain, and a rash that began 3 days ago. He reports that the symptoms started suddenly, and he has been feeling progressively worse. The patient has a history of recent travel to a tropical region known for outbreaks of viral infections. A physical examination reveals a maculopapular rash on his trunk and limbs. Laboratory investigations show leukopenia and elevated liver enzymes. Given the clinical presentation and recent travel history, the attending physician suspects a viral infection, possibly a flavivirus or alphavirus, and decides to send clinical samples for virus cultivation and characterization to confirm the diagnosis.

(a) What types of clinical samples would be appropriate for virus cultivation in this case, and why?
(b) Describe the steps involved in the virus cultivation process from these clinical samples.
(c) What cell culture systems would be suitable for the isolation of viruses suspected in this patient, and what are the criteria for selecting these systems?
(d) Once a virus is successfully cultivated, outline the methods that can be used for its characterization and identification.
(e) Discuss the biosafety considerations that must be considered during the cultivation and characterization of the virus from these clinical samples.
(f) How would the identification and characterization of the virus impact the clinical management of the patient?

9.2 A 50-year-old male presents to the emergency department with a fever, fatigue, dry cough, and difficulty breathing. He also reports experiencing confusion and headaches over the past two days. The patient has a history of traveling to East Asia and chronic obstructive pulmonary disease (COPD) and is currently on medication. He mentions that he recently attended a large family gathering. Physical examination reveals crackles in the lungs upon auscultation, and the patient is slightly disoriented.
Given the symptoms, the physician suspects a viral infection, possibly a respiratory virus such as the Middle

East respiratory syndrome coronavirus (MERS-CoV) or SARS-CoV-2. To confirm the diagnosis, the physician decides to send clinical samples for virus cultivation and characterization.

(a) What criteria should be used to prioritize testing for viral respiratory pathogens in this patient, considering his symptoms and history?
(b) Describe the role of virus cultivation in confirming a diagnosis of MERS-CoV or SARS-CoV-2. What advantages does it offer over molecular testing methods like PCR?
(c) What are the ethical considerations when handling and processing potentially infectious clinical samples, especially with viruses known to cause outbreaks?
(d) What steps would you take if the virus does not grow in the initial cell culture attempts? How would you troubleshoot the cultivation process?
(e) How can virus cultivation and subsequent characterization contribute to understanding the epidemiology and transmission dynamics of MERS-CoV or SARS-CoV-2 in the community?
(f) If a novel respiratory virus is suspected, what additional steps might be taken in the characterization process to identify and study this new virus?

9.3 A 60-year-old male patient presents to the hospital with symptoms of fever, severe cough, shortness of breath, and muscle pain that began a week ago. The patient reports that his symptoms have been progressively worsening. He also mentions that a few of his family members and neighbors have recently had similar symptoms. Physical examination reveals bilateral crackles upon lung auscultation, and chest X-rays show diffuse interstitial infiltrates. The attending physician suspects a viral infection, possibly a respiratory virus, and decides to use immunoassays to detect and characterize the virus.

Questions:

(a) What type of immunoassays would be appropriate for detecting and characterizing the suspected respiratory virus in this patient, and why?
(b) Which clinical samples would be ideal for performing these immunoassays, and at what stage of the illness should they be collected?
(c) Explain the principle behind the immunoassays used for detecting respiratory viral infections. How can these tests help differentiate between different respiratory viruses, such as RSV, influenza virus, and human metapneumovirus?
(d) Discuss the advantages and limitations of using immunoassays for the diagnosis and characterization of respiratory viral infections. What are the potential challenges in interpreting the results?
(e) How would the identification and characterization of the specific respiratory virus impact the clinical management and treatment of the patient, as well as infection control measures?

Objectives

9.1 **Which of the following methods is most used to determine the specific tropism of a virus during its cultivation in cell culture?**
(a) Hemagglutination inhibition assay
(b) Immunofluorescence microscopy
(c) Plaque assay
(d) Neutralization test

9.2 **Which factor is least likely to affect the outcome of a plaque assay?**
(a) The concentration of the virus in the inoculum
(b) The type of cell line used for the assay and the type of virus
(c) The duration of the incubation period
(d) The temperature at which the cell culture is maintained

9.3 **When characterizing a newly isolated virus from a clinical sample, which of the following techniques is most appropriate for determining the viral envelope structure?**
(a) Transmission electron microscopy
(b) Western blotting
(c) ELISA
(d) RT-PCR

9.4 **Which of the following methods would be least effective in distinguishing between two closely related viruses that differ primarily in their surface glycoproteins?**
(a) Hemagglutination inhibition assay
(b) Immunofluorescence microscopy
(c) PCR amplification of viral RNA
(d) ELISA for surface proteins

9.5 **What is the primary advantage of using primary cell cultures over continuous cell lines for virus isolation and characterization?**
(a) Higher susceptibility to a wider range of viruses
(b) Consistency in cell behavior over prolonged periods
(c) Easier to maintain and propagate
(d) Ability to provide a more uniform environment for viral replication

9.6 **In the process of characterizing a virus isolated from a clinical sample, which of the following methods would provide the most definitive evidence of viral replication?**
(a) Detection of viral proteins using immunoassays
(b) Observing cytopathic effect in cell culture
(c) Quantification of viral RNA by RT-PCR
(d) Measurement of virus-induced hemagglutination

9.7 **Which of the following is the primary purpose of virus cultivation from clinical samples?**
(a) To develop vaccines
(b) To identify bacterial contaminants
(c) To detect and identify viral pathogens
(d) To produce antibiotics

9.8 **Which cell culture system is commonly used for the cultivation of viruses from respiratory samples?**
(a) Hep-2 cells
(b) Vero cells
(c) Madin–Darby canine kidney (MDCK) cells
(d) HeLa cells

9.9 Which of the following clinical samples is most appropriate for detecting HSV via virus cultivation?
(a) Blood
(b) CSF
(c) Throat swab
(d) Vesicular fluid

9.10 What is the role of immunofluorescence in the characterization of viruses in cell culture?
(a) To detect bacterial contamination
(b) To observe cytopathic effects
(c) To identify viral antigens in infected cells
(d) To stain cell nuclei

9.11 Which of the following techniques is used to quantify the relative number of infectious virus particles in a sample?
(a) Hemagglutination assay
(b) Polymerase chain reaction (PCR)
(c) Plaque assay
(d) Western blot

9.12 In virus isolation, which biosafety level is generally required for handling highly pathogenic viruses like Ebola?
(a) Biosafety Level 1 (BSL-1)
(b) Biosafety Level 2 (BSL-2)
(c) Biosafety Level 3 (BSL-3)
(d) Biosafety Level 4 (BSL-4)

9.13 What is a key indicator of successful virus cultivation in cell culture?
(a) Bacterial growth on the culture medium
(b) Presence of viral RNA in the cell culture supernatant
(c) Observation of cytopathic effect
(d) Change in color of the culture medium

9.14 Which method is commonly used to identify a virus after successful cultivation?
(a) Gram staining
(b) ELISA
(c) NGS
(d) Mass spectrometry

9.15 What is the purpose of using a plaque assay in virus characterization?
(a) To determine the viral genome sequence
(b) To measure the ability of antibodies to neutralize viral infectivity
(c) To estimate the total number of viral particles
(d) To identify viral proteins

References

1. Uyeki, T.M., et al., *Clinical practice guidelines by the Infectious Diseases Society of America: 2018 update on diagnosis, treatment, chemoprophylaxis, and institutional outbreak management of seasonal influenzaa*. Clin. Infect. Dis., 2019. **68**(6): p. e1–e47.
2. WHO, *Guidance on regulations for the transport of infectious substances 2021–2022*, G.W.H. Organization, Editor. 2021.
3. prevention, U.S.C.f.D.c.a. Smallpox specimen packaging and shipping. 01.05.2024 [cited 2025]; Available from: https://www.cdc.gov/smallpox/hcp/laboratories/specimen-packing.html.
4. IATA. *Infectious substances shipping regulations (ISSR)*. 06.08.2025 [cited 2025]; Available from: https://www.iata.org/en/publications/manuals/dangerous-goods/.
5. IATA, *Dangerous Goods Regulations*. 66th Edn. 2025.

6. Goodman, J.L. and Stevens, J.G., *Passage of herpes simplex virus type 1 on chick embryo fibroblasts confers virulence for chick embryos*. Virus Res., 1986. **5**(2–3): p. 191–200.
7. Rodgers, F.G., *Growth of herpes simplex virus types 1 and 2 in tissues of fertile hens' eggs in ovo and in vitro*. Br. J. Exp. Pathol., 1981. **62**(3): p. 317–322.
8. Brauer, R. and Chen, P., *Influenza virus propagation in embryonated chicken eggs*. J. Visualized Exp., 2015. **97**.
9. Manders, T.T.M., et al., *Success rates of inoculation of the various compartments of embryonated chicken eggs at different incubation days*. Avian Pathol., 2021. **50**(1): p. 61–77.
10. Henle, G. and McDougall, C.L., *Mumps meningo-encephalitis. Isolation in chick embryos of virus from spinal fluid of a patient*. Proc. Soc. Exp. Biol. Med., 1947. **66**(1): p. 209–211.
11. Henle, G., Henle, W., et al., *Isolation of mumps virus from human beings with induced apparent or inapparent infections*. J. Exp. Med., 1948. **88**(2): p. 223–232.
12. Manso, P.P., et al., *Kinetic study of yellow fever 17DD viral infection in Gallus gallus domesticus Embryos*. PLoS One, 2016. **11**(5): e0155041.
13. Ambagala, A., et al., *Susceptibility of chicken embryos, sheep, cattle, pigs, and chickens to Zika virus infection*. Front. Vet. Sci., 2020. **7**: p. 23.
14. Gilhare, V.R., et al., *Pock forming ability of fowl pox virus isolated from layer chicken and its adaptation in chicken embryo fibroblast cell culture*. Vet. World, 2015. **8**(3): p. 245–250.
15. Burnet, F.M., *The growth of viruses on the chorioallantois of the chick embryo*, in *Handbuch der Virusforschung: Erste Hälfte*, R. Doerr et al., Editors. 1938, Springer Berlin Heidelberg: Berlin, Heidelberg. p. 419–446.
16. Agency, U.A.a.P.H *Procedure for the Isolation and Detection of Viruses from Avian Species using Embryonated Eggs*. 2020; Available from: chrome-extension://efaidnbmnnnibpcajpcglclefindmkaj/`https://science.vla.gov.uk/flu-lab-net/Documents/english/protocol_IsolationEmbryonatedEggs.pdf`.
17. Olasunkanmi, O.I., et al., *Novel antiviral activity of ethyl 3-hydroxyhexanoate against coxsackievirus B infection*. Front. Microbiol., 2022. **13**: 875485.
18. ThermoFisher. *Useful numbers for cell culture*. 2025. Useful cell culture numbers for various sizes of dishes and flasks. [cited 2025]; Available from: `https://www.thermofisher.com/us/en/home/references/gibco-cell-culture-basics/cell-culture-protocols/cell-culture-useful-numbers.html`.
19. Zhu, D., et al., *Systemic delivery of fusogenic membrane glycoprotein-expressing neural stem cells to selectively kill tumor cells*. Mol. Ther., 2013. **21**(8): p. 1621–1630.
20. Zhang, H., et al., *Exploring therapeutic targets from spreading patterns against respiratory syncytial virus*. FASEB J., 2025. **39**(14): e70858.
21. Burton, C. and Bartee, E., *Syncytia formation in oncolytic virotherapy*. Mol. Ther.Oncolytics, 2019. **15**: p. 131–139.
22. Dolnik, O., Gerresheim, G.K., and Biedenkopf, N., *New perspectives on the biogenesis of viral inclusion bodies in negative-sense RNA virus infections*. Cells, 2021. **10**(6): p. 1460.
23. Hoenen, T., et al., *Inclusion bodies are a site of ebolavirus replication*. J. Virol., 2012. **86**(21): p. 11779–11788.
24. Lahaye, X., et al., *Functional characterization of Negri bodies (NBs) in rabies virus-infected cells: evidence that NBs are sites of viral transcription and replication*. J. Virol., 2009. **83**(16): p. 7948–7958.
25. Nikolic, J., et al., *Negri bodies are viral factories with properties of liquid organelles*. Nat.Commun., 2017. **8**(1): p. 58.
26. Murphy, F.A. and Whitfield, S.G., *Morphology and morphogenesis of arenaviruses*. Bull. World Health Organ., 1975. **52**(4–6): p. 409–419.

27. Gilden, D., et al., *Review: the neurobiology of varicella zoster virus infection*. Neuropathol. Appl. Neurobiol., 2011. **37**(5): p. 441–463.
28. Riedel, S., *Smallpox and biological warfare: a disease revisited*. Proc. (Bayl. Univ. Med. Cent.), 2005. **18**(1): p. 13–20.
29. Mattes, F.M., et al., *Histopathological detection of owl's eye inclusions is still specific for cytomegalovirus in the era of human herpesviruses 6 and 7*. J. Clin. Pathol., 2000. **53**(8): p. 612–614.
30. Klotz O, Belt TH. The pathology of the liver in Yellow Fiver. Am. J. Pathol. 1930: 6(6):663-688.1. PMID: 19969934; PMCID: PMC 2007356. https://pubmed.ncbi.nlm.nih.gov/19969934/.
31. Sasaki, S. and Ludwig, H., *In borna disease virus infected rabbit neurons 100 nm particle structures accumulate at areas of Joest-Degen inclusion bodies*. Zentralbl. Veterinarmed. B, 1993. **40**(4): p. 291–297.
32. Eaves, G. and Flewett, T.H., *The structure of fowl-pox inclusions (Bollinger bodies)*. J. Hyg. (Lond.), 1955. **53**(1): p. 102–105.
33. Adair, B.M. and Curran, W., *Studies on the cytopathology of bovine adenoviruses in calf testis and calf kidney cell cultures*. Vet. Microbiol., 1979. **3**(4): p. 241–254.
34. Rao, K., et al., *Molluscum contagiosum*. J. Oral. Maxillofac. Pathol., 2013. **17**(1): p. 146–147.
35. Porter, C.D., et al., *Molluscum contagiosum virus*, in *Molecular and Cell Biology of Sexually Transmitted Diseases*, D. Wright and L. Archard, Editors. 1992, Springer: Netherlands: Dordrecht. p. 233–257.
36. Moise, A., et al., *Ulcer of Lipschutz, a rare and unknown cause of genital ulceration*. Facts Views Vis. Obgyn., 2018. **10**(1): p. 55–57.
37. Ebert, M.H. and Otsuka, M., *Virus diseases of skin, with special reference to elementary and inclusion bodies: I. Variola-vaccinia and molluscum contagiosum*. Arch. Dermatol. Syphilol., 1943. **48**(6): p. 635–649.
38. Blinzinger, K., Luh, S., and Anzil, A.P., *The neuronal Torres bodies in experimental yellow fever encephalomyelitis. An additional example of the heterogeneous nature of Cowdry type A intranuclear inclusions*. Arch. Psychiatr. Nervenkr. (1970), 1976. **221**(3): p. 199–212.
39. Leland, D.S. and Ginocchio, C.C., *Role of cell culture for virus detection in the age of technology*. Clin. Microbiol. Rev., 2007. **20**(1): p. 49–78.
40. Duc-Nguyen, H., *Hemadsorption of mumps virus examined by light and electron microscopy*. J. Virol., 1968. **2**(5): p. 494–506.
41. Kistler, G.S., et al., *Electron microscopic studies on rubella infectied human tissue cultures*. Schweiz. Med. Wochenschr., 1967. **97**(42): p. 1377–1382.
42. Levy, N., *The use of animal as models: ethical considerations*. Int. J. Stroke, 2012. **7**(5): p. 440–442.
43. Baumans, V. and Van Loo, P.L.P., *How to improve housing conditions of laboratory animals: the possibilities of environmental refinement*. Vet. J., 2013. **195**(1): p. 24–32.
44. Noroozi, F., et al., *Alpha-pinene ameliorates liver fibrosis by suppressing oxidative stress, inflammation, and the TGF-β/Smad3 signaling pathway*. Iran J. Basic Med. Sci., 2025. **28**(4): p. 451–460.
45. Stefan, C.P., et al., *Comparison of illumina and oxford nanopore sequencing technologies for pathogen detection from clinical matrices using molecular inversion probes*. J. Mol. Diagn., 2022. **24**(4): p. 395–405.
46. Bejaoui, S., et al., *Comparison of Illumina and Oxford Nanopore sequencing data quality for Clostridioides difficile genome analysis and their application for epidemiological surveillance*. BMC Genomics, 2025. **26**(1): p. 92.
47. Medicine, N.L.o *Basic local alignment search tool*. 06.08.2025 [cited 2025]; Available from: https://blast.ncbi.nlm.nih.gov/Blast.cgi.
48. Wadas, I. and Domingues, I., *Systematic review of phylogenetic analysis techniques for RNA viruses using bioinformatics*. Int. J. Mol. Sci., 2025. **26**(5): p. 2180.

10 Public Health, Epidemiology, and Control of Viral Infections in Tropical Countries

10.1 Introduction to the Pathogenesis of Virus Infection

Viral pathogenesis refers to the biological mechanisms by which viruses cause disease in the host. This includes the sequence of events from viral entry, replication, and spread within the body, to the host immune response and the resulting tissue damage or clinical manifestations. Understanding the pathogenesis of virus infections is crucial to addressing the public health burden of viral diseases, particularly in tropical countries with unique environmental, socioeconomic, and infrastructural factors that can influence transmission rates. For example, in tropical regions, viral infections such as dengue, Lassa, Ebola, chikungunya, yellow fever, hepatitis viruses, and emerging zoonotic viruses display distinct patterns of pathogenesis and are shaped by these factors, which may contribute to the emergence, severity of illness, as well as complicate diagnosis, treatment, and disease control measures [1]. In this chapter, an overview of viral infections of public health importance in tropical countries is presented. The epidemiological patterns, modes of transmission, and major risk factors that contribute to the persistence and spread of these infections are examined.

10.1.1 Viral Infections of the Respiratory Tract

The respiratory tract is the primary portal of entry for viral pathogens [2]. It is constantly exposed to the external environment, making it a frequent target for airborne and droplet-borne viruses. Respiratory viruses are transmitted via this route, usually through human-to-human or animal-to-human transmission [3], and are primarily transmitted through two main mode: direct and indirect transmission.

The direct transmission includes inhalation of virus-laden aerosols or droplets expelled during sneezing, coughing, breathing, or talking [4], while the indirect transmission involves contact with contaminated surfaces (fomites) or touching the mouth, nose, or eyes after exposure to infected respiratory secretions.

Infection with this type of virus spreads with great speed, poses a significant public health challenge globally, and is among the leading causes of mortality and morbidity, particularly in children, the elderly, and immunocompromised individuals [5, 6]. Their impact is especially pronounced in densely populated areas and regions with limited access to healthcare infrastructure.

10.1.1.1 Common Viral Etiology

Common respiratory viruses include coronaviruses, enteroviruses, influenza viruses, rhinoviruses, parainfluenza, metapneumoviruses, respiratory syncytial viruses (RSVs), and adenoviruses (see Table 10.1). Most respiratory tract infection starts with an infection of airway epithelial cells [7] (RSV and adenoviruses), ciliated cells (Human parainfluenza viruses) [8], or scattered ciliated cells of the nose and non-ciliated cells of the nasopharynx (Human rhinoviruses) in the upper respiratory tract [9], causing the viral titer to reach a high level in the nasal secretion rapidly. In most cases, viral respiratory diseases may have a very short incubation period, 2–7 days, and viruses may develop spontaneous mutation, developing into new strains capable of leading to new epidemics. Risk factors associated with development of disease state are listed in Table 10.2.

10.1.1.2 Symptoms

Principal clinical syndromes of onset of disease include the common cold, coughing, fever, asthma, croup, acute otitis media, nasal flaring, wheezing, laryngitis, and sinusitis. Respiratory viruses, such as severe acute respiratory syndrome coronavirus 2 (SARS-CoV-2), influenza, and RSV, are well known to infect the lower respiratory tract and can cause bronchitis, bronchiolitis, gastrointestinal symptoms, and pneumonia [10]. Viremia and systemic dissemination may occur in some patients [11].

10.1.1.3 Laboratory Diagnosis

Diagnosis of respiratory viral infections, in most tropical countries, is primarily through clinical manifestations. Laboratory investigations could further support this. Based on the availability of test facilities, tests that can be carried out in most laboratories include serological detection (serum) and polymerase chain reaction (PCR) (sawb). Advanced diagnosis includes culture, and sequencing can be carried out in reference laboratories.

Table 10.1 Summary of the pathogenesis of some respiratory viruses.

Virus	Disease	Incubation time	Interaction with host defense	Pathophysiology	Phase of infection
Influenza virus	Influenza (flu)	1–4 days	The virus triggers an innate immune response, including interferon production, but may evade immunity by antigenic drift/shift.	Infects respiratory epithelial cells, leading to necrosis of ciliated cells, inflammation, fever, cough, and myalgia.	Acute infection
Respiratory syncytial virus (RSV)	Bronchiolitis, pneumonia	4–6 days	Inhibits type I interferon production, recruits immune cells to infected airways, and causes cytokine dysregulation.	Infects bronchiolar epithelium, causing mucus buildup, airway obstruction, and necrosis.	Acute infection, primarily in infants
Rhinovirus	Common cold	1–3 days	Activates innate immune pathways (e.g., TLRs), but weakly induces long-lasting adaptive immunity.	Infects the upper respiratory tract, causing mild inflammation, rhinorrhea, sneezing, and congestion.	Acute infection
Adenovirus	Pharyngitis, conjunctivitis, pneumonia	5–8 days	Activates both innate and adaptive immune responses but may persist in lymphoid tissues.	Infects epithelial cells of the respiratory tract, eyes, or gastrointestinal tract, causing inflammation and cell damage.	Acute or persistent infection
Parainfluenza virus	Croup, bronchiolitis, pneumonia	2–6 days	Elicits a strong innate immune response, including neutrophil infiltration, but lacks long-lasting immunity.	Infects the larynx and bronchi, causing airway swelling (croup), bronchiolitis, and inflammation, leading to cough and stridor.	Acute infection

(*Continued*)

Table 10.1 (Continued)

Virus	Disease	Incubation time	Interaction with host defense	Pathophysiology	Phase of infection
Coronavirus (SARS-CoV-2, etc.)	Common cold, COVID-19, MERS, SARS	2–14 days	Suppresses type I interferon response, induces excessive inflammatory cytokines (cytokine storm in severe cases).	Infects respiratory epithelial cells, causing widespread inflammation, alveolar damage, and ARDS in severe COVID-19 cases.	Acute infection. COVID-19 may have long-term effects
Human metapneumovirus (HMPV)	Bronchiolitis, pneumonia	3–6 days	Induces a strong innate immune response but lacks a robust memory T-cell response.	Infects airway epithelial cells, leading to mucus buildup, inflammation, and bronchiolar obstruction.	Acute infection, especially in children
Bocavirus	Upper and lower respiratory tract infections	5–9 days	Weakly activates the immune response, allowing for prolonged viral shedding.	Causes mild respiratory infections but can lead to severe lower respiratory infections in immunocompromised individuals.	Acute infection, prolonged shedding in some cases
Enterovirus (e.g., EV-D68)	Respiratory infections (runny nose, cough)	3–6 days	Elicits both innate and adaptive immune responses, though cross-immunity is limited between strains.	Infects the upper and lower respiratory tracts, causing cold-like symptoms or severe respiratory illness in children.	Acute infection, with potential for outbreaks
Measles virus	Measles (can lead to severe respiratory symptoms like pneumonia)	10–14 days	Inhibits innate immune responses, particularly interferon, allowing widespread viral replication.	Infects respiratory epithelial cells, spreads systemically, causing maculopapular rash, fever, and potential pneumonia.	Acute infection

Influenza virus	Croup (more common in parainfluenza viruses), bronchitis, pneumonia	2–6 days	Weak innate immune responses and incomplete adaptive immunity lead to recurrent infections.	Infects respiratory epithelial cells, causing airway obstruction, bronchiolitis, and pneumonia in severe cases.	Acute infection and reinfections are common
Epstein–Barr virus (EBV)	Infectious mononucleosis (can cause respiratory symptoms)	4–6 weeks	Evades immune detection by residing in B-cells; a strong CD8+ T-cell response leads to symptoms.	Infects B-cells and epithelial cells, causing systemic symptoms like fever, sore throat, and enlarged lymph nodes.	Latent infection with reactivation potential
Cytomegalovirus (CMV)	Pneumonia (especially in immunocompromised patients)	3–12 weeks	Persists in host cells and evades immune response via multiple immune-evasion mechanisms.	Infects a variety of cell types, causing latent infections, but reactivation in immunocompromised individuals can lead to pneumonia.	Latent infection with reactivation potential
Herpes simplex virus (HSV)	Rarely causes respiratory infections, but can cause severe pneumonia in immunocompromised patients	2–12 days	Evades the immune response by residing in nerve cells; reactivates during immune suppression.	Causes latent infection, leading to severe pneumonia in immunocompromised individuals upon reactivation.	Latent infection with potential reactivation
Varicella-zoster virus (VZV)	Chickenpox, shingles (can lead to pneumonia in severe cases)	10–21 days	Evades immune detection by establishing latency in sensory ganglia; reactivates as shingles later in life.	Causes primary infection (chickenpox), with the latent virus residing in nerves, leading to shingles and possible pneumonia in severe cases.	Latent infection with reactivation potential

Table 10.2 Risk factors associated with respiratory disease.

Adults	
■ Smoking: Smoking damages the respiratory tract and impairs the immune response, making smokers more susceptible to viral infections.	■ Crowded populations: Living in close quarters, such as in dormitories, nursing homes, or crowded urban areas, increases the risk of transmission of respiratory viruses
■ Poor hygiene	■ Underlining medical conditions like cardiac, respiratory, or kidney disease, and allergies
■ Stress	■ Pregnancy
■ Occupational exposure: Healthcare workers, caregivers, and those who work in public settings are at a higher risk due to frequent exposure to infected individuals.	
■ Poor immunization status: Not being vaccinated against preventable respiratory viruses, such as influenza, increases the risk of contracting and spreading these infections.	■ Seasonality: Many viral respiratory infections are more common in the fall and winter months when people are more likely to be indoors and near each other
■ Travel history: Traveling, especially to areas with a high prevalence of certain respiratory viruses, can increase the risk of infection.	■ Chronic health conditions: People with chronic respiratory conditions like asthma, chronic obstructive pulmonary disease (COPD), or cardiovascular diseases are more vulnerable to severe outcomes from viral respiratory infections
Neonates and infants	
■ Low birth weight: Premature infants and those with low birth weight have underdeveloped lungs and immune systems, making them more susceptible to respiratory infections.	■ Low gestational age
■ Contact with contaminated mechanical ventilators: Infants who require mechanical ventilation, particularly in neonatal intensive care units (NICUs), are at increased risk of contracting respiratory infections from contaminated equipment.	■ Inadequate breastfeeding: Breastfeeding provides essential antibodies and nutrients that help protect against infections. Inadequate breastfeeding can deprive neonates of this natural immunity, increasing their risk of respiratory illnesses
■ High Clinical Risk Index for Babies (CRIB) score	■ Pollution: Indoor and outdoor air pollution associated with the use of biofuels
■ Educational status: Poor parental educational status	■ Poor immunization status: Children who are not fully vaccinated are more susceptible to preventable respiratory infections, such as influenza and pertussis
■ Attendance in a daycare center	■ Close contact with smoking parents
■ Living in a rural area with poor hygiene	■ Nutritional status

10.1.2 Viral Meningitis and Encephalitis

10.1.2.1 Common Viral Etiology

Viral meningitis is the most common form of meningitis in most developed countries. In most tropical countries, however, meningococcal meningitis accounts for most meningitis cases, especially in sub-Saharan Africa. Although viral meningitis may be prevalent in tropical countries, it often goes unreported. It typically resolves without treatment (without associated encephalitis), and the associated morbidity is generally low [12]. However, in some cases, it can be life-threatening and may require hospitalization and emergency treatment, particularly in children under 1 year of age. All suspected cases of meningitis should be referred to a hospital for evaluation.

Most cases of viral meningitis are caused by non-polio enteroviruses [13]. However, infections with viruses such as cytomegalovirus (CMV) (in immunocompromised individuals) [14], herpes simplex viruses (HSV 1 and 2) [15], varicella-zoster virus (VZV), human immunodeficiency virus (HIV) (usually bacterial-associated or cryptococcal meningitis) [16], and Epstein–Barr virus (EBV) should also be considered in patients suspected to have viral meningitis [17]. Since the introduction of the measles, mumps, and rubella (MMR) vaccine, cases of mumps and measles meningitis and encephalitis have been nearly eliminated. However, in unvaccinated or partially vaccinated individuals, mumps and measles viruses remain potential causes. Additionally, for patients with a travel history to certain regions in the Americas, West Asia, Australia, Africa, or mainland Europe, infections with West Nile virus (WNV), tick-borne encephalitis viruses, or Saint Louis encephalitis virus should be considered (see Table 10.3).

10.1.2.2 Symptoms of Viral Meningitis

Most cases of viral meningitis are self-limiting; however, symptoms may persist in some individuals. Early symptoms include fever, loss of appetite, headache, irritability, lethargy, or sleepiness. Most people with mild symptoms usually get better within 7–10 days. However, it is important to see a doctor immediately once a case of meningitis or encephalitis is suspected. In some cases, symptoms can become complicated, and a prolonged delay in treatment can lead to a stiff neck or body, severe headache with nausea, vomiting, confusion, or even death. Risk factors associated with development of disease state are listed in Table 10.4.

10.1.2.3 Laboratory Diagnosis

Aside from the clinical symptoms, lumbar puncture and analysis of cerebrospinal fluid are primary for the diagnosis of viral meningitis, as well as to exclude bacterial meningitis. Identification of the specific viral cause using serological tests like enzyme-linked immunosorbent assay (ELISA), PCR-based test and rapid diagnostic kits in blood samples is beneficial and enhances patients' prognosis and reduces the use of antibiotics. This can be carried out in district laboratories. Where available, virus isolation from throat and rectal swabs should be considered. PCR tests, sequencing, and microscopy techniques like immunofluorescence and electron microscopy can also be used for identification.

Table 10.3 Summary of the pathogenesis of viral meningitis and encephalitis.

Virus	Disease	Incubation time	Interaction with host defense	Pathophysiology	Phase of infection
Enterovirus (e.g., EV-A71, coxsackievirus)	Aseptic meningitis, encephalitis	3–7 days	Activates both innate and adaptive immune responses; it can invade the CNS through the blood–brain barrier.	Infects neurons and glial cells in the CNS, leading to inflammation and neuronal damage in meningitis/encephalitis.	Acute infection, typically self-limiting
Herpes simplex virus (HSV-1, HSV-2)	Encephalitis, meningitis	2–12 days	Evades the immune response by establishing latency in neurons; can reactivate during immunosuppression.	HSV-1 primarily causes encephalitis by infecting the temporal lobes, leading to necrosis and hemorrhage. HSV-2 causes meningitis.	Acute infection, with potential for reactivation
Varicella-zoster virus (VZV)	Encephalitis, meningitis	10–21 days	Establishes latency in sensory ganglia and reactivates during immune suppression, leading to viral spread to the CNS.	Primary infection causes chickenpox; reactivation can lead to encephalitis with inflammation and neuronal damage.	Latent infection with potential reactivation
Mumps virus	Viral (aseptic) meningitis, encephalitis	12–25 days	Suppresses interferon response and inhibits host protein synthesis, enabling immune evasion.	Infects the parotid glands and can spread to the CNS, leading to inflammation of the meninges and/or brain tissue.	Acute infection, primarily in children
West Nile virus (WNV)	Encephalitis, meningitis	2–14 days	Evades immune detection by infecting dendritic cells and macrophages; it induces a strong inflammatory response in the CNS.	Infects neurons and glial cells, leading to neuroinflammation, neuronal death, and potentially long-lasting neurological deficits.	Acute infection, but it can cause long-term effects.
Japanese encephalitis virus	Encephalitis	5–15 days	Elicits a strong innate immune response but may evade through suppression of type I interferon signaling in some cases.	Infects neurons, leading to severe neuroinflammation, neuronal apoptosis, and CNS dysfunction.	Acute infection, with high mortality in severe cases

Rabies virus	Encephalitis	1–3 months (up to years)	Evades immune detection by slow transport along peripheral nerves to the CNS, avoiding an early immune response.	Infects neurons, leading to inflammation, neural dysfunction, paralysis, and ultimately death without treatment.	Acute, progressive infection, fatal without treatment
Measles virus	Subacute sclerosing panencephalitis (SSPE), acute encephalitis	10–14 days for acute	Evades immune response by impairing T-cell function; it induces a strong immune response in the CNS in SSPE cases.	Acute measles may lead to encephalitis, while SSPE develops years after infection, causing chronic inflammation and demyelination.	Acute infection, with potential for chronic phase
Cytomegalovirus (CMV)	Encephalitis, meningitis (especially in immunocompromised patients)	3–12 weeks	Persists in host cells by evading immune detection, particularly in immunocompromised patients.	Infects various cell types, leading to inflammation in the meninges or brain in immunosuppressed individuals.	Latent infection, with reactivation potential
Epstein–Barr virus (EBV)	Meningitis, encephalitis (rarely)	4–6 weeks	EBV establishes a latent infection in B-cells; immune response, particularly $CD8^+$ T cells, can lead to CNS inflammation.	Infects B-cells and epithelial cells, and in rare cases, it causes neuroinflammation, leading to meningitis or encephalitis.	Latent infection with rare reactivation
Lymphocytic choriomeningitis virus (LCMV)	Viral meningitis, encephalitis	1–2 weeks	Stimulates strong immune responses, particularly T-cells, leading to CNS inflammation.	Infects meninges and CNS, causing inflammation and immune-mediated damage to neurons.	Acute infection, rare chronic complications
Zika virus	Meningitis, encephalitis (especially in neonates)	3–14 days	Elicits a strong innate immune response, including interferon production, but may cause immune evasion in some cases.	Infects neural progenitor cells in the developing brain, leading to microcephaly in neonates and CNS inflammation in adults.	Acute infection, with potential long-term neurological effects
Enterovirus 71 (EV-A71)	Hand, foot, and mouth disease can cause encephalitis	3–5 days	Elicits strong innate immune responses but may invade the CNS by crossing the blood–brain barrier.	Infects neurons, leading to inflammation, neuronal damage, and paralysis in severe cases.	Acute infection, especially in young children

Table 10.4 Risk factors associated with viral meningitis.

Adults or children	
Incomplete or unvaccinated: Individuals who have not received or completed their vaccinations, particularly for measles, mumps, and rubella (MMR), are at a higher risk of contracting viral meningitis caused by these viruses.	**Pregnancy/unsanitary conditions**
Compromised immunity	**Overcrowded environments:** Living or working in crowded conditions, such as schools, daycare centers, or refugee camps, facilitates the spread of viral infections, including those that cause viral meningitis.
Poor sanitation: Living in or being exposed to environments with inadequate sanitation increases the risk of viral transmission, particularly for enteroviruses, which are common causes of viral meningitis.	**Travel history:** Recent travel to regions where specific viruses are endemic, such as WNVs in parts of the Americas or tick-borne encephalitis in Europe and Asia, can increase the risk of contracting viral meningitis.
Age: Infants, particularly those under 1 year of age, are at a higher risk of developing viral meningitis due to their immature immune systems. Elderly individuals may also be at increased risk due to age-related immune decline.	**Weather and seasons:** Certain viral infections are more common during specific seasons or climates. For example, enteroviruses, a common cause of viral meningitis, tend to be more prevalent in the summer and fall.
Exposure to infected mosquitoes or ticks: Certain viruses that cause viral meningitis, such as West Nile virus (WNV) and tick-borne encephalitis viruses, are transmitted through the bites of infected mosquitoes or ticks. People living in or traveling to areas where these vectors are common are at higher risk.	**Contact with contaminated droppings or urine:** Exposure to rodent droppings or urine, which can carry certain viruses like lymphocytic choriomeningitis virus (LCMV), increases the risk of viral meningitis.

10.1.3 Viral Infection of the Eyes

Limited information exists on the population incidence of infectious viral infections in tropical areas; data available are focused on bacterial pathogens. Few reports that addressed the incidence and prevalence of virus infection of the eyes implicated herpesviruses as a major cause of keratitis in sub-Saharan Africa. However, other possible viruses are listed below (see Table 10.5).

10.1.3.1 Common Viral Etiology

Conjunctivitis Localized virus infection of the eyes without systemic manifestations usually results mainly from infection with adenoviruses (keratoconjunctivitis) and sometimes

Table 10.5 Summary of the pathogenesis of viral infection of the eyes.

Virus	Disease	Incubation time	Interaction with host defense	Pathophysiology	Phase of infection
Adenovirus	Viral conjunctivitis; keratoconjunctivitis	5–8 days	Induces a strong innate immune response, including interferon production, but may evade adaptive immune responses.	Infects epithelial cells of the conjunctiva, leading to inflammation, redness, discharge, and sometimes corneal involvement.	Acute infection
Herpes simplex virus (HSV)	Herpetic keratitis	2–12 days	Evades the immune system by establishing latency in sensory neurons, with periodic reactivation causing recurrent eye infections.	Infects corneal epithelial cells, causing ulceration, inflammation, and potential scarring (can lead to vision loss).	Recurrent/ latent infection
Varicella-zoster virus (VZV)	Herpes zoster ophthalmicus (shingles)	10–21 days	Establishes latency in sensory ganglia; reactivates later in life, especially during immune suppression or stress.	Reactivates in the ophthalmic branch of the trigeminal nerve, causing severe inflammation, pain, and potential vision loss.	Latent infection with reactivation potential
Cytomegalovirus (CMV)	CMV retinitis (mostly in immunocompromised, and organ transplant patients)	3–12 weeks	Evades immune detection by establishing a latent infection, particularly in immunocompromised individuals (e.g., HIV patients).	Infects retinal cells, causing inflammation and necrosis of retinal tissue, leading to vision impairment or blindness.	Latent infection with reactivation potential
Enterovirus (e.g., coxsackievirus A24, EV-70)	Acute hemorrhagic conjunctivitis (commonly known as pink eye)	1–2 days	Induces robust innate immune responses, triggering inflammation and immune cell infiltration into the conjunctiva.	Rapid infection of conjunctival epithelial cells causes hemorrhaging, conjunctival swelling, and redness.	Acute infection
Measles virus	Measles-associated conjunctivitis	10–14 days	Disrupts immune responses by suppressing interferon production; also induces immunosuppression during the acute phase.	Affects the conjunctiva along with systemic symptoms like rash and fever, and can lead to complications like keratitis and blindness.	Acute infection

(*Continued*)

Table 10.5 (Continued)

Virus	Disease	Incubation time	Interaction with host defense	Pathophysiology	Phase of infection
Zika virus	Zika-associated conjunctivitis	3–14 days	Activates innate immune response but may evade adaptive immunity in some cases.	Infects conjunctival cells, causing redness, eye pain, and mild discharge. Can be associated with systemic symptoms.	Acute infection
Epstein–Barr virus (EBV)	Conjunctivitis in infectious mononucleosis	4–6 weeks	Evades immune response by residing in B-cells; triggers strong T-cell-mediated immune response, leading to systemic symptoms.	Mild conjunctival inflammation is associated with mononucleosis symptoms like fever, fatigue, and sore throat.	Latent infection with reactivation potential
Rubella virus	Rubella-associated conjunctivitis (mild pink eye)	14–21 days	Evoke a weakened immune response in conjunctival tissue but a more significant immune response systemically.	Mild conjunctivitis is associated with systemic rubella infection (rash, fever). In severe cases, it may affect the cornea.	Acute infection
Human papillomavirus (HPV)	Conjunctival papilloma	Variable (weeks to months)	Evades immune detection by altering the host cell cycle and immune evasion mechanisms.	Causes benign conjunctival tumors (papillomas) due to abnormal cell proliferation in the conjunctiva.	Chronic infection
Chikungunya virus	Chikungunya-associated conjunctivitis	3–7 days	Induces a strong innate immune response and inflammatory cytokine production.	Infects the conjunctiva, causing inflammation, redness, and discharge. Associated with systemic arthralgia and fever.	Acute infection
Mumps virus	Mumps-associated conjunctivitis	14–25 days	Suppresses interferon response, which allows systemic spread, including to the conjunctiva.	Involves inflammation of the conjunctival tissue along with classic mumps symptoms like parotitis and fever.	Acute infection

enteroviruses. Epidemic keratoconjunctivitis is associated with serotypes 8, 19, and 37 of adenovirus [18]. Infection with the Ebola virus and SARS-CoV-2 can cause bilateral conjunctival hyperemia [19]. Usually, HSV-1 can cause a potentially serious type of eye infection called herpes simplex eye infections (neurotrophic keratitis) that leads to impaired corneal nerve function [20, 21]. Although essentially associated with skin conditions, symptomatic ocular infection by VZV can also occur in most patients who are not promptly treated with the appropriate antivirals. VZV infection of the first branch of the trigeminal nerve can cause shingles involving the eye or the surrounding area (herpes zoster ophthalmicus) and can progress to keratitis and uveitis [22]. Commonly reported in immunocompromised patients, infection with either VZV, HSV-1, or CMV can also cause retinitis in immunocompetent individuals. Some arboviruses, like non-congenital or congenital Zika virus infection, may cause macular or optic atrophy (see Figure 10.1) [23, 24].

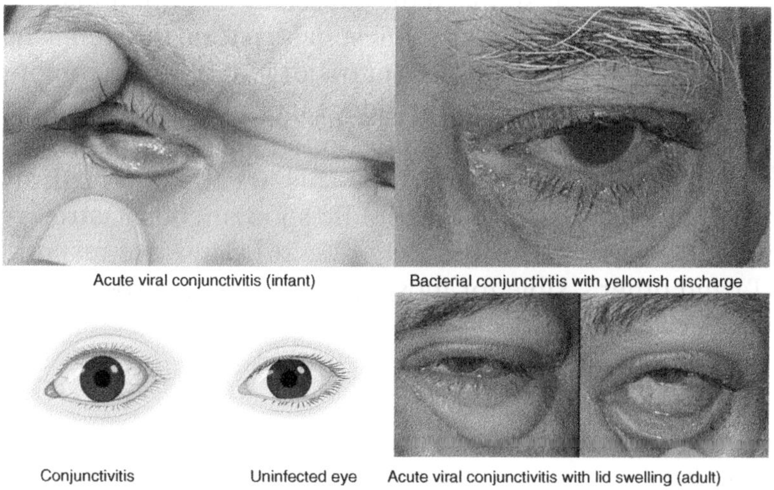

Feature	Bacterial Conjunctivitis	Viral Conjunctivitis
Cause	Bacteria such as *Staphylococcus aureus, Streptococcus pneumoniae, Haemophilus influenzae,* and *Moraxella*	Viruses, most commonly adenovirus; can also be caused by herpes simplex virus (HSV), enteroviruses
Discharge	Thick, yellow, green, or white, typically more abundant	Watery or mucus-like, clear to slightly cloudy
Eye redness	Common, often with more localized redness	Common, usually with more diffuse redness across the eye
Affected eyes	Often starts in one eye and may spread to the other	Usually begins in one eye, may spread to the other
Swelling	Eyelid swelling can occur, usually mild	Mild to moderate swelling round the eye
Pain	Gritty feeling, discomfort, but not severe pain	Gritty sensation, mild discomfort, or irritation
Contagiousness	Moderately contagious through direct contact with discharge	Highly contagious, especially through respiratory droplets or contact with contaminated surfaces
Associated Symptoms	Possible ear infection or other bacterial infection	Often associated with cold-like symptoms (sore throat, fever, cough)

Figure 10.1 Cases of eye flu.
Source: Photo by Dr Hashem Abu Serhan.

10.1.3.2 Symptoms

Unlike conjunctivitis of bacterial origin, which usually presents with purulent or mucopurulent discharges, viral conjunctivitis typically presents with bilateral watery discharges and follicles. Other symptoms include extreme redness of the eyes (especially in one eye only), sore eyes, photophobia, itchy eyes, a feeling of grittiness/burning (which is often described by patients as being painful), or swelling of the eyelids, a feeling of "something" in the eye. Symptoms of viral conjunctivitis (pink eyes) can be mistaken for eye herpes. Risk factors associated with development of disease state are listed in Table 10.6.

10.1.3.3 Laboratory Diagnosis

The prevalence of viral infection of the eyes is mainly based on clinical diagnosis. Nucleic acid amplification tests (NAAT) and other rapid, immunodiagnostic tests can be useful, especially when the inflammation is severe. For proper and correct differential diagnosis of viral conjunctivitis and eye herpes, a culture sample-eye swab test is essential.

10.1.4 Viral Infection of the Oral Cavity and Digestive System

10.1.4.1 Viral Etiology

Viral Gastroenteritis This is also known as stomach flu, caused by the inflammation of the gastrointestinal tract and the stomach. It is a disease of major public health importance worldwide, especially in developing countries and resource-poor settings. Norovirus and rotavirus are the two leading causes of sporadic spontaneous gastroenteritis globally [25, 26]. They are responsible for most cases of viral gastroenteritis

Table 10.6 Risk factors associated with virus infection of the eyes.

Adults	
■ **Poor hygienic habits:** Inadequate hand washing and touching the eyes with unclean hands can facilitate the transfer of viruses to the eye, leading to infections like viral conjunctivitis or keratitis.	■ **Contaminated personal articles:** Sharing or using contaminated personal items such as towels, pillowcases, or eye makeup can transmit viruses to the eye, increasing the risk of infections.
■ **Contact lens misuse:** Improper handling, cleaning, or prolonged use of contact lenses can lead to eye infections. Contact lenses can harbor viruses and bacteria if not maintained correctly.	■ **Immune compromise**
■ **Weather** (viral conjunctivitis peaks in the summer)	■ Previous history of herpes infection (reactivation can be triggered)
■ Using infected tissues and towels on the face and eyes	■ Using infected eye makeup
Neonates and infants	
■ Maternal infections are harbored in the mother's birth canal and congenital infection.	■ Mechanical ventilation

morbidity and mortality in children, who may have multiple (six or more) infections in a year. Rotavirus infection can be life-threatening, especially in children <5 years old [27]. Viral gastroenteritis is less commonly reported among adults, possibly due to acquired immunity because of previous exposure. Severity may increase in immunocompromised individuals, and co-infection with bacteria can contribute to accelerated disease progression.

Other viruses associated with gastroenteritis include calicivirus, astroviruses, and enteric adenoviruses [29]. Adenovirus species F (serotypes 40 and 41) and D have been identified in fecal samples of Tunisian and Kenyan children with diarrhea [30, 31].

Ulcerative Disease An ulcerative disease of the gastrointestinal tract is characterized by inflammation and lesions, sores, or ulcers of the gastrointestinal mucosa because of a break in the lining of the digestive tract. Peptic ulcer and ulcerative colitis are common ulcerative diseases of the human gastrointestinal tract. Infection with *Helicobacter pylori* and the use of nonsteroidal anti-inflammatory drugs (NSAIDs) are considered the main risk of developing peptic ulcer disease. Recent evidence has reported a possible involvement of herpesvirus infection. Five common herpesviruses have been associated with infection of the gastrointestinal tract. Among these, two major herpesviruses, CMV and HSV, are associated with colitis, an ulcerative disease of the gastrointestinal tract [32, 33]. Although several viruses can be found in the gastrointestinal tract, possibly due to their mode of infection and pathogenesis, they have not been primarily associated with disease of the gut.

CMV infection has been implicated in several cases of steroid-refractory colitis among African populations. Human papillomavirus (HPV), EBV, and HSV-8 have been implicated in benign and malignant proliferative diseases of the gastrointestinal tract, especially among immunocompromised patients [34, 35]. Usually, these viruses can infect the intestinal tissues and remain latent. The virus may later reactivate several years after initial infection, often triggered by immunosuppression, to cause diseases. There is a need for a growing awareness of the possible importance of the virus as a cause or contributory factor in a range of gastrointestinal ulcer disease conditions in some populations.

Viral Infection of the Oral Cavity Oral HPV infections are typically transmitted through oral-to-oral or genital-to-oral contact [36]. While most cases are asymptomatic and clear on their own, certain strains of HPV can lead to more severe health outcomes. There have been several reports of oral papilloma linked to HPV infection [37]. Though these are rare, some subtypes of HPV (HPV 16, 18, others are 31 and 33) can cause head and neck cancer, including the base of the tonsil and the base of the tongue [38]. See Table 10.7 and Figure 10.2. Risk factors associated with development of disease state are listed in Table 10.8.

10.1.4.2 Laboratory Diagnosis

The most widely used diagnostic techniques for the identification of viruses in fresh fecal samples, 10% clarified stool sample (without preservative) or rectal swabs are immunoassay – ELISA andlatex agglutination (for adenovirus, enterovirus, rotavirus,

Table 10.7 Summary of the pathogenesis of viral infection of the gastrointestinal tracts.

Virus	Disease	Incubation time	Interaction with host defense	Pathophysiology	Phase of infection
Rotavirus	Gastroenteritis (severe diarrhea, especially in infants)	1–3 days	Evades innate immune responses by non structural protein 1(NSP1)-mediated inhibition of interferon signaling.	Infects mature enterocytes in the small intestine, causing villous atrophy and malabsorption, leading to diarrhea.	Acute infection, primarily in children
Norovirus	Viral gastroenteritis (diarrhea, vomiting)	12–48 hours	Highly infectious; rapidly spreads through the fecal–oral route and aerosolized vomitus. Weak adaptive immune response.	Infects enterocytes, causing epithelial cell damage and inflammation, leading to nausea, vomiting, and watery diarrhea.	Acute infection, often self-limiting
Adenovirus (types 40, 41)	Gastroenteritis, diarrhea (primarily in young children)	3–10 days	Triggers innate immune responses and inflammation but lacks long-lasting immunity in the gut.	Infects intestinal epithelial cells, leading to inflammation and disruption of water absorption in the intestines.	Acute infection, primarily in infants and children
Astrovirus	Gastroenteritis (mild diarrhea in children)	1–4 days	Induces limited immune responses, with weak activation of both innate and adaptive pathways.	Infects intestinal epithelial cells, leading to mild inflammation and diarrhea without severe tissue damage.	Acute infection
Sapovirus	Gastroenteritis (diarrhea, vomiting)	1–4 days	Induces both innate and adaptive immune responses, but limited immunity leads to potential reinfections.	Infects enterocytes, causing cell death and inflammation, resulting in watery diarrhea and vomiting.	Acute infection, reinfections are possible
Enteric adenovirus (types 40, 41)	Gastroenteritis, diarrhea	3–10 days	Elicits both innate and adaptive immune responses but can persist in lymphoid tissues.	Infections enter enterocytes, causing damage to the intestinal lining, leading to diarrhea and dehydration.	Acute infection, persistent infection in some cases

Hepatitis A virus (HAV)	Hepatitis, jaundice, mild diarrhea	15–50 days	Strong innate immune response, including interferon production, activates adaptive immune response with liver involvement.	Replicates in hepatocytes, leading to liver inflammation, jaundice, and mild gastrointestinal symptoms.	Acute infection, lifelong immunity post-infection
Hepatitis E virus (HEV)	Hepatitis, gastrointestinal symptoms	2–9 weeks	Evades immune response via suppression of T-cell activation; causes severe disease in immunocompromised or pregnant patients.	Infects hepatocytes, leading to liver inflammation, jaundice, and gastrointestinal symptoms (nausea, vomiting, and diarrhea).	Acute infection, chronic infection in immunocompromised patients
Enterovirus (non-polio types)	Gastroenteritis, hand-foot-and-mouth disease	3–6 days	Activates both innate and adaptive immune responses, but immune evasion by the virus allows for widespread replication.	Infects mucosal surfaces of the gut and sometimes the respiratory tract, leading to diarrhea, hand-foot-mouth lesions, and fever.	Acute infection
Poliovirus	Poliomyelitis (may start with mild gastrointestinal symptoms)	6–20 days	Initially infects the gut, evades immune responses, and spreads to the CNS in some cases, leading to paralysis.	Infects the intestinal mucosa and spreads via blood to the CNS, causing paralysis in a small percentage of cases.	Acute infection, with potential for chronic CNS involvement
Coronaviruses (some strains)	Gastroenteritis (in addition to respiratory symptoms)	2–14 days	Suppresses innate immune responses, including interferon, and may cause excessive cytokine release in severe cases.	Infects the gastrointestinal epithelium, causing inflammation and diarrhea; in some strains, it can affect both the respiratory and GI tracts.	Acute infection, potential for severe outcomes

(*Continued*)

Table 10.7 (Continued)

Virus	Disease	Incubation time	Interaction with host defense	Pathophysiology	Phase of infection
Torovirus	Gastroenteritis, mild diarrhea	3–5 days	Induces a moderate immune response but may persist in the host for longer periods.	Infects epithelial cells of the gastrointestinal tract, leading to mild diarrhea and dehydration in some cases.	Acute infection, possibly prolonged shedding
Coxsackievirus A and B	Hand-foot-and-mouth disease, gastroenteritis	3–6 days	Activates both innate and adaptive immune responses, and can spread to other organs (e.g., the heart in the coxsackievirus B).	Infects mucosal surfaces of the gastrointestinal tract, leading to diarrhea, fever, and in some cases systemic symptoms.	Acute infection, potential systemic involvement
Calicivirus	Gastroenteritis (like norovirus)	1–2 days	Causes a strong innate immune response, but reinfections are common due to antigenic variation.	Infects the epithelial cells of the gastrointestinal tract, causing rapid-onset diarrhea, vomiting, and dehydration.	Acute infection and re-infection are common.

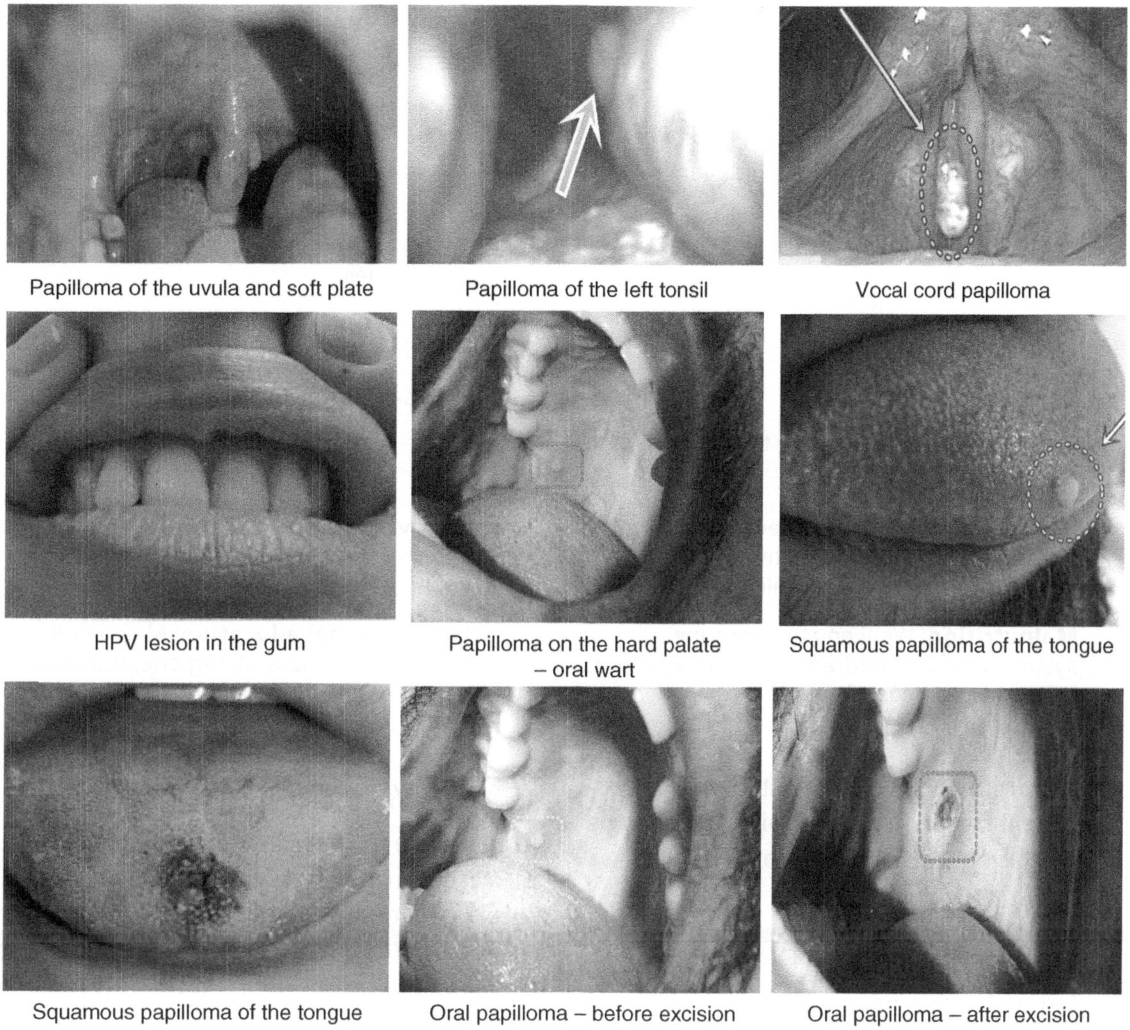

Figure 10.2 Cases of oral papilloma infections.
Source: Photo by Dr. Murat Enöz.

norovirus, and astrovirus) – and reversetranscription PCR (for coronavirus, astrovirus, rotavirus, enterovirus, and calicivirus) or PCR (foradenovirus and norovirus). Most members of this group can be detected by electron microscopic examination (a laboratory technique that is available only at a few large centers in tropical countries). The culture of some of these viruses might be difficult. Acute- and convalescent-phase serum samples can be collected for serological tests. For papilloma infection, the lesion can be clinically identified. Testing for HPV in the oral cavity is not routinely done but may be

Table 10.8 Risk factors associated with virus infection of the oral cavity and digestive system.

Adults	
■ Eating or drinking contaminated food or water. Poor food safety practices and unclean water sources are significant contributors.	■ Contaminated personal articles
■ Poor food-handling	■ Compromised immunity and nutritional deficiencies.
■ Age and vaccination status	■ Previous history of herpes infection (reactivation can be triggered)
■ Using infected utensils	■ Poor sanitation and hygiene
■ Seasons (vary from country to country)	
Children	
■ **Lack of exclusive breastfeeding (in infants):** Breast milk provides protective antibodies; lack of it increases vulnerability to infections	■ **Poor vaccination status:** Not receiving vaccines (e.g., rotavirus vaccine) increases the risk of infection
■ **Malnutrition:** Weakens the immune system, making children more susceptible to infections and complications	■ **Low maternal education:** May result in poor awareness of hygiene, nutrition, and preventive care for children

used in certain cases of suspected oral or oropharyngeal cancer. Swabs or biopsies may be taken to detect the presence of HPV DNA.

10.1.5 Viral Infection of the Skeletal System

Virus infection of almost all the organs has been described, except for the infection of the bone. Physicians, infectious disease specialists, and microbiologists have focused virtually on septic bone infections. Request for tests for the etiology of bone infections and diseases, in routine laboratory tests, due to viral infection is generally rare. There are two types of bone and joint involvement in viral infection: osteomyelitis and arthritis (see Table 10.9).

10.1.5.1 Viral Etiology

Osteomyelitis Osteomyelitis is the inflammation or swelling of the bone, which can occur because of an infection from another part of the body that has spread to the bone. It can start in the bone, usually due to injuries that are exposed to germs. Although viral infection of the bone is somewhat rare, it has been reported in the past. Osteomyelitis variolosa is an infection of bone and joints by smallpox virus [39]. It is associated mostly with the infection of the elbows, wrists, ankles, hands, and feet, capable of forming different kinds of bone deformities. Osteomyelitis of the skull (calvaria) as a complication of VZV infection has been reported in developed countries. In the recent past, several cases of bony sequelae

Table 10.9 Summary of pathogenesis of viral infection of the skeletal system.

Virus	Disease	Incubation time	Interaction with host defense	Pathophysiology	Phase of infection
Parvovirus B19	Erythema infectiosum (Fifth disease), arthropathy, viral arthritis	4–14 days	Suppresses erythropoiesis and evades immune detection by infecting erythroid progenitor cells in the bone marrow.	Causes joint pain and inflammation, especially in adults, affecting the knees, wrists, and fingers.	Acute infection, with prolonged joint symptoms in some cases
Chikungunya virus	Chikungunya fever, a viral arthritis	3–7 days	Elicits strong innate and adaptive immune responses, but chronic joint pain may result from prolonged inflammation.	Infects muscle and joint tissues, causing severe joint pain and swelling, which can persist for months to years.	Acute phase, with chronic arthralgia possible
Dengue virus (DENV)	Dengue fever, dengue arthritis	4–10 days	Induces strong immune activation, including a cytokine storm; subsequent infections may worsen due to antibody-dependent enhancement.	Causes muscle and joint pain (breakbone fever), leading to inflammation and fluid leakage, sometimes affecting the skeletal system.	Acute infection, risk of severe dengue with subsequent infections
Ross River virus	Ross River fever, viral arthritis	7–10 days	Activates a strong immune response, including type I interferon, but chronic joint inflammation may persist in some cases.	Infects joints and muscles, causing prolonged arthritis-like symptoms, joint swelling, and pain, especially in the knees, wrists, and fingers.	Acute infection, with chronic joint pain possible
Rubella virus	Rubella (German measles), viral arthritis in adults	2–3 weeks	Induces both humoral and cell-mediated immune responses; immune complexes can deposit in joints, leading to arthritis.	Causes mild fever and rash, but in adults, particularly women, it can lead to joint inflammation in the hands, knees, and wrists.	Acute infection, joint symptoms may persist briefly
Hepatitis C virus (HCV)	Hepatitis C, cryoglobulinemia, polyarthritis	2–26 weeks	Chronic infection with persistent immune activation leads to cryoglobulin formation and immune complex deposition in joints.	Chronic infection can lead to polyarthritis, cryoglobulinemia, and skeletal complications due to immune complex deposition.	Chronic infection, with potential long-term joint issues

(Continued)

Table 10.9 (Continued)

Virus	Disease	Incubation time	Interaction with host defense	Pathophysiology	Phase of infection
Zika virus	Zika fever, viral arthritis	3–14 days	Activates both innate and adaptive immune responses; immune-mediated joint inflammation may occur, though rare.	Mild symptoms like rash and fever are common, but joint pain and swelling can affect the wrists, hands, and feet.	Acute infection, joint symptoms are generally mild
Epstein–Barr virus (EBV)	Infectious mononucleosis, viral arthritis	4–6 weeks	Evades immune detection by infecting B-cells; triggers an immune response that can lead to immune-mediated joint inflammation.	Typically causes fatigue, sore throat, and swollen lymph nodes, but may also cause viral arthritis in rare cases.	Latent infection, with possible arthritis flare-ups
Human immunodeficiency virus (HIV)	HIV infection, HIV-associated arthropathy	Variable (months to years)	Suppresses the immune system, leading to chronic inflammation and immune dysregulation, which can affect joints.	Chronic immune activation and inflammation can lead to joint pain, stiffness, and arthritis-like symptoms in the later stages of the disease.	Chronic infection, with long-term joint complications possible
Coxsackievirus B	Viral myositis, myalgia, pleurodynia (Bornholm disease). Infection of the bone marrow.	2–5 days	Elicits an immune response that leads to muscle and joint inflammation, primarily targeting muscle tissues.	Causes inflammation of skeletal muscles, leading to severe chest pain (pleurodynia) and sometimes joint pain or swelling.	Acute infection, with potential for recurrent pain
Human T-lymphotropic virus (HTLV-1)	HTLV-1-associated myelopathy, arthropathy	Months to years	Chronic immune activation due to viral persistence; HTLV-1 infection of T-cells may lead to immune dysregulation in joints.	Can lead to chronic joint pain, muscle weakness, and skeletal complications due to immune system involvement in chronic infections.	Chronic infection and joint issues develop over time

of smallpox and a history of childhood fever with rashes have been reported in India and Hong Kong [39, 40]. During an epidemic of smallpox in Nigeria in 1957, about 34 cases of osteomyelitis variolosa were reported in children, mainly between the ages of 7 months and 7 years [41]. No case has been reported so far post-eradication era of smallpox in most countries. Although the pathogenesis is unclear, osteoblast-like cells are susceptible to Zika virus infection [42]. Some tropical African countries like Senegal, Nigeria, Angola, the Central African Republic, and Uganda that have reported cases of isolation of the Zika virus should have raised concern about possible bone infections.

Viral Arthritis Viral arthritis is inflammation of the joints from a viral infection. Infection from a broad range of viruses can result in arthritis and arthralgia. Viruses, such as parvovirus B19, mumps virus, hepatitis A, C virus (HCV-associated cryoglobulinemia), and rubella virus, can infect joints in people of any age. Enterovirus infections like coxsackievirus A9, B2, B3, B4, and B6, and echovirus, Human T-lymphotropic virus (HTLV-1) (in Japan), HCV-associated cryoglobulinemia, and parvovirus B19 (in young children) cause most cases of nonspecific viral arthritis [43]. Human herpesviruses like EBV and CMV, as well as other viruses like chikungunya virus, O'nyong-nyong virus, found in the joints, bone and joint lesions, or synovial fluid, can cause nonspecific inflammation usually associated with arthralgia or arthritis [44, 45]. HIV infection in African children has been associated with arthritis [46]. For many patients, viral arthritis is self-limiting because of an efficient elimination of the organism by the immune system; others may experience significant systemic disease. Rather than the direct damage to the cartilage tissue, viral arthritis can also be explained by virus-induced autoimmunity (with the presence of autoantibodies, like rheumatoid factor and antinuclear antibody) and resulting inflammatory responses [47]. Risk factors associated with development of disease state are listed in Table 10.10.

10.1.5.2 Laboratory Diagnosis

The most common diagnosis is radiological and imaging evaluation. Routine laboratory findings are nonspecific; they can be carried out to confirm the underlying infection. These may include antibody tests that detect disease-specific antibodies. PCR tests to detect disease-specific viral DNA. Full blood count to assess neutrophils and lymphocytes, and an elevated erythrocyte sedimentation rate (ESR). Joint fluids with raised white counts. The incubation period of the virus should be put into consideration to assess the state of viremia.

10.1.6 Viral Infection of the Genitourinary Tract

Primary viral infections of the genitourinary tracts are usually due to direct contact of the virus with the mucosal lining of the tract, sexually transmitted, or viremia. Although it

Table 10.10 Risk factors associated with virus infection of the skeletal system.

Adults	
Bone marrow transplantation and preexisting arthritis. Individuals who have undergone bone marrow transplantation are at an increased risk due to the potential for immunosuppression. Similarly, those with preexisting arthritis are more susceptible to viral infections that can exacerbate their condition or lead to viral arthritis.	Immunosuppression, e.g., HIV and immunosuppressive therapy. Immunocompromised individuals, including those with HIV or those receiving immunosuppressive therapy (e.g., for autoimmune diseases or after organ transplantation), are more vulnerable to viral infections that can affect the skeletal system.
Healthcare workers	Travelers from the endemic areas
Intravenous drug abusers: Intravenous drug users are at risk of blood-borne viral infections, such as HIV and hepatitis B or C, which can lead to complications affecting the skeletal system, including viral arthritis or osteomyelitis.	Previous history of herpes infection (reactivation can be triggered)
Trauma, including intra-articular injections	Previous central intravenous catheterization (sternoclavicular)

is often caused by bacterial pathogens, viral infections of the genitourinary tract can occur even in an immunocompetent host. Viral diseases of the genitourinary tract vary broadly from hemorrhagic cystitis of adenovirus, Burkitt's lymphoma-associated (BK) virus, and CMV to cellular dysplasia affected by HPV infection and epididymoorchitis of the mumps virus (see Table 10.11).

10.1.6.1 Viral Etiology

Viral Urethritis HPV and HSV, especially type 2 (HSV-2), are the common sexually transmitted infections causing disease burden in sub-Saharan tropical Africa [48] and also in many other tropical regions, including parts of Southeast Asia, Latin America, and the Caribbean [49]. HPV accounts for the most co-existence of genitourinary infection and ulcer disease cases. HSV-2 is responsible for an increasingly high proportion of cases of urethritis (inflammatory disease of the urethra) worldwide. Viral orchitis and epididymitis are associated with testicular or scrotal pain and swelling. Having epididymitis can cause orchitis. Viral orchitis rarely occurs; when it happens, it is usually related to mumps virus infection. Other viruses, such as VZV, CMV, and rubella virus, can cause orchitis. Coxsackievirus A6, a member of the human enterovirus species A, has been identified in epididymal fluid [50]. Viral epididymitis should be suspected in young men who present with testicular pain, between the ages

Table 10.11 Summary of pathogenesis of viral infection of the genitourinary tracts.

Virus	Disease	Incubation time	Interaction with host defense	Pathophysiology	Phase of infection
Human Papillomavirus (HPV)	Genital warts, cervical cancer, penile cancer, anal cancer, vaginal cancer	3 weeks to several months	Elicits local immune responses; some strains evade immunity, leading to persistent infections.	Infects epithelial cells, causing hyperplasia; certain high-risk types can integrate into host DNA, leading to malignancy.	Acute infection with potential for chronicity
Herpes simplex virus (HSV)	Genital herpes, urethritis, vulvovaginitis, cervicitis	4–7 days	Elicits a strong local immune response but can establish latency in sensory ganglia.	Infects mucosal epithelium, causing vesicular lesions; recurrent infections occur due to reactivation from latency.	Acute infection with latent reactivation potential
Human immunodeficiency virus (HIV)	HIV/AIDS, opportunistic infections affecting the genitourinary tract	2–4 weeks or more	Initially activates innate immune responses; chronic infection leads to immune system depletion.	Infects $CD4^{+}$ T-cells, resulting in progressive immune dysfunction; allows for opportunistic infections.	Chronic infection with acute and asymptomatic phases
Cytomegalovirus (CMV)	CMV-related urethritis, nephritis (in immunocompromised individuals)	3–12 weeks	Evades immune response by residing in host cells; can reactivate in immunocompromised individuals.	Infects a variety of cells, causing inflammation; can lead to organ dysfunction, particularly in immunocompromised hosts.	Latent infection with potential reactivation
Epstein–Barr virus (EBV)	Posttransplant lymphoproliferative disorder (PTLD) in the urinary tract	4–6 weeks	Induces both innate and adaptive immune responses, and can lead to lymphoproliferative disorders in immunocompromised hosts.	Can transform B-cells, leading to uncontrolled growth and potentially affecting urinary tract tissues posttransplant.	Acute infection, the potential for chronicity in immunocompromised individuals

(*Continued*)

Table 10.11 (Continued)

Virus	Disease	Incubation time	Interaction with host defense	Pathophysiology	Phase of infection
Adenovirus	Hemorrhagic cystitis, urethritis, vulvovaginitis (less common)	5–10 days	Elicits a strong innate immune response; generally self-limiting, but can lead to significant symptoms.	Infects epithelial cells of the urinary tract, leading to inflammation and potential bleeding in the bladder.	Acute infection, usually self-limiting
Zika virus	Zika-associated genitourinary symptoms, urethritis	3–14 days	Elicits an innate immune response; potential for sexual transmission and complications during pregnancy.	Infects various cell types, including endothelial cells; can cause inflammation and affect fetal development.	Acute infection, potential long-term effects in pregnancy
Molluscum Contagiosum virus	Molluscum contagiosum (genital lesions)	2–8 weeks	Elicits a local immune response, but lesions may persist due to immune evasion.	Causes benign skin tumors; lesions may be more persistent in immunocompromised individuals.	Acute infection with potential for chronicity

of 15 and 35, with a recent history of hand, foot, and mouth disease (HFMD)-related coxsackievirus A6 infection.

Hemorrhagic Cystitis Hemorrhagic cystitis, a diffuse inflammatory condition of the urinary bladder, previously thought to be a complication of chemotherapy, is also caused by viral infections [51]. BK virus, adenovirus, and CMV are important pathogenic infections involved in hemorrhagic cystitis [52], especially in children and patients transplanted with stem cells and solid organ kidneys. Some of these viruses may cause pyelonephritis [53]. Though viral causes of pyelonephritis (inflammation of the kidney) are rare, when they occur, it is often due to a viral infection of the kidney from the bladder or is carried in the bloodstream (viremia).

Vulvovaginitis Vulvovaginitis (inflammation of the vulva and vagina) is a common condition that affects women and girls of all ages, which is mostly spread through sexual contact. HSV or HPV (types 6, 11, 16, and 18) is a common cause of vulvovaginitis. These viruses cause painful lesions or sores (HSV), usually visible on the vulva or vagina (occasionally visible sores may not be present), or white to gray colored painful warts (HPV) that grow in the vagina, vulva, rectum, or groin (occasionally visible warts may not be present). Koilocytes may appear in squamous epithelial cells that are infected with HPV (see Figure 10.3) [54]. HSV and HPV can also cause viral cervicitis, endometritis, and anogenital warts in sexually active homosexual men [55, 56]. Other risk factors associated with development of disease state are listed in Table 10.12.

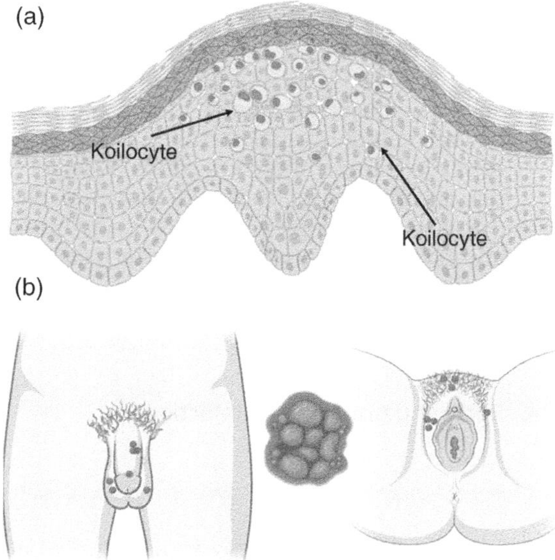

Figure 10.3 (a) Representative illustration of Epidermis containing numerous koilocytes. Koilocytes are squamous epithelial cells that are infected with the human papillomavirus (HPV). (b) Illustration of where genital herpes occurs and what the blisters may look like. Common in HSV-1 and 2.

10.1.6.2 Laboratory Diagnosis

Generally, early diagnosis and treatment of genitourinary tract infections may prevent significant morbidity. First-pass urine or urethral swab, urethral smears, epididymal fluid, semen samples, cervical smears, and fluids from fine-needle aspirates of the swollen testis can be submitted for viral culture (where possible). Cytology examination (CMV and BK virus) by characteristic changes on urine cytology for the diagnosis of orchitis and epididymitis. A Pap test can be carried out if infection with HPV is

Table 10.12 Risk factors.

Adults	
Sexual behaviors: Engaging in unprotected sex, having multiple sexual partners, or having a history of sexually transmitted infections (STIs) increases the risk of contracting viruses like human papillomavirus (HPV), herpes simplex virus (HSV), and human immunodeficiency virus (HIV), which can cause genitourinary infections.	Immunosuppression, e.g., HIV and immunosuppressive therapy
Congenital abnormality of the urinary tract	Surgery that involves the genitals or urinary tract
Recurring urinary tract infections	Poor immunization against some viruses

suspected. Other laboratory diagnoses, such as molecular-based techniques like real-time PCR, are often the method of choice because they allow for the quantification of viral load in blood and urine. Adenoviral and HSV and most infections can be detected by using conventional PCR and real-time PCR. A complete blood count can be done to determine infections. Direct immunofluorescence of the virus protein and ELISA can be used to detect the presence of the pathogen (antigen) in spun body fluid (or cells).

10.1.7 Viral Infection of the Skin

Skin diseases, such as allergic reactions, infections, skin appendages, and pigmentation, are generally considered more prevalent in the rural areas of some countries. Common viral skin diseases are chickenpox or varicella (infectious diseases of childhood that cause an itchy, spotty rash on the skin); shingles (results from a reactivation of the virus); measles (highly contagious, causing total-body skin rash, common among children of school age); molluscum contagiosum (presenting as a single or multiple raised, pearl-like papules or bumps on the skin); mumps (cause a flat, red rash usually starts from the face and spreads to other parts of the body); dengue skin rash (maculopapular or macular confluent rash over the face that spread to the thorax, and flexor surfaces, with islands of skin-sparing); roseola infantum (also known as sixth disease, a high fever followed by a rash usually presented as cutaneous erythematous almond-shaped macules and papules on the neck, trunk, and proximal extremities in infants or very young children); monkeypox (produces pox-like lesions on the skin); and cowpox. Kaposi's sarcoma and viral warts/verruca (a benign lesion) are mainly associated with HIV infection [57]. Smallpox (a contagious disease caused by the variola virus) has been eradicated. Skin problems are among the most frequent medical presentations in returned travelers, and transmission

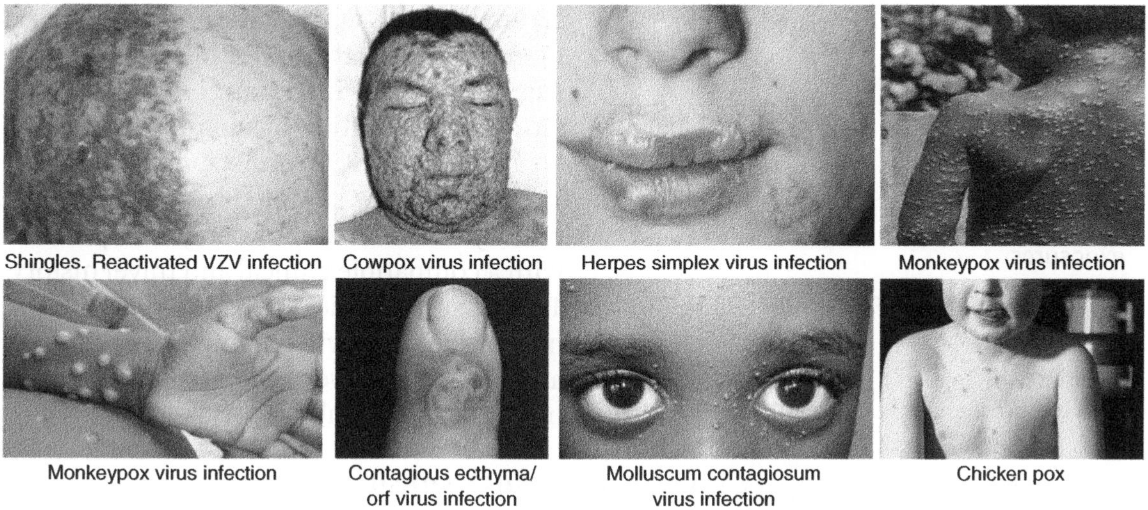

Figure 10.4 Representative images showing examples of viral infections affecting the skin. Photo by Dr Hans R. Gelderblom and Google/Nigeria Center for Disease Control.

may occur by direct skin-to-skin contact with infected surfaces, like furniture. Bed linen may also serve as a contact surface capable of harboring infectious agents. It can spread through coughing and sneezing (measles and mumps), inhalation of saliva, or respiratory droplets of an infected person (see Figure 10.4).

10.1.7.1 Viral Etiology

- Dengue skin rash is caused by a virus of the *Flavivirus* family – dengue virus (1–4 serotypes), a virus transmitted by *Aedes* spp. Mosquitoes are often found in urban settings. Dengue disease is endemic in more than 100 countries. Burkina Faso and Kenya reported localized outbreaks of probable cases in 2016 and 2020.
- Infection with Zika virus is associated with measles-like (morbilliform) maculopapular erythematous rash [58]. Other arboviruses associated with viral skin disease are West Nile, O'nyong-nyong, and Sindbis viruses, especially at the inoculation site [59, 60].
- Molluscum contagiosum is a common viral skin infection caused by the molluscum contagiosum virus (a poxvirus). It is prevalent in tropical and subtropical regions, particularly in countries like Papua New Guinea, Mali, Nigeria, and Fiji.
- Monkeypox, closely related to smallpox, is caused by the monkeypox virus. It was first discovered in the Democratic Republic of Congo and is primarily confined to the rainforests of central and western Africa [61], though cases have spread to countries like the United States and Sudan.
- The VZV (a member of the herpesvirus family) causes varicella (chickenpox)

Table 10.13 Risk factors associated with viral skin diseases.

Exposure history: Human contact, occupational, recreational, animal/insect, and school exposures	Immunosuppression, e.g., HIV
Travel history: Location and duration of travel	Vaccination status
Contact with an infected host	Overcrowded environment

and herpes zoster (shingles). Another herpesvirus, HSV-6 or -7, is responsible for roseola infantum.

- Other viruses that can cause skin diseases include the measles virus, rubella virus, variola virus (smallpox), EBV, CMV, cowpox virus, HPV, and parapox virus (*Orf virus infection*) [62].
- HIV infection can result in acute retroviral syndrome, which often presents as a generalized skin rash with pink or red macules or papules.
- Viral infections such as HBV, EBV, enteroviruses, echoviruses, RSV, and SARS-CoV-2 (Coronavirus disease 2019 [COVID-19]) can cause papular acrodermatitis or viral exanthems in children [63].

Risk factors associated with development of disease state are listed in Table 10.13.

10.1.7.2 Laboratory Diagnosis

Diagnosis includes antigen and antibody detection tests, as well as PCR assays; immunoassays detecting antibodies and antigens in the appropriate clinical setting. Acute-phase serum obtained early (days after fever onset) in the illness, swabs, and skin samples could suggest a recent infection. For the dengue virus, determination of the presence of the virus by RT-PCR or nonstructural protein 1 antigen in the sample is considered laboratory confirmation of dengue in patients with a compatible clinical and travel history. Tzanck smear (chickenpox skin test and the herpes skin test) will enhance identification and can help prevent the further spread of infection to the public.

Tzanck smear

A Tzanck smear is used primarily for the diagnosis of herpesvirus infections, including both HSV-1 and HSV-2, and VZV infections (chickenpox and shingles) [64]. A Tzanck smear is typically performed on a sample taken from the base of a skin vesicle, blister, or ulcer. The smear can help identify characteristic multinucleated giant cells, which are a hallmark of these viral infections. This test aids in diagnosing and managing these conditions, potentially preventing further spread to the public through early detection and treatment.

10.1.8 Virus Infection of the Circulatory System

The two primary heart infections caused by viruses are pericarditis (infection or inflammation of the pericardium – the innermost layer of the heart) and myocarditis (infection or inflammation of the myocardium – the middle muscular layer of the heart). Endocarditis, the third type of infection that affects the inner lining of the heart, is less commonly caused by viruses. Viral infection is the most common cause of myocarditis. Viral myocarditis begins with the direct invasion and replication of the infectious agent within or around the myocardium. Infected cardiac cells may produce proinflammatory cytokines to fight the virus. Viruses that avoid elimination begin to replicate, producing viral proteins that can cause direct myocardial injury (see Table 10.14). A defective or excessive innate immune response may cause further damage to the cardiac cells.

10.1.8.1 Viral Etiology

Pericarditis Viruses are believed to be the cause of around 85% of pericarditis cases in developed countries. Commonly known viruses include coxsackievirus, particularly coxsackievirus B, which is one of the most common viral causes of pericarditis. Herpesviruses such as CMV (especially in immunocompromised individuals) and EBV (known for causing mononucleosis, can also lead to pericarditis [65]. HSV is occasionally associated with pericarditis [66]. Others include echovirus, adenovirus, parvovirus B19, and HIV [67, 68].

Myocarditis Generally, the most common causes of viral myocarditis are adenoviruses and coxsackieviruses (Figure 10.5). The etiology and estimated incidence of acute viral myocarditis are largely undefined in some tropical areas. Parvovirus B-19, EBV, and HSV-6 were characterized as the main causes of viral acute myocarditis in other places. HIV-related acute myocarditis has been reported. Other viruses capable of causing myocarditis are CMV, influenza B and A, and HCV. Cases of myocarditis have been associated with COVID-19 infection [69–71].

10.1.8.2 Laboratory Diagnosis

Diagnosis of viral myocarditis is important at the early stages of viral infection. The gold standard for the diagnosis of myocarditis, when necessary, is a histopathology from an endomyocardial biopsy. This might help confirm a viral infection and the type of virus, but done in extreme cases. PCR to detect viral DNA/RNA in the blood will be helpful. Serological markers are often not available or reliable during the acute phase of the disease. Other tests like complete blood count, ESR, and C-reactive protein can reveal signs of a viral infection, but might not be indicative of a viral infection. Risk factors associated with development of disease state are listed in Table 10.15.

10.1.9 Virus Infection of the Central Nervous System

Viruses are a common cause of central nervous system infection. Following infection, the virus replicates in susceptible cells and spreads via peripheral nerve fibers or the bloodstream to the central nervous system. Although some viruses can affect the CNS, many of them are not primarily neurotropic. While some appear to be (rabies virus and poliovirus), others affect nervous cells secondarily. Viruses spread to the central nervous system either through neuronal and axonal transport- peripheral nerve fibers, or via the bloodstream. The viruses that pass through the bloodstream encounter the impenetrable blood–brain and or blood–CSF

Table 10.14 Overview of the pathogenesis of some virus infections of the circulatory system.

Virus	Disease	Incubation time	Interaction with host defense	Pathophysiology	Phase of infection
Dengue virus	Dengue fever, dengue hemorrhagic fever	4–10 days	Elicits both humoral and cellular immune responses; antibody-dependent enhancement can worsen the disease.	Infects endothelial cells and immune cells, leading to increased vascular permeability and hemorrhage.	Acute infection with potential for severe forms
West Nile virus	West Nile fever, a neuroinvasive disease	2–14 days	Initial innate immune response: can lead to neurological complications if it crosses the blood–brain barrier.	Infects neurons and causes inflammation, leading to encephalitis or meningitis in severe cases.	Acute infection, potential chronic neurological issues
Zika virus	Zika fever, associated with congenital microcephaly	3–14 days	Elicits both innate and adaptive immune responses; potential for sexual transmission.	Infects various cell types, including endothelial cells, potentially leading to microcephaly in fetuses.	Acute infection, with long-term effects possible
HIV	HIV/AIDS	2–4 weeks	Elicits an initial strong immune response but progresses to chronic immune dysfunction.	Targets $CD4^+$ T-cells, leading to progressive immune deficiency and increased susceptibility to opportunistic infections.	Chronic infection with acute and asymptomatic phases
Cytomegalovirus (CMV)	CMV infection, associated with immunocompromised states	3–12 weeks	Evades immune response; can remain latent and reactivate in immunocompromised individuals.	Infects endothelial cells and leukocytes, causing vasculitis and end-organ damage in severe cases.	Latent infection with potential for reactivation
Ebola virus	Ebola virus disease	2–21 days	Elicits a strong immune response but can lead to immune dysregulation and severe systemic effects.	Causes widespread endothelial cell infection, leading to severe hemorrhagic manifestations and multi-organ failure.	Acute infection with a high mortality rate
Marburg virus	Marburg virus disease	2–21 days	Similar immune response dynamics to Ebola often lead to severe hemorrhagic fever.	Infects endothelial cells, causing significant vascular leakage and coagulopathy, leading to hemorrhage.	Acute infection with a high mortality rate

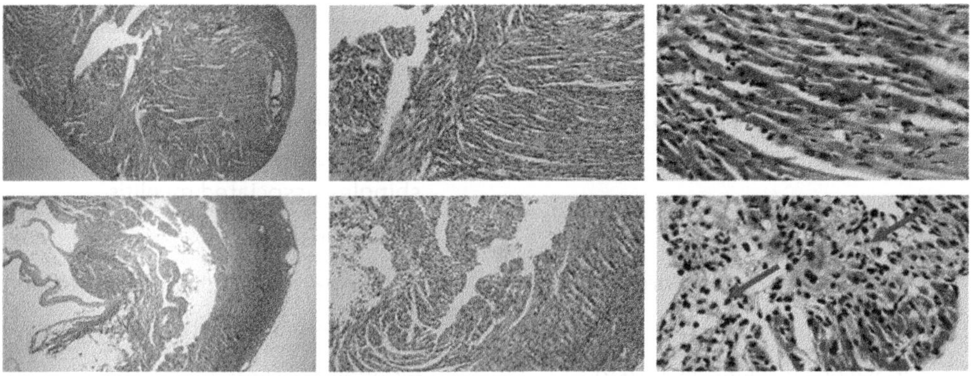

Figure 10.5 Viral infection in suckling mice, demonstrating inflammation of the heart muscle (myocarditis).

Source: Ref. [72]/American Society For Microbiology.

Table 10.15 Risk factors with virus infection of the circulatory system.

Hemodialysis for kidney disease	Immunosuppression, e.g., HIV
Personal history of endocarditis	IV drug abuse
Age: Children are more susceptible to enteroviral myocarditis	

barrier (choroid plexus). HIV-1, WNV, human T-cell leukemia virus, Zika virus, and Ebola virus can cross the blood–brain barrier [73, 74]. Chikungunya virus may pass through the blood–CSF barriers through the choroid plexus to infect ependymal cells [75]. Some viruses use either the blood–brain barrier or the neural spread. Examples of such are rabies and HSV-1. Is the most common cause of sporadic encephalitis in adults, which causes severe encephalitis cases [76, 77]. Viral destruction of these barriers can be both cause and effect of viral diseases. Congenital, infant, and childhood infection of neuroteratogenic or neurotropic viruses, especially during the peak of neuronal cell proliferation, migration, and developmental phases, can cause slight or obvious changes in brain maturation and functions. Development of neurological defects may appear later in life, even following asymptomatic infections in newborns. Although usually a long process, some people may eventually fully recover from encephalitis (see Table 10.16). Other risk factors associated with development of disease state are listed in Table 10.17.

Zika virus infection can cause neurological complications, particularly microcephaly in newborns and Guillain–Barré syndrome in adults [78]. The virus crosses the blood–brain barrier and infects neural progenitor cells, leading to cell death and impaired brain development in fetuses [79]. This is especially devastating in pregnant women, where Zika virus can result in congenital Zika syndrome, a group of severe birth defects, including brain malformations (see Figure 10.6).

Table 10.16 Viral etiology associated with viral infection of the central nervous system.

Virus	Disease
Herpes simplex virus (HSV)	Herpes simplex encephalitis
Varicella-zoster virus (VZV)	Varicella-zoster encephalitis, meningitis, shingles-associated myelitis
Rabies virus	Rabies encephalitis
Enteroviruses (e.g., poliovirus)	Poliomyelitis (polio), aseptic meningitis
West Nile virus	West Nile encephalitis, meningitis
Japanese encephalitis virus (JEV)	Japanese encephalitis
Zika virus	Zika virus-associated microcephaly, meningoencephalitis, and Guillain–Barre syndrome
Eastern equine encephalitis virus (EEEV)	Eastern equine encephalitis
Western equine encephalitis virus (WEEV)	Western equine encephalitis
St. Louis encephalitis virus (SLEV)	St. Louis encephalitis
Mumps virus	Mumps meningitis, encephalitis
Measles virus	Subacute sclerosing panencephalitis (SSPE)
Cytomegalovirus (CMV)	CMV encephalitis, ventriculitis, especially in immunocompromised patients
Epstein–Barr virus (EBV)	Epstein–Barr virus-associated encephalitis
Human immunodeficiency virus (HIV)	HIV-associated neurocognitive disorders (HAND), HIV encephalopathy
Lymphocytic choriomeningitis virus (LCMV)	Lymphocytic choriomeningitis
Arboviruses (e.g., dengue virus)	Dengue-associated encephalitis, meningoencephalitis
Tick-borne encephalitis virus (TBEV)	Tick-borne encephalitis
Hendra virus	Hendra virus encephalitis
Nipah virus	Nipah virus encephalitis
La Crosse virus	La Crosse encephalitis
Influenza virus	Influenza-associated encephalopathy/encephalitis

Table 10.17 Risk factors with virus infection of the central nervous system.

Older adults	Immunosuppression, e.g., HIV
Children under the age of 1	Vaccination status
Mother-to-child transmission	

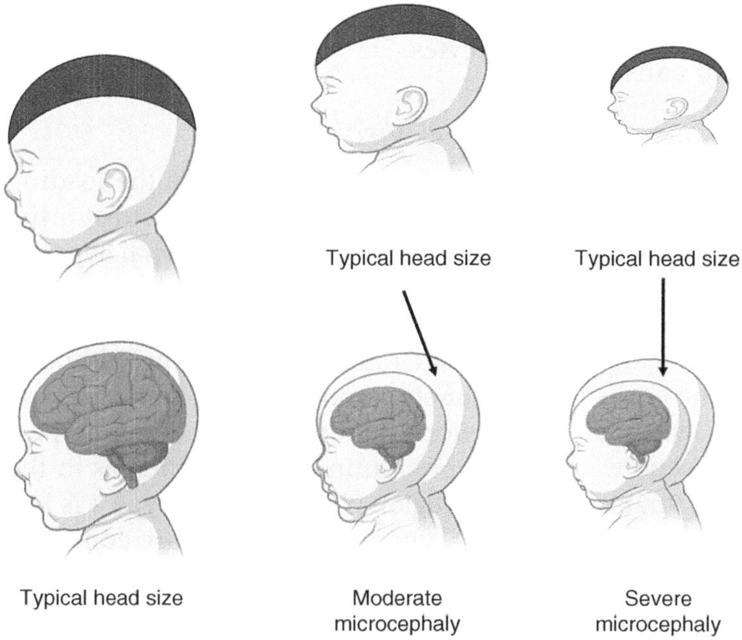

Figure 10.6 Illustration of infants exhibiting moderate to severe microcephaly linked to maternal Zika virus infection, in contrast to a healthy newborn.
Source: With permission of BioRender.

Other viruses include WNV [80] (in North and West Africa, South Africa, and Madagascar), Ebola virus (in the Democratic Republic of the Congo and South Sudan), and chikungunya virus (in Mozambique, Tanzania, and Kenya) are considered to be neuroinvasive viruses that cause neurological complications. In Malawi, a wide range of neurotropic viruses, in particular adenovirus infection, occur among children with suspected non-bacterial CNS infections. Other viruses are mumps, HSV-6, CMV, HSV-1, and enterovirus. Japanese encephalitis virus (JEV), a mosquito-borne flavivirus, remains a significant public health concern in India and many other Asian countries, including Nepal, Bangladesh, Vietnam, Thailand, China, and Indonesia, where it is a leading cause of viral encephalitis [81].

10.1.9.1 Laboratory Diagnosis

PCR can be done to identify the virus genome from postmortem brain biopsy samples (rarely performed when the patient is alive because of the high risk of complications). Blood, urine, swabs, and cerebrospinal fluid (CSF) (taken during the first 3 days after the onset of symptoms) can be collected for cells, protein, and glucose estimation and can be tested by PCR for virus nucleic acid detection. For serological tests, cross-reactivity may occur. This may be a result of secondary infections (or previous infections) by co-circulating viruses of the same family. Careful interpretation of results that considers available clinical and epidemiological information is suggested. A complete blood test can reveal signs of a viral infection.

10.1.10 Viral Hepatitis

Viral hepatitis is an infection that causes liver inflammation and damage. The liver involvement in viral infections can be caused by hepatotropic viruses that replicate in the liver and non-hepatotropic viruses that primarily target other tissues. Hepatitis A, B, C, D, and E are typically hepatotropic viruses. The hepatitis A and E viruses can cause acute, or short-term, infections and can be self-limiting depending on the host's immunity [82]. Hepatitis B, C, and D viruses can cause both acute and chronic infections. Chronic infection occurs if the host is unable to eliminate the virus [83]. This can lead to severe complications like liver cirrhosis, liver failure (fulminant hepatitis), and liver cancer (see Figure 10.7 and Table 10.18). The consequence of the hepatitis is thought to be a result of a cellular immune response directed against viral antigens scattered in hepatocytes.

10.1.10.1 Viral Etiology

Hepatitis B and C viruses are the leading causes of hepatitis and bear a substantial portion of the global burden of viral hepatitis (Table 10.19). Africa has the highest number of HBsAg-positive individuals. Largely, genotypes 1 and 4 are predominating, although other genotypes are present. In the central and eastern regions of Africa, genotype 4 is more predominant [84]. Whereas in the western region of Africa, genotype 4 is less frequent, genotypes 1 and 2 are common. All genotypes of HCV are found in Southern Africa, except genotype 6, which is primarily found in Southeast Asia.

HBV, HCV, and Hepatitis D can cause acute and chronic infection and spread through

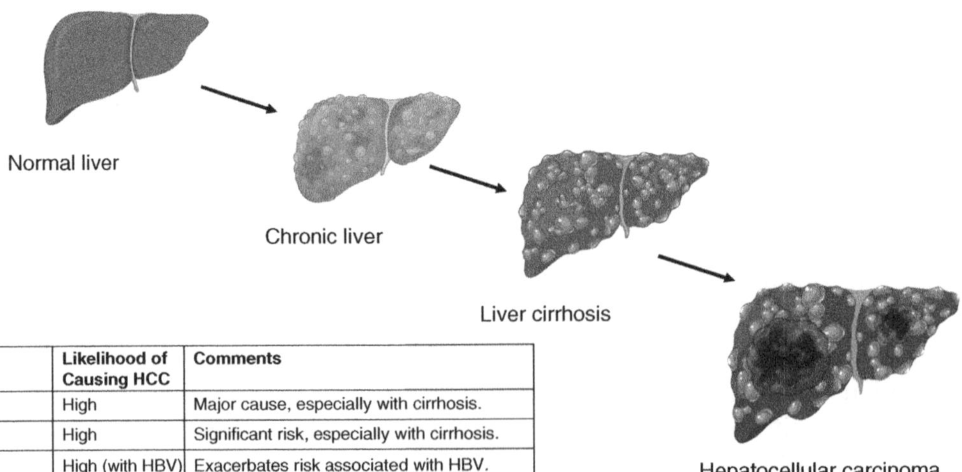

Hepatitis Virus	Likelihood of Causing HCC	Comments
HBV	High	Major cause, especially with cirrhosis.
HCV	High	Significant risk, especially with cirrhosis.
HDV	High (with HBV)	Exacerbates risk associated with HBV.
HEV	Low	Rarely associated with HCC, mostly in chronic cases.
Other hepatitis viruses	Very low	Minimal association with HCC.

Figure 10.7 Stages in the development of hepatocellular carcinoma (HCC).

Table 10.18 Overview of the pathogenesis of viral hepatitis.

Virus	Disease	Incubation time	Interaction with host defense	Pathophysiology	Phase of infection
Hepatitis A virus (HAV)	Hepatitis A	15–50 days	Elicits a strong humoral immune response; immunity is generally lifelong postinfection.	Infects hepatocytes, leading to liver inflammation and necrosis; often self-limiting.	Acute infection typically resolves within weeks
Hepatitis B virus (HBV)	Hepatitis B	30–180 days	Elicits both humoral and cellular immune responses; chronic infection can evade immune surveillance.	Infects hepatocytes, leading to liver inflammation; can cause chronic infection, cirrhosis, and hepatocellular carcinoma.	Acute and chronic phases; chronic infection can last for years
Hepatitis C virus (HCV)	Hepatitis C	14–180 days	Elicits a variable immune response; chronic infection can lead to immune evasion and liver damage.	Infects hepatocytes, causing chronic inflammation, fibrosis, and potentially cirrhosis and hepatocellular carcinoma.	Acute infection often progresses to a chronic phase
Hepatitis D virus (HDV)	Hepatitis D (requires HBV for infection)	2–8 weeks	Requires co-infection or superinfection with HBV; can lead to more severe liver disease.	Causes severe liver inflammation and increases the risk of cirrhosis and liver failure when co-infected with HBV.	Acute and chronic phases exacerbate HBV infection
Hepatitis E virus (HEV)	Hepatitis E	2–10 weeks	Elicits a strong humoral immune response; typically, self-limiting in healthy individuals.	Infects hepatocytes, leading to liver inflammation; can be severe in pregnant women.	Acute infection usually resolves within weeks

contact with contaminated blood and blood products. Hepatitis A and E viruses most often cause acute hepatitis, which can become chronic. They spread through contact with uncooked food or contaminated water (fecal–oral route). Non-A-E hepatitis (also known as hepatitis X), caused by herpesviruses – CMV, EBV, HSV, parvovirus, adenovirus, and influenza viruses – is most often acute but can become chronic [85, 86].

10.1.10.2 Laboratory Diagnosis

Hepatitis virus-infected individuals can be asymptomatic for decades; hence, routine diagnostic screening is highly recommended for all individuals, especially people at high

Table 10.19 Risk factors associated with viral hepatitis.

Hepatitis B and C	Hepatitis A and D
People born in countries with higher HBV prevalence	Poor sanitation
IV drug abuse	Men who have sex with men
Household and sexual contacts of people with hepatitis B	International travelers
Mother-to-child transmission	Homeless people
Healthcare, emergency medical, and public safety personnel	Men who have sex with men
Hemodialysis patients	IV drug abuse
Immunosuppression, e.g., HIV	Household and sexual contacts of people with hepatitis B

risk and pregnant women. Rapid diagnostic tests (point-of-care tests) are an acceptable alternative to laboratory-based immunoassays (invariably ELISA) and nucleic acid-based tests. Though expensive and not easily accessible, HBV DNA testing is essential to identify an individual's viral load and genotype. Assessment of Aspartate Aminotransferase (AST) to Platelet Ratio Index (APRI) score and alpha-fetoprotein levels can be used as biomarkers of fulminant hepatitis and fibrosis.

10.2 Geographical Distribution and Patterns of Virus Transmission

10.2.1 Tropical Vs. Temperate Regions

Tropical regions: Areas close to the equator, such as some parts of Africa, South America, and Southeast Asia, often experience higher incidence rates of vector-borne viral diseases like dengue fever, Zika virus, and chikungunya [87, 88]. This is due to the presence of vectors like mosquitoes, which thrive in tropical climates.

Temperate regions: Regions with seasonal temperature variations, such as North America and Europe, may see seasonal outbreaks of respiratory viruses like influenza and coronaviruses, which can peak in colder months [89].

10.2.2 Urban Vs. Rural Areas

Urban areas: Higher population density in cities can facilitate the rapid spread of viruses [90]. For example, influenza and norovirus outbreaks are common in urban settings due to close human contact.

Rural areas: Rural areas may have less frequent outbreaks but can experience sporadic cases of zoonotic viruses (e.g., hantavirus, Ebola) due to closer interactions with wildlife [91].

10.2.3 Regions with Endemic Viruses

1. **Sub-Saharan Africa**
   - **HIV/AIDS:** Sub-Saharan Africa has the highest burden of HIV/AIDS in the world. The virus is endemic in the region due to several factors, such as limited access to healthcare and high rates of co-infections like tuberculosis [92].
   - **Yellow fever:** This mosquito-borne virus is endemic in West and Central Africa. Vaccination campaigns are critical in controlling outbreaks in these areas [93].
   - **Lassa fever:** Endemic in West Africa, particularly in countries like Nigeria, Sierra Leone, and Liberia, Lassa fever is spread through contact with the urine or feces of infected rodents.
2. **Southeast Asia**
   - **Dengue fever:** This mosquito-borne virus is endemic in Southeast Asia, particularly in countries like India, Thailand, Indonesia, and the Philippines [94]. The region's tropical climate and urbanization contribute to the high prevalence of dengue.
   - **Hepatitis B virus:** Hepatitis B is endemic in several parts of Southeast Asia [95]. Vertical transmission (from mother to child) and early childhood infections are common in the region, contributing to high prevalence rates.
   - **Japanese encephalitis:** This virus, transmitted by mosquitoes, is endemic in rural areas of Southeast Asia and parts of the Western Pacific [96]. It primarily affects children, with outbreaks often occurring during the rainy season.
3. **South America**
   - **Dengue fever:** Like Southeast Asia, dengue fever is also endemic in parts of South America, including Brazil, Colombia, and Venezuela [97]. The region experiences regular outbreaks, especially during the rainy season.
   - **Chikungunya:** This virus, transmitted by Aedes mosquitoes, is endemic in parts of South America, including Brazil and Venezuela [98]. It causes fever and severe joint pain.
   - **Zika virus:** Endemic in parts of Central and South America, Zika virus gained global attention during the 2015–2016 outbreak [99]. It is transmitted by mosquitoes and can cause birth defects if contracted during pregnancy.
4. **Middle East and North Africa**
   - **Hepatitis C:** Egypt has one of the highest prevalence rates of hepatitis C in the world [100]. The virus became less endemic due to historical public health campaigns against the use of non-sterile needles.
   - **Middle East Respiratory Syndrome (MERS):** MERS is caused by a coronavirus, is endemic in the Arabian Peninsula, particularly in Saudi Arabia. It is transmitted from camels to humans, with sporadic outbreaks occurring since its discovery in 2012 [101].

5. **Central Asia**
   - **Crimean–Congo hemorrhagic fever (CCHF):** Endemic in parts of Central Asia, including Kazakhstan, Uzbekistan, and Tajikistan [102], CCHF is a tick-borne virus that can cause severe hemorrhagic fever.
   - **Hepatitis B and D:** Hepatitis B is endemic in many Central Asian countries, with co-infection with the hepatitis D virus (HDV) also prevalent in these regions.
6. **Western Pacific**
   - **Hepatitis B:** Some countries in the Western Pacific, such as China, Taiwan, and the Pacific Islands, have high rates of hepatitis B infection [103]. Perinatal transmission is common, contributing to chronic cases.
   - **Dengue fever:** Endemic in many parts of the Western Pacific, dengue fever is a significant public health concern in countries like the Philippines and Malaysia.
7. **Central and Eastern Europe**
   - **Tick-borne encephalitis (TBE):** Endemic in parts of Central and Eastern Europe, including Russia, Latvia, and the Czech Republic [104], TBE is transmitted by ticks and affects the central nervous system.
8. **South Asia**
   - **Hepatitis E:** Hepatitis E is endemic in parts of South Asia, particularly in India **[105]**. The virus is primarily transmitted through contaminated water and is a significant cause of viral hepatitis in the region.
   - **Dengue fever:** Dengue is also endemic in South Asia, with regular outbreaks occurring in countries like India, Sri Lanka, and Bangladesh.

10.2.4 Epidemic Vs. Pandemic

Epidemic: A sudden increase in the number of cases of an infection (viral)/disease in a specific area. Examples include the Ebola outbreak in West Africa (2014–2016) and the Zika virus epidemic in the Americas (2015–2016).

Pandemic: A global outbreak affecting multiple countries or continents. The COVID-19 pandemic is caused by the SARS-CoV-2 virus.

10.3 Modes of Transmission of Viral Infections

Virus infection can be transmitted in a variety of ways, such as through sexual contact (e.g., HIV), fecal–oral and foodborne (e.g., Poliovirus), airborne (e.g., coronavirus), touch, animal bites (e.g., rabies virus), contaminated needles, and sharp objects. Also, viruses can be transmitted via the parenteral route (e.g., blood transfusion, tattooing, and tissue transplants) and vertical routes (through intrauterine, human milk, and perinatal) (see Figure 10.8).

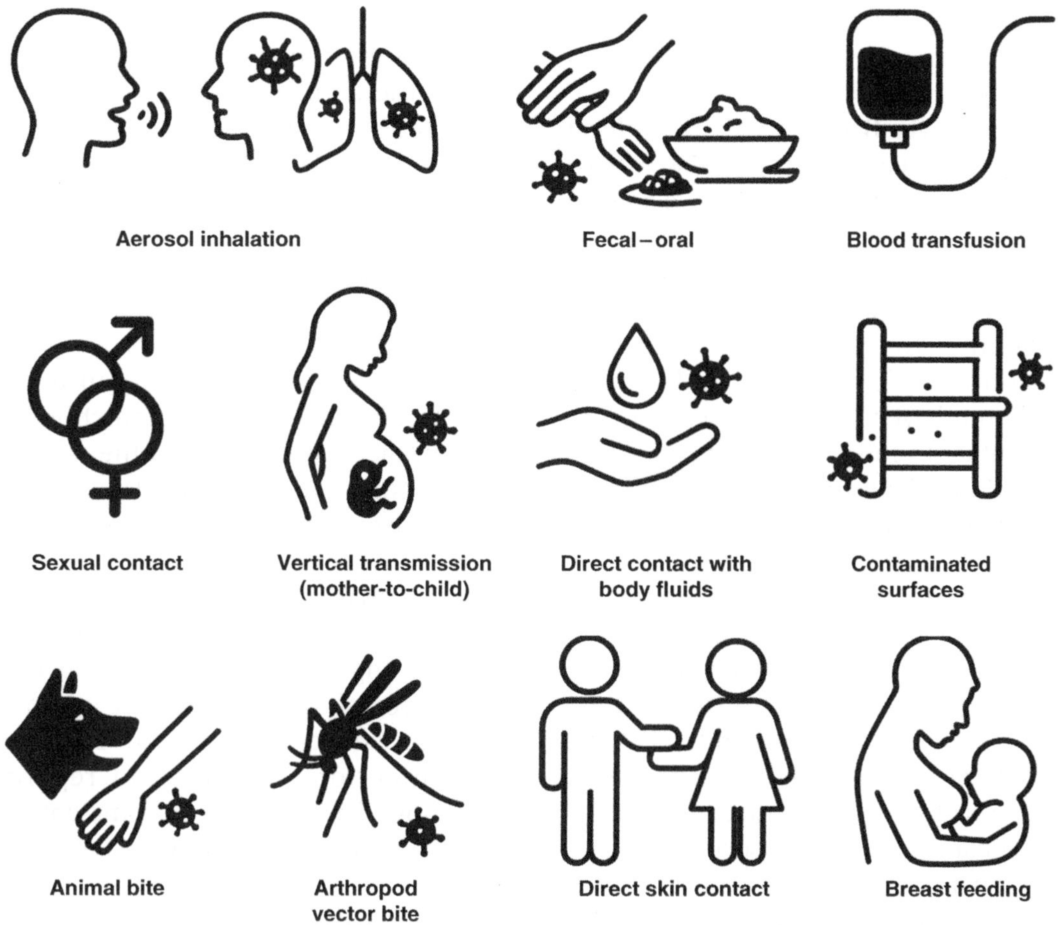

Figure 10.8 Routes of viral transmission.

10.3.1 Sexually Transmitted Viral Diseases

These are infections that spread through contact with body fluids and tissues during sexual intercourse. They can also be blood-borne. Examples include:

- **HIV** is a virus that attacks the immune cells of an infected individual. Progression of infection, most times when not treated, results in a decreased body's ability to fight against common diseases and eventually could lead to acquired immune deficiency syndrome (AIDS). HIV is a major public health concern worldwide. It is transmitted mainly through unprotected sexual intercourse with an infected individual. HIV-infected mother-to-child transmission (MTCT), transfusion of unscreened blood and blood products, and prick or cut with an HIV-contaminated needle or sharp object are other modes of transmission of HIV.

- **Hepatitis B infection** is a blood-borne disease, and it is also a sexually transmitted disease. It is a global public health challenge that accounts for millions of chronic carriers worldwide. It is commonly transmitted from mother to child and through close contact with infected individuals. Though vertical transmission from mother to child is common, most early childhood transmissions are horizontal. There is a high burden of HIV/HBV co-infection in some countries.
- **Human papillomavirus** is a sexually transmitted infection that causes cervical cancer. It is common among younger age groups because of early sexual activity. It is also the most common type of viral infection of the reproductive tract. Some types of HPV can cause genital or ano-genital warts in heterosexual men.
- **Genital herpes** is a common sexually transmitted infection caused by HSV-1 or HSV-2. The virus causes painful small blisters or cold sores that appear in the genital and surrounding areas. Genital herpes usually spread by having anal, oral, or vaginal sex. After primary infection, the virus can remain dormant in the body and can reactivate several years later.

10.3.2 Fecal–Oral and Foodborne Viral Infections

The fecal–oral route of transmission of viral infection implies the ingestion of virus particles that are shed in the feces of an infected person. This usually occurs when an infected individual has poor personal hygiene and spreads the virus by shaking hands, touching surfaces, contaminating drinking water, and preparing food. Fecal–oral viral infections cause mild to severe diseases such as watery diarrhea, hepatitis, vomiting, paralysis, and nausea. Examples of viruses that can be transmitted through this means are:

- **Rotavirus** is a common cause of gastroenteritis in babies and children in tropical countries, and they are often misdiagnosed or undiagnosed. It causes severe diarrhea that can lead to severe dehydration.
- **Norovirus** is one of the leading causes of gastroenteritis worldwide and among all age groups. However, the report shows that there are unreported sporadic cases of human enterovirus infection. Information regarding norovirus is scarce in some countries because routine norovirus surveillance is limited.
- **Hepatitis A virus** is found in the stool and blood of infected individuals. It is very contagious, and it usually has a short-term acute infection and does not usually become chronic. Most infected individuals are asymptomatic, and up to 15% of infected individuals experience recurrent infection within 6 months of infection. Symptoms include diarrhea, nausea, vomiting, and yellow skin or eyes.
- **Poliovirus and coxsackievirus.** They are transmitted through the fecal–oral or oral–oral route, especially through contamination of water or food. Although any age group can contract

the virus, children younger than 5 years are more likely to contract the virus and show symptoms. The transmission is very contagious, and it is a major public health concern. The global effort to eradicate wild-type poliovirus through vaccination has been successful so far. For example, in August 2020, the World Health Organization announced the eradication of wild-type poliovirus in Africa.

10.3.3 Airborne and Respiratory Viral Infections

Respiratory viruses affect the lungs and airways and are mainly airborne. These viruses are most commonly transmitted via inhalation of droplets containing viral particles through direct or indirect contact with infected individuals. They can also be transmitted through contact with the mucus membranes by hands contaminated with droplets of viral particles. They are one of the leading causes of infection among all age groups, and the severity ranges from mild to severe infection. Examples of viruses that can be spread via air droplets include:

- **Coronaviruses** are a group of viruses that cause respiratory tract infections in animals and humans, with the severity of infection ranging from mild to severe. Mild illness shows symptoms that include a common cold, muscle pain, sore throat, diarrhea, and cough. Some strains of coronavirus can cause severe illness with high fatality. Viral respiratory diseases caused by coronavirus include COVID-19 (that caused a pandemic in 2020), severe acute respiratory syndrome (SARS); the 2002–2004 epidemic outbreak, but no case has been reported since 2004, and MERS, which was first reported in Saudi Arabia in 2012, and it spread to other countries. Although no clinical case was reported in Africa, several studies have reported the presence of antibodies against MERS-CoV among domestic donkeys in several tropical countries.
- **Influenza (flu) viruses** are responsible for seasonal flu in the United States every year. They are detected all year round in the United States, but are most common during fall and winter. However, the impact of influenza infection, in some countries, is not well documented, making the need for more surveillance. The disease may have contributed significantly to mobility (especially among children) and mortality on the continent.
- **RSV** infection is a common respiratory virus that usually causes both mild cold-like symptoms (upper respiratory infection) and pneumonia or bronchiolitis (lower respiratory infection) among individuals of all age groups. Most infected individual recovers, but it is more severe in infants and the elderly, and they are at higher risk of infection. In tropical and semi-tropical countries, RSV infections are frequently associated with rainy rather than cold seasons.
- **Rhinovirus** is the common cause of the common cold. Human rhinovirus is spread by aerosolized droplets,

direct or indirect human-to-human contact, or contact with surfaces contaminated by virus droplets. Transmission is common among close contacts in families, schools, and daycares. It is generally believed that it is self-limiting and mild. It is increasingly recognized as an important cause of morbidity among children. Critical insights into the epidemiology and disease dynamics show that it can affect a patient's quality of life and could lead to more severe diseases like bronchiolitis and pneumonia. It is a major trigger of acute wheezing in young children.

10.3.4 Animal-Associated (Zoonotic) Viral Infections

Zoonotic virus diseases are caused by the transmission of viruses from animals to humans. Generally, some people interact with animals in their daily lives as companions, for food, sports, and means of transportation. Most times, because of the connection of humans with animals, humans become infected with the virus, unaware, when they come in contact with infected animals. Common examples of zoonotic infections are bat-associated transmissible infections (e.g., Ebola virus, coronavirus, rabies virus infection) or ape-associated transmissible infections (e.g., Herpes B virus, monkeypox virus, Orf virus infection).

- **Direct contact with the animals:** This occurs when a susceptible individual has direct human-to-animal contact with an infected animal. Examples include bites (rabies virus) or scratches. Others include coming in contact with droppings, mucus, and body fluids of infected animals.
- **Indirect contact with animals:** This occurs when there is no direct animal-to-human contact. Usually, humans get infected by meeting objects or surfaces where roaming animals have lived or touched.

10.3.5 Vector-Borne Transmission of Virus Infections

Vector-borne transmission refers to the process by which viruses are transmitted to humans or animals through arthropods such as mosquitoes, ticks, or fleas. These vectors act as carriers, acquiring the virus from an infected host and passing it on to another, often without becoming ill themselves. Viruses transmitted by vectors, such as dengue, Zika, chikungunya, and WNV, pose significant public health challenges worldwide, particularly in tropical and subtropical regions. The complexity of vector-borne transmission is influenced by several factors. This includes the biology of the virus, the ecology of the vector, environmental conditions, and human activity. Understanding this mode of transmission is essential for developing strategies for control, prevention, and management of viral infections that are often associated with outbreaks and epidemics.

10.3.5.1 Mosquitoes

Aedes Species

- **Dengue virus:** Dengue fever is caused by four closely related dengue viruses (DENV 1–4), transmitted primarily by *Aedes aegypti* (which bite mostly during the day) and *Aedes albopictus* mosquitoes [106]. It is endemic in tropical and subtropical regions worldwide.
- **Zika virus:** Zika virus, also transmitted by Aedes mosquitoes (*A. aegypti* and *A. albopictus*), gained global attention during the 2015–2016 outbreak in the Americas [107, 108]. It can cause birth defects when transmitted to pregnant women.
- **Chikungunya virus:** Also spread by *A. aegypti and A. albopictus* [109], chikungunya virus causes fever and severe joint pain. Outbreaks have occurred in Africa, Asia, and the Americas.

Culex Species

- **WNV:** Culex mosquitoes transmit WNV, which can cause encephalitis or meningitis in severe cases [110]. It is found in Africa, Europe, the Middle East, and North America.
- **Japanese encephalitis virus (JEV):** JEV, transmitted by Culex mosquitoes (particularly *Culex tritaeniorhynchus*), is endemic in rural areas of Asia and the Western Pacific, primarily affecting children [111].

10.3.5.2 Ticks

Ixodes Species

- **Tick-borne encephalitis virus (TBEV):** TBEV is transmitted by Ixodes ticks (mainly *Ixodes ricinus* and *Ixodes persulcatus*) and is endemic in parts of Europe and Asia. It can cause severe neurological disease.
- **Crimean–Congo hemorrhagic fever virus (CCHFV):** CCHFV is spread by Hyalomma ticks and is found in Africa, Asia, the Middle East, and Eastern Europe (in Europe, especially the species *Hyalomma. marginatum* and *Hyalomma. lusitanicum*) [112]. It causes severe hemorrhagic fever. It can also be transmitted by exposure to infected blood, body fluids, and tissues of viremic animals or humans.

Rhipicephalus Species

- **Kyasanur forest disease virus (KFDV):** This virus is transmitted by ticks and is endemic to certain forested areas in India, causing fever, headache, and bleeding [113].

10.3.5.3 Sandflies

Phlebotomus Species

Sandfly fever virus (phlebotomus fever): Transmitted by sandflies, also known as "three-day fever," prevalent in Mediterranean regions, the Middle East, and parts of Asia, and has been reported in some African countries [114].

Geographically, Toscana, Sicilian, and Naples sandfly fever viruses are predominantly found across Mediterranean regions and northern Africa. Meanwhile, Germiston virus (GFV), originally discovered in South African, can be plausible in other African countries including Sudan, Senegal, Central African Republic, Nigeria and Benin, due to sandfly virus ecology, while Karimabad virus (or antibodies against it) has been reported in humans in Central Asia and some parts of the Middle East, including Sudan, South Sudan and Egypt. [115].

Mechanisms of Transmission

Biological transmission: *In this type of transmission, the virus must undergo replication or development within the vector before it can be transmitted to the next host. For example, the dengue virus replicates within the Aedes mosquito before being transmitted through a bite.*

Mechanical transmission: *This is less common for viruses but occurs when a vector physically carries the virus from one host to another without biological development.*

10.3.6 Vertical Transmission (Mother-to-Child)

Vertical transmission, also known as MTCT, occurs when a virus is passed from an infected mother to her child during pregnancy, childbirth, or breastfeeding. This mode of transmission is significant because it can lead to serious health consequences for the newborn, including congenital infections, chronic diseases, and developmental disorders (see Table 10.20).

10.4 Surveillance and Reporting of Virus Infection

Monitoring viral infections involves tracking the spread, evolution, and impact of viruses within populations. This is crucial for public health, allowing for timely interventions and the development of strategies to control outbreaks. Several methods and strategies can be applied to identify and monitor the spread of viral infection. This includes:

10.4.1 Surveillance Systems

- **Passive surveillance:** This type of surveillance collects data from healthcare providers and laboratories who report cases of viral infections. It relies on existing data and is often used for routine monitoring.
- **Active surveillance:** This involves proactive data collection, often by contacting healthcare facilities or conducting field visits. It is more resource-intensive but provides more accurate real-time data.
- **Sentinel surveillance:** Uses specific cohorts such as selected healthcare facilities, individuals (volunteers or paid), or laboratories to monitor specific areas or diseases. These sites act as early warning systems for outbreaks.

10.4.2 Epidemiological Methods

- **Contact tracing:** Ideal contact tracing is a critical public health tool used to prevent the spread of infectious diseases by identifying, notifying, and monitoring individuals who have been in close contact with an infected person [116]. Its primary goal is to break the chain of transmission by ensuring early detection and intervention. Contact tracing involves identifying infected individuals (index

Table 10.20 Examples of some viral diseases that can be transmitted by vertical transmission.

Virus	Transmission pathways	Impact on the infant	Prevention/ management
HIV	Pregnancy, childbirth, breastfeeding	Risk of developing AIDS, severe immunodeficiency	Antiretroviral therapy (ART), elective cesarean, and avoid breastfeeding
HBV (hepatitis B)	Childbirth (exposure to blood and body fluids)	Chronic hepatitis B, risk of liver cirrhosis and cancer	Hepatitis B vaccine, HBIG within 12 hours of birth
HCV (hepatitis C)	Childbirth	Chronic hepatitis C, potential liver disease	Regular monitoring, no current vaccine
CMV (Cytomegalovirus)	Pregnancy, childbirth, breastfeeding	Hearing loss, vision impairment, and developmental delays	Good hygiene, antiviral treatments
Zika virus	Pregnancy	Congenital Zika syndrome, microcephaly, neurological disorders	Mosquito bite prevention, contraception, and no current vaccine
Rubella virus	Pregnancy (especially the first trimester)	Congenital rubella syndrome, heart defects, hearing loss, cataracts	MMR vaccination before pregnancy
HSV (Herpes simplex virus)	Childbirth (active genital herpes), possible during delivery	Neonatal herpes, encephalitis, and severe infections	Antiviral treatment, cesarean if active outbreak
VZV (Varicella-zoster virus)	Pregnancy, particularly the first or second trimester	Congenital varicella syndrome, limb abnormalities, neonatal varicella	Varicella vaccination before pregnancy

cases), tracking their recent contacts, and advising these contacts to isolate, get tested, or receive treatment depending on the disease in question. This strategy is particularly effective in controlling outbreaks of highly contagious diseases.

- **Case–control studies:** Case–control studies are an essential observational study design used in public health research to investigate the causes of diseases and identify risk factors associated with various health outcomes. In this type of study, two groups of individuals are compared: cases, who have the disease or condition of interest, and controls, who do not. Researchers then look retrospectively to determine whether exposure to a particular risk factor or variable was more common among the cases than the controls.

- **Cohort studies:** Cohort studies are a fundamental research design in public health used to investigate the causes of disease and establish links between risk factors and health outcomes [117]. In a cohort study, a group of individuals sharing a common characteristic (the cohort) is followed over time to assess how certain exposures (e.g., lifestyle, environmental factors, or medical interventions) influence the development of specific outcomes such as disease, mortality, or health-related conditions. These studies are particularly valuable for understanding the natural history of diseases, estimating the incidence of outcomes, and identifying potential causal relationships. In public health and epidemiology, cohort studies help track the incidence, risk factors, and long-term outcomes associated with viral infections.

10.4.3 Environmental Monitoring

- **Wastewater surveillance:** This can help to detect viral particles in sewage systems, providing early warning of community spread [118]. Wastewater surveillance offers several advantages in public health monitoring. It provides real-time, population-wide data without the need for individual testing, offering insights into the prevalence of infections at a community level. This approach has proven particularly useful for tracking asymptomatic cases and viral transmission trends over time, guiding public health responses, and resource allocation.
- **Air and surface sampling:** Air sampling involves collecting airborne particles or aerosols that may contain viral particles [119]. This can be done using various devices like impactors, filters, or liquid impingers that capture viral particles from the air, allowing for subsequent laboratory analysis. Air sampling helps to assess viral presence in enclosed spaces, such as hospitals, public transport systems, and workplaces, contributing to the development of ventilation guidelines and air filtration measures to reduce viral spread.

10.4.4 Digital and Syndromic Surveillance

- **Digital surveillance** refers to the use of digital tools and data sources, such as social media, online search trends, electronic health records (EHRs), and mobile applications, to monitor and track the spread of diseases in real time. This method allows public health authorities to quickly identify outbreaks, monitor disease trends, and respond more rapidly than traditional surveillance methods. Digital surveillance leverages vast amounts of data from nontraditional sources, providing a broader, more immediate view of emerging health threats. Analyzing social media platforms for mentions of symptoms or outbreaks also helps in providing information on the disease. Digital surveillance can utilize mobile apps and wearable devices to track symptoms and potential exposure.

- **Syndromic surveillance**, on the other hand, focuses on collecting and analyzing health-related data based on clinical symptoms, or "syndromes," before a specific diagnosis is made. It relies on symptom patterns reported in hospitals, emergency departments, and pharmacies to detect early signals of potential outbreaks. This type of surveillance is particularly useful for monitoring diseases with common symptoms, such as respiratory infections or gastrointestinal illnesses, and can be enhanced by digital tools for faster data collection and analysis.

10.4.5 Mathematical Modeling

Mathematical modeling plays a critical role in understanding and predicting the dynamics of viral infections [120]. These models use mathematical equations to simulate the spread of viruses within populations, helping scientists and public health officials predict outbreaks, understand the impact of interventions, and plan responses. By incorporating variables such as transmission rates, recovery rates, population behavior, immunity, and environmental factors, these models provide valuable insights into how a virus might behave under different scenarios.

An example is the Susceptible-Infectious-Recovered (SIR) model [121], which categorizes populations into compartments based on their disease status. More complex models, like Susceptible-Exposed-Infectious-Recovered (SEIR), account for latent periods, making them useful for diseases with incubation times, such as COVID-19.

- **Predictive models:** Use data from various sources to predict the spread and impact of viral infections, guiding public health interventions.
- **Transmission dynamics models:** Analyze how a virus spreads within a population, helping to identify key factors driving an outbreak.

10.5 Public Health Strategies for Prevention and Control of Viral Diseases

Public health strategies for the prevention and control of viral infections are essential for safeguarding populations and reducing the impact of outbreaks. These strategies encompass a range of interventions, each targeting different aspects of disease transmission and management.

10.5.1 Vaccination Programs

Vaccination is one of the most effective public health strategies for preventing viral infections. Vaccines stimulate the immune system to recognize and combat specific viruses, reducing the incidence and severity of disease. Importance of vaccination are:

- **Herd immunity:** Widespread vaccination leads to herd immunity, which is based on the principle that a large portion of the population becomes immune, thus limiting the spread of the virus [122]. This protects those who cannot be vaccinated (such as immunocompromised individuals) and are vaccine-hesitant.

- **Eradication of diseases:** Successful vaccination programs have led to the eradication of diseases like smallpox and have significantly reduced the prevalence of others, such as polio and measles.
- **Impact on public health:** Vaccination reduces healthcare costs by preventing disease, reduces the burden on healthcare systems, and improves overall public health.

The threshold for herd immunity is the proportion of the population that must be immune to an infectious disease (through vaccination or previous infection) to prevent its spread within the community [123]. This threshold varies depending on the contagiousness of the disease, which is often represented by the basic reproductive number, R_0 (the average number of secondary cases generated by one infected individual in a fully susceptible population) [124].

The formula to estimate the herd immunity threshold is:

$$H = 1 - \frac{1}{R_0}$$

Here are some examples of herd immunity thresholds for various diseases:

1. **Measles:** Measles is highly contagious, with an R_0 of approximately 12–18 [125]. The herd immunity threshold for measles is therefore around 92–95%. This means that at least 92–95% of the population needs to be immune to prevent sustained transmission.
2. **Polio:** Polio has an R_0 of approximately 5–7, leading to a herd immunity threshold of around 80–86% [126].
3. **COVID-19 (SARS-CoV-2):** The herd immunity threshold for COVID-19 varies depending on the specific variant and its transmissibility. Initially, it was estimated to be around 60–70% for earlier strains (with an R_0 of about 2–3) [127]. However, more transmissible variants (like Delta and Omicron, with higher R_0 values) require higher thresholds, possibly 80% or more.
4. **Influenza:** Influenza has a lower R_0, typically ranging from 1.5 to 2, which means the herd immunity threshold is around 33–50%. However, because of the virus's ability to mutate and the varying effectiveness of flu vaccines, maintaining herd immunity is challenging.
5. **Rubella:** Rubella has an R_0 of 5–7, so its herd immunity threshold is around 80–86% [128].

10.5.2 Hygiene and Sanitation Practices

- **Hand hygiene:** Regular handwashing with soap and water is a simple yet effective measure to prevent the spread of many viruses, including those causing respiratory and gastrointestinal infections.
- **Disinfection:** Proper cleaning and disinfection of surfaces, particularly in

healthcare settings and public spaces, helps reduce the transmission of viruses.

- **Safe water and sanitation:** Ensuring access to clean water and proper sanitation facilities reduces the spread of waterborne and fecal–oral transmitted viruses, such as those causing hepatitis A and rotavirus infections.
- **Food safety:** Adhering to food hygiene practices prevents foodborne viral infections, such as norovirus.

10.5.3 Public Awareness and Education Campaigns

- **Education on preventive measures:** Public health campaigns educate communities on the importance of vaccination, hygiene practices, and other preventive measures. This knowledge empowers individuals to take proactive steps in preventing infections.
- **Behavioral change:** Campaigns aim to change behaviors that contribute to the spread of viruses, such as encouraging the use of condoms to prevent sexually transmitted infections or promoting mask-wearing during respiratory outbreaks.
- **Combating misinformation:** Public education efforts also counteract misinformation and vaccine hesitancy, which can undermine public health efforts.
- **Outreach to vulnerable populations:** Targeted campaigns to reach high-risk groups, such as those with chronic conditions or living in crowded conditions, to ensure they are informed and protected.

10.5.4 Quarantine and Isolation Measures

Quarantine: Involves restricting the movement of individuals who may have been exposed to a virus but are not yet symptomatic. This helps prevent potential spread while monitoring for signs of infection.

Isolation: Refers to the separation of individuals who are infected from those who are healthy to prevent transmission. Isolation can occur in healthcare settings or at home, depending on the severity of the infection.

These measures are particularly effective during outbreaks of highly contagious diseases, such as COVID-19 or Ebola, where controlling person-to-person transmission is critical. However, implementing quarantine and isolation can be challenging due to the need for public compliance, provision of essential services to those affected, and managing the economic and social impacts.

10.5.5 Vector Control Programs

- **Targeting disease vectors:** Vector control programs focus on controlling the organisms that transmit viruses, such as mosquitoes, ticks, and flies. This is crucial for preventing diseases like malaria, dengue, Zika, and yellow fever.
- **Environmental management:** Reducing breeding sites for vectors (e.g., eliminating standing water for mosquitoes) is a key strategy in vector control.

- **Insecticide use:** Spraying insecticides in areas with high vector populations can reduce the risk of transmission, although careful management is needed to avoid resistance and environmental harm.
- **Use of protective measures:** Encouraging the use of bed nets, insect repellents, and protective clothing helps individuals protect themselves from vector-borne diseases.
- **Biological control:** Introducing natural predators of vectors or using genetically modified organisms to reduce vector populations are emerging strategies in vector control.

10.6 "One Health" Approach to Controlling Viral Diseases

The "One Health" approach is a collaborative, multisectoral strategy that recognizes the interconnectedness of human health, animal health, and environmental health. This perspective is particularly relevant in controlling the emergence and spread of pathogenic diseases, such as viral diseases, and many other pathogens that can be transmitted between humans and animals (zoonoses), and are influenced by environmental factors [129]. An overview of how the One Health approach can effectively control viral diseases in tropical countries includes:

1. **Integrated surveillance and monitoring**
   - **Zoonotic disease surveillance:** Improving the implementation of surveillance systems that monitor animal populations for emerging viral threats helps detect and respond to outbreaks before they spill over into human populations. This includes tracking wildlife, livestock, and domestic animals.
   - **Environmental monitoring:** Assessing environmental factors that contribute to the emergence and spread of viruses, such as climate change, habitat destruction, and biodiversity loss, can provide insights into potential outbreaks.
2. **Cross-sector collaboration**
   - **Interdisciplinary teams:** Engaging professionals from veterinary medicine, human medicine, environmental science, and public health to work together enhances the understanding of viral diseases and improves response strategies.
   - **Data sharing:** Creating platforms for sharing data across sectors allows for a more comprehensive view of viral threats and facilitates timely decision-making.
3. **Preventive measures and risk management**
   - **Vaccination programs:** Improving and implementing vaccination programs for both humans and animals can further significantly reduce the transmission of viral diseases. For instance, vaccinating domestic animals can prevent zoonotic diseases like rabies and influenza.
   - **Biosafety and biosecurity:** Establishing and enforcing biosafety measures in laboratories and biosecurity protocols in agricultural

practices can minimize the risk of viral outbreaks from accidental releases or bioterrorism.

4. **Public health education and awareness**
   - **Community engagement:** Educating communities about the risks of viral diseases, transmission pathways, and prevention strategies fosters resilience and encourages compliance with public health measures.
   - **Behavioral change initiatives:** Promoting behaviors that reduce contact with wildlife and improve hygiene practices can mitigate the risk of zoonotic viral infections.
5. **Research and development**
   - **Viral pathogen research:** Improving research on viral pathogens and their interactions with hosts (both humans and animals) will further help identify potential risk factors and inform the development of vaccines and treatments.
   - **Innovative technologies:** Leveraging technologies such as genomic sequencing and bioinformatics can enhance the understanding of viral evolution and assist in outbreak prediction and management.
6. **Environmental health considerations**
   - **Habitat conservation:** Protecting natural habitats and biodiversity is crucial for reducing human–wildlife interactions that could lead to the spillover of viral pathogens. Conservation efforts can help maintain ecological balance and limit the spread of diseases.
   - **Impact of climate change:** Understanding how climate change affects the distribution and behavior of animal reservoirs can help predict and mitigate potential outbreaks of viral diseases.
7. **Global cooperation and governance**
   - **International collaboration:** The One Health approach emphasizes the need for global cooperation to address viral diseases that cross borders. Initiatives like the Global Health Security Agenda (GHSA) promote collaboration between countries to strengthen health systems and improve response capabilities.
   - **Policy development:** Creating policies that integrate health systems across sectors, such as zoonotic disease prevention, environmental protection, and public health, can lead to more effective control of viral diseases.

10.7 Challenges in the Control of Viral Infections

Controlling viral infections presents numerous challenges, especially in a rapidly changing global landscape. These challenges stem from the dynamic nature of viruses, human behavior, and the complexities of implementing public health measures across diverse populations. Below are some of the key challenges in the control of viral infections.

10.7.1 Emerging and Re-emerging Viral Infections

- **Emerging infections:** New viral infections continue to emerge, often from zoonotic transmission, where viruses jump from animals to humans (see Table 10.21).
- **Re-emerging infections:** Some viral infections, thought to be under control, may resurge due to factors like waning immunity, changes in human behavior, or mutations in the virus. Diseases such as measles and dengue have seen a resurgence in recent years.

Emerging and re-emerging viruses may initially go undetected, delaying response efforts. The lack of preparedness, insufficient surveillance, and limited understanding of the new virus's behavior can exacerbate the challenge of controlling outbreaks.

Table 10.21 Examples of newly emerged viruses in the last 20 years.

Viral infection	Date of emergence	Description
COVID-19 (SARS-CoV-2)	Dec 2019	The novel coronavirus is causing the global COVID-19 pandemic.
Monkeypox (Clade IIb – Global Outbreak)	May 2022	Widespread human-to-human transmission in non-endemic countries.
MERS-CoV (Middle East Respiratory Syndrome Coronavirus)	2012	Coronavirus causes severe respiratory illness, primarily in the Middle East.
Zika virus (major outbreak)	2015–2016	Mosquito-borne viruses are linked to birth defects like microcephaly.
H5N1 avian influenza	2003–2004	Highly pathogenic avian influenza virus with high mortality.
Chikungunya virus (Indian Ocean outbreak)	2004	Mosquito-borne virus causes fever and severe joint pain.
Severe fever with thrombocytopenia syndrome virus (SFTSV)	2009–2011	Tick-borne virus infection causes severe fever and multi-organ failure.
Lujo virus	2008	Rare Arenavirus causes viral hemorrhagic fever with high mortality.
Bourbon virus	2014	Tick-borne viruses have been identified in the United States, causing severe illness.
Heartland virus	2009–2012	Tick-borne virus causing severe febrile illness in the United States.

10.7.2 Impact of Travel and Globalization

- **Rapid spread of viruses:** Increased global travel and trade facilitate the rapid spread of viral infections across borders. An infection can move from one part of the world to another within hours, as seen with COVID-19.
- **Urbanization and population density:** The growth of largely populated urban areas creates environments where viruses can spread more easily [130]. In such settings, controlling outbreaks becomes more complex due to the sheer number of people and the interconnectedness of cities.
- **Global supply chains:** The globalized nature of supply chains means that disruptions caused by viral outbreaks in one region can have far-reaching impacts, affecting the availability of medical supplies and resources needed for outbreak control [131, 132].
- **Coordination across borders:** Effective control requires international cooperation, but differences in public health infrastructure, policies, and resources between countries can hinder coordinated efforts.

10.7.3 Antiviral Resistance and Challenges in Vaccine Development

- **Antiviral resistance:** Long-term use or misuse of antiviral medications can lead to the development of resistance, where viruses evolve to resist treatment [133]. This makes managing viral infections, such as HIV or influenza, more difficult, as standard treatments may become less effective.
- **Viral mutation:** Many viruses, particularly RNA viruses like influenza and HIV, mutate rapidly, which can render existing vaccines less effective. This necessitates frequent updates to vaccines or the development of entirely new ones.
- **Long development times:** Developing a vaccine is a lengthy process, often taking years to ensure safety, efficacy, and mass production. This time lag can be critical during fast-moving outbreaks.
- **Vaccine hesitancy:** Public skepticism and misinformation about vaccines can lead to lower vaccination rates, undermining efforts to control viral infections.
- **Supply chain and distribution:** Even after a vaccine is developed, challenges remain in manufacturing, distributing, and administering the vaccine, especially in low-resource settings.

10.7.4 Socioeconomic and Cultural Factors Affecting Control Measures

- **Access to healthcare:** In some low- and middle-income countries, limited access to healthcare services, vaccines, and antiviral treatments hinders efforts to control viral infections. Inadequate healthcare infrastructure can lead to delayed diagnosis and treatment.

- **Economic constraints:** The costs associated with controlling viral infections, including surveillance, vaccination, and treatment programs, can be relatively unaffordable, particularly in resource-poor settings. Economic downturns caused by pandemics can further strain public health systems [134].
- **Low socioeconomic status:** This has contributed to the increased incidence rate and spread of viral infections in developing countries. Factors like unemployment, poverty, malnutrition, poor drainage, sewage disposal, poor personal hygiene, and unsafe water systems are the leading factors that can contribute to the spread and endemicity of viral infections [134, 135]. For example, as access to safe water is increasing, there is a decrease in the seroprevalence of the hepatitis A virus.
- **Cultural practices:** Certain cultural practices and beliefs may conflict with recommended public health measures, such as vaccination or quarantine [136, 137]. For example, traditional burial practices during the Ebola outbreak may have contributed to the spread of the virus.
- **Education and awareness:** In some regions, low levels of health literacy and education can impede the adoption of preventive measures, such as hygiene practices or vaccination [138, 139]. Public health campaigns may struggle to reach certain communities.
- **Political instability:** In areas affected by conflict or political instability, public health measures may be difficult to implement. Infrastructure may be damaged, and populations may be displaced, making it challenging to conduct surveillance, deliver healthcare, or carry out vaccination campaigns.

10.7.5 Climate Change and Geographical Factors

Climate-induced changes in temperature and rainfall across countries could increase the rate of virus infection and death. Warmer temperatures can help viral insect vectors like mosquitoes breed and mature faster [140, 141]. Whereas prolonged rainfall can increase the breeding of mosquitoes by providing stagnant ponds of water, which aids egg laying. Geographical factors can result in difficulties for health workers in reaching out to villages and remote areas for vaccination or awareness.

10.8 Global Health Initiatives

Global health initiatives are crucial in managing and controlling viral infections worldwide. These efforts involve guidelines from organizations like the World Health Organization (WHO) and the National Centers for Disease Control (CDC), international collaborations, and successful public health interventions.

10.8.1 Overview of the Activities of the WHO and the National Center for Disease Control of Viral Infection Control

10.8.1.1 World Health Organization (WHO)

- **Standards and recommendations:** WHO provides comprehensive guidelines and recommendations for controlling viral infections. These include best practices for surveillance, diagnosis, treatment, and prevention.
- **Infection prevention and control (IPC):** WHO guidelines emphasize the importance of IPC measures in healthcare settings to prevent the spread of infections. These include protocols for hand hygiene, use of personal protective equipment (PPE), and safe handling of infectious materials.
- **Emergency response:** WHO coordinates international responses to viral outbreaks, providing technical support and guidance to affected countries. This includes outbreak preparedness plans, response strategies, and resource mobilization.
- **Global health security:** WHO promotes the strengthening of global health security by encouraging countries to implement the International Health Regulations (IHR), which require nations to report and respond to public health emergencies.

10.8.1.2 National Center for Disease Control (CDC)

- **Surveillance and reporting:** National CDCs develop guidelines for local and national surveillance of viral infections. They provide protocols for case reporting, data collection, and analysis to monitor and control outbreaks.
- **Vaccination programs:** CDCs establish vaccination guidelines, including recommendations for routine immunizations and outbreak-specific vaccines. They also oversee vaccine distribution and monitoring.
- **Public health interventions:** CDC guidelines cover a range of public health interventions, including quarantine and isolation measures, travel advisories, and community education.
- **Research and development:** CDCs often lead or collaborate on research initiatives to develop new diagnostics, treatments, and vaccines for emerging viral infections.

10.8.2 International Collaborations and Efforts to Control Viral Outbreaks

10.8.2.1 Global Initiatives

- **Global Outbreak Alert and Response Network (GOARN):** Coordinated by WHO, GOARN brings together international partners to respond to infectious disease outbreaks. It provides technical assistance, mobilizes resources, and facilitates information exchange.

- **Global fund:** The Global Fund provides financial support for the prevention, treatment, and care of major diseases, including HIV/AIDS, tuberculosis, and malaria. It collaborates with countries to strengthen health systems and response capabilities.
- **Coalition for Epidemic Preparedness Innovations (CEPI):** CEPI focuses on accelerating the development of vaccines for emerging infectious diseases and improving global preparedness for future outbreaks.

10.8.2.2 Regional and Bilateral Collaborations

- **African Centers for Disease Control and Prevention (Africa CDC):** Africa CDC (`https://africacdc.org`) plays a pivotal role in supporting African Union member states by strengthening health systems, enhancing laboratory capacity, and coordinating outbreak preparedness and response. It has been instrumental in responding to viral threats such as Ebola, COVID-19, Lassa fever, and yellow fever. The organization also facilitates regional training and the implementation of the Integrated Disease Surveillance and Response (IDSR) strategy.
- **Southeast Asia Regional Office of the World Health Organization (WHO-SEARO):** WHO-SEARO (`https://www.who.int/southeastasia`) provides technical and operational support to Southeast Asian countries, helping them to address tropical viral infections such as Japanese encephalitis, dengue, Nipah virus, and chikungunya. It also supports cross-border surveillance and early warning systems.
- **Regional Economic Communities (RECs):** This is an organization of the African Union (`https://au.int/en/recs`) including organizations such as the Economic Community of West African States (ECOWAS), and East African Community (EAC). These play important roles in promoting regional health strategies, harmonizing disease surveillance systems, and enabling joint procurement of vaccines and diagnostics for viral disease control.
- **Bilateral agreements:** Countries often collaborate bilaterally to address specific health threats, share resources, and provide technical assistance.
- **Academic and research collaborations:** Universities and research institutes across tropical regions often engage in collaborative research with international partners to advance virological research, vaccine development, and the understanding of viral pathogenesis in tropical settings.

10.9 Viruses as Bioterrorism Agents: Implications for Biosecurity in Tropical Countries

Bioterrorism is the deliberate use of biological agents such as viruses, bacteria, fungi, or toxins by individuals or groups aiming to cause harm or intimidation to cause illness, death, or disruption [142]. With viruses emerging as significant threats due to their potential for rapid spread and high mortality rates, the implications of using viruses as bioterrorism agents are high. Therefore, it is crucial to develop robust biosecurity measures to protect public health and national security.

1. Characteristics of viruses as potential bioterrorism agents [143]
   - **Infectivity:** Many viruses, such as smallpox and Ebola, capable of being used as bioweapons, are highly infectious and can spread easily through respiratory droplets, direct contact, or contaminated surfaces.
   - **Transmission:** Many viruses, capable of being used as bioweapons, can spread rapidly within populations. For example, airborne viruses like influenza and coronaviruses can transmit from person to person with relative ease, leading to widespread outbreaks.
   - **Latency and incubation:** Some viruses have incubation periods during which infected individuals show no symptoms but can still transmit the virus. This allows for the silent spread of the pathogen before detection.
   - **Lethality:** Viruses can cause severe diseases with high mortality rates, making them effective for instilling fear and chaos.
   - **Stability in the environment:** Some viruses can remain viable in the environment for extended periods, enhancing their potential for use as weapons.
   - **Challenges in detection and diagnosis:** Rapid and accurate diagnosis can be difficult, allowing for a window of opportunity for the virus to spread before public health measures are enacted.
2. **Historical Context**
   - **Past incidents:** There have been instances of bioterrorism using viruses, such as the deliberate spread of anthrax in the 2001 attacks in the United States [144]. While not viral, this incident highlighted the vulnerabilities in public health responses to biological threats.
   - **Potential targets:** Viruses like smallpox, which has been eradicated but still exists in laboratory stocks, are considered potential bioweapons. Other viral agents, such as the Marburg and Ebola viruses, are also concerning due to their lethality. Tropical countries should also consider wild-type poliovirus as a potential bioterrorism agent.
3. **Implications for biosecurity**
   - **Enhanced surveillance:** Countries must implement robust surveillance systems to detect unusual patterns of illness or outbreaks that may indicate bioterrorism.
   - **Research and development:** Investing in research to develop vaccines, antiviral treatments, and rapid diagnostic tools is essential to counter potential viral threats.
   - **Laboratory security:** Biosecurity measures in laboratories working with dangerous pathogens must be stringent, including access

controls, training, and emergency response plans.

- **Public health preparedness:** Developing and testing emergency response plans for potential viral outbreaks is critical for minimizing public health impacts.

4. **International cooperation**
   - **Global health security:** Collaboration among nations is essential to share intelligence on potential bioterrorism threats and to coordinate responses to viral outbreaks.
   - **Treaties and agreements:** All countries should endeavor to be part of international treaties, such as the Biological Weapons Convention (BWC), which aims to prevent the development and use of biological weapons, including viruses.
5. **Public awareness and education**
   - **Community resilience:** Educating the public about the risks of bioterrorism and the importance of biosecurity can enhance community resilience and preparedness.
   - **Crisis communication:** If there is a case of bioterrorism, effective communication strategies are necessary to inform the public during a viral outbreak, minimizing panic and ensuring compliance with health advisories.

10.10 Historical Outbreaks and Lessons Learned

Knowing some cases and historical outbreaks provides valuable insights into the epidemiology, pathogenesis, and control of viral infections. By examining specific instances, scientists and public health officials can identify risk factors, develop effective prevention and control strategies, and learn from past mistakes (see Table 10.22).

Table 10.22 Examples of historical outbreaks.

Viral outbreak	Key points	Lessons learned
HIV/AIDS pandemic	**Virus:** Human immunodeficiency virus (HIV) **Disease:** Acquired immunodeficiency syndrome (AIDS) **Date of emergence**: Early 1980s **Impact:** A global pandemic with millions of cases and deaths; it impacts the immune system, leading to increased susceptibility to other infections. **Global response:** Initial slow response due to stigma; significant progress with antiretroviral therapy (ART), transforming HIV into a manageable chronic condition. **Treatment:** ART has improved life expectancy and quality of life.	**Stigma reduction:** Addressing stigma and discrimination is crucial for effective public health interventions. **Community engagement:** Involving affected communities helps tailor interventions and improve outcomes. **Research and innovation:** Ongoing research and development of new treatments and prevention methods are essential for continued progress.

Table 10.22 (Continued)

Viral outbreak	Key points	Lessons learned
Ebola virus outbreaks	**Virus:** Ebola virus **Disease:** Ebola virus disease (EVD) **Date of significant outbreak:** 2014–2016 (West Africa) **Impact:** One of the largest and deadliest outbreak with over 28,000 cases and 11,000 deaths in tropical Africa **Response:** Involvement of international healthcare workers, the establishment of treatment centers, and the development of an experimental vaccine. **Containment:** Aggressive contact tracing, community engagement, and isolation of cases were key.	**Rapid response:** An early and coordinated international response is crucial for controlling outbreaks. **Infrastructure:** Strengthening local healthcare infrastructure and capacity is essential for managing future outbreaks. **Vaccine development:** Investing in vaccine research and having pre-positioned vaccines can significantly enhance outbreak response.
COVID-19 pandemic	**Virus:** SARS-CoV-2 **Date of emergence:** December 2019 **Disease:** COVID-19 **Impact:** Global pandemic with millions of cases and deaths; caused widespread disruption to daily life, economies, and healthcare systems. **Global spread:** Rapid spread due to global travel and urbanization. **Response:** Public health measures included social distancing, mask-wearing, lockdowns, and rapid vaccine development. **Vaccine development:** mRNA vaccines (e.g., Pfizer-BioNTech, Moderna) and other platforms have been developed swiftly.	**Preparedness:** Having pandemic response plans, stockpiles of essential supplies, and robust surveillance systems is critical. **Data sharing:** Transparent sharing of data helps understand the virus and develop effective interventions. **Equity:** Ensuring equitable access to vaccines and treatments across populations is crucial for global health security.

Notable historical outbreaks

- **The 1918 influenza pandemic:** This pandemic, often referred to as the "Spanish Flu," was one of the deadliest in human history [145]. It infected an estimated 500 million people worldwide and resulted in an estimated 50 million deaths.
- **Polio epidemics:** Polio was a major public health concern in the 20th century, causing paralysis and even death in children and young adults [146]. The development of the polio vaccine in the 1950s and 1960s led to a significant decline in polio cases worldwide.
- **HIV/AIDS epidemic:** The HIV/AIDS epidemic began in the 1980s and continues to be a major global health challenge. The virus is transmitted

through bodily fluids and can lead to severe immune system impairment.

- **SARS and MERS outbreaks:** SARS and MERS caused significant outbreaks in the early 21st century. These outbreaks highlighted the potential for emerging viral infections to cause global pandemics.
- **The Zika virus outbreak in the Americas (2015–2016):** The Zika virus outbreak in the Americas was associated with increased cases of microcephaly, a birth defect [147]. This outbreak raised concerns about the potential long-term health consequences of viral infections.

10.11 Study Questions

10.1 A 12-year-old student presents to the school's health center with a fever, headache, sore throat, and fatigue that started a few days ago. He also reports swollen lymph nodes in his neck. On examination, the healthcare provider notices a mild rash on his chest. The student mentions that several of his classmates have had similar symptoms over the past 2 weeks. His medical history reveals that he had not received vaccinations during childhood due to personal and parental preferences.

Given the symptoms, the healthcare provider suspects a viral infection, possibly a re-emerging virus due to low vaccination rates in the community. Blood tests confirm the presence of antibodies against the measles virus, indicating a recent infection.

(a) What are the key epidemiological factors contributing to the resurgence of measles in this school setting?

(b) Identify and describe the primary modes of transmission for measles. How do these modes of transmission influence the spread of the virus in a school environment?

(c) What public health measures should be implemented immediately to control the spread of measles in this school?

(d) How can vaccination campaigns be utilized as a long-term strategy to prevent future outbreaks of measles in the school and surrounding community?

(e) Discuss the potential public health implications if this measles outbreak is not controlled effectively. What impact could it have on the broader community beyond the school?

(f) What role does epidemiological surveillance play in identifying and managing outbreaks of common viral infections such as measles?

(g) Considering this scenario, what communication strategies should be employed to inform students, staff, and the community about the outbreak and preventive measures?

10.2 A 45-year-old man presents to the emergency department with a sudden onset of fever, headache, vomiting, and confusion. He recently returned from a rural area where he was involved in flood relief work. Physical examination reveals neck stiffness and a positive Brudzinski's sign, raising concern for meningitis. Initial laboratory results show elevated white blood cell count and slightly increased protein in cerebrospinal fluid (CSF). A rapid diagnostic test confirms the presence of WNV in the patient's serum.

The local health department is notified, and an investigation reveals that several cases of WNV have been reported in the area, correlating with an increase in mosquito populations following recent flooding. There is concern about the spread of the virus among the population, especially since the rainy season is ongoing, providing favorable breeding conditions for mosquitoes.

(a) What are the key factors that have likely contributed to the outbreak of WNV in this area?
(b) Identify the primary vector responsible for the transmission of the WNV. How does the ecology of this vector influence the spread of WNV in flood-affected areas?
(c) What immediate public health interventions should be prioritized to control the spread of WNV in the affected community?
(d) Discuss the role of environmental management in preventing future outbreaks of WNV, particularly in flood-prone areas. What specific actions can be taken?
(e) How can public health education be utilized to reduce the risk of WNV transmission among the community? What key messages should be communicated?
(f) Explain the importance of surveillance and monitoring in managing WNV outbreaks. What specific data should be collected and analyzed?
(g) Consider the potential long-term impacts of the WNV on the affected population. What measures can be taken to support individuals who have been infected?

Objectives

10.1 **Which of the following is NOT part of the primary objectives of virologists in the context of viral infections?**
(a) To diagnose individual patients
(b) To identify the causative agents of infections
(c) To determine the distribution of infections in populations
(d) To develop vaccines

10.2 **What is the most effective method to prevent the spread of influenza in a community setting?**
(a) Hand hygiene
(b) Isolation of infected individuals
(c) Vaccination
(d) Use of antiviral medications

10.3 **Which of the following viral infections is primarily transmitted through a fecal–oral route?**
(a) Hepatitis B
(b) Hepatitis C
(c) Hepatitis A
(d) HIV

10.4 **In the control of a measles outbreak, what is the recommended percentage of population immunity required to achieve herd immunity?**
(a) 50%
(b) 60%

(c) 70%
(d) 95%

10.5 **Which of the following is a key feature of a successful public health surveillance system?**
(a) High specificity with low sensitivity
(b) Ability to detect outbreaks early
(c) Infrequent reporting intervals
(d) Focus solely on infectious diseases

10.6 **What is the primary goal of quarantine as a public health measure during a viral outbreak?**
(a) To treat infected individuals
(b) To reduce the number of susceptible individuals
(c) To prevent the spread of infection to healthy individuals
(d) To vaccinate the population

10.7 **Which viral infection has been successfully eradicated through global vaccination efforts?**
(a) Polio
(b) Smallpox
(c) Measles
(d) Chickenpox

10.8 **What is the role of contact tracing in managing viral outbreaks?**
(a) To identify and isolate cases of infection
(b) To administer vaccines
(c) To provide antiviral treatment
(d) To conduct laboratory testing

10.9 **Which of the following is an example of a vector-borne viral infection?**
(a) HIV
(b) Influenza
(c) Dengue fever
(d) Hepatitis B

References

1. Zumla, A. and Ustianowski, A., *Tropical diseases: definition, geographic distribution, transmission, and classification*. Infect. Dis. Clin. North Am., 2012. **26**(2): p. 195–205.
2. White, D.O. and Brown, L.E., *Respiratory viruses*, in *Encyclopedia of Virology*. 1999. p. 1488–1496. doi: 10.1006/rwvi.1999.0247 Epub 2004 Jun 17. PMCID: PMC7173581. https://pmc.ncbi.nlm.nih.gov/articles/PMC7173581/.
3. Kutter, J.S., et al., *Transmission routes of respiratory viruses among humans*. Curr. Opin. Virol., 2018. **28**: p. 142–151.
4. Dhand, R. and Li, J., *Coughs and sneezes: their role in transmission of respiratory viral infections, including SARS-CoV-2*. Am. J. Respir. Crit. Care Med., 2020. **202**(5): p. 651–659.
5. Watson, A. and Wilkinson, T.M.A., *Respiratory viral infections in the elderly*. Ther. Adv. Respir. Dis., 2021. **15**: 1753466621995050.
6. Troeger, C., et al., *Estimates of the global, regional, and national morbidity, mortality, and aetiologies of lower respiratory infections in 195 countries, 1990–2016: a systematic analysis for the Global Burden of Disease Study 2016*. Lancet Infect. Dis., 2018. **18**(11): p. 1191–1210.
7. Rijsbergen, L.C., et al., *In vitro modelling of respiratory virus infections in human airway epithelial cells – a systematic review*. Front. Immunol., 2021. **12**: 683002.
8. Zhang, L., et al., *Infection of ciliated cells by human parainfluenza virus type 3 in an in vitro model of human airway epithelium*. J. Virol., 2005. **79**(2): p. 1113–1124.
9. Martinez-Giron, R., Van Woerden, H.C., and Martínez-Torre, C., *Ciliated nasal epithelial cells damage and human rhinovirus infection: cytological findings*. Acta Biomed., 2020. **91**(1): p. 146–147.

10. Thompson, C. and Zambon, M., *Influenza, respiratory syncytial virus and SARS*. Medicine (Abingdon), 2009. **37**(12): p. 679–685.
11. Hagman, K., et al., *Prevalence and clinical relevance of viraemia in viral respiratory tract infections: a systematic review*. Lancet Microbe, 2025. **6**(2): p. 100967.
12. Anon, *Viral meningitis*. Paediatr. Child Health, 1998. **3**(6): p. 437.
13. Kohil, A., et al., *Viral meningitis: an overview*. Arch. Virol, 2021. **166**(2): p. 335–345.
14. Hsu, C.H., et al., *Clinical presentation of cytomegalovirus meningoencephalitis: a retrospective study of 12 adult patients with a variety of immunocompromised conditions*. BMC Infect. Dis., 2025. **25**(1): p. 575.
15. Silwal, S., et al., *A case of herpes simplex virus meningitis in an immunocompromised individual: avoiding common diagnostic pitfalls*. Cureus, 2023. **15**(7): e42242.
16. McHale Thomas, C., et al., *Diagnosis and management of cryptococcal meningitis in HIV-infected adults*. Clin. Microbiol. Rev., 2023. **36**(4): e00156-22.
17. Wang, Y., et al., *Intracranial Epstein-Barr virus infection appearing as an unusual case of meningitis in an immunocompetent woman: a case report*. J. Int. Med. Res., 2020. **48**(2): 300060520903215.
18. Ishiko, H. and Aoki, K., *Spread of epidemic keratoconjunctivitis due to a novel serotype of human adenovirus in Japan*. J. Clin. Microbiol., 2009. **47**(8): p. 2678–2679.
19. Bavinger, J.C., Shantha, J.G., and Yeh, S., *Ebola, COVID-19, and emerging infectious disease: lessons learned and future preparedness*. Curr. Opin. Ophthalmol., 2020. **31**(5): p. 416–422.
20. Bagga, B., et al., *Herpes simplex infection of the eye: an introduction*. Community Eye Health, 2020. **33**(108): p. 68–70.
21. Chucair-Elliott, A.J., Zheng, M., and Carr, D.J., *Degeneration and regeneration of corneal nerves in response to HSV-1 infection*. Invest. Ophthalmol. Visual Sci., 2015. **56**(2): p. 1097–1107.
22. Litt, J., et al., *Herpes zoster Ophthalmicus: presentation, complications, treatment, and prevention*. Infect. Dis. Ther., 2024. **13**(7): p. 1439–1459.
23. Zaidi, M.B., et al., *Non-congenital severe ocular complications of Zika virus infection*. JMM Case Rep., 2018. **5**(6): e005152.
24. Ventura, C.V. and Ventura, L.O., *Ophthalmologic manifestations associated with Zika virus infection*. Pediatrics, 2018. **141**(Suppl 2): p. S161–s166.
25. Ballard, S.B., et al., *Tropical and travel-associated norovirus: current concepts*. Curr. Opin. Infect. Dis., 2015. **28**(5): p. 408–416.
26. Oldak, E., et al., *Norovirus and Rotavirus--two major causative agents of sporadic viral gastroenteritis in hospitalized Polish children*. Adv. Med. Sci., 2009. **54**(2): p. 183–186.
27. Nguyen, T.V., et al., *Diarrhea caused by rotavirus in children less than 5 years of age in Hanoi, Vietnam*. J. Clin. Microbiol., 2004. **42**(12): p. 5745–5750.
28. Hellysaz, A., et al., *Viral gastroenteritis: sickness symptoms and behavioral responses*. mBio, 2023. **14**(2): e0356722.
29. Glass, R.I. and Bresee, J.S., *Astroviruses, enteric adenoviruses, and other gastroenteritis viral infections*, in *Tropical Infectious Diseases: Principles, Pathogens and Practice*, R.L. Guerrant, D.H. Walker, and P.F. Weller, Editors. 2011, Elsevier. p. 416–419. 10.1016/B978-0-7020-3935-5.00063-X. Epub 2011 Apr 29
30. Magwalivha, M., et al., *High prevalence of species D human adenoviruses in fecal specimens from Urban Kenyan children with diarrhea*. J. Med. Virol., 2010. **82**(1): p. 77–84.
31. Sdiri-Loulizi, K., et al., *Molecular epidemiology of human astrovirus and adenovirus serotypes 40/41 strains related to acute diarrhea in Tunisian children*. J. Med. Virol., 2009. **81**(11): p. 1895–1902.

32. Leal, T., et al., *Colitis due to cytomegalovirus and herpes simplex type 2 as a complication of a first presentation of inflammatory bowel disease*. GE Port. J. Gastroenterol., 2022. **29**(1): p. 56–60.
33. Hock, R.A., et al., *A rare case of herpes simplex virus and cytomegalovirus dual infection inducing unremitting ulcerative colitis*. Cureus, 2023. **15**(9): e45166.
34. Jagirdhar, G.S.K., et al., *Gastrointestinal tract and viral pathogens*. World J. Virol., 2023. **12**(3): p. 136–150.
35. Gupta, I., et al., *High-risk human papillomaviruses and Epstein–Barr virus in colorectal cancer and their association with clinicopathological status*. Pathogens, 2020. **9**(6): p. 452.
36. Wierzbicka, M., San Giorgi, M.R.M., and Dikkers, F.G., *Transmission and clearance of human papillomavirus infection in the oral cavity and its role in oropharyngeal carcinoma - a review*. Rev. Med. Virol., 2023. **33**(1): e2337.
37. Andrei, E.C., et al., *Oral papillomatosis: its relation with human papilloma virus infection and local immunity-an update*. Medicina (Kaunas), 2022. **58**(8): p. 1103.
38. Michaud, D.S., et al., *High-risk HPV types and head and neck cancer*. Int. J. Cancer, 2014. **135**(7): p. 1653–1661.
39. Balaji, D., *Osteomyelitis variolosa: a case report*. J. Orthop. Surg. (Hong Kong), 2011. **19**(1): p. 120–122.
40. Tang, J., et al., *Osteomyelitis variolosa, an issue inherited from the past: case report and systematic review*. Orphanet J. Rare Dis., 2021. **16**(1): p. 354.
41. Crawford, R.R., Hodson, C.M., and Errickson, D., *Guidance for the identification of bony lesions related to smallpox*. Int. J. Paleopathol., 2024. **44**: p. 65–77.
42. Mumtaz, N., et al., *Zika virus infection perturbs osteoblast function*. Sci. Rep., 2018. **8**(1): p. 16975.
43. Marks, M. and Marks, J.L., *Viral arthritis*. Clin. Med. (Lond), 2016. **16**(2): p. 129–134.
44. Mejía, C.-R. and López-Vélez, R., *Tropical arthritogenic alphaviruses*. Reumatol. Clíni. (Eng. Ed.), 2018. **14**(2): p. 97–105.
45. Marks, M. and Marks, J.L., *Viral arthritis*. Clin. Med., 2016. **16**(2): p. 129–134.
46. Chinniah, K., et al., *Arthritis in association with human immunodeficiency virus infection in Black African children: causal or coincidental?* Rheumatology, 2005. **44**(7): p. 915–920.
47. Sundaresan, B., et al., *The role of viral infections in the onset of autoimmune diseases*. Viruses, 2023. **15**(3): p. 782.
48. Harfouche, M., et al., *Epidemiology of herpes simplex virus type 2 in sub-Saharan Africa: systematic review, meta-analyses, and meta-regressions*. EClinicalMedicine, 2021. **35**: 100876.
49. Liu, C., et al., *Association between herpes simplex virus type 2 and high-risk human papillomavirus infections: a population study of the National Health and Nutrition examination survey, 2009–2016*. J. Infect. Dis., 2025. **231**(4): p. e650–e658.
50. Vuorinen, T., et al., *Epididymitis caused by coxsackievirus A6 in association with hand, foot, and mouth disease*. J. Clin. Microbiol., 2014. **52**(12): p. 4412–4413.
51. Manikandan, R., Kumar, S., and Dorairajan, L.N., *Hemorrhagic cystitis: a challenge to the urologist*. Indian J. Urol., 2010. **26**(2): p. 159–166.
52. Paduch, D.A., *Viral lower urinary tract infections*. Curr. Urol. Rep., 2007. **8**(4): p. 324–335.
53. Singh, H.K. and Nickeleit, V., *Kidney disease caused by viral infections*. Curr.Diagn. Pathol., 2004. **10**(1): p. 11–21.
54. Krawczyk, E., et al., *Koilocytosis: a cooperative interaction between the human papillomavirus E5 and E6 oncoproteins*. Am. J. Pathol., 2008. **173**(3): p. 682–688.

55. Guidry, J.T. and Scott, R.S., *The interaction between human papillomavirus and other viruses*. Virus Res., 2017. **231**: p. 139–147.
56. Sausen, D.G., et al., *Herpes simplex virus, human papillomavirus, and cervical cancer: overview, relationship, and treatment implications*. Cancers (Basel), 2023. **15**(14): p. 3692.
57. Angeletti, P.C., Zhang, L., and Wood, C., *The viral etiology of AIDS-associated malignancies*. Adv. Pharmacol., 2008. **56**: p. 509–557.
58. Cosano-Quero, A., et al., *Zika virus: cutaneous manifestations in 3 patients*. Actas Dermo-Sifiliogr. (Engl. Ed.), 2018. **109**(3): p. e13–e16.
59. Colpitts, T.M., et al., *West Nile virus: biology, transmission, and human infection*. Clin. Microbiol. Rev., 2012. **25**(4): p. 635–648.
60. Avci, O. and Ertam, I., *Viral infections of the face*. Clin. Dermatol., 2014. **32**(6): p. 715–733.
61. Kantele, A., et al., *Emerging diseases-the monkeypox epidemic in the Democratic Republic of the Congo*. Clin. Microbiol. Infect., 2016. **22**(8). p. 658–659.
62. Andreani, J., et al., *Human infection with orf virus and description of its whole genome, France, 2017*. Emerg. Infect. Dis., 2019. **25**(12): p. 2197–2204.
63. Swali, R.N., Lee, E.B., and Adams, J.L., *Gianotti-crosti syndrome in the setting of recent coronavirus disease-19 infection*. Pediatr. Dermatol., 2021. **38**(3): p. 629–631.
64. Oranje, A.P. and Folkers, E., *The Tzanck smear: old, but still of inestimable value*. Pediatr. Dermatol., 1988. **5**(2): p. 127–129.
65. Golzarian, H., et al., *A case of Epstein-Barr virus-associated pericarditis progressing to complete atrioventricular block and cardiac arrest*. HeartRhythm Case Rep., 2023. **9**(10): p. 709–714.
66. Yamanaka, T., et al., *Pericarditis caused by herpes zoster*. J. Cardiol. Cases, 2019. **19**(3): p. 77–80.
67. Rerkpattanapipat, P., et al., *Cardiac manifestations of acquired immunodeficiency syndrome*. Arch. Intern. Med., 2000. **160**(5): p. 602–608.
68. Seishima, M., et al., *Pericarditis and pleuritis associated with human parvovirus B19 infection in a systemic lupus erythematosus patient*. Mod. Rheumatol., 2010. **20**(6): p. 617–620.
69. Magno Palmeira, M., et al., *Heart failure due to cytomegalovirus myocarditis in immunocompetent young adults: a case report*. BMC Res. Notes, 2016. **9**: p. 391.
70. Matsumori, A., et al., *Myocarditis and heart failure associated with hepatitis C virus infection*. J. Card Fail, 2006. **12**(4): p. 293–298.
71. Baral, N., et al., *Influenza myocarditis: a literature review*. Cureus, 2020. **12**(12): e12007.
72. Olasunkanmi Oluwatayo, I., et al., *Antiviral activity of trans-hexenoic acid against coxsackievirus B and enterovirus A71*. Antimicrob. Agents Chemother., 2023. **67**(3): e00868-22.
73. Sagui, E., et al., *Severe Ebola virus infection with encephalopathy: evidence for direct virus involvement*. Clin. Infect. Dis., 2015. **61**(10): p. 1627–1628.
74. Winkelmann, E.R., Luo, H., and Wang, T., *West Nile virus infection in the central nervous system*. F1000Res., 2016. **5**: p. 5.
75. Couderc, T. and Lecuit, M., *Focus on Chikungunya pathophysiology in human and animal models*. Microbes Infect., 2009. **11**(14): p. 1197–1205.
76. Rozenberg, F., *Chapter 122 – Acute viral encephalitis*, in *Handbook of Clinical Neurology*, O. Dulac, M. Lassonde, and H.B. Sarnat, Editors. 2013, Elsevier. p. 1171–1181.
77. Duerlund, L.S., Nielsen, H., and Bodilsen, J., *Current epidemiology of infectious encephalitis: a narrative review*. Clin. Microbiol. Infect., 2025. **31**(4): p. 515–521.

78. Krauer, F., et al., *Zika virus infection as a cause of congenital brain abnormalities and Guillain-Barré syndrome: systematic review*. PLoS Med., 2017. **14**(1): e1002203.
79. Muñoz, L.S., Parra, B., and Pardo, C.A., *Neurological implications of Zika virus infection in adults*. J. Infect. Dis., 2017. **216**(suppl_10): p. S897–s905.
80. Fulton, C.D.M., et al., *Long-term, West Nile virus-induced neurological changes: a comparison of patients and rodent models*. Brain Behav. Immun. Health, 2020. **7**: 100105.
81. Mulvey, P., et al., *The ecology and evolution of Japanese encephalitis virus*. Pathogens, 2021. **10**(12): p. 1534.
82. Sayed, I.M., *Dual infection of hepatitis a virus and hepatitis E virus- what is known?* Viruses, 2023. **15**(2): p. 298.
83. Riaz, M., et al., *An overview of triple infection with hepatitis B, C and D viruses*. Virol. j., 2011. **8**: p. 368.
84. Iacob, S., et al., *Hepatitis B prevalence and referral rates in vulnerable populations undergoing community-based screening-results from the LIVE(RO)2 program*. Viruses, 2024. **16**(8): p. 1318.
85. Tapper, E.B. and Curry, M.P., *Hepatitis caused by other viruses*, in *Handbook of Liver Disease*. 2018, Elsevier. p. 78–83. 10.1016/B978-0-323-47874-8.00006-7https://pmc.ncbi.nlm.nih.gov/articles/PMC7152265/.Epub 2017 Sep 8. PMCID: PMC7152265.
86. Rodriguez-Frias, F., Rando-Segura, A., and Quer, J., *Solved the enigma of pediatric severe acute hepatitis of unknown origin?* Front. Cell. Infect. Microbiol., 2023. **13**: p. 1175996.
87. Maneerattanasak, S., et al., *Prevalence of dengue, Zika, and chikungunya virus infections among mosquitoes in Asia: a systematic review and meta-analysis*. Int. J. Infect. Dis., 2024. **148**: 107226.
88. Lim, A., et al., *The overlapping global distribution of dengue, chikungunya, Zika and yellow fever*. Nat. Commun., 2025. **16**(1): p. 3418.
89. Quintero-Salgado, E., et al., *Seasonal shifts in influenza, respiratory syncytial virus, and other respiratory viruses after the COVID-19 pandemic: an eight-year retrospective study in Jalisco, Mexico*. Viruses, 2024. **16**(12): p. 1892.
90. Khavarian-Garmsir, A.R., Sharifi, A., and Moradpour, N., *Are high-density districts more vulnerable to the COVID-19 pandemic?* Sustainable Cities Soc., 2021. **70**: 102911.
91. Worsley-Tonks, K.E.L., et al., *Strengthening global health security by improving disease surveillance in remote rural areas of low-income and middle-income countries*. Lancet Glob. Health, 2022. **10**(4): p. e579–e584.
92. Moyo, E., et al., *Key populations and Sub-Saharan Africa's HIV response*. Front. Public Health, 2023. **11**: 1079990.
93. Mensah, E.A., et al., *A proposed One Health approach to control yellow fever outbreaks in Uganda*. One Health Outlook, 2024. **6**(1): p. 9.
94. Wiyono, L., et al., *Dengue and COVID-19 infections in the ASEAN region: a concurrent outbreak of viral diseases*. Epidemiol. Health, 2021. **43**: e2021070.
95. Kaewdech, A., Charatcharoenwitthaya, P., and Piratvisuth, T., *Asian perspective on hepatitis B virus and hepatitis C virus elimination*. Viruses, 2024. **17**(1): p. 34.
96. Arimurti, A.R.R., et al., *The alarming spread of Japanese encephalitis: a growing public health concern*. Open Vet. J., 2025. **15**(4): p. 1505–1519.
97. Lessa, C.L.S., et al., *Dengue as a disease threatening global health: a narrative review focusing on Latin America and Brazil*. Trop. Med. Infect. Dis., 2023. **8**(5): p. 241.
98. Cunha, M.S., et al., *Chikungunya virus: an emergent arbovirus to the South American continent and a continuous threat to the world*. Front. Microbiol., 2020. **11**: p. 1297.

99. Mittal, R., et al., *Zika virus: an emerging global health threat*. Front. Cell. Infect. Microbiol., 2017. **7**: p. 486.
100. Mohamoud, Y.A., et al., *The epidemiology of hepatitis C virus in Egypt: a systematic review and data synthesis*. BMC Infect. Dis., 2013. **13**: p. 288.
101. Hajjar, S.A., Memish, Z.A., and McIntosh, K., *Middle East respiratory syndrome coronavirus (MERS-CoV): a perpetual challenge*. Ann. Saudi Med., 2013. **33**(5): p. 427–436.
102. Nurmakhanov, T., et al., *Crimean-Congo haemorrhagic fever virus in Kazakhstan (1948–2013)*. Int. J. Infect. Dis., 2015. **38**: p. 19–23.
103. Hennessey, K., et al., *Hepatitis B control in the World Health Organization's Western Pacific Region: targets, strategies, status*. Vaccine, 2013. **31**: p. J85–J92.
104. Chrdle, A., Chmelík, V., and Růžek, D., *Tick-borne encephalitis: what travelers should know when visiting an endemic country*. Hum. Vaccines Immunother., 2016. **12**(10): p. 2694–2699.
105. Aggarwal, R., *Hepatitis E: the endemic perspective*. Clin. Liver. Dis. (Hoboken), 2013. **2**(6): p. 240–244.
106. Pourzangiabadi, M., et al., *Dengue virus: etiology, epidemiology, pathobiology, and developments in diagnosis and control – a comprehensive review*. Infect., Genet. Evol., 2025. **127**: 105710.
107. Ciota, A.T., et al., *Effects of Zika virus strain and Aedes mosquito species on vector competence*. Emerg. Infect. Dis., 2017. **23**(7): p. 1110–1117.
108. Gomes, E.O., et al., *Detection of Zika virus in Aedes aegypti and Aedes albopictus mosquitoes collected in urban forest fragments in the Brazilian Amazon*. Viruses, 2023. **15**(6): p. 1356.
109. Vega-Rúa, A., et al., *High level of vector competence of Aedes aegypti and Aedes albopictus from ten American countries as a crucial factor in the spread of Chikungunya virus*. J. Virol., 2014. **88**(11): p. 6294–6306.
110. Molaei, G., et al., *Host feeding patterns of Culex mosquitoes and West Nile virus transmission, northeastern United States*. Emerg. Infect. Dis., 2006. **12**(3): p. 468–474.
111. Van den Eynde, C., et al., *Japanese encephalitis virus interaction with mosquitoes: a review of vector competence, vector capacity and mosquito immunity*. Pathogens, 2022. **11**(3): p. 317.
112. Hawman, D.W. and Feldmann, H., *Crimean–Congo haemorrhagic fever virus*. Nat. Rev. Microbiol., 2023. **21**(7): p. 463–477.
113. Munivenkatappa, A., et al., *Clinical & epidemiological significance of Kyasanur forest disease*. Indian J. Med. Res., 2018. **148**(2): p. 145–150.
114. Izri, A., et al., *Sandfly fever Sicilian virus, Algeria*. Emerg. Infect. Dis., 2008. **14**(5): p. 795–797.
115. Tchouassi, D.P., et al., *Sand fly-associated phlebovirus with evidence of neutralizing antibodies in humans, Kenya*. Emerg. Infect. Dis., 2019. **25**(4): p. 681–690.
116. Hossain, A.D., et al., *Effectiveness of contact tracing in the control of infectious diseases: a systematic review*. Lancet Public Health, 2022. **7**(3): p. e259–e273.
117. Song, J.W. and Chung, K.C., *Observational studies: cohort and case-control studies*. Plast Reconstr. Surg., 2010. **126**(6): p. 2234–2242.
118. Carmo Dos Santos, M., et al., *Wastewater surveillance for viral pathogens: a tool for public health*. Heliyon, 2024. **10**(13): e33873.
119. Verreault, D., Moineau, S., and Duchaine, C., *Methods for sampling of airborne viruses*. Microbiol. Mol. Biol. Rev., 2008. **72**(3): p. 413–444.
120. Arinaminpathy, N., Metcalf, C.J.E., and Grenfell, B.T., *Viral dynamics and mathematical models*, in *Viral*

Infections of Humans. 2014, Springer. p. 81–96 10.1007/978-1-4899-7448-8_5. PMCID: PMC7119994. https://pmc.ncbi.nlm.nih.gov/articles/PMC7119994/.

121. Trejo, I. and Hengartner, N.W., *A modified susceptible-infected-recovered model for observed under-reported incidence data*. PLoS One, 2022. **17**(2): e0263047.
122. Law, K.B., et al., *Modelling infectious diseases with herd immunity in a randomly mixed population*. Sci. Rep., 2021. **11**(1): p. 20574.
123. Yadegari, I., Omidi, M., and Smith, S.R., *The herd-immunity threshold must be updated for multi-vaccine strategies and multiple variants*. Sci. Rep., 2021. **11**(1): p. 22970.
124. Plans-Rubió, P., *Percentages of vaccination coverage required to establish herd immunity against SARS-CoV-2*. Vaccines (Basel), 2022. **10**(5): p. 736.
125. Guerra, F.M., et al., *The basic reproduction number (R(0)) of measles: a systematic review*. Lancet Infect. Dis., 2017. **17**(12): p. e420–e428.
126. Shaw, C.L. and Kennedy, D.A., *What the reproductive number R(0) can and cannot tell us about COVID-19 dynamics*. Theor. Popul Biol., 2021. **137**: p. 2–9.
127. Manathunga, S.S., Abeyagunawardena, I.A., and Dharmaratne, S.D., *A comparison of transmissibility of SARS-CoV-2 variants of concern*. Virol. j., 2023. **20**(1): p. 59.
128. Otani, N., et al., *Changes in the epidemiology of rubella: the influence of vaccine-introducing methods and COVID-19*. Vaccines (Basel), 2023. **11**(8): p. 1358.
129. Horefti, E., *The importance of the one health concept in combating zoonoses*. Pathogens, 2023. **12**(8): p. 977.
130. Reyes, R., et al., *Urbanization and infectious diseases: general principles, historical perspectives, and contemporary challenges*, in *Challenges in Infectious Diseases*. Vol. 19. 2012. p. 123–146. 10.1007/978-1-4614-4496-1_4. PMCID: PMC7119955. https://pmc.ncbi.nlm.nih.gov/articles/PMC7119955/ eCollection 2013.
131. Kazancoglu, Y., et al., *Impact of epidemic outbreaks (COVID-19) on global supply chains: a case of trade between Turkey and China*. Socio-Econ. Plan. Sci., 2023. **85**: 101494.
132. Guan, D., et al., *Global supply-chain effects of COVID-19 control measures*. Nat. Human Behav., 2020. **4**(6): p. 577–587.
133. Dinata, R., Baindara, P., and Mandal, S.M., *Evolution of antiviral drug resistance in SARS-CoV-2*. Virus, 2025. **17**(5): 722.
134. Tosam, M.J., Ambe, J.R., and Chi, P.C., *Global emerging pathogens, poverty and vulnerability: an ethical analysis*, in *Socio-cultural Dimensions of Emerging Infectious Diseases in Africa*. 2019. p. 243–253. 10.1007/978-3-030-17474-3_18. PMCID: PMC7123202. https://pmc.ncbi.nlm.nih.gov/articles/PMC7123202/.
135. Sahasranaman, A. and Jensen, H.J., *Poverty in the time of epidemic: a modelling perspective*. PLoS One, 2020. **15**(11): e0242042.
136. Lucas, T., et al., *Justice beliefs and cultural values predict support for COVID-19 vaccination and quarantine behavioral mandates: a multilevel cross-national study*. Transl. Behav. Med., 2022. **12**(2): p. 284–290.
137. Keselman, A., et al., *Cognitive and cultural factors that affect general vaccination and COVID-19 vaccination attitudes*. Vaccines (Basel), 2022. **11**(1): p. 94.
138. Vasileia, E., et al., *Health literacy and acceptance of COVID-19 preventive measures and vaccination in the European Union: a scoping review*. Health Lit. Res. Pract., 2025. **9**(1): p. e46–e55.

139. Silva, M.J. and Santos, P., *The impact of health literacy on knowledge and attitudes towards preventive strategies against COVID-19: a cross-sectional study*. Int. J. Environ. Res. Public Health, 2021. **18**(10): 5421.
140. Liao, H., et al., *Climate change, its impact on emerging infectious diseases and new technologies to combat the challenge*. Emerg. Microbes Infect., 2024. **13**(1): 2356143.
141. He, Y., et al., *Viral respiratory infections in a rapidly changing climate: the need to prepare for the next pandemic*. EBioMedicine, 2023. **93**: 104593.
142. Jansen, H.J., et al., *Biological warfare, bioterrorism, and biocrime*. Clin. Microbiol. Infect., 2014. **20**(6): p. 488–496.
143. Meyer, R.F. and Morse, S.A., *Viruses and bioterrorism*, in *Encyclopedia of Virology*. 2008. p. 406–411. 10.1016/B978-012374410-4.00549-5 Epub 2008 Jul 30. PMCID: PMC7149760. https://pmc.ncbi.nlm.nih.gov/articles/PMC7149760/.
144. Hughes, J.M. and Gerberding, J.L., *Anthrax bioterrorism: lessons learned and future directions*. Emerg. Infect. Dis., 2002. **8**(10): p. 1013–1014.
145. Morens, D.M., Taubenberger, J.K., and Fauci, A.S., *A centenary tale of two pandemics: the 1918 influenza pandemic and COVID-19, part I*. Am. J. Public Health, 2021. **111**(6): p. 1086–1094.
146. Nathanson, N. and Kew, O.M., *From emergence to eradication: the epidemiology of poliomyelitis deconstructed*. Am. J. Epidemiol., 2010. **172**(11): p. 1213–1229.
147. Brady, O.J., et al., *The association between Zika virus infection and microcephaly in Brazil 2015–2017: an observational analysis of over 4 million births*. PLoS Med., 2019. **16**(3): e1002755.

Index

A

B

C

H

I

J

K

L

M

Printed and bound by CPI Group (UK) Ltd, Croydon, CR0 4YY
17/09/2026
14964238-0001